I0883278

Late Roman Combat Tactics

Late Roman Combat Tactics

Ilkka Syvänne

Medallion of Justinian
source: Diehl

Pen & Sword
MILITARY

First published in Great Britain in 2024 by
Pen & Sword Military
An imprint of Pen & Sword Books Limited
Yorkshire – Philadelphia

Copyright © Ilkka Syvänne 2024

ISBN 978 1 52679 395 9

The right of Ilkka Syvänne to be identified as
Author of this Work has been asserted by him in accordance
with the Copyright, Designs and Patents Act 1988.

A CIP catalogue record for this book is
available from the British Library

All rights reserved. No part of this book may be reproduced or transmitted in any form or by any means, electronic or mechanical including photocopying, recording or by any information storage and retrieval system, without permission from the Publisher in writing.

Typeset by Mac Style
Printed in the UK by CPI Group (UK) Ltd, Croydon, CR0 4YY.

MIX
Paper | Supporting responsible forestry
FSC® C013604

Pen & Sword Books Limited incorporates the imprints of After the Battle, Atlas, Archaeology, Aviation, Discovery, Family History, Fiction, History, Maritime, Military, Military Classics, Politics, Select, Transport, True Crime, Air World, Frontline Publishing, Leo Cooper, Remember When, Seaforth Publishing, The Praetorian Press, Wharncliffe Local History, Wharncliffe Transport, Wharncliffe True Crime and White Owl.

For a complete list of Pen & Sword titles please contact

PEN & SWORD BOOKS LIMITED
47 Church Street, Barnsley, South Yorkshire, S70 2AS, England
E-mail: enquiries@pen-and-sword.co.uk
Website: www.pen-and-sword.co.uk
or
PEN AND SWORD BOOKS
1950 Lawrence Rd, Havertown, PA 19083, USA
E-mail: uspen-and-sword@casematepublishers.com
Website: www.penandswordbooks.com

Dedicated to the memory of the pioneers of the trade, the great Byzantinists Alphonse Dain and George T. Dennis

Contents

Acknowledgements

First of all, I would like to thank the Commissioning Editor Philip Sidnell for accepting the proposal for this book. Special thanks are also due to Matt Jones, Tara Moran, and other staff of Pen & Sword for their stellar work and for the outstanding support they gave the author. Ian Hughes also deserves a thanks for his part in the editing work.

I would also like to thank Perry Gray for his perceptive comments regarding the content of the book and Dario Wielec for the photos of horses which I used as models for some of my vector drawings in the text portion of the book. Jyrki Halme deserves once again particular gratitude from me for his re-enacting photos. I appreciate the efforts of all contributors greatly. If there are any mistakes left, those are the sole responsibility of the author. My warmest thanks are also due to my wife Sini and children Ari and Nanna for their patience.

List of Plates

List of Maps, Diagrams and Battle Formations

Introduction

The intention of this book is to analyze how the Late Romans fought, in particular at the tactical level. It complements the already published eight volume series *The Military History of Late Rome* by analyzing the tactics, face of battle and other factors of combat in greater detail than was possible in the narrative.

The endnotes are kept at a minimum so as not to distract from the actual analysis. Excluding a short summary of the sources I have also kept to a minimum the analyses of the sources because there exists expert literature devoted to this subject. These include also my own previous studies and in particular *The Age of Hippotoxotai* (Tampere 2004).

I have usually used the commonly used transliterations, but not slavishly and not always consistently. For example, I have used the F instead of the PH when transliterating Greek military terms. Similarly, the words στρατηγός and πύκνωσις (and other words with long vowels) should be transliterated *stratêgos* and *pyknôsis/puknôsis* respectively, but I have not included the long vowel every time in the transliteration of the Greek words because the common practice is to leave those out. I have basically included those only inconsistently. I have also occasionally used the word *strategos* as equivalent of the general or commander of the army despite the fact that during the Late Roman period the commander in charge could officially be the emperor, *magister militum*, *dux* or some other title holder.

All illustrations, drawings, maps and diagrams etc. have been drawn and prepared by the author unless stated otherwise. I have used the *Barrington Atlas* as the principal source for the Maps.

The aim of this book is to analyze Roman combat tactics with particular focus on pitched battles during the so-called Late Roman period, which is usually dated to cover the years 284–641. The primary objective of this study is to reconstruct and understand the land combat tactics, and secondarily to assess their effectiveness. The battle tactics are further divided into the sub questions of campaign and battle tactics, and cavalry and infantry tactics. The siege tactics, naval tactics and other fighting methods are dealt with only insofar as it is necessary for the understanding of land warfare. The reason for this is that those will be studied in separate studies later. The land tactics cannot be understood in a vacuum, which means that the study also includes discussions of military organization, intelligence gathering, logistics, strategy and types of enemies.

This monograph aims to achieve its aims through the use of the qualitative method. In other words, the period combat tactics are reconstructed on the basis of the combined information provided by the narrative sources, military treatises and other sources. Of these the period military manuals form the most valuable source of information. It is thanks to the existence of these military treatises that historians can see through the eyes of the contemporaries.

As I noted in my doctoral dissertation, the tactical effectiveness of the armed forces derived from many sources. The effectiveness of an army cannot be calculated solely on the basis of the outcome of a battle, because the victorious army could achieve success by mere superiority of numbers. The effectiveness of the army results from a combination of good training methods; quality of troops and officers; indoctrination of the soldiers; amount of money spent on the soldiers; ability of the commanders to cooperate with each other; ability to manoeuvre and operate in different types of terrain and weather; sound intelligence gathering methods; good theoretical foundations based on past experience; sound criteria for the choosing of the types of engagement; efficient and adaptable fighting methods; ability of the different units and arms of service to coordinate their actions; and readiness to learn from mistakes and from enemies.[1]

The best sources for the assessment of the effectiveness of the late Roman armed forces are the military manuals of which the most important is the late sixth century *Strategikon* of Maurice.[2] When the detailed information drawn from these is analyzed in conjunction with other period evidence, the most important of which are the narrative histories of Ammianus Marcellinus, Procopius, Agathias and Theophylact Simocattes, one can arrive at a reconstruction of period combat tactics and their effectiveness which is as accurate as one can obtain. Ultimately it is the combat tactics and so-called decisive battles that decided the fate of the Mediterranean and humanity in general. This study therefore aims at nothing less than to provide the ultimate answer for the question of why Rome survived the many challenges facing it during the Late Roman period and why it survived even the crushing defeat on the battlefield of Yarmuk in 634.

Abbreviations

Cav.	Cavalry
DMS	*De militari scientia*
DRB	*De rebus bellicis*
DRM	*De re militari*
DV	*De velitatione*
HC	Heavy cavalry
HI	Heavy infantry
Inf.	Infantry
JE	John of Ephesus
LC	Light cavalry
LI	Light infantry
MHLR	*Military History of Late Rome* by Ilkka Syvänne
Or.	Orations
PE	*Excerpts of Polyaenus*
PKA	*Peri katastaseôs aplêktou, De castrametatione*
PPE	*Peri politikês epistêmetes*
REF1	*The Roman Eastern Frontier and the Persian Wars (AD 226–363)*
RH	*Rhetorica militaris*
STR	*Strategikon*
UT	Urbicius, *Tactica, Taktikon*

Chapter One

The Military Manuals and Combat Tactics in ca. 211–641[1]

1.1. Graeco-Roman heritage

The most important sources for reconstructing the period combat methods are the military manuals and in this we are fortunate because we possess several dating from the era under investigation. However, we still need to begin our search with the treatises that date from the period before our era because these combat systems represent the stage in development that existed at the beginning of our era. Furthermore, as will be made clear, most of the earlier treatises retained their importance at least until the advent of the firearms, and in some cases even beyond, so that we still see men fighting in ranks and files during the nineteenth century. It was because of this that the East Romans (and others) kept on producing new copies of these just like they did of their newer treatises. For example, John Lydus *(Mag. 1.47)*, who wrote during the reign of Justinian, names the following military theorists: Celsus (1st cent. BC, not extant); Paterus (2nd cent. AD, not extant); Catiline (not extant); Cato (3rd–2nd cent. BC, not extant); Frontinus (1st–2nd cent. AD; a fragment of *De Officio Militari* in Lydus and *Stratagems*); Renatus (Vegetius); Aelian (2nd cent. AD); Arrian (1st–2nd cent. AD); Aeneas (4th cent. BC); Patro (not extant); Apollodoros (2nd cent. AD); and the emperor Julian (*Military Engines*, not extant). As can be seen, quite a few of these are no longer extant, but the list given by Lydus is by no means complete, because we know that there existed treatises which are no longer extant by Pyrrhus, Alexandros, Clearchos, Pausanias, Evangelos, Polybios, Eupolemos, Ifikrates, Posidonios, Dorylaos, Theodoridas, and Bryon. In addition to which, there are treatises which are still extant by: Biton (3rd cent. BC); Philon of Byzantion (3rd cent. BC); Athenaios (1st cent. BC); Heron of Alexandria (2nd–1st cent. BC); Asklepiodotos/Asclepiodotus (1st cent. BC); Onesandros/Onasander (2nd cent. AD); Polyainos/Polyaenus (2nd cent. AD); Julius Africanus (3rd cent. AD); and other fragments. Of these the treatises of the Greek texts of Onasander, Arrian, Aelian, Polyaenus and Julius Africanus were to have great influence on the development of the East Roman combat methods, while the Latin texts of Frontinus and Renatus Vegetius were to have greater influence on the Latin-speaking west. The military treatises of Aelian and Arrian in particular had great influence on infantry tactics, because both were good sources for the various tactical uses of the Macedonian phalanx. The only real difference between the Roman and Macedonian phalanx was that the Romans usually used shorter spears (*hasta, kontarion*, a spear usually ca. 2.5–3.74m in length) than the Macedonians (*sarissa*, a pike usually ca. 5–6m in length) had used. This is already in evidence in the *Kestoi* of Julianus Africanus and Arrian's *Ektaxis kata Alanon*.

1.2. Arrian and Frontinus[2]

The texts of Arrian (mainly *Ektaxis kata Alanon* and *Taktika*) and Frontinus (*Stratagems* and a fragment of *De Officio Militari* in Lydus *Mag.* 3.3) are particularly valuable because both were practising soldiers who described actual Roman combat practices. When their information is combined with other sources we get a good picture of how the Roman armed forces fought during the second century AD. The picture that emerges from these treatises closely resembles the Late Roman armed forces of the third and fourth centuries.

Both authors demonstrate in no uncertain terms that the Romans deployed their footmen as phalanxes, even if Frontinus's text proves that the Romans were also using earlier methods (manipular and cohortal arrays) alongside it. It is quite probable that these methods could also be used on a smaller scale during the Late Roman period when the Romans employed smaller units in local conflicts. The infantry formations in Arrian's *Tactics* consisted of open, close (*pyknosis*) and locked shields (*synaspismos*) orders. Arrian (11.4–12.6) equated the *synaspismos* (interlocked-shields) with the Roman *testudo*-formation which he stated could be square, circular or oblong in shape. Arrian's description of the infantry tactics included most of the variants that we can find not only in the Hellenistic military treatises and in practical use among the Roman armies and also in the military manuals of Frontinus and Vegetius. These included the lateral oblong (*plagia*) phalanx; column (*orthia*) phalanx (= 'tower formation', called in Greek *pyrgos* and in Latin as *turris*);[3] oblique (*loxe*) phalanx; two-fronted (*amfistomos*) phalanx; two-fronted double phalanx (*amfistomos difalaggia*); each-other-facing (*antistomos*) phalanx; each-other-facing-double-phalanx (*antistomos difalaggia*); marching column of two phalanxes side by side or behind each other in which the file leaders were on the same side (*homoiostomos difalaggia*), phalanx with files of light-armed placed between heavy infantry files; *hypotaxis*-formation, in which the light-armed were placed slightly behind or slightly in front of the edges of the phalanx in an *epikampios*-formation (half-square); interjection of men from the rear to the front (*parembole*); marching formations of one, two, three or four fronted formations; wedge (*embolon*); hollow wedge (*koilembolos*); rectangular-hollow-formation (*plaision*, hollow oblong in which the light infantry was placed in the middle of the heavy); hollow square 'brick' (*plinthion*; square); double outflanking (*hyperfalangesis*); outflanking (*hyperkerasis*).

Naturally, Arrian includes also a discussion of the cavalry unit formations. The cavalry unit orders were the open, close, tortoise (horses brought side-to-side so far forward and so close that each rider in the front could cover with his shield the head of the horse behind him) and ranks partially interjected into ranks (rhomboid). Arrian (16.13–4) opposed the tightening of the cavalry formation (*pyknosis*) too much by bringing the rear rank horses partially between those in front of them, because, in his opinion, when the horses were massed and pressed together it only panicked the horses – this means that he did not approve the same practice with the smaller 28 horsemen rhomboid. The unit formations for combat were: square (*tetragônos*), oblong (*heteromêkês*), rhomboid (28 horsemen *rhomboiedês*), wedge (*embolos*) and depthless one rank (*abathês taxis*). In addition to this, it is known that the Romans used the larger 128 horsemen rank-and-file rhomboids, because these were used by the Armenian and Parthian mounted archers that served as auxiliaries. Most of this followed the traditional Hellenistic material which can be found in Asclepiodotus (7.1–11) and Aelian (Matthew ed. 18–21), but Arrian's referral to the depthless, one-rank formation (17.5) is unique to him and probably reflects contemporary Roman practices or comes from some unknown earlier treatise. The one-rank formation was to be used only when raiding or trampling enemy underfoot because it was very disadvantageous for battles. We learn from Arrian that from the reign of Hadrian onwards, Roman cavalry was expected to be equally adept at long range combat (mounted archery with crossbows and composite bows), medium range with javelins, and at close quarters with spears (*lancea, lancea pugnatoria*, *xyston,* one-handed Gallic *contus*, and the two-handed Sarmatian *contus*, which is a pike with a length of ca. 3–4m), javelins, swords and axes. Arrian also makes it clear that the cavalry could vary the amount of protective equipment according to the situation, so that at times the horses were armoured while at other times they were not. Frontinus's text (*Stratagems* 2.3.23) also makes clear that the Romans expected their horsemen to be ready to dismount and fight as infantry. The cavalry was therefore purpose-built for each task so that the equipment of each unit and individual in combat reflected either the intended use of the unit or the position of each individual trooper in the combat formation. In other words, it was not the Huns who caused the Romans to become mounted archers – the Sarmatians, Alans and Parthians had a far greater role in this. We also learn from Arrian's texts that the Roman infantry was expected to use the *contus* (in the *Strategikon* known as *kontarion*) and kneeling version of the *foulkon/testudo/chelone* (tortoise, one version of the *synaspismos*, interlocked-shield order) and the phalanx array against cavalry. However, it should be noted that Arrian's influence on ancient and medieval warfare was profound, not only because of his military treatises but even more so because of his *Campaigns of Alexander the Great*. The military campaigns of Alexander as described by Arrian were idolized by generation after generation of military leaders.

In his *De Officio Militari*, Frontinus positioned the overall commander of the army (either *Kaisar/Caesar/autokrator/*emperor, or *hyparchos/praefectus*) in the centre of the army, the *hipparchos* (cavalry commander) or *hyparchos* (*praefectus*) on the left wing, while the right wing was placed under the so-called *praitôres* (*praetores*, praetors, governors) and *lêgatoi* (*legati*, legates). The *praetores* were considered *strategoi* (generals) and the *legati* their *presbeutai* (lieutenants, second-in-command, *hypostrategoi*, vicars). Even if the commander could always choose his place as the situation required, the positioning of the overall commander in the centre was also the standard positioning pattern of commanders during the Late Roman era. Vegetius placed the overall commander on the right wing, but it is clear that he had derived this instruction from some Republican-era treatise.

The combat formations listed by Frontinus in his *Stratagems* include the triple line of either maniples or cohorts for the Romans, and a triple line of phalanxes for their enemies (*triplex acies*, a variation of the Greek *trifalaggia*, but the latter of which could also mean *epikampios*); the crescent (*menoeides*); the convex (*kurte/kyrte*); the *epikampios emprosthia* (forward-angled half-square, or rather its variant,

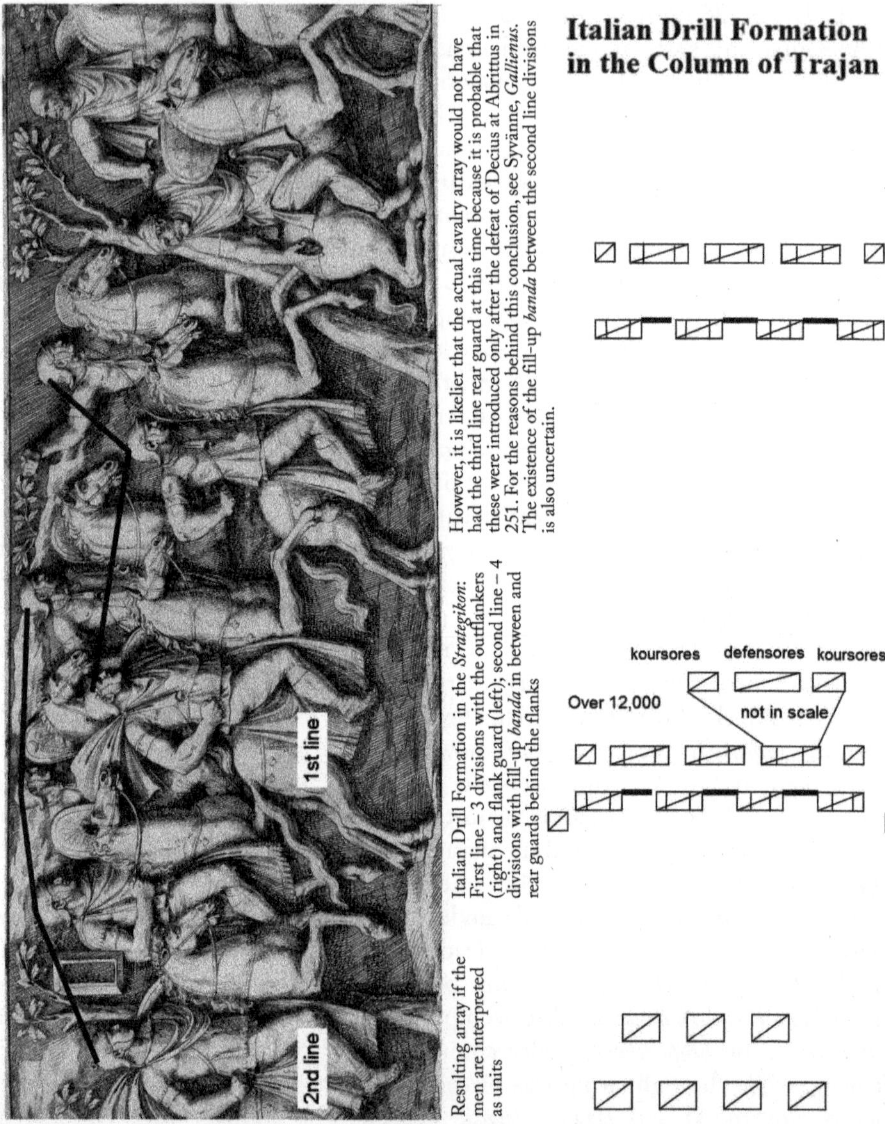

which was to send wings forward against the enemy's flanks); the *epikampios opisthia* (rearward-angled half-square, *trifalaggia*); the double line either of legionaries, or with light-armed (auxiliaries) in front and legionaries behind (*difalaggia*); the cavalry in front and infantry behind; left and right oblique formations (*oblique acie*, a variation of the *loxê falagx*/column); the hollow square/oblong (*plinthion*/square, *plaision*/oblong); the phalanx (*plagia falagx*); and the modification of the arrays according to the terrain (e.g. array placed against a river, or in a defile, in which case some of the men could be placed on the heights). Most of these formations can also be found in the military treatises depicting Macedonian tactics (Asclepiodotus, Arrian and Aelian).

Alan and Illyrikian Drills in the Column of Trajan

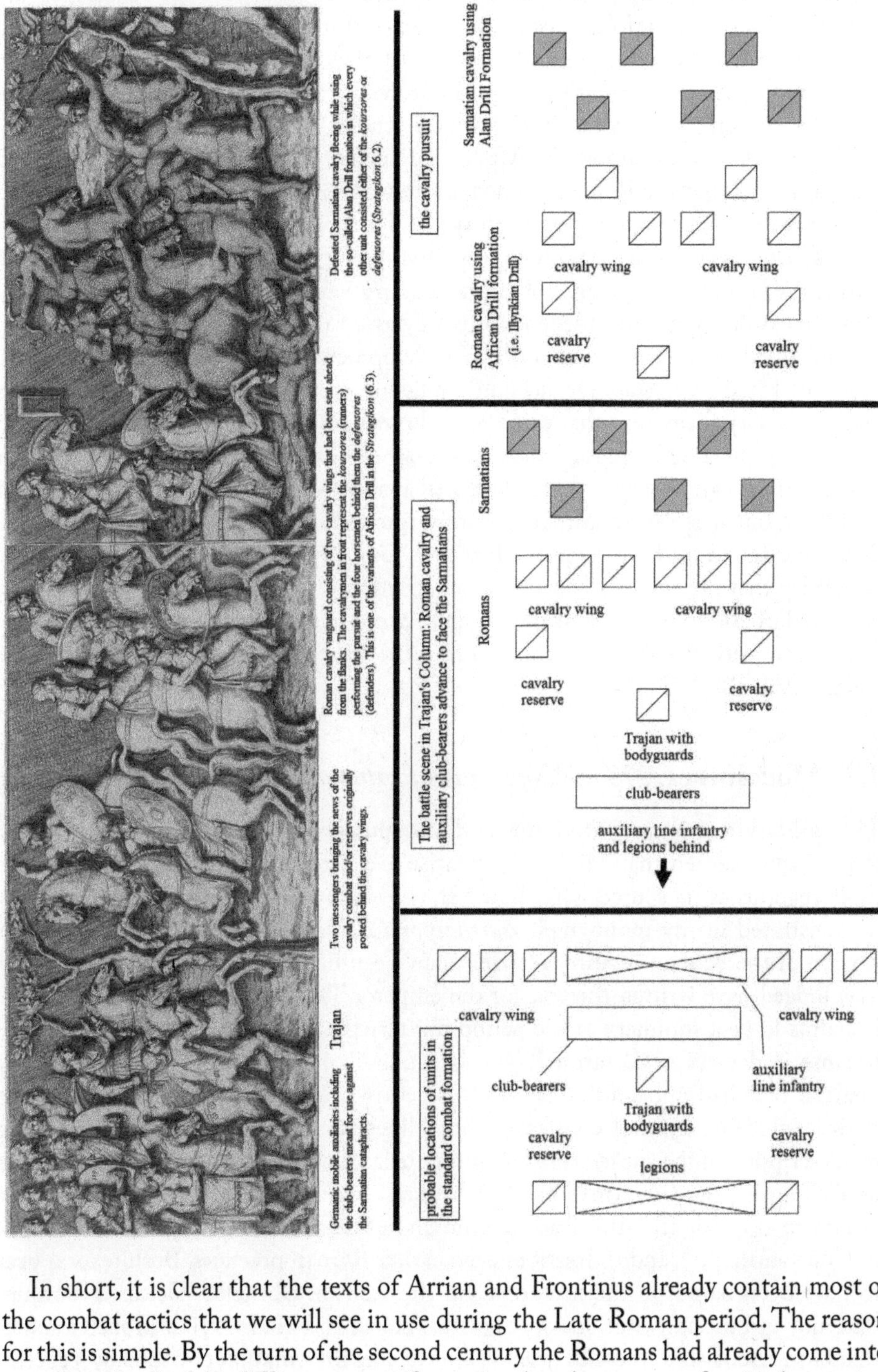

In short, it is clear that the texts of Arrian and Frontinus already contain most of the combat tactics that we will see in use during the Late Roman period. The reason for this is simple. By the turn of the second century the Romans had already come into contact with all of the different types of enemies that they were to face in the coming centuries. The Romans were fast learners, so they adopted their tactics immediately. The only details missing from the military treatises of Arrian and Frontinus are referrals to the cavalry formations, but we learn from Arrian's *Ektaxis kata Alanôn*, and from the narrative sources and works of art, that the Romans typically deployed their cavalry with reserves early on. We also learn from these sources that the Romans were already using cavalry tactics that we can find in the *Strategikon* (6).

These include for example the African, Illyrikian, Alan, Scythian and Italian drill formations, so that the Romans were definitely using the African, Illyrian and Alan cavalry drill by the mid-second century AD. Furthermore, the Romans could also use a separate cavalry army without any infantry support, as happened for example during the German campaigns of Germanicus in 13–16 AD, even if it was more typical that the cavalry operated independently only as a vanguard for the combined army that included the infantry element. This development reached its apogee during the reigns of Philip the Arab, Decius and Gallienus, so that we see all three emperors using a cavalry formation that came to be known as the Italian Drill Formation in the *Strategikon*. The basic formation of the array was in use by the turn of the first century AD, as can be seen from the attached analyses of two reliefs from the Column of Trajan, but it is probable that the array reached its final shape (three divisions: flank guards and outflankers in the first line; four divisions with fill-up *banda* (flags of cavalry that presumably had the paper strength of ca. 500 horsemen, but which consisted of 200 to 400 men at the time the *Strategikon* was written) between them; a third line with two divisions) only during the reign of Gallienus, hence the name Italian Drill Formation.[4]

1.3. Modestus ca. 275 and Vegetius's *Epitoma Rei Militaris* ca. 390–450[5]

The military treatises of Modestus and Vegetius pose a problem for any analysis. It has been claimed that Modestus's treatise is a fifteenth century forgery that has used Vegetius as its source. This, however, is by no means certain, as I have already demonstrated in my monograph *Aurelian and Probus*. Modestus presents a better version of the Roman infantry phalanx than Vegetius, with the implication that he may indeed have written the text for the emperor Tacitus in 275 as the text claims. It claims to be a summary of the period military practices. By and large this is true, because it depicts a *pila*-armed (*pila* is a plural of *pilum* heavy javelin) infantry phalanx that had apparently been introduced by Septimius Severus alongside the traditional legionary array consisting of two lines of cohorts. The only thing missing is a description of the spear-armed infantry phalanx, which we know that the Romans used alongside the *pila*-armed phalanx. Vegetius wrote his treatise in stages for some unknown emperor after the death of Gratian but before 450 (a copy attested to exist in Constantinople), and it describes even earlier Roman practices. Both texts appear to have been based on the texts of Cato the Censor, Cornelius Celsus, Frontinus, Paternus and the constitutions of Augustus, Trajan and Hadrian, but in such manner that both refer also to these in period context. The information provided by Vegetius (ca. 390–450) is discussed here together with Modestus (275) because its information has clearly been derived from the same source(s) as Modestus's material – hence its data is relevant also for the late third century.

Both treatises are valuable as sources of Roman training methods and also for their descriptions of basic infantry combat methods before the fourth century, but which also remained similar after. Both treatises make it clear that the standard Roman combat formations were the single line, double line, wedge and hollow square. Vegetius adds to this list the circle (*orbis*) formation. The text of Modestus represents a return to the traditional Roman combined arms tactic (infantry with cavalry) during the reign of Aurelian (270–5), so that the cavalry was no longer the dominant arm of service as it had been under Gallienus. The probable reason for the change is that Aurelian lost a significant portion of his cavalry forces very early during his reign at the battle of Placentia in 271. This, however, does not mean that the Romans would have completely abandoned the use of cavalry independently of infantry, only a change in the typical usage of different arms of service when the Romans fought major battles.[6]

Both treatises include two different legionary battle formations, the first for the older legions, which consisted of two lines of cohorts (with cavalry on the flanks), the second for the legionary array used by the new legions, which consisted of a shallow six deep infantry phalanx with cavalry posted on the flanks and reserves of infantry and cavalry. Modestus (18) and Vegetius (3.16) add further details to the cavalry wings by noting that the *loricati* and *contati* heavy cavalry were placed next to the infantry array and mounted archers further out.

The legionary phalanx of Modestus and Vegetius consisted of six ranks. The first two ranks consisted of the heavy-armed legionaries (shield, armour, helmet, spear/javelin, sword), but Vegetius (3.14) suggests that the second rank also included armoured archers (i.e. multipurpose heavy infantry). The third and fourth ranks consisted of the light-armed, in that the third rank consisted of archers and javelineers, while the fourth rank consisted of javelineers or men armed with *martiobarbuli/mattiobarbuli/plumbatae* (lead-weighted darts). The fifth and sixth ranks once again consisted of so-called heavy armed men. The fifth rank consisted of the *ballistarii/balistarii* (probably users of *manuballista* or *arcuballista*), *funditores* (slingers), *tragularii* (in this case probably users of the *tragula*-javelin), *fustibalatores* (staff-sling users).[7] In other words, the fifth rankers consisted of the heavy armed men who were actually equipped as light-armed to make the array two-fronted if the enemy approached from the rear. The sixth rank, the *triarii*, consisted of strong soldiers who were equipped with *scutum* shields and every type of defensive and offensive equipment. The Roman unit structures were based on the tent group called *contubernium*, which consisted of ten men of whom eight were fighting men deployed as a single file for combat, who were supported by one recruit (*tiro*) and one servant. In a cavalry unit the elite units had fewer fighting men with a corresponding number of squires/servants so that the total remained ten per tent group.

In the example given of the legionary phalanx by Modestus and Vegetius, the eight-man file of heavy infantry (the *contubernium*) had been divided into two files of four men, meaning that it had later been strengthened with two light-armed footmen. The missile equipment of the second and fifth ranks of the heavy infantry reflect the fact that a quarter to a third of the heavy armed footmen had been trained as archers ever since the second century BC. This also means that in a unit that had four ranks of legionaries the Romans could deploy one of the ranks as archers if there were no light-armed specialists available. The file of eight men was divisible by two, which means that the Romans could vary the depth of their formation (4, 8, 16, or 32 ranks of men) in the same manner as in the Hellenistic model. When one remembers that the regular Roman cohort had 480 footmen (divided as 3 maniples of 160 men), Modestus's cohort of 560 footmen (five centuries of 111 men plus supernumeraries), and the Hellenistic *pentekosirarchia* had 512 footmen (divided into two 256 *syntagmata*, each consisting of two 128-men *taxeis*, sing. *taxis*), the actual difference in sizes was not particularly significant when one takes into account the fact that in practice the size of the unit was not necessarily the same as its paper strength. Most importantly, all units were divisible by the numbers two and four. This means that the new unit types that were introduced in the third and fourth centuries, which had unit structures modelled after the Hellenistic model (units as follows: 8, 16, 32, 64, 128, 256, 512, 1024, 2048, 4096, 8192, and 16384), could be used alongside the older units without any problems.

In addition to this, Vegetius adds other formations to this list from some other source not used by Modestus, but which was clearly an old treatise just like the rest of the sources used by both Modestus and Vegetius: 1) left and right oblique formations (*loxe* in Hellenistic terminology), 2) two forward-angled formations (wings sent forward; wings sent forward while light-armed advance to protect them; in Hellenistic terminology this represents the *epikampios emprosthia* formation); 3) outflanking the enemy by sending the right wing forward, while the rest remained behind while lengthening the width (compare with the *Peri strategikes* tactical sections; 10–11 below); and 4) the use of the terrain. Vegetius's military treatise (3.23) contains the first referral to the use of the mixed formation (cavalry placed between infantry units) in a military treatise. We do not know if this reflected the material Vegetius was using or his own judgement derived from the sources he had read. Vegetius recommended the posting of the *catafracti equites* either in front of the footmen or between them for use against enemy infantry. What is certain is that the Romans were using this tactic by the fourth century, which implies that Vegetius may have derived these instructions from some unknown third century or earlier text. Even if there were earlier precedents for the use of this array (e.g. by the Achaemenid Persians and Seleucids), it is probable that the Romans had copied the formation from the Sasanian Persians, together with their *clibanarii* cavalry and extra-large cavalry wedge, in the course of the third century wars with them.[8]

As regards cavalry warfare, Vegetius's text (2.14) stressed both the melee skills and skill with the bow equally, with the implication that it represents the middle stage in the development of the importance of mounted archery in Roman military thinking from Arrian to Maurice. The ending of Book III (3.26) shows nicely how the period cavalry had been influenced by the Persians, Huns and Alans and was not in need of reform when Vegetius wrote his treatise in about 390–450 – all three nations are likely to have only increased the importance of mounted archery, which is reflected in Vegetius's mutual emphasis on the use of spears and bows. Vegetius is also valuable for his descriptions of siege warfare and naval warfare. Siege warfare remained much the same throughout the era, but Vegetius's description of naval warfare held true only for as long as the Romans preferred the combination of ramming and missile fire. The ever increasing use of the ram-spur and spurs by the early-sixth century lessened the importance of ramming in naval warfare. However, until then his advice can be seen to reflect the actual naval tactics in use.

The narrative sources confirm the use of the *plagia falagx* (lateral phalanx), *antistomos difalaggia*, *hyperfalaggêsis*, *epikampios emprosthia* (forward-angled half-square), *menoeides* (crescent), *plinthion/plaision* (hollow square/oblong), *embolos* (wedge), *koilembolos* (hollow wedge), *triplex acies/trifalaggia* (triple phalanx/line), *kyrte* (convex), and the continued use of the cavalry formations with reserves (i.e. the different size variants of the Italian Drill Formation) during the period 284 until 363.[9] This means that the Romans continued to use all of the tactics described in the previous treatises by Arrian, Frontinus, and Modestus, and also by those that used earlier material like Vegetius did or which can be found in such Hellenistic treatises as those by Asclepiodotus and Aelian.

1.4. *De Rebus Bellicis* by anonymous ca. 365–366[10]

The military treatise known with the title *De Rebus Bellicis* was written for two emperors (brothers Valentinian and Valens) by an anonymous author in about 365–6. The *terminus ante quam* for the date of the treatise is the year 367/8, because it is known that it was then that Valens put into effect policies advocated by the treatise. It is also probable that the treatise was written by someone belonging to the staff of Valentinian for Valens. The best piece of evidence for this is that the author considered formal education unnecessary (Preface 4). What counted in his opinion were intellectual abilities. Valens lacked good education and could not even understand Greek.[11] The anonymous author also states (Preface 10) that many of the remedies suggested by him were not unknown to the closest friends of the emperors, and that the author had just collected and assembled those from all sources. This implies that many of the suggested reforms had indeed been discussed in the imperial circles and that it had been the author who had collected them while adding his 'own inventions' so that the emperors (esp. Valens) could consult those more easily. In this context it is of particular note that Ammianus (30.9.4) claims that the emperor Valentinian was himself a painter and modeller and had himself invented many kinds of new arms – it is therefore not too farfetched to suggest that some of the inventions presented in the *DRB* were actually the handiwork of Valentinian himself and that these had been collected together by the anonymous author and then sent to Valens as a set of instructions from his brother. It is possible that Valentinian had painted at least some of the original illustrations himself.[12]

The stated aim of the treatise was to improve the military capabilities of the Roman Empire through financial and military reforms and through inventions. The author claimed that the financial problems facing the emperors had been caused by: 1) Constantine's release of the temple treasuries into circulation that had made the rich richer and the poor poorer (the rich violated the rights of the poor with impunity); 2) fraud (debasement of coins), committed by mint workers that had resulted in inflation; and 3) corruption of governors that had exhausted the taxpayers. The author proposed to remedy these and other problems by: 1) limiting public grants to the administrators; 2) placing all workers of the mint on an island; 3) appointing upright governors; 4) forcing the soldiers to retire when they were still capable of tilling the land; 5) introducing military reserves in the form of reservists who received lower salaries; 6) reducing the taxes paid in kind by the farmers by 50 per cent; 7) refortifying the frontiers to provide security for the farmers; 8) and by introducing various new military machines and equipment meant to make the army more efficient.[13]

Many of these suggested reforms were actually implemented by Valens and Valentinian, which lends further credence to the claims that the unknown author's suggestions reflected the ideas presented in the imperial courts by both the emperors themselves and by their courtiers, and that he merely collected these ideas for the emperors' use.[14] Similarly, it is clear that the fortifying of the frontiers with forts and towers about a mile apart against the barbarians suggested by the *De Rebus Bellicis* reflected the aspirations of both emperors. The imperial policy under the brothers was to isolate the Roman Empire from its neighbours through legislation, by fortifying and/or refortifying the borders, and by stopping the tribute payments in return for treaties when possible.[15]

As noted above, the *DRB* (7.1ff.) also claims that the effectiveness of the Roman armed forces could be improved by introducing improvements to the military equipment and machines and tactics. These suggestions are simultaneously either very impractical, or traditional and highly practical. For example, the *ballista quadrotis* is a wheeled vehicle drawn by two horses, manned by two men, powered by a windlass instead of torsion, and turned easily around to face any of the four directions by means of slots. It is difficult to see what the novelty of this machine was, unless it was the turning mechanism itself. Whatever the claimed novelty, it is certain that it was very practical and various kinds of wagon/cart-mounted ballistae continued to be used until the advent of gunpowder.

The following list of military equipment found in the treatise follows closely my analysis in *MHLR 361–395* (Vol. 2):

The *tichodifrus* was a mobile protective cover used to protect advancing ballistae and men in formation. It could also be used in sieges as a cover. It is impossible to detect what was the novelty about this type of movable cover (possibly the nails, spears and tridents covering it), but similar equipment was certainly used before and after, at least in sieges.

The *clipeocentrus* was a small round shield (*clipeus/parma*) studded with nails. The author envisaged two uses for it: 1) by soldiers; 2) as an attachment to a *tichodifrus* or another device. Shields like this had been used for ages so it is very difficult to see any innovation in this.

The *tribulata* (*plumbata*-dart, with a caltrop attached) and *plumbata mamillata* (a dart with lead weight attached to increase its penetrating power) appear similar to standard darts used by the Romans, so it is unknown what was unique about these (possibly a different type of shaft and/or head?).

The author advocated the use of various types of cataphracted (scale-armour) scythed chariots (*currodrepanus, currodrepanus singularis, currodrepanus clipeatus*) for use in pursuit of the enemy. This smacks of theory, but not conclusively, because of the periodical referrals to the use of scythed chariots in warfare. Their inclusion in, for example, the works of Vegetius and Leonardo da Vinci suggest the ancient, medieval and renaissance militaries experimented with these kinds of chariots. What is notable about the intended uses of the chariots and scythed chariots is that all treatises (Julius Africanus's *Kestoi*, the *DRB* and Vegetius) intended these as counter measures against the Persians. On the basis of Julius Africanus's *Kestoi* and Vegetius's *Epitoma rei militaris* (also known as *De re militari*) it is clear that scythed chariots were not only used for pursuit but also as anti-elephant devices. None of these experiments appear to have been long-lasting, because war chariots required very level terrain which made them impractical (a waste of resources).

There is one piece of equipment, however, that the Polish Hussars found very practical (probably a defence against lassoes) which are the wings attached to the riders' back, as shown in the illustration *currodrepanus clipeatus*.[16] On the basis of the actual usage of the other tactics (wagon laager together with hollow square/oblong formations and *carroballistae*) depicted in the *DRB* during the period from the late fourth century until the early sixth century, and on the basis of the inclusion of scythed chariots both in the *DRB* and Vegetius, I would actually suggest that there is a very high probability that the Romans did indeed experiment with these during

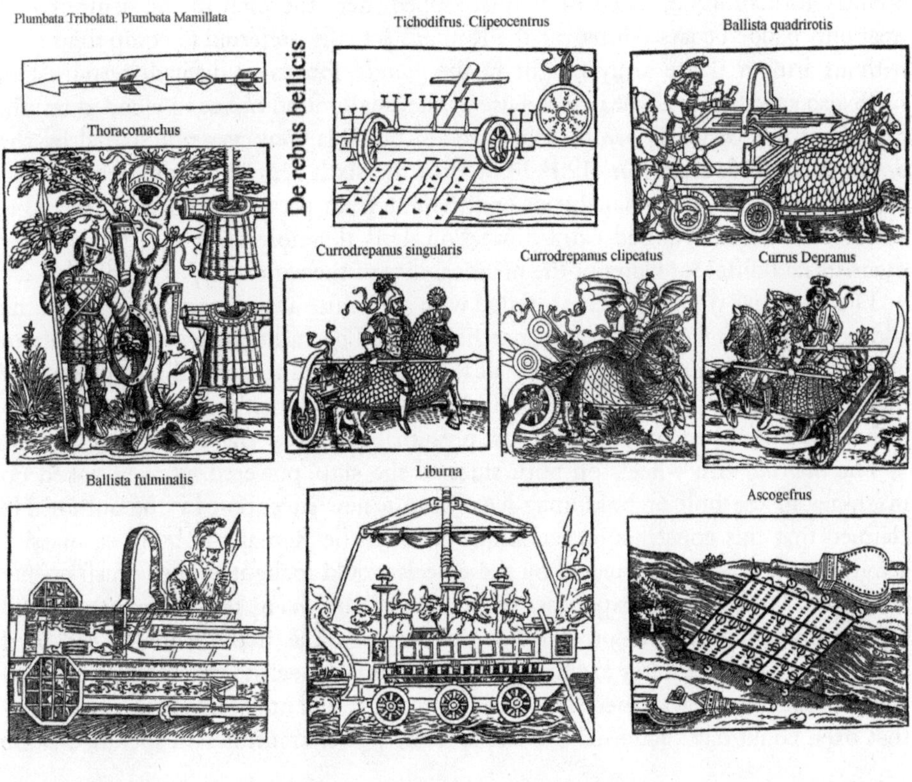

their wars against the Persians during the very same period. The unfortunate fact is that this is not visible from our narrative sources because these give us actual details of combat very sparingly.

In the case of the *thoracomachus* (a defensive padded undergarment worn underneath *lorica* or *cliuanus* known as *subarmalis*), the author actually admitted that it was a very old invention. He just advocated the wearing of garments on top of the *thoracomachus* to protect it from rain. In the opinion of the *DRB*, a foot soldier was ready for combat when he wore the *thoracomachus*, *socci*-boots, *ocreae*-metal greaves, *galea*-helmet, *scutum* (presumably the regular large *scutum*; the author (19.1) recommended the use of a *clipeus* i.e. the *clipeocentrus* for use in wooded terrain), *gladius*, and *lanceae*-lances (the extant illustrations have just a single *lancea/hasta*). This was all that was needed to fight against *miles* (footman or soldier in general). The only unique feature in this is that the author appears to prefer the use of lighter equipment (no armour on top of the padded garment) over heavy. This indeed appears to have been the preferred type of equipment after the writing of the *DRB* because Vegetius (1.20) criticizes this. Vegetius claims that the Roman infantry had been protected by armour and helmets until the reign of Gratian and that it was because of this lack of armour that the Goths had defeated the Romans with multitudes of archers.

As already noted, the *DBR* appears to have reflected the policies adopted by Valentinian and Valens. They were short of cash, so they had to economize by not equipping all of the new recruits with armour. It was not really necessary for all the men in the infantry phalanx to be armoured. If the front, rear and flanks of the infantry formation consisted of fully equipped men, the men in the centre could wear only padded coats. Moreover, the Romans actually preferred to equip their men without armour if the army fought in the woods, forests or difficult terrain. The *DRB* also specifically advocated the use of the small round *clipeus* in wooded terrain because it was easier to use in dense woods, but this view was not shared by the *Strategikon* of Maurice (*Str.* 12.B.20.7–8) in the sixth century. Maurice advocated the use of larger than regular shields in difficult terrain, presumably because the non-armoured soldiers required extra protection. It is therefore clear that the military theorists had different views of the practicability of the various types of shields.

The *ascogeryfi* were calf-skins inflated with air for use as pontoon bridges with the addition of iron stakes, ropes, and goat-hair mats. The author stated that only about 50 pack-horses and some men were all that was needed to carry these. The use of inflated calf-skins to build a pontoon bridge was an old invention, and all one can say is that the author just favoured their use presumably because these were easy to carry.

The *liburna* with wheels on both sides of the ship, powered by oxen yoked on machines in the hull or hold, may have been a new invention by the author. He claimed that this construction technique enabled the Romans to build so massive a warship that the speed created by the wheels would make it easy to crush enemy warships. Modern era experiments conducted between the seventeenth and nineteenth centuries have proven that this was possible, if the Romans had used horses instead of oxen. For example, Prince Rupert's wheeled vessel outstripped the Royal Barge rowed by 16 men in 1683. However, I would not preclude the possibility that oxen could have achieved the same, if the power transmission included some features that increased the speed of the wheels progressively from one wheel to another. Nevertheless, we still do not know for certain if the Romans ever used this type of ship in combat.[17] All that the modern experiments prove is that it would certainly have performed as explained.

The most important section of the treatise (19) concerns the tactics that the author envisaged for Roman armies when they were to use the above-mentioned pieces of equipment, and instructions for the use of the civilian paramilitary forces together with the fortifications for the protection of the frontiers and borders. The author stated that the *dux* (general) was to array his soldiers (presumably the 6,000-man legion) in three *agmina* with intervals in between so that each had no more than 2,000 troops. Notably, this is the way how Theodosius I the Great deployed his army at Poetovio. It should also be noted that it was possible for each of the separate 2,000 to 6,000 strong units in a single large hollow square/oblong to be deployed as separate hollow oblongs or double-fronted phalanxes to give each unit an all-round defensive perimeter if the enemy managed to break inside the massive hollow square/oblong.[18]

On the basis of the *DRB*, it is therefore likely that after ca. 366 the Roman infantry phalanxes consisted of the 2,000 men *moirai* (each deployed as a hollow oblong) that were grouped together to form 6,000 men *mere* (legions). The author advised that when the army was operating in icy regions, with the implication that he expected the army to operate also during winter (note how a similar order by the emperor Maurice led to his downfall in 602), the soldiers were to wear the *thoracomachus* with the other relevant pieces of equipment (see above) so that they could withstand both the cold and missiles. The soldiers were also to use the *clipei* (small round shields), because the dense woods would make a larger shield cumbersome. The extant works of art from the fourth century confirm that the Romans indeed increasingly used the small and medium sized round shields rather than the large oval/rectangular *scutum*, which does suggest the probability that the above instructions reflected reality. The *ascogefyri* were to be used for the crossings of rivers. In short, the extant evidence suggests that the instructions of the *DRB* (lighter equipment etc.) reflected reality.

On the basis of the analysis of the Persian war under Julian and Jovian in 363, the *DRB* instructed the Romans to oppose the Persians with the infantry square formations and a greater quantity of military equipment. The *tichodifrus*, each manned by two men, was to be used to protect the ballistae carts/wagons and the *scutati* (shield-bearers). In other words, the author thought that the Roman heavy infantry *scutati* (= *skoutatoi* = shield-bearers) and *carroballistae* required additional protective measures against the Persians. After all, during the 363 war with Persia, the Persians had managed to penetrate inside the Roman hollow square marching formation, which had left the *carroballistae* which were inside vulnerable. The *tichodifrus* protected the *scutati* in two different manners: the shield protected them against missiles, while its spikes protected them against cavalry and elephants. We do not know whether the Romans used these movable shields, but what we do know is that they used caltrops against enemy cavalry and that Urbicius (at the turn of the sixth century)[19] also recommended the use of a sort of Spanish Riders to protect the Roman hollow square similarly from the effects of the cavalry charge. It is also of note that the anonymous tenth century *De re militari* (G.T. Dennis's *Campaign Tactics*) includes a childlike drawing of a wagon that may represent a device similar to the *tichodifrus* in the *DRB*. See image 4 below.

It is worthwhile to compare the attitudes of the *DRB* and Vegetius towards the use of scythed chariots in warfare. The *DRB* recommended the use of scythed chariots for pursuit when the enemy turned to flight. Vegetius's view (3.24) was that the sickle-bearing chariots were next to useless, because they rarely found a level field and even a single wounded or fallen horse made the chariot useless, which was easy to achieve with caltrops. This, however, was not the entire truth, because Vegetius also saw a use for scythed chariots. He considered the chariots drawn by cataphracted horses with two fully armoured men (i.e. the *currodrepanus* of the *DRB*) armed with *sarisae* to be very useful as a counter-measure against elephants. This may mean that Vegetius criticized the *DRB*-based experiment made by Valens, but at the same time it may also mean that the Romans had used scythed chariots with great success against Persian elephants. Regrettably, in the absence of detailed information about the Roman campaigns against the Persians after 366 we do not know the reason

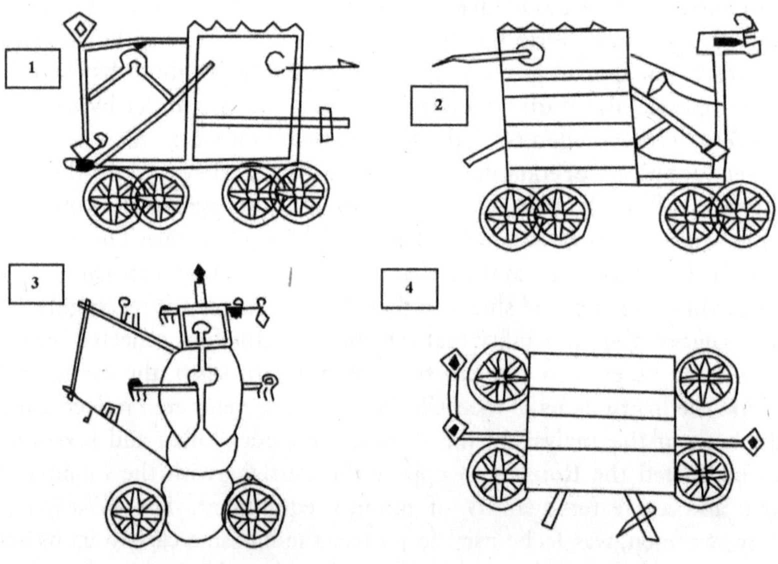

Ballistae wagons/carts and a possible tenth century *"tichodifrus"*

Source: DRM, Vatican Library, Cod.Gr.1164, f.238v, Rome (Dennis, 1985, "Campaign Tactics", 261).

The childlike drawings of military machines from the anonymous tenth century *DRM* demonstrate the remarkable continuity of the Roman military methods throughout the centuries. The figures 1 and 2 are clearly wagon-mounted ballistae and the figure 3 is a cart-mounted ballista, but the figure 4 is less easy to explain. However, since the figure 4 was to be used in the field together with the ballistae wagons and carts and because it bears some resemblance to the *tichodifrus* of the *DRB*, it is in my opinion likely to be the tenth century equivalent of the *tichodifrus*, a device used to protect the ballistae wagons and men operating these.

for Vegetius's comments. It is possible that Vegetius drew these conclusions on the basis of much earlier ancient material but did not see any practical relevance for his day. However, on balance the circumstantial evidence does suggest that the inclusion of scythed chariots in both the *DRB* and Vegetius had a contemporary relevance, so that the sickle-bearing chariots were introduced into the Roman army thanks to the *DRB* and this then resulted in the comments of Vegetius in the *Epitoma rei militaris/De re militari*. In short, there is every reason to believe that the Romans used scythed chariots at least when operating against the Persians from ca. 366 until the fifth century.

Another important tactical detail in the *DRB* refers to situations in which the army could not be deployed for combat. In that case the author instructed the Romans to array their field artillery and machines around the packed formation so that the enemy could not threaten its flanks. With this in mind the *DRB* instructed the general to take twice as many beasts of burden as usual to secure an adequate supply of animals to transport the *tormenta* (field artillery) because it was particularly important to possess a reserve of these to replace those exhausted or killed. The Romans had a long history of the use of field artillery to protect their army and these arguments were sound. The instruction to increase the numbers of beasts of burden was a particularly relevant reform in the aftermath of Julian's failed Persian campaign, which failed in particular because Julian lost too many of his beasts of burden. In sum, the author recommended the use of a variant version of the wagon laager (*carrago, karagos*) which we find the Romans using from this date onwards until the sixth century, and an increasing of the number of beasts of burden as a security measure. For an analysis of these different types of hollow square/oblong and wagon laager formations, see Chapter 9 on infantry battle formations.

Since the *DRB* (20) reflected the policies that Valentinian and Valens adopted it is not surprising to find out that the treatise promoted the building of effective fortifications around the empire. The *DRB* required the building of a continuous line of forts (*castella*) at one mile intervals with strong walls and towers. In order to save money, the forts were to be built by local landowners, and these were to be garrisoned by the *vigiles* (guards = i.e. civilian *burgarii)* and the countryside was to be patrolled by *agrarii* (a paramilitary force of peasants?). The emperors implemented a watered down version of this, because the intervals between the forts and guard towers only rarely reached the recommended minimum distance. It was presumably either too costly or unnecessary to build as many forts as recommended.

In sum, the extant evidence suggests that these regulations (lighter equipment, hollow squares, mobile artillery and machines, strengthening of the frontiers with fortifications etc.) became the military doctrine followed by the Roman emperors from 366 onwards. This doctrine was an expression of the lessons learned in the course of the previous 30 years of wars, and in particular the reflection of the defeat in the war against the Persians in 363. The narrative sources together with the texts of Urbicius and Syrianus Magister (see below) prove that the different versions of the hollow square/oblong (*plinthion, plaision*) and wagon laager (*carrago*) became the standard combat methods for the period from ca. 366 until the 530s. The only other combat methods that can be found in the narrative sources are the standard *plagia falagx* (lateral phalanx), mixed formation (cavalry posted between footmen) and the Italian Drill Formation for cavalry. During this period the commanders appear to have varied the infantry battle formations as follows: 1) if the infantry forces consisted of veterans, the commander could use either the lateral or mixed phalanx; 2) if the infantry consisted of veterans and the situation required an all-round defensive perimeter the commander used the hollow oblong/square; 3) if the quality of the infantry was poor or consisted of untested men, the commander used either the traditional wagon laager, or a wagon laager with extra safety measures, or the wagon laager described by the *DRB* (ballistae carts and footmen protected by the *tichodifri*).[20]

1.5. *Definitiones* (second to third century) = *Hermeneia* (before sixth century)[21]

There also exists a glossary of military terms called *Hermeneia*, which may date from the Late Roman period. The *Hermeneia* is not an original treatise, because it contains the earlier text known as the *Definitiones*. The *Definitiones* (also known as *Glossarium militare* or *Lexicon militare*) in its turn is merely a lexicon of military terms (63 entries with additional subentries) based on earlier military treatises such as Arrian, Aelian, Asclepiodotus and other Hellenistic treatises. According to Dain (1967, 332), the entries 29 and 59 of the *Definitiones* are unique to this lexicon. This, however, is only partially true. We find corresponding information at 29 (*parembole* vs. *parentaxis*) also in Asclepiodotus (6.1), and similarly the referral to the *leptysmos* (thinning of the phalanx by dividing the file of sixteen into two files of eight) at 63 is almost a word-for-word copy of Aelian's text (Matthew ed. 49). In short, there is nothing unique in this text. Notably, the *Peri strategikes* (25.13–5) warned against the thinning of the depth too much in such cases, which is not found in either *Definitiones/Hermeneia* or Aelian. Bury dates the *Hermeneia* to Justinian's reign, but Dain leaves the exact date open. After all, the *Hermeneia* is merely a copy of the earlier glossary in the *Definitiones*. The principal value of the text for this study lies solely in the preservation of the military lexicon of phalanx warfare which remained the same. The exact dating of these two similar treatises is therefore meaningless.

1.6. Urbicius: *Epigrams, Taktikon, Epitedeuma* ca. 505–18[22]

A man known as Urbicius wrote two tactical treatises the *Taktikon* and the *Epitedeuma*, and two epigrams during the reign of Anastasius (491–518).[23] These three texts are of the greatest importance for the analysis of Roman combat doctrine and tactics at the turn of the sixth century. The epigrams were a preface to the *Taktikon* and the *Epitedeuma*. The *Taktikon* in its turn is an epitome of Arrian's *Taktikon* (second century AD) while the *Epitedeuma* presents the author's invention that could be used to protect the infantry against barbarian cavalry charge. My own view is that Urbicius was a *magister militum per Orientem* (*stratēlatēs tēs Anatolēs*) and *patricius* as the later sources claim. Like so many other military appointees of Anastasius, he lacked military experience which he sought to remedy through the study of earlier military treatises. The results of this study are visible in the treatises he wrote.

Urbicius sought to improve the general standard of the period infantry with his treatise that summarized Arrian's treatise on the Macedonian phalanx. Urbicius basically lists all of the typical Hellenistic unit names and sizes for phalangites/hoplites (16,384 heavy-armed), *psiloi* (8,192 light-armed) and cavalry (4,096 horsemen), which were based on the structures of 16, 32, 64, 128, 256, 512, 1024, 2048, 4096, 8192, and 16384; unit orders (*pyknosis*, shields rim-to-rim; *synaspismos*, shields interlocked; *chelône/testudo*, a formation which was either square or round in shape, the equivalent of the *foulkon*); commands (turning, wheeling, about turn, counter-marches, reinforcing etc.); and various types of phalanx formations. The combat formations included: 1) regular Macedonian *plagia* phalanx; 2) *prostaxis* (one *stifos* of 4,096 men of light-armed men placed forward on each of the flanks of the phalanx, a sort of crescent array, which was probably used by Narses at Taginae/Busta Gallorum in 552); 3) *entaxis* (insertion of light-armed between the files of hoplites called by Asclepiodotus *parentaxis*); 4 *hypotaxis* (the light-armed placed behind the flanks of the phalanx – Arrian's original text likened it to the *epikampios* = *epikampios opisthia*); and 5) hollow square or round *chelône/testudo* (tortoise), with wagon mounted ballistae on the basis of the *Epitedeuma*. The baggage train, the *touldon/skeuoforoi*, was to be arrayed as follows: 1) if it was expected that the enemy would be in front, the baggage train was to be placed behind the army; 2) if it was expected that the enemy was to the right, then the baggage train was to be on the left of the army; 3) if it was expected that the enemy was to the left, then the baggage train was to be on the right of the army, 4) if the enemy was suspected of threatening all sides, then the baggage train was to be placed inside (i.e. inside a hollow square/oblong or circle or double phalanx).

However, Urbicius knew that inexperienced footmen were not really ready to use the complex manoeuvres represented by the *Taktikon*/Arrian so he recommended in the *Epitedeuma* the use of the new anti-cavalry devices invented by him alongside the wagon mounted ballistae (*carroballistae, ballistroforoi hamaxai*) and the hollow square formation to negate the moral impetus of the barbarian cavalry charge. He sought to negate the psychological impact of the impetuous barbarian cavalry charge (the *impetus*) in the minds of the soldiers by building around it an encircling fence (this was his new invention) outside the infantry square at such a distance that the enemy arrows would be unable to reach the soldiers. When this took place the ballistae wagons (*carroballistae, ballistroforoi hamaxai*) on the outer edges of the square began to shoot at the attacking enemy horses. The ballistae outranged the bows significantly (ballistae range was three bowshots) so this was expected to work well. The combination of the ballistae and fence would result in the death of the foremost barbarians so that these would pile up in front of the fence and also on the nails of the fence. The following ranks of barbarian cavalry would only pile up in the same place.

1.7. *De scientia politica dialogus / Peri politikes epistemes* (*On Political Science*) ca. 530–550[24]

The surviving portions of the *On Political Science* (*Peri politikes epistemes/De scientia politica dialogus*) traditionally attributed to Peter the Patrician also contain a short treatise dealing with military matters. The treatise is usually dated to a period before 532, but I prefer to date it to the period from the late 540s to the 550s because the importance of having both infantry and cavalry present would have been the target of particular attention at a time when Procopius felt it necessary to defend the use of cavalry in his *Wars* and Narses had achieved great advantage from the use of infantry. Furthermore, the treatise concentrates on the use of the lateral phalanx (*plagia falagx*) and cavalry, with the implication that its information is more relevant to the period when the infantry was expected to perform better than when Urbicius wrote his treatise. Most of the treatise deals with the relationship between the civilian population and the armed forces. The most important piece of information in the treatise concerns the infantry. The training scheme shows that the heavy armed footmen were expected to fight like ancient hoplites by using shields and *doru*-spears. The text stressed the value of the infantry in combat. The infantry was deployed with cavalry wings, separate flank guard cavalry next to them and with reserves of infantry and cavalry where needed. On its own the cavalry was deployed as two lines. In short, the treatise shows the Romans using infantry phalanxes in which the men were armed like hoplites and it also shows the Romans using their cavalry and infantry with reserves. The unknown author lamented the period preference for cavalry because in his opinion infantry and cavalry were complementary and equally important for success. This demonstrates nicely the sixth-century prejudices of the period military men against which the unknown author and others spoke.

1.8. Syrianus Magister: *Peri strategikes/strategias, Naumachica* and *Rhetorica militaris* (ca. 545–52)[25]

Modern consensus opinion among historians is that Syrianus Magister (Syrianos Magistros) is the author of three military treatises entitled *Peri strategias/strategikes* (also known as *De re strategica*), *Naumachica* and *Rhetorica militaris*. Even if we cannot know this for certain, it is quite probable that this is the case. The name Syrianus comes from the *Naumachica*. The other two treatises were previously credited to anonymous authors.

Syrianus's treatise is a compilation of earlier materials and personal observations and instructions. Most of the modern authors suggest that Syrianus used the lost naval treatise of Aeneas Tacticus (ca. 350–340s BC) as his source for the *Naumachica*. This is possible, but equally well he could have used the lost naval treatise of Aelian (Devine ed. p.43, 2.1 on p.45; Matthew ed. 2), which modern research has established to be his principal source for the infantry tactics. The *Rhetorica militaris* was modelled after the theoretical handbook on oratory written by Hermogenes (ca. 160–230 AD), but the actual contents are the work of Syrianus himself. The *Peri strategikes* is a compilation of material drawn from Aelian, Apollodorus, *De arcus usu* and some other unknown military and political treatises which Syrianus has updated with his own material and streamlined to form up a unified text.

None of the three treatises contain any specific dating information, which means that Syrianus's treatise has to be dated on the basis of its contents. The inclusion of the name Belisarius gives us the date *post quam*, and Leo VI the Wise's use of Syrianus's treatise as a source gives us the *ante quam* date.[26] The modern consensus opinion among the historians is that Syrianus wrote his treatise during the so-called Middle Byzantine Era, so that the ninth century is seen as the likeliest era.[27] However, I do not share this view. I am inclined to accept the sixth century dating that H. Köchly and W. Rüstow (1855), and G.T. Dennis (1985) gave to the *Peri strategikes*, which is obviously to be extended to cover the rest of the treatises now credited to Syrianus. In fact, this is also the dating scheme preferred by Constantin Zuckerman (1990, 216), who suggests that Syrianus wrote his treatise in the late sixth-early seventh century. Indeed, the latest possible dating for the treatises would be the early- to late-seventh century for the reasons given below. In short, the likeliest date for the *Peri strategikes* is the reign of Justinian after the Empire had recovered from the plague and now faced a naval threat (specifically the years 545–552), or the early part of the reign of Justin II.

It has also been suggested that the different parts of the *Peri strategikes* were written at different periods of time. This suggestion was first made by Köchly and Rüstow. On the basis of the fact that the *Peri strategikes* books 1–32 were longer and more detailed than books 33–47, they suggested that those were not in their original form, but had been shortened by a later epitomizer. This has caused D. Lee and J. Shepard (29–30) to suggest that these books represent an independent work by an unknown author who then added these books to the original text during the Middle Byzantine period. Neither of these arguments is valid, because the *Peri strategikes* is a compilation which the author has updated with his own observations. The known sources for books 1–32 are Aelian and Apollodorus while it is conjectured that the last books, 44–47, have been borrowed from an unknown archery treatise which has received the name *De arcus usu* in research literature.

In short, it is probable that all of the books contain at least some information that has been borrowed from some other treatise or treatises. As will be made clear below, one of the sources used was clearly a treatise which contained similar instructions for the arraying of the infantry phalanx with cavalry (PST 35), as can be found in Modestus (18) and Vegetius (3.16), and which was no longer valid after the *Strategikon* was published. The use of the different sources obviously influenced the length of the material contained in each book, which in its turn means that the more concise presentation of the material in the books 33–47 cannot be taken to mean that these were either later abridgments or books added to the original by some unknown author.

I will first go through the arguments in favour of Middle Byzantine dating for the *Peri strategikes*, together with counter arguments, after which I will add material which supports the sixth-century dating. Philip Rance (2007, 707–19) has reiterated the standard arguments in favour of the sixth century dating and against it, after which (719ff.) he added new arguments in favour of the Middle Byzantine dating.

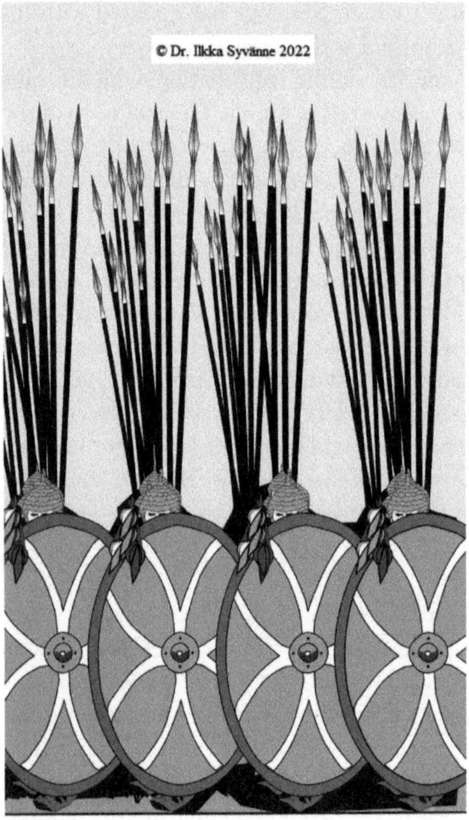

ABOVE RIGHT: Note the *hasta*-spear, sword, hoplite-shield with protruding shield-boss, scale-armour, pteruges, helmet and the arrow-quiver. The shield with the spike could have been used by front-rankers in phalanx formation. This suggests the use of multipurpose troops and phalanx formation later described by Syrianus Magister in the *Peri Strategikes* (16, 27, 36).
BELOW RIGHT: A 'Trojan' archer from the same manuscript. Note the use of scale-armour also by the archers.
SOURCE: Vergilius Romanus, origin probably western from the 4th or 5th century.

ABOVE LEFT:
The infantry phalanx of the *Peri strategikes* (16) frontal view with files ca. 63 cm wide. The spears of the first four ranks reached the front. The front rankers had extra large shields with a spike attached to the shield boss for shield bashes and all men had conical helmets with a spike on the top for head buts.

BOTTOM:
The infantry phalanx of the *Peri strategikes* (16). The hoplites are also depicted as multipurpose footmen with bows and arrows as expected by the *Peri Strategikes* (36, see also *MHLR 6*, 342).

Large infantry shields in art

1) An infantryman in the Altar of Domitius Ahenobarbus (1st cent. BC). Drawn after Bishop and Coulston. Note the size of the *scutum*-shield.
2) A footman in the Glanum stele. 1st century BC. Author's drawings. Note the baldric and horned helmets that predate the late Roman era helmets depicted in the Arch of Constantine the Great by more than three centuries.
3) Heroes duelling, a wedding chest, Breton, 12th century. Yet another example of the use of large shields in the areas formerly belonging to the Roman Empire. Drawn after the photo of the painting in Nicolle (1999, 109). It is probable that this was the typical size also for the late Roman shields in situations in which the ancient author referred to the use of extra large shields.
4) Soldiers from the Sienese army at the battle of Sinalunga/Val di Chiena (1363) in the painting of Lippo Vanni (ca. 1370). Drawn by author with some details like decorations left out. The painting shows nicely how the large rectangular shields remained in use until High Middle Ages in Italy.
5) Late Roman era large oval shields in the Geneva silver missorium depicting either Valentinian I (364-375, possible), or Valentinian II (375-392; the least likely), or Valentinian III (425-455, possible). Author's drawing.
6) Late Roman soldiers with large oval shields in the Ravenna mosaic (ca. 500).

Rance has noted that the following arguments do not provide decisive evidence for the dating: 1) the referrals to triumphs (these were used at least until the eleventh century); 2) Belisarius (well-known person who could also be used as an example by later authors); 3) *Arabes* using ambushes (the name *Arabes* was more typical for the Middle Byzantine authors, but can be found e.g. in the text of Evagrius); 4) *kataphraktoi* (typical classification for heavy cavalry later, but can be found in John Lydus); 5) Persians as enemies (this name was also used later to mean Arabs, Turks and others); 6) current emperor inciting enemies against enemies (Justinian famous for this, but typical for most of the Roman emperors); and 7) the prominence of archery (typical also after the sixth century). In addition to this, Rance demonstrates that the absence of the elephants from the *Peri strategikes* is not relevant for the dating, because elephants were not mentioned in the *Strategikon* either. However, he notes that the inclusion of the *Naumachica/Naumachia* in the companion points to the post-649 date of composition, when in his opinion instructions for naval warfare would have been more relevant than in the sixth century – the *Strategikon* includes only a section on riverine warfare. This, however, is a mistake because the sixth century Romans certainly faced serious naval threats in the form of the Vandals and Ostrogoths, both of which could easily have led Syrianus to the conclusion that the inclusion of the naval treatise was also necessary. The need for the inclusion of the naval treatise in the compendium was particularly relevant during the years 545–52[28] because it was during those years that Totila built a sizable fleet. For example, the navy which he took to Sicily consisted of 400 warships and a large fleet of cargo ships. The situation had changed by the time the *Strategikon* was written, hence only the inclusion of the riverine warfare in that treatise.

Philip Rance (2007, 719–37) adds to the above his own evidence for the Middle Byzantine dating of Syrianus's treatise. This consists of the following: 1) Syrianus's description of the marching camp contains instructions to build around the camp a palisade of shields and spears (shield-barrier *skoutarôma*, with spears resting on shields in such a manner that these point towards the enemy) which is unknown from any other source before the tenth century; 2) the use of extra-large shields with a diameter of at least seven *spithamai*; 3) the protection of the hooves of the cavalry against caltrops with iron plates; 4) in the codex *Ambrosianus Graecus* 139 only the texts of Syrianus have not been updated into period Greek; 5) Syrianus's text is unusual because it was produced uncorrupted without any mistakes during the majuscule period of writing.

As regards Philip Rance's first addition to the criteria for dating, I agree with him that there appear not to be any references to the shield-spear *skoutarôma* in the late Roman literature. However, in this context I add the cautionary remark that the absence of this from the literature does not necessarily mean that it was not used earlier. After all, if Syrianus is dated to the sixth century, it can be used as evidence that the *skoutaroma* was used in the sixth century. For example when analysing the evidence we need to keep in mind that there existed a late Roman treatise on castramentation which was later used as a source by Syrianus, Maurice, Leo (*Taktika*) indirectly and by the unknown authors of the *De re militari* and *Apparatus bellicus*.[29] All of these treatises contain some differences and omissions, even though all of them have used the same late Roman source. It is easy to see why Maurice would have left out the details concerning the shield-barrier, because he produced only a short summary of how to build a marching camp when the wagons followed the army (Str. 12.B.22). In this case, the wagons were parked around the four-sided camp, with the ditch dug outside the wagons and the earth thrown up on the inner side with caltrops and pits placed outside. In other words, the wagons were used as a barricade and protection. The *Strategikon* (12.B.20) fails to describe what type of extra defence was added to the marching camp when the army was not accompanied by wagons, as happened when the army traversed in difficult terrain. One can easily imagine that in such situations the footmen would have used their shields and spears as *skoutarôma*. In this context, it is of particular note that the *Peri strategikes* (27–9) does not mention the wagons in the context of the camp, which means that Syrianus was describing an army which was not accompanied by wagons. This means that we can use the *Peri strategikes* as evidence for the sixth century use of the *skoutarôma*. It also pays to note that the name derives from the Latin word *scutum* (*skouton*), which suggests an earlier origin for the device.

As noted, Rance (2007, 723–9) draws attention to the extra-large size of the shield in the *Peri Strategikes* and claims that there does not exist any evidence for such in the sixth century sources. According to the *Peri strategikes* the front rankers in the phalanx formation were to possess shields with at least a seven *spithamai* diameter, so that these had a spike of four *daktyloi* (7.8cm) in the boss for attack. If the *spithame* (span) is interpreted according to the royal span (*spithame basilike*) of 23.4cm, this results in a shield of ca.164cm diameter. He points out that the resulting size is unwieldy and points out, like most commentators have, that it is probable that this refers to the height of the shield. On the basis of this Rance then draws attention to the fact that the *foulkon* (tortoise, *testudo*) combat formation described by the *Strategikon* would be impossible with this shield. As a further point of evidence he notes that Leo's *Taktika*, which has used the *Strategikon* as its source, has updated this information with period material so that the *Taktika* includes a referral to *thureoi* shields, and other shields called *peltai*, and still other shields of iron, round and polished. The *thureoi* shields were large, oblong and sometimes four-sided (*tetragonos*) so that these covered the entire body of a man. Rance notes that Leo is the first Byzantine military writer to describe this shield and suggests that he added this into his revision of the *Strategikon* from Onasander, because his text contains a similar description. Rance notes that the large man-sized shields (four-sided tapering towards the bottom, triangular, kite, tear and almond-shaped) can also be found in the tenth-century *Sylloge tacticorum*, *Praecepta militaria* and Nikephoros Ouranos's *Taktika*. He claims that the large shield in the *Peri strategikes* does not find any parallels in late antiquity, but finds its closest match in the Middle Byzantine period.

This can be demonstrated to be incorrect on three grounds. Firstly, Tim Dawson (2007) has proven that the late Romans and 'Byzantine Romans' used several different variants of the *spithame*, of which the royal span (*spithame basilike/vasilike*) of 23.4cm was only one of the versions. The other two ways to measure the *spithame* were the common span (*spithame koine*) of 19.5cm, and the equation of *spithame* with a *dichas* (8 *daktyls* = 15.6cm). This means that the seven *spithamai* shield has archaeological matches (largest Dura-Europos shield height 118cm, width 97cm[30]) for the known shield sizes noted by Tim Dawson if we use either of these two smaller versions of the span (height with the *spithame koine* is at least 136cm; height with the *spithame/dikhas* at least 109cm). However, this is not even the entire truth of the matter, because the Dura-Europos finds also include a wood and rawhide shield with measurements of height more than 155cm and width 78cm (James, 186–7). The shield in question is usually identified as Persian and the Persians were certainly known for such large shields: Procopius notes that the Goths used large *thureos*-type shields effectively against Roman arrows at the siege of Rome and likened them to the large Persian shields. Most importantly, we have a direct reference for the Roman use of such large shields in the *Iohannidos* of Flavius Cresconius Corippus (4.553–63), who not only refers to the wall of *scuta* covering the bodies of the soldiers but notes that only the tops of the conical helmets, *bibennis*-battle-axes and *hasta*-spears were visible. In this context it is particularly important also to note that the *Peri Strategikes* 16.27–8 specifically stated that the infantry helmet should have a spike (*xifos* = a sword, here a spike) three *daktyloi* long (5.85cm) usable for head butts. This is in full agreement with the details provided by Corippus above for the Roman infantry phalanx under Tarasis and John Troglita at the Battle of Antonia Castra in 547.[31] In sum, there exist both literary and archaeological evidence even for the largest possible variant the *spithame vasilike*, which the *Peri strategikes* gives as a shield height of ca. 164cm. This was the man-sized *thureos*-shield which we can find later in the text of Leo's *Taktika*. The shape of this shield, however, is unfortunately unknown because we possess evidence for both the rectangular and oval large shields. In the accompanying illustration on p.20 I have made the guess that the shield would have been oval but it is equally possible that it was actually rectangular.

Secondly, and equally importantly, the *Strategikon* includes shields of several different sizes, which Rance fails to recognise. Here it suffices to note that in the context of fighting in difficult terrain, the *Strategikon* (12.B.20.7–9)[32] demanded that the lightly-equipped *skoutatoi* (shield-bearers without armour and helmets) were required to use larger shields for their protection. This means that in the open terrain the *skoutatoi* used smaller *skouton/scutum*-shields, with the implication that the referrals to the interlocking of the shields in the *foulkon* array cannot be used as evidence against the use of larger shields during the sixth century. Furthermore, the comparative for 'larger' (*meizosi*) implies that the Romans used even larger shields than the ones used in difficult terrain. These would have been the *thureioi*. The *Strategikon* also required that the light-armed *psiloi* used smaller shields than usual in difficult terrain, with the implication that there also existed a class of larger shields for lightly-equipped troops which is unlikely to have been as large as the regular *scutum* used by the *skoutatoi* in open terrain. The requirement to use a larger than normal *scutum* in difficult terrain is peculiar, because previously the men had been required to use smaller shields when they fought in such terrain.

Philip Rance also sees the requirement to use iron plates as protection for hooves (PST 17) as something that would suggest a Middle Byzantine date because there exists a reference to metal protection for hooves in an eleventh-century historical work.[33] This, however, is not conclusive as he also notes himself.

As Rance states, this is an argument from silence in a situation in which it was typical for the 'Byzantine' authors to leave out the details of mundane equestrian paraphernalia. The metal protection for hooves had been needed as long as armies used caltrops. The so-called horse slippers used by the Romans undoubtedly had metal protection for hooves. As regards the other evidence for the period cavalry tactics in the *Peri strategikes*, the use of the gallop in charge and looser or tighter cavalry orders (horses from the rear rank placed between the horses of the front rank) do not provide secure dating because, as will be made clear in the proper place later, these tactical variations were used during different periods. However, it should be noted that the use of the gallop (vs. canter or trot in the *Strategikon*) would be particularly relevant in the context of cavalry warfare as practised by Belisarius – in fact Belisarius's cavalry tactics are the best fit for the use of the gallop in combat. See *MHLR Vol.6* with index 'Belisarius'.

Rance also notes that the only text which has not been updated into contemporary Greek in the *codex Ambrosianus graecus 139* is Syrianus. In his opinion, this implies that its Greek was more contemporary than the rest of the texts. Rance actually answers this point himself. He implies that Syrianus's text was in need for modernising. Why this did not happen is unknown, but it certainly does not prove that the text was later than the sixth century. The same concerns Rance's last point, which was that Syrianus's text is unusual because it was produced in uncorrupted form without any mistakes during the majuscule period of writing, after which it was transliterated into minuscule without signs of errors. Unlike the claim by Rance, this does prove that the text would not have been written in the sixth century. It is quite possible that the copyists were particularly diligent and in fact the great personal interest of Constantine Porphyrogennetos in this text may explain that.

It has also been claimed that the texts of Syrianus must belong to the Middle Byzantine era on the basis of the instructions of Constantine Porphyrogennetos in *De ceremoniis* (p. 467 in the Moffat and Tall edition) for the emperor to take the treatise of Syrianus with him on a campaign is completely irrelevant, because Constantine also demands the emperor should take the text of Polyaenus/Polyainos with him on campaign. Polyaenus wrote during the second century.

I will now turn to the details that in my opinion prove that the text of Syrianus belongs to the sixth century and in particular to the period from ca. 545–565 (the period 545–552 being likeliest), but conceding that the early successful portion of the reign of Justin II (ca.565–570/2) could also be considered possible. Syrianus includes almost all of the tactical phalanx combat variants (see below) and manoeuvres (doubling, turning, wheeling, deepening and widening of the frontage, counter-marches by rank and file) that one can find in the Hellenistic military treatises, but his specific referrals to the period tactics enable one to pinpoint his treatise to the sixth century before the publication of the *Strategikon* of Maurice. However, there are several differences in the tactical details between Syrianus (*PST* 31–2, 34–7) and Aelian (the alleged source for his information) which will be summarized below. This suggests that Syrianus may have drawn some of his information from some other source(s), or that he has misunderstood Aelian, or that he has altered the source(s) he has used on the basis of his own opinion or experience or period practices, or that all or some of these alternatives are true. As will be made clear below, it is certain that Syrianus used also some other sources for his description of phalangial combat.

Furthermore, it is clear that Syrianus was not a mere copyist. For example, he criticized Apollodorus's pontoon bridge on the basis of his personal experience and he did not blindly reproduce the Hellenistic unit structures (16,384 *hoplitai*; 8192 *psiloi*; 4096 *hippeis*). He merely reproduced the unit terms for the hoplites from 16 men *lochos* up to 4,096 men *falaggarchia* (*lochos, dilochia, terarchia, taxiarchia, syntagma, pentakosiarchia, chiliarchia, merarchia, falaggarchia*) and its structure (ranks, files, front, rear guard, *protostates, epistates* etc.). The descriptions of the cavalry phalanx (see pp.149–51), the statement that in the hoplite phalanx only the spears of the first four ranks reached the front, and the referrals to the multipurpose footmen (spears, swords and bows) demonstrate that Syrianus was describing period practices. Similarly, when Syrianus (*PST* 15.1–5) introduced the phalanx to the readers, he stated that the phalanx could assume different shapes; *kuklos/kyklos* (circle), *rhombos* (rhombus), *rhomboeides* (rhomboid), *embolos* (wedge), *koilembolos* (hollow wedge) and many others. It is of note that Syrianus does not explain what he meant by *rhombos* (rhombus) and *rhomboeides* (rhomboid) and neither does his text explain the *kyklos*/circle array.[34] These were included only as background information because Syrianus (*PST* 15.1–10) chose not to discuss these as so few people had any practical knowledge of tactics – his stated goal was to discuss only those things that his public was familiar with.

According to Syrianus (*PST* 15.11–20), the period Romans were using only hollow square (*plinthion*) and oblong (*plaision* and *difalaggia* with *psiloi* posted between those) formations while his stated goal was that they should also use the other phalanx variants.[35] This is the exact opposite of what Urbicius set out to do. He described the other tactical formations, but stated that in practice the green recruits were capable of using only the hollow square/wagon laager formation with the addition of his new security measures. Furthermore, there is also another significant difference between Syrianus and Urbicius. Urbicius's hollow square was actually a moving wagon fortress (*carrago*) in which at least some of the wagons were actually *carroballistae* (cart-mounted ballistae), and Urbicius was not even satisfied with these protective measures – he proposed to improve it with an 'anti-cavalry fence'. Syrianus's *plinthion* and *plaision* were not protected by wagons, *carroballistae* or anti-cavalry devices. His *plinthion* and *plaision* were formed by *hoplitai* and *psiloi*. This change fits perfectly the improved conditions of infantry during the reign of Justinian. The footmen were no longer green recruits, but regulars who did not need the additional safety measures, the wagons, *carroballistae* and anti-cavalry devices. Syrianus wanted to take the next step forward by having the infantry manoeuvre more offensively.

The *Strategikon* of Maurice[36] (or at least its extant version) took this development even further. In the *Strategikon*, the hollow square (together with the double phalanx with the light infantry placed between) was restricted only to retreating and for marching camps. In other words, Maurice did not envisage the use of the hollow square/oblong as an infantry combat formation even if during his reign it was still used in practice for these purposes – although it should be noted that in the instances in which the Romans used the hollow square during his reign the army was either retreating or defending itself against cavalry forces (*MHLR* 7, 307–8, 311–2). In short, the instructions in the *Peri strategikes* fit perfectly the situation facing the Romans between Urbicius and Maurice.

Peri Strategikes (31–2, 34–8): **the use of phalangial battle formations:**
1) *Orthia falagx* (column) was to be used only when marching and not in combat. In contrast, Aelian (Matthew ed. 35) refers to its use in combat by stating that the *falagx orthia* was opposed by *koilembolos* (a hollow wedge that had a V-shape). In the same context, Aelian (Matthew ed. 35, Devine ed. 36.8) notes that the *koilembolos* was opposed by a *trifalaggia*, so that the flank phalanxes opposed the wings of the *koilembolos* and the centre one remained behind as a reserve. It is therefore unlikely that Syrianus was using Aelian in this case.

2) *PST* 31–2: *plagia falagx* (lateral) was to be used against enemy attempts to outflank or encircle. This is implied by the shape of the formation but not explicitly stated in Aelian. Similarly, according to Syrianus, the *plagia falagx* was used when the Romans intended to outflank (*hyperkerasis*) or encircle (*hyperfalaggesis*). Both types of encirclements (*hyperkerasis* and *hyperfalaggesis*) can be found in Aelian (Matthew ed. 49). The *hyperkerasis* (encirclement of one flank) was used when the Romans outnumbered the enemy. The *hyperfalaggesis* (double outflanking) was to be used when the Romans outnumbered the enemy significantly. If the Romans possessed the same number of men and the enemy attempted to outflank, the Romans were to lengthen their line to equal width. If the Romans had fewer men than the enemy there were two options: 1) if the outflanking enemy force consisted of infantry, then this was to be opposed by the Roman infantry posted on the flanks (this implies the presence of infantry reserves on the flanks, because only reserves could be used in this manner – the referral to the infantry on the flanks in the context of facing outflanking enemy infantry implies that there were also cavalry reserves on the flanks which was not used for this); and 2) if the encircling enemy force consisted of cavalry, then the Romans were to use caltrops while firing arrows and javelins at them (this once again implies the presence of reserves on the flanks that could shoot arrows and javelins when the enemy cavalry ran into the caltrops). Both of these instructions on how to deal with outflanking attacks imply that there were infantry and cavalry reserves behind the flanks. This is once again indicative of the Justinianic dating for the treatise, because it was during his reign that we find the Romans employing either no reserves at all (e.g. at the Battle of Callinicum, see Syvänne, *MHLR 6*, 67–70) or reserves only on the flanks but not in the middle (e.g. the battles of Narses in Italy, see Syvänne, *MHLR 6*, 340–53).

Syrianus (*PST* 34) later presents somewhat similar instructions, which clearly represent either his own input or has been drawn from some other source. Syrianus advised the Roman commander to oppose each enemy phalanx (one, two, or more phalanxes) with his own phalanx (one, two, or more phalanxes). If the enemy used a single phalanx and the Romans had a numerical superiority, the commander was to widen the formation and not deepen it, because the increased depth did not necessarily help those in front. The commander was not to make the formation too shallow when he extended the width of the formation to obtain a single or double envelopment. If the Romans had twice as many men as the enemy, or even more, the commander was to use the double phalanx. See also section 11 below. The double phalanx was to be used for the double encirclement of the enemy (the front phalanx could be divided to outflank the enemy while the support phalanx advanced forward, or the support phalanx could be divided and sent to the flanks of the enemy; Syrianus fails to state which he means) in such a manner that the Romans left open a route for the enemy to flee, the idea being to induce them to flee and not fight. If the enemy used several phalanxes advancing from different directions, then the Roman commander was to oppose each of these with his own. As noted by Syrianus, this could result in the use of the *plinthion* or *plaision* formation (see also section 9 below). The other alternatives were obviously double and triple phalanxes, so that the triple phalanx could look like an *epikampios* (half-square).

The *Peri strategikes 35* gives us yet another version of the *plagia falagx* for a situation in which it faced the enemy only from one direction. It has also clearly been drawn from yet another source. The *Peri strategikes* states that the infantry phalanx stood in the middle and the cavalry on the wings so that the *katafraktoi* cavalry stood next to the infantry and the light cavalry further out. The light cavalry was used to outflank and pursue the enemy while the *katafraktoi* were used to protect them and the infantry. If the enemy fled, the light cavalry pursued while the *katafraktoi* and infantry phalanx followed at steady pace and served as a place of refuge for the light cavalry if needed. The light infantry was placed according to the situation, so that these were sometimes placed in the rear of the phalanx, sometimes on the flanks, and sometimes in front (this took place when the enemy used cavalry). Excluding the instructions concerning the light-armed, we find this same array in Modestus (18) and Vegetius (3.16) who stated that the *loricati* and *contati* heavy cavalry was placed next to the infantry array and mounted archers further out. Notably we know that a copy of Vegetius was produced in Constantinople in 450, so it is possible that Syrianus could have used Vegetius as his source. This array was relevant only when one did not divide all of the cavalry forces into *koursôres* (cavalry skirmishers/pursuers) and *defensôres* (cavalry defenders in close order) in the manner described by the *Strategikon*, which means that it was relevant only before Maurice introduced these standardized structures into the Roman army. The *koursores/defensores* system was certainly used before Maurice,[37] but it is clear that prior to his reign there was far more variety in combat doctrine and units than after he had streamlined and standarized these. It is therefore obvious that Syrianus wrote his treatise before the reign of Maurice and that he has either used Vegetius or some other source containing similar information when describing this array. In the context of ambushes, Syrianus (*PST* 40.35–8) also recommends as a precaution that one should send another group of cavalry to support the pursuing cavalry. Once again this instruction was relevant only when the *koursores* and *defensores* system was not the official military doctrine.

Notably, the *Peri strategikes* mentions the so-called Mixed Array of the *Strategikon* (*taxis symmiktos*) at Chapter 35, but does not recommend its use. This is particularly relevant as Syrianus warned against the use of this array because once the cavalry had advanced forward from the middle of the formation the horsemen could not retire back into the same place in the infantry array without danger. This warning against the use of the Mixed Array was particularly relevant in the aftermath of Belisarius's defeat at the battle of Callinicum in 531 (see *MHLR* 6, 67–70), but which appears to have been forgotten by the time the *Strategikon* was written. This once again points to the reign of Justinian as the likeliest time when Syrianus wrote his treatise.

3) *Loxe falagx* (oblique) was used to break up or disorder either the enemy's close order troops or their wing. If the enemy used *loxe falagx*, the Romans turned their array so that the battle became a battle between two *plagia* phalanxes. Aelian (Matthew ed. 29) only describes the oblique array, not its use.

4) In the *prostaxis/protaxis* formation, files advanced from the phalanx to harass the enemy so that some of the file leaders were in advance of the other files. The idea was to break up the cohesion of the enemy array. The actual name for this array was the *peplegmene falagx* (woven or saw array, Vegetius's *serra*), which Syrianus has failed to understand. A variation has the light-armed *psiloi* (in particular slingers) in front to oppose enemy cavalry. It is probable that Syrianus has not fully understood the differences between the *prostaxis*, *protaxis*, *hypotaxis* and *peplegmene falagx*. See also the *hypotaxis* below. Aelian (Matthew ed. 23, 30) separates the *prostaxis/protaxis* from each other. The *prostaxis* meant the placing of soldiers on one or both flanks in front of the phalanx (fourth, fifth and sixth formations of Vegetius). The *protaxis* meant the placing of the light-armed in front. In other words, there are clear differences between the two texts.

Syrianus's misunderstanding of the *protaxis/prostaxis* and the omission of the *peplegmene falagx* (woven/saw phalanx) could be taken as the saw array being no longer used. Vegetius (ca.390–450) included the saw array (*serra*, 3.17, 3.19), but Urbicius did not include it in his *Taktikon* (505–18) and neither did Maurice in his *Strategikon* (593). On the basis of this, one could perhaps conclude that the array was no longer used in the sixth century, but this would be pushing the evidence too far, because it is clear that Syrianus has confused the *peplegmene* array with the *prostaxis* and *protaxis* and also with the *koilembolos* (hollow wedge) and *embolos* (wedge). See below. The reason for the confusion is that in the case of *peplegmene* some of the men were indeed *prostaxis/protaxis* (forward) position. Furthermore, Urbicius's source, Arrian, also failed to mention this tactical formation and Maurice's text includes only those aspects of infantry warfare that he considered necessary (e.g. Maurice does not provide any detailed discussion of the hollow square formation, even though it was one of the main infantry formations in use). We see a description of this array in Asclepiodotus (11.7), Aelian (Matthew ed. 48) and *Byzantine Interpolations of Aelian* (Devine ed. 48.1–3, Dain ed. M1–3) and in the *Hermeneia/Definitiones* (Köchly & Rüstow ed.57, p.232; Montfaucon, p.512), with the implication that the knowledge of the *peplegmene* array was retained in the Roman armed forces as long as these texts were copied and read. Syrianus just failed to realize that he was describing this formation. Aelian provides us with the best description of the aim of this array. In the *peplegmene* formation men were sent forward so that the array looked like a saw

– on the basis of Asclepiodotus, this was the equivalent of the ancient chequerboard formation or alternatively of units posted like a saw in which different units advanced and retreated. According to Aelian, this array was used in particular against the hollow oblong (*plaision*) with the intention of inducing the men in the *plaision* to break up their order so that some of the units advanced against the projecting parts of the saw which the reserve units in the *peplegmene* array anticipated. The descriptions of Vegetius and Syrianus resemble this, but they do not mention that the array would have been designed in particular against the hollow oblong formation.

5) In the *entaxis*, *psiloi* archers and javelineers were inserted into the files of the *hoplitai* before the phalanx assumed close order, so that the resulting array had one file of *psiloi* between each file of *hoplitai* – the slingers were not deployed outside this formation. This was to be used when the enemy consisted of cavalry. The same solution could also be used to make the array 'thicker' (i.e. close order). Aelian (Matthew ed. 23, 30) only describes the formation, but says nothing about its use in combat. Asclepiodotus (6.1) has a different term, *parentaxis*, for this same formation so is definitely not the source. The *Definitiones/Hermeneia* (29) states that the *parentaxis* meant that files of *hoplitai* were inserted into files of *psiloi*, or files of *psiloi* into files of *hoplitai*. When (*PST* 36) the Romans used multipurpose footmen (these used bows against the approaching cavalry), the entire infantry phalanx obviously put the approaching cavalry under heavy missile bombardment.

6) *Epitaxis* (light-armed *psiloi* posted behind the rear guards of the phalanx, i.e. behind the hoplite phalanx) was used when the infantry phalanx was not deep so that the light-armed could provide effective support for the infantry phalanx. Aelian (Matthew ed. 23, 30) only describes the array, but does not say anything about its intention.

7) *Hypotaxis* (light-armed posted off to both sides of the phalanx) was meant to be used against enemy cavalry when the enemy used an infantry centre with cavalry wings. The slingers were considered particularly effective in this case. Aelian (Matthew ed. 23, 30) only describes the structure (the light infantry posted on the flanks either slightly forward or back of the phalanx so that it looked like a triple-gate) but says nothing about the purpose of the array. Arrian (*Techne Taktike* 26.7) equates *hypotaxis* with the *epikampios* (forward or rearward angled half-square), but this is not found in the *PST*.

8) *Parembole* meant the strengthening of the men posted in forward position (*protaxis, prostaxis*), with men posted behind pushing forward through the intervals (*PST* 31.23–5). Aelian (Matthew ed. 30) provides us with similar information. Syrianus (32.39–41) adds to this that the Romans could also strengthen those posted in front by doubling their number by inserting men between the files (this made the array denser and was the actual meaning of *parembole*) or either in width (reinforcements sent to the flank or flanks of those in front) or depth (reinforcements posted as rear ranks). Asclepiodotus's text (6.1) and *Definitiones/Hermeneia* (29) provide an additional explanation for the term which is not found in Syrianus or Aelian. According to these two treatises, *parembole* actually meant the interjection of the same type of soldiers beside one another so that hoplites reinforced hoplites or light-armed reinforced light-armed. This shows nicely that the different ancient authors had quite different ideas of the meaning of the term.

9) *Plinthion* (hollow square), *plaision* (hollow oblong) and *heteromêkês falagx* (double phalanx with light infantry placed between flanks or in *plaision*) were used when the Romans were unable to engage the enemy in every place they appeared. Aelian (Matthew ed. 35, 41, 48 *plinthion*, *plaision*, *difalaggia*, *tetrafalaggia*) obviously includes similar instructions but with fewer details. According to Aelian, the *plinthion* was used against an *homoiostomos* phalanx (file of 16 in front with another file of 16 following, i.e. two 16 deep phalanxes one after another). This presumably means that the *plinthion* was used against the double phalanx (*difalaggia*). According to Aelian, the best formation against the *plaision* was the *peplegmene* (see above).

10) According to Syrianus, *antistomos* (file-leaders opposite each other, here with the meaning of *koilembolos*-hollow-wedge, i.e. a unified double phalanx, συνημμένη διφαλαγγιᾳ), *amfistomos* (file-leaders on the opposite sides here with the meaning of the *embolon*-wedge, i.e. another version of the unified double phalanx, συνημμένη διφαλαγγιᾳ) and *heterostomos* (file leaders in the first column on one side and in the second column on the opposite side) phalanxes were the best counter measures against the *tetragonos falagx* (*plinthion*) and *heteromêkês falagx* (*plaision* and *difalaggia* with *psiloi* on the flanks). Additionally, Syrianus notes that the hollow-wedge and wedge were both used when the intention was to break up the enemy formation. The *peplegmene* array used against the *plaision* could obviously consist of both hollow wedges and wedges depending on what point of the array one looked at, so the inclusion of both in this context corresponds with the above description of *prostaxis/protaxis* and *peplegmene*. The *heterostomos* phalanx can be found in Aelian (Matthew ed. 42), but neither Syrianus nor Aelian describe how it was to be used in practice. One may make the educated guess that Syrianus expected that the right-facing marching column advanced to the left of the *plaision* so that its front faced the enemy's right flank, while the left facing column advanced either to the right so that its front faced the enemy's left flank (i.e. the Romans attacked both sides of the enemy *plaision*; see also section 11 below) or alternatively that it was wheeled so that its front faced the enemy's long frontal side (i.e. the Roman array looked like the letter L), and vice versa if the left facing column advanced right. In Aelian (Matthew ed. *koilembolos* 35, *embolon* 47; Devine ed. 36–4–5, Devine ed. 47.1–5, Dain ed. L1–5) the infantry hollow wedge was used against the infantry column formation and the infantry wedge against cavalry arrayed in square formation. In other words, there are once again clear differences in the material.

11) Syrianus (*PST* 32.72–8) instructed that the divided double phalanx (διῃρημένη διφαλαγγιᾳ) was to be used either for attacking the enemy frontally, or for attacking the enemy's flank and front simultaneously. Aelian (Matthew ed. 35, *difalaggia* – only mentioned not described) has no such information, but if one assumes that what Syrianus had in mind actually means Aelian's *difalaggia peristomos* (Matthew ed. 40) then we have something that resembles this. According to Aelian, the phalanx was opened up obliquely from the middle so that one phalanx marched obliquely towards the left and the other towards the right (which resulted in an open hollow wedge formation) to surround the opposing square formation. As we can see this is not exactly the same instruction that we find in Syrianus.

12) Syrianus's treatise (*PST* 36) contains a unique method to be used by infantry against cavalry. According to the *Peri strategikes* (16, 27, 36), the Romans possessed multipurpose heavy infantry which was armed with spears, swords and bows. The use of this type of multipurpose force can also be found in Vegetius and the *Strategikon* (12.B.9). According to the *Strategikon*, if there were more than 24,000 footmen present, the commander was to deploy half of the footmen as *psiloi* (light-armed) foot archers, but if he had fewer than 24,000 footmen, he was to assign a third as foot archers. This means that the heavy infantry was also trained as archers, so that the commander could vary the number of foot archers according to the needs of the moment. However, in contrast to Syrianus, in the *Strategikon* the men chosen as foot archers were always lightly equipped (*STR* 12.B.5), meaning that they were not multipurpose footmen in the same sense as in the *Peri strategikes*. The use of this type of heavy-armed multipurpose footman for the reign of Justinian is confirmed by Procopius (see *MHLR 6*, 342). This is yet another piece of evidence for the Justinianic dating of the text. According to Syrianus, it was useful to hide the presence of the multipurpose footmen from the enemy cavalry, so he advised the commander to post cavalry in front as a screen. When the enemy then advanced close, the Romans should withdraw their cavalry to the flanks so that the enemy cavalry suddenly faced the Roman infantry phalanx in front of them. The *Strategikon* (12.A.7) has similar instructions for hiding the presence of the infantry *epikampios opisthia* behind the cavalry screen. Syrianus instructed that when the enemy cavalry then approached the multipurpose footmen, the three ranks rested their spears on the ground, the first two ranks aimed their bowshots at the horses and the third (and others with bows behind them) shot at a higher angle so that the arrows dropped on the enemy from above. The idea was that the enemy horsemen had to choose whether to protect themselves or their horses with their shields. When the enemy approach had been slowed down by the barrage of arrows, the three front ranks picked up their spears and with the fourth rank (*PST* 16.31–9) formed a wall of spears and engaged the enemy. The ranks behind them (*PST* 16.45–7) used their spears, javelins and other weapons as missiles, if opportunity for that arose. There is nothing comparable to this in Aelian or other Hellenistic treatises.

13) If the enemy unexpectedly concentrated such a force that the Romans would be defeated but could not avoid battle, Syrianus (*PST* 37) instructed the Roman commander to occupy some high ground and then fight back by using slings and throwing rocks if the enemy attempted to climb. The next night the Romans were to retreat. If the enemy pursued, the Romans were to resort to the same solution. There are no equivalent instructions in the Hellenistic treatises, excepting those that contain stratagems. If it was impossible to retreat to higher ground, then Syrianus advised the commander to use a ruse. The commander was to draw his army in two to three phalanxes facing the enemy. Into the intervals were to be inserted two to three ranks so that the phalanx would appear as one continuous line equal in width to the enemy. The phalanx was to be posted so that the front ranks were on higher ground and the rear ranks on lower to cover the stratagem. When the enemy approached, the phalanx was moved forward a short distance so that the ranks in the intervals became level with the rear of the formation. According to the *Peri strategikes*, the surprised enemies who faced the intervals would not dare to enter those because they feared being caught up in a crossfire so that in practice in such situations the enemy could fight only against the forward posted phalanxes.

14) True to his defensive character Syrianus (*PST* 38) instructed the Romans also to prepare themselves for the possibility of defeat. This was done by posting selected courageous veteran horsemen (not too many because these did not participate in the battle) about two to three miles behind the phalanx in some concealed terrain (behind hills or trees, or in a wet or dry river bed) if possible. If these men saw their own men in retreat, they were to ride out and cut off the pursuit by showing themselves.[38] In open terrain, these horsemen were posted three or four miles away and they did the same by showing themselves to the enemy. Syrianus also instructed those soldiers who protected the rear during retreat to carry caltrops which they were to scatter during the retreat to delay the pursuers. This was an ancient defensive tactic, which the Roman cavalry employed for example in the battle of Nisibis in 217 (see Syvänne, *Caracalla*, 286). Syrianus's source for this information is not known, but it could be a work of history or a treatise on ruses.

In sum, it is quite clear that the sixth century dating remains the most convincing one, with references to Belisarius and circumstances that were relevant only to the mid-to-late sixth century. It is also of note that the description of the cavalry tactics fits particularly well the period when Belisarius was in charge.[39] Similarly, the use of spear- and bow-armed multipurpose footmen in the *Peri strategikes* (27.11–22, 36), very much in evidence during Justinian's reign (see e.g. *MHLR Vol.6*, 342), makes the reign of Justinian or its immediate aftermath the most likely candidate.[40] The referral to the fact that only the spears of the first four ranks protruded outside the formation similarly point to the sixth century dating (*PST* 16.31–53).

There is even more evidence for the sixth century dating in the sections not directly related to unit tactics. The references to the Persians as enemies, tower structure in the forts, present day prosperity (!) and the building of cities on level ground in the *Peri strategikes* (10–12, 41.12–5) also suggest a sixth century dating. In the context of river crossings, Syrianus (*PST* 19.1–11) notices that some rivers like the Danube (deep) and Euphrates (shallow) required different kinds of methods when one intended to cross them. This instruction was hardly relevant for the period 641–886! The Roman Empire that Syrianus saw before his eyes was a prosperous one that built cities far away from the border (i.e. the enemy was not located at the Taurus Range) and which operated across the Danube and Euphrates.

Other evidence in the *Rhetorica militaris* and *Naumachica* also point to a sixth century dating. The enemies that the *Rhetorica militaris* mentions by name are the 'barbarians', Persians and Scythians (e.g. 39.8–9/39.6–7). The referrals to fighting for Christianity against the infidels who seek to subject these to their own religion (e.g. 9.3, 10.1–2, 36.11, 37.6/37.7) do not necessarily refer to the Muslims, because the Persians in particular were already waging a religious war against the Romans. It was because of this that Chosroes I targeted the city of Edessa in 540 (it was claimed that Christ protected the city) and it was because of this that Chosroes captured the True Cross of the city of Apameia/Apamea during the same campaign. The religious nature of the war against Persia became even more pronounced after the Persians captured Jerusalem and seized its True Cross in 614. There is no specific mention of the Muslims as enemies in any of the three texts. Most importantly, it is clear that the naval treatise *Naumachica* of Syrianus was written before the introduction of Greek Fire, because there is no mention of this in the treatise. This means that the latest period when the *Naumachica* could have been written is before Kallinikos of Heliopolis introduced the weapon in about 672–3 (Theophanes AM 6164–5) or very soon after it. However, the period from ca. 545 to 552 when the Ostrogothic navy posed a very formidable threat for the Romans operating in Italy remains far likelier than a seventh century date.

Modern historians have not found a consensus view of the quality of the military knowledge in these three texts. Some see the author as an armchair theorist while others think that he was a man with military experience. On the basis of his texts it is certain that he had at least some personal knowledge of military engineering and crossing of rivers, and I would therefore suggest that Syrianus was indeed a man with military experience who just used ancient military treatises as his sources like most of the authors of these texts did. The simple reason for this is that these retained their relevance as long as the weapons remained the same.

1.9. *The Strategikon* (ca. 593)[41]

The principal source for the tactics of the Late Roman period is the sixth-century *Strategikon* of Maurice. The reasons for this are: 1) it was intended for practical use; 2) it is based on Roman military traditions; and 3) it includes material which represents also recent developments in warfare.

Maurice states in the *prefatio* that he wrote the treatise as an introduction to warfare because the Romans had neglected their armed forces and the period Roman commanders did not understand even basic principles. It would be easy to think that this is a *topos*, but the evidence proves Maurice correct. The level of knowledge of military matters varied greatly among commanders. For example, the great commander Belisarius and footmen under him did not know that the ancients had used the *salpigx* (trumpet) with two different strains to urge the men forward or to retreat as needed – it was then Procopius who advised Belisarius to employ this system. Similarly, the emperor Heraclius was quite inadequately taught in military matters prior to the winter of 621–2, when he finally took time to study earlier military treatises. This implies that the situation was improved only temporarily during the reign of Maurice, so that Roman military leadership was once again in need of revision studies after the last generals of Maurice's generation retired.[42]

Of particular note is also the fact that Maurice stated in the *prefatio* that his treatise was meant only as an elementary handbook to correct the deficiencies in the military knowledge of his generals (*strategoi*). It was based on the earlier ancient treatises and on Maurice's personal experience. It was meant as an introduction to military science, which the generals were to broaden with a study of the ancient treatises. In short, Maurice claimed no originality, because his text was after all based on Roman military traditions and as already noted most of the things described by the *Strategikon* were already in use by the second century.

Most modern researchers think that we can use the internal information to pinpoint the time period when the treatise achieved its final form. I agree with this view. The *Strategikon* was written after the Avars surprised the Roman cavalry at Heracleia in 592, but before the downfall of Maurice. In short, the text reached its final form between the years 592 and 602. However, there is still no consensus among the scholars regarding the exact date of publication and its author. Depending on the individual, the manual has been variously attributed to Urbicius, Maurice, Heraclius, Justinian (the son of Germanus, Justinian's nephew), Philippicus (Maurice's brother-in-law), and even to the imperial circles in the 590s.

As stated in my doctoral dissertation in 2004, I see no reason to challenge the traditional view that the author of the *Strategikon* was the emperor Maurice – most of the manuscripts name him the author. The author of the text clearly imparts a presence of authority when he instructed the general (*strategos*) to act as instructed. It is also of note that the author expected the officers below the general to be literate, because the general distributed his commands to them in writing.

The treatise appears to consist of two military policy statements (= military doctrine) of the emperor Maurice for two different wars. The original version consisted of the first eleven books (Book 11 ends with an epilogue) or less, because it is possible that Maurice inserted additional material at a later stage. This section deals primarily with cavalry tactics (Books 1–7) and matters relating specifically to the eastern front (Books 2.1, 2.6, 3.8, 7.B.2, 9–10). It is clear that this first half of the *Strategikon* was partially rewritten and updated when the final form of the treatise was produced. It is probable that this first half of the treatise was produced to correct the problems within the Roman eastern army that became evident as a result of two defeats it suffered in 582–3. The first of the instances is the defeat of Roman cavalry under John Mystacon at the Battle of Nymphius in 582. The defeat was caused by the use of a single cavalry line. It is therefore not surprising to find out that the treatise warned against the use of a single cavalry line and that it took into account the lessons learned as a result of the contacts made with the Avars by the previous emperor Tiberius and general Bonus – the Roman cavalry formation needed reserves such as were present in the Italian Drill formation. The defeat of the Romans at the siege of Akbas in 583, resulting from the inadequate use of scouts, also found its way into this treatise, hence the instruction to use adequate numbers of scouts and guards during sieges.

As a result of the defeats at Nymphius and Akbas, Philippicus, the *comes excubitorum*, replaced John Mystacon as *strategos*. This happened either during the spring or summer of 584. All the subsequent military operations on the eastern front reflect the instructions of the *Strategikon*. In sum, it is very likely that the first half of the treatise was produced as a military policy statement by the emperor Maurice at this juncture to ensure successful operations under Philippicus. Philippicus and others close to the emperor are likely to have contributed to this treatise. Therefore, it is probable that this portion of the manual was written during the winter of 583–4. However, it is still possible that the *Strategikon* also contains a number of other layers of dates of composition. A good example of this is that Book 7 summarizes all of the necessary information regarding the use of cavalry, which may imply that the treatise ended at this point at some other time in the history of the existence of the treatise, or possibly that it was just copied into the *Strategikon* from some other pre-existing treatise.

The second half of the *Strategikon* consists of the 12th Book which is divided into four separate treatises, the first two of which deal with the infantry tactics (12.A-B) particularly relevant for the Balkan theatre of war and of two short sections (12.C-D) dealing with the marching camp and hunting as a form of military exercise. The 12th Book, and the updating and streamlining of Books 1–11, was clearly made after 592 as a response to the needs in the Balkans. It is unlikely to be a coincidence that the 12th Book pays particular attention to fighting in difficult terrain with infantry when the Romans invading forces faced the Slavs there. Similarly, the instructions for the use of the *epikampios opisthia* against cavalry betray a similar tendency to calibrate the treatise to the conditions prevailing in the Balkans, in this case against the Avars. It is probable that the 11th Book, dealing with different types of enemies, was also added to the original treatise at this same juncture, because the 7th book (7.A.Pr.) already contains instructions for the cavalry on how to deal with different types of enemies. In conclusion, the 12th Book and the 11th Book (in particular the sections of the 11th Book dealing with the Slavs and Avars) again suggest that Maurice made a military policy statement about the military methodology to be used in the Balkans.

As noted in 2004, Maurice's role in the writing of these instructions is certain, because it was Maurice himself who stubbornly insisted upon the need of the army to winter in enemy territory or in the Balkans despite stiff resistance from the soldiers. The earliest order for the army to winter in the Balkans was made by Maurice in 593. This allows us to pinpoint the writing of the final version of the *Strategikon* to this same year. It is clear that Maurice was assisted by other military men such as Philippicus, Comentiolus, Narses and Petrus (Maurice's brother) and possibly also by Priscus and others, but it is still certain that the final form of the treatise was Maurice's own choice. This is clear from the fact that Comentiolus, Priscus and Petrus were all readier to listen to the complaints of the soldiers against wintering in the Balkans than Maurice. The final version of the *Strategikon* of Maurice was the military doctrine that the generals and soldiers were expected to follow both in the Balkans and in the East and elsewhere in the Roman Empire – Book 11 contained specific instructions also on how to deal with the Lombards, Franks and other 'Light-Haired Peoples'. It should be noted, though, that the version of the *Strategikon* that has come down to us is not necessarily the original text of Maurice because, for example, Book 12 seems to be incomplete.

The *Strategikon* of Maurice is a compilation of earlier and contemporary material that reflected Roman military traditions. The sources for the treatise consisted of independent earlier works that were revised and updated by Maurice and his inner circle. The earlier texts used by Maurice consisted at least of the following: Book 7 of the *Codex Theodosianus* for military laws; the *Taktika* of Onasander; some unidentified earlier military treatise(s) containing Hellenistic and Roman military traditions; and probably some official military pamphlets and drill manuals. It is obvious that the cavalry (Books 1.1–5, 2–7) and infantry (12.A-B) treatises were drawn from some earlier military treatises or drill manuals or their combination. For example, it is clear that the Alan, Scythian, African and Italian cavalry drills were copied from earlier treatises. In fact, one can detect the use of the Alan and Italian drills on the Column of Trajan.[43] The treatise also betrays the influences of the Huns, Avars, Slavs, Heruls, Persians and others. Most of the direct borrowings from the earlier known treatises are among the military maxims (Book 8.A-B). Some of these actually contradict what is stated in the main sections of the text, but all the same these are still relevant for the reconstruction of Late Roman tactics because Maurice clearly considered these relevant enough to include in his treatise.

In spite of the apparent novelty of some of the instructions in the *Strategikon* (e.g. the instructions on how to deal with different types of enemies), it needs to be stressed that the combat methods and tactics in the *Strategikon* were traditional. All of them had earlier precedents and the apparent novelty of some of the sections (e.g. the one on enemies) is likely to be a reflection of the fact that we no longer possess the treatises that Maurice used as his sources. The only real novelty of the treatise appears to be the fact that the emperor saw it necessary to exercise greater control of his armed forces and its combat doctrine than was the case before. Maurice clearly sought to impart some uniformity into his armed forces. The command structure was simplified and combat methods were streamlined. Henceforth, Roman armed forces were expected to follow the same combat doctrine regardless of where the forces were posted. The *Strategikon* forms the basis of our understanding of the military methods of the sixth- and seventh-century armies. However, thanks to its very detailed information, use of earlier treatises, and preservation of the older Roman military traditions, the treatise is also of the greatest importance for an understanding of the earlier periods as well.

As we shall see, the *Strategikon* included almost all of the combat methods used by the Romans during the Late Roman period. These include the Italian Drill cavalry formation, the lateral phalanx, double phalanx, *epikampios opisthia*, hollow square, mixed formation, 'another' formation, column formation and wagon laager (but only as a form of marching camp and not as a marching formation). However, there are also some notable omissions, which include the oblique formations, hollow wedge and opening up of the formation (*antistomos difalaggia*). These can be found from the other extant Late Roman period treatises, so it is clear that these were also used in this period even if Maurice did not consider these necessary for his infantry forces.

It is also of note that Maurice did not see the hollow square/oblong (including its variant the double phalanx with light infantry flanks) as a battle formation, but rather as a formation that was used during retreat or marching. This means that Maurice expected his infantry to be of so high a quality that this was not necessary.

The heavy infantry *skoutatos* (shield-bearer) of the *Strategikon* was equipped with Gothic-style tunics, Gothic shoes, mantlets, a Herulian sword (*spathion*), *kontarion*-spear, plumed helmet, a sling, darts, and *skoutarion*-shield. All should wear *zaba*-armour (i.e. *lorikion*), but if that was not possible then at least the first two ranks. Naturally the padded under armour (*subarmalis*) was also used, even if Maurice does not mention it. The *skoutatoi* were also to wear iron or wooden greaves, or at least the first and last in each file (see for example the *skoutatos* on p.13 of the colour plates). When the *skoutatoi* were equipped for difficult terrain, they did not wear armour or helmets and they changed their long spear into shorter spears or javelins and preferably used the 'larger shield'. The light infantry *psiloi* were equipped with slings, darts, bows, small shields, and arrow-guides (short darts of various shapes that gave longer range and which the enemy could not reuse). Those *psiloi* who could not use bows were to use small javelins or Slavic spears.

Each cavalryman was to have a hooded coat of mail reaching his ankles, an Avar type neck-guard, a helmet with a small plume, a bow, a bow-case, spare bow strings, a quiver with a cover (30 or 40 arrows), two cavalry lances of the Avar type (ca. 3.74m *kontarion*) with a leather thong in the middle and pennon (removed before combat), extra-large cloak or hooded mantle, a lasso, a saddle bag with three to four days' rations, and a sword. The padded under-amour was obviously also used. The horses of the front rankers and officers were to be equipped with protective iron or quilted/padded armour for their heads and breasts, or Avar-type armour. The young foreigners who were unskilled with bow were allowed to use only lances and shields. All horses were to have two iron stirrups, a saddle-cloth, good quality saddles and bridles. With the exception of the iron stirrups (even if the Romans had been using these ever since the first century) and the recent copying of particular types of equipment (Herulian swords, Avar equipment) from the enemies, the regular multipurpose Roman cavalryman that Maurice describes was already in use by the time Arrian wrote his treatise, but before Maurice the Romans had not reformed all of their cavalry forces to follow the same training and tactical doctrine.

Despite the adoption of some specific types of gear from their enemies, the basic types still remained the same as they had been during the second century: heavy, medium and light infantry; multipurpose regular cavalry and foreigners. For the types of forces envisaged by Maurice, see the Plates section.

1.10. *De militari scientia* (mid-to-late seventh century)[44]

De militari scientia is a short text or fragment that exists only in one manuscript, the *codex Laurentianus graecus*, in which it follows immediately after the *Strategikon*. It was first published in 1880 by K.K. Müller under the title 'Ein griechisches Fragment über Kriegswesen'. The beginning of the text is missing. It was therefore named *De militari scientia* by Alphonse Dain. The latest definitive edition with Italian translation was produced by Immacolata Eramo in 2018. All references in the text are to this edition. The extant text contains eighteen chapters of varying topics, but most of the details only concern cavalry combat, which was particularly relevant during the era when the Romans were on the defensive against the Muslims. It has been established beyond doubt that *De militari scientia* has either used the *Strategikon* as its source or alternatively that both texts used the same source. The author and the date of composition are unknown, but the most likely date for its composition is mid-to-late seventh century.[45] The most valuable portion of the text concerns the instructions and commands (*DMS* 10–11) given to the three frontline *mere*-divisions, which are not specifically mentioned in the *Strategikon*. The *DMS* shows in no uncertain terms that each of the three (left, centre, and right) divisions (pl. *mere*, sing. *meros*) was expected to support each other, so that if the enemy in front of any of the divisions fled they were expected to support the division beside them if it had not been equally successful. It is this in particular which suggests the possibility that the *DMS* may have used the same source as the *Strategikon* rather than the *Strategikon* as its source.

1.11. Aelian, early second century and *The Byzantine Interpolation of Aelian*, sixth to tenth century[46]

Aelian's *Taktike theoria/Tactica theoria* from the reign of Trajan is the most important of the surviving Hellenistic tactical manuals on the basis of its influence upon subsequent military thinkers. The Romans continuously produced new copies and interpolations of the text, a number of which are still extant as later copies. Some of these were obviously copies of the original treatise of Aelian, while others are considered to be Roman interpolations. Alphonse Dain placed the earliest version of the known interpolations of Aelian (known as the *Byzantine Interpolated Recension*) to the sixth century. Dain's argument was based on the vocabulary and comparison of the treatises between the sixth and tenth centuries. A.M Devine agrees with this view, in that he too considered the extant version of the *Interpolated Recension of Aelian* to date from the middle of the tenth century, but suggests that it was based on an earlier recension that the 'Byzantine scholiasts' attempted to elucidate by incorporating additional material from other Hellenistic tactical manuals, now lost. The existence of the *Corpus Élianique* (Dain's term) around the fourth and fifth centuries supports this conclusion. However, it is clear that the extant text as it survives in the manuscripts has additions dating from the late-ninth or tenth century. The most obvious of these later additions are the referrals to the habits of the Bulgars, Turks, and Pechenegs (Patzinakians).

The instructions of both Aelian and the Interpolated versions of his text retained their relevance during the Late Roman period because Roman tactical doctrine was based on the Hellenistic model of the infantry phalanx with its light-infantry and cavalry – in fact, Aelian retained its relevance until the nineteenth century as a description of how rank-and-file formations were to be manoeuvred. The copying, interpolation and use of Aelian during the Late Roman period also attest to its continued relevance. In short, the treatise of Aelian with its different variants retained its relevance for this period. I do not include here a list of the different tactical methods because Aelian's treatise included almost all of the tactical methods mentioned above and those will be analysed in greater detail in the context of battle formations.

1.12. Polyaenus/Polyaionos ca. 161 and *The Excerpts of Polyaenus*, sixth century–ca. 850[47]

We know the exact date for Polyaenus's *Stratagems* (*Strategica, Polyainou stratēgēmatōn*) because he dedicated it to the Emperors Marcus Aurelius and Lucius Verus on the occasion of the Parthian War (AD 161–166). Polyaenus's treatise retained its relevance during later periods. The best evidence for this is that in the tenth century the emperor Constantine (Konstantinos) Porphyrogennetos named Polyaenus and Syrianus Magister as the two treatises that an emperor should take on campaign. The use of the stratagems was highly valued in Graeco-Roman military thinking, so this is by no means surprising. It has been speculated that Constantine referred to the *Byzantine Excerpts of the Stratagems* (*Hypotheseis*) or its subsequent later copies and adaptations rather than to the original text of Polyaenus. This is quite probable but not conclusively so.

However, from the point of view of our research – the Late Roman period – the treatise that interests us is the *Excerpts of Polyaenus*. The *Excerpts of Polyaenus* date from before c. 850, and in my opinion the likeliest date is the sixth century on the basis of the use of Roman military terminology and the description of tactics. The obvious problem with this dating is that those would still fit almost any date before 850 but after ca. 500. As the other studies of the text have already demonstrated, it is ultimately impossible to date this text with any certainty.

In my opinion the following details (after my doctoral dissertation) taken from the *Excerpts of Polyaenus* suggest a sixth century dating for the text. These include the details of training, the singling out of cowards, the use of spies, encampments, single combats, the time and place of battle, exhortations, deception of numbers, and the use of wind, sun and terrain (*Excerpts*, 3, 5, 7, 13–16). Of special importance is the special attention given to measures to be used against cavalry, which singles out the use of caltrops, trenches, horse breakers or empty amphorae and the flames of fire (*Excerpts*, 37), all of which were particularly relevant for the fifth- and sixth-centuries when the Romans faced the cavalry-based enemies of the Goths, Vandals, Alans, Huns, Lombards, Gepids, Heruls and Persians.

The tactical formations in the *Excerpts* are based upon variations of the phalanx formation. In other words, these are not of great help in determining the time period for the treatise, excepting those instances which find particular relevance for the Late Roman period. For example, there is a description of the phalanx awaiting the enemy's attack in stationary disposition (17.1); the sudden doubling of the length/width of the line by inserting the second rank men to outflank the enemy (17.2); the use of *epikampios emprosthia* (forward angled-formation, 17.3); the description of the importance of the maintenance of a continuous line of shields (*synaspismos*), without which the phalanx would be a useless crowd of individuals (18.4); and the structure of the phalanx having the first three ranks point their spears with spears of such length that these present an even front while the men behind hold their spears upright (18.8). As can be seen, the phalanx is clearly indebted to the Hellenistic military theorists, which retained its importance throughout the period under discussion. However, there is still one possible clue in it that could be used to pinpoint the time period. It is the referral to length of the spears. The theoretical recommendation of having the spears present an even front can also be found from the *Peri strategikes* (16.40–43), Aelian (Matthew ed. 14) and Asclepiodotus (5.2), which means that this may not be relevant for the dating, but the referral to the length of the spears could be. According to the *Excerpts*, only the first three ranks were expected to point their spears out of the formation, which means that the spears were shorter than the pikes of the Macedonian phalanx – this length corresponds with the length of Late Roman spears. The *Strategikon* (12.A.7.51–8) expected that the spears of either three or four men protruded outside the formation when the Roman infantry adopted the kneeling version against the enemy cavalry. This implies that the spear length in the *Strategikon* and *Excerpts* was roughly the same. However, there are two major problems with this argument too, which are: 1) Arrian in his *Ektaxis kata Alanôn* (16–7, 26) describes the Romans using similar length *kontoi* (the spears of the four ranks protruding in front of the phalanx); and 2) the *Strategikon* became the Bible of the Roman military until the tenth century, which means that its description fits the circumstances described by the *Excerpts*, even if the text dates from the ninth century.

The other tactical concepts put forth by the *Excerpts of Polyaenus* were the various methods of outflanking that involved: a feigned retreat or the sending of soldiers beforehand to outflank the enemy (19); the admittance of a deep enemy wedge formation by opening up the ranks (32.3); the use of terrain and feigned retreat to put the enemy cavalry between the infantry archers in crescent formation (36.3); and ambushes in general (36). The recommended formation of retreat, the double-faced phalanx with the baggage in the middle and cavalry and peltasts behind (46.5), and the use of prisoners as shields (42.2). were also typical sixth-century tactical concepts. The sections dealing with sieges and naval tactics also show similarities with sixth-century practices. Regardless, the problem remains that these same instructions would also be relevant to other periods too. The inner logic of military methods has remained the same throughout millennia when it is the question of using stratagems and ruses to fool the enemy. Therefore, the instructions in the *Excerpts* are simply Polyaenus's original text rearranged and relevant to this period of study only in a very general way.

1.13. *Apparatus Bellicus*, tenth century AD

The so-called *Apparatus Bellicus*, written by an unknown author in the tenth century, contains information that is relevant for the Late Roman period. The first 30 chapters of the *Apparatus Bellicus* are borrowed from the *Kestoi* of Julius Africanus (ca.232–4). The remaining chapters, from 31 to 77, are borrowed from *De arcus usu*, Syrianos's *Peri Strategias/Strategikes*, Aeneas/Aineias, Polyeanus/Polyainos, Aelian/Ailianos and some unknown ancient treatises. It is the last mentioned sources that interest us here, because this part of the text contains a version of the *epikampios opisthia* which dates from the Late Roman period but is earlier than the one described in the *Strategikon*. An analysis of this information can be found in Chapter 9.6. Dain suggested that Chapters 31–59 were borrowed from some unknown collection of earlier military material. Dain named this treatise *Exercitationes*. Constantin Zuckerman has analyzed Chapters 44–7 which deal with infantry and cavalry tactics and marching camps. Friedrich Lammert attributes these chapters to Julius Africanus, but Zuckerman opposes this because Julius Africanus would not have made so many mistakes in the terminology. I agree with this, but with the caveat that one of the versions used by the *Apparatus Bellicus* could still have been borrowed from the *Kestoi* so that the unknown compiler then inserted into it material from another treatise that contained similar material. In my opinion it is probable that the extant description of the *epikampios opisthia* actually contains two layers of ancient sources which were then compiled by the unknown author as one text.[48]

Chapter Two

Background[1]

2.1. Military structures: The types of units and their commanders

The Late Roman era did not signify a break with the past and it did not stop developing. Therefore, Late Roman land combat units consisted of multiple different types, some of which were created earlier during the Principate and Dominate periods (ca. 30 BC–AD 284) and some of which were created in the course of the following 357 years (284–641) that makes up the so-called Late Roman period. In the sixth century, the two basic types of regular units consisted of those created prior to the third century or during the early-third century (*legions, alae, turmae, auxilia, numeri*) and the new types of units created in the third and fourth centuries (*vexillationes, cunei, auxilia palatina, scholae*), but the mixed infantry and cavalry units (e.g. the legions and *cohors equitata*) were divided into separate infantry and cavalry units. The new legions created after the reign of Constantine the Great consisted solely of infantry and appear to have been usually smaller in size than the previous legions, possibly because of corruption. What is certain is that the legions had a paper strength of about 6,000 footmen at least until the reign of Diocletian and that these were usually smaller after his reign. In addition to this, there existed irregularly-sized detachments of both cavalry and infantry drawn from the legions or from auxiliaries (means both *auxilia* and *numeri*) that were originally named *vexillationes*, but which became to be known with the names *arithmoi/numeri* (an unspecified number of men, not to be confused with the units carrying these names) or *drouggos* (throng) when *vexillationes/vexillarii* came to mean the cavalry units drawn originally from their mother units, the legions. The land forces were obviously supported by the naval forces, which included also units that were used in land warfare. These included specialist naval legions and rowers and sailors trained in land combat. In addition to this, the naval forces included artillery and siege engineering specialists who were used in support of the land forces, in particular during sieges. The civilian paramilitary forces (e.g. police forces) and militias were also used in a military capacity when needed.

In addition to this, the Romans employed *foederati*, who were basically mercenaries who fought on the basis of a treaty (*foedus*) made with the Roman Empire. Originally the *foederati* had consisted only of foreigners hired on the basis of individual treaties, but from the reigns of Honorius and Arcadius onwards the *foederati/foideratoi* could also include natives, meaning that, in the Eastern half of the empire, the *foederati* became part of the regular army and by the sixth century the only real difference between the *foederati* and the regular army were their terms of service. In east Rome, the overall commander of the *foederati* was the *comes foederatorum*, a title which is attested for the first time in the early-fifth century.

In addition to this, the armed forces included foreign allies under their own leaders serving only in some specific section of the Roman frontier. By the sixth century, these had received the name *symmachoi* (allies) to separate them from the *foederati* that now belonged to the regular forces.

The Late Roman imperial bodyguard units consisted of the *praetoriani* and *equites singulares Augusti* (both disbanded in 312), *scholae, protectores domestici, excubitores* (created during the reign of Leo I), and possibly also of the private *bucellarii* (eaters of *bucellatum* hard tack) of the emperor and such enigmatic units as the *optimates* (the best). The inclusion of the *optimates* among the bodyguards is uncertain, because these may have belonged to the praesental forces, just like some of the *bucellarii* units that we find in the *Strategikon*.

The late-fourth and fifth centuries saw the birth of a new type of unit, the private retinues called *bucellarii* (hard tack eaters after the hard tack *bucellum*) employed by wealthy individuals. The Roman generals had always had bodyguards, which were usually chosen from regular units. These were called *excubitores*-guards (not to be confused with the Late Roman bodyguard unit), or as *singulares* usually chosen from the auxiliaries to serve as bodyguards of governors. The wealthy Roman commanders had even used gladiators in this capacity, but the large scale employment of *bucellarii* was still a new phenomenon, which had probably been copied from German warlords who employed personal retainers. The typical employers of the *bucellarii* were the emperor, members of the imperial family, senators and high ranking officers. The wealthiest individuals could employ thousands of them. The retinues of Bonifatius, Aetius and Belisarius played very significant, if not decisive, roles in the wars of the fifth and sixth centuries. Major households like this had a separate commander (*efestôs tê oikia, maior domus*), treasurer (*optio*), officers (*doryforoi*), and privates (*hypaspistai*). Their owners, in particular Belisarius, were also in the habit of using the officers of these retinues as commanders of separate armies, detachments, divisions, or as *hypostrategos*. Regardless of their private nature, the *bucellarii* essentially still served as regular soldiers. They swore an oath both to their employer and to the emperor. It was because of this that the emperor could order the senators and their *bucellarii* to the east to fight against the Persians or he could confiscate the *bucellarii* and hand them over to some other employer.

During the Late Roman era the land forces, regardless of their type, were divided into units posted along the frontiers (*limitanei*, *ripenses*), the regional field armies (*comitatenses*) and the central reserve consisting of the palatine forces (both the bodyguard units and other 'praesental/palatine' or federate military units posted nearby) at the immediate disposal of the emperor. The strategic disposition of the forces was therefore based on the concept of defence in depth. In practice, individual units from frontier, regional field armies and from praesental armies could also serve in other places than their regulation army for prolonged periods of time. During the tetrarchy the regional field armies served under the four emperors, but these were eventually placed under regional *magistri militum* (masters of soldiers). The central reserve at the immediate disposal of the emperor was initially under either a *praefectus praetoriani* (Praetorian Prefect) or a *magister officiorum*, but after the *praetoriani* were disbanded the Praetorian Prefects gradually lost their military functions, so that they were primarily in charge of the logistical side of military affairs. In fact, the larger field armies could have separate praesental prefects to take care of their logistical needs, while the office as such was further divided into regional praefectures. The functions of the Praetorian Prefect were overtaken by the already-mentioned *magister officiorum* (commanded the bodyguards) and *comes domesticorum*, and by the *magister equitum* (Master of the Horse) and the *magister peditum* (Master of the Foot),[2] who commanded the land forces making up the field army of the emperor. The renaming of many generals as *magistri militum*, the creation of honorary *magistri*, and the creation of new field armies each under a regional *magister militum* resulted in additional changes, so that when the empire was then divided into eastern and western halves the forces at the immediate disposal of the western emperor consisted of his bodyguards (officially under a *magister officiorum* or *comes/comites domesticorum*) and of those under the *magister peditum* and *magister equitum*, while those at the disposal of the eastern emperor consisted of his bodyguards officially still under *magister officiorum* (or *comes/comites domesticorum)* and of those under the *magistri militum in praesenti* (two praesental Masters of Soldiers). At some point in time the *scholae* received a separate commander called the *comes scholariorum* and the new unit of *excubitores* a commander called the *comes excubitorum*. By then all of the regional field armies (*comitatenses*) were under *magistri militum* of their own, while those belonging to the frontier armies (*limitanei*) were either under *duces* (dukes) or *comites* (counts), but so that the frontier armies were subordinated to the regional *magistri*.

The bodyguard units of the emperor served initially under the *praefectus praetoriani* and *magister officiorum* (Master of Offices), but when the former lost its purely military functions, the *magister officiorum* became the overall commander of all imperial bodyguard units until the appearance of the *comes excubitorum* with his new unit of bodyguards, the *excubitores*, during the latter half of the fifth century. The *comes excubitorum* was in practice independent from the *magister officiorum* so that this position became the most important military office by the sixth century. The emperors Justin I, Tiberius II and Maurice had all served in this office before becoming emperors. In practice, the forces serving under the *magister officiorum* were usually placed under his subordinate commanders, which included (depending on the time period) either the *comes domesticorum* (Count of the Domestics), or the *comes domesticorum equitum* (Count of the Cavalry Domestics), the *comes domesticorum peditum* (Count of the Infantry Domestics), and the *comes scholariorum*. The division of the office of the *comes domesticorum* into the offices of the *comes domesticorum equitum* and *comes domesticorum peditum* meant the use of members of the unit of the *protectores/protectores domestici* as officers of both cavalry units (included e.g. the *scholae*) and infantry units.[3]

The late-sixth century and early-seventh centuries saw further changes in the organization, which was the first step in the direction of uniting all different types of forces under the same unified structure. Either Tiberius II or Maurice introduced the exarchates into Italy and Africa. The office of exarch (*exarchus*) united both the civilian and military administrations under this one person, so that it was easier for him to organize the upkeep of his armed forces. The military titles became Hellenized and uniform during the reigns of Maurice and Heraclius, so that even when the officers retained their Latin titles (*magister, vicarius, comes, dux, tribunus, centurion* etc.) these started to be changed to the Greek titles *strategos* (general, commander of the army), *hypostrategos* (*vicarius*, lieutenant general, second-in-command), *merarches* (commander of ca. 6,000–7,000 men), *moirarches* (commander of ca. 2,000–3,000 men, also known as *taxiarches* when in command of the *optimates* or *doux/dux*, or *chiliarches*), *drouggarios* (commander of an irregular number of troops, but at this time usually in charge of ca. 5,000 men), *archôn* (commander of a *bandon* or *tagma* or *arithmos*, who held the title of *tribunus* or *comes*), and *hekatontarches* (centurion, commander of 100 men).

The second step in the unification and simplification process was the creation of the military *themata* (themes) under Heraclius during the years 613–22. This resulted from the withdrawal of the armies to Asia Minor when the Persians conquered most of the Middle East during those years. The field army and *limitanei* of Armenia became the *thema* of *Armeniakon*. The praesental armies, *bucellarii*, and *optimates* became the *thema* of *Opsikion*. In the eighth century, the *bucellarii* and *optimates* were divided into the separate themes of *Boukellarion* and *Optimaton*. Even if the *thema* of *Thrakesion* is mentioned in the sources for the first time only in 741/2, it is possible that the core of the future *thema* of *Thrakesion* in Asia Minor was created after 622 when Heraclius transferred forces from Thrace to Asia Minor. With the exception of the *thema* of *Opsikion*, whose head was *komes opsikiou*, the commander of each theme was a *strategos*.

The creation of the *themata* removed the distinction between the praesental forces, *comitatenses*, *limitanei* and *foederati*. The principal subordinates of the *strategos* were the officers called *tourmarchai* (*tourmarches* was also roughly the equivalent of the *merarches*), and *drouggarioi/droungarioi* (typically in command of 5,000 men each, i.e. they were basically *merarchai*). In the case of the *thema* of *Opsikion*, the principal subordinates were probably called *domestikoi*. The *tourmarchai* commanded the *banda/tagmata* (flags of 200–400 men) of some particular area assigned to them, while the *drouggarioi* commanded irregularly sized grouping of *banda/tagmata*.

During the Late Roman period the command structure had some unified aspects, while there were also many different variations. At the top of the military hierarchy were always the emperors so that *Augustus* was the senior emperor and *Caesar* the junior emperor. Then there were the unofficial de facto rulers of the empire, such 'generalissimos' as Stilicho, Aetius, Aspar, Ricimer and Odoacer, who did not become emperors themselves but ruled in practice. They had various different official military titles, in addition to which they added the title *patricius* to lend them additional informal authority. In due course of time *patricius* saw an inflation in use, so that most of the generals came to possess this title. In practice the standing of the 'generalissimos' *patricii* vis-à-vis emperor was dependent on their *bucellarii* and personal control of the armed forces which the emperors were too weak to challenge.

In the official hierarchy just below the emperors were the generals who were given supreme command of some large section of the frontier. In the third century Philip the Arab's brother Priscus, Odanaethus of Palmyra and Maximian (before being appointed as emperor) were early examples of this phenomenon, but it did not end with them. For example, the above-mentioned generalissimos acted in this manner. In the sixth century the same practice continued. Belisarius was named *strategos autokrator* (roughly *dux imperator*) successively for the African and Italian wars, so that he had temporary imperial powers to fight and settle the wars. The special commands for individual large theatres of war which continued during the latter half of the sixth century belong to this same category. These commanders were called *exarchai* in Italy and Africa and *strategoi* elsewhere. However, during the Late Roman period, the typical official titles for the generals and senior officers were in descending order of seniority the *magistri militum* (several different grades depending on time period and place), *comites, duces, praefecti,* and *tribuni.* All of these titles saw inflation in use, so that, for example, the title *magister* could be used by commanders who had very few soldiers under their command. It was presumably largely thanks to this inflation of titles that the command structure was simplified and unified first by Maurice and then by Heraclius.

2.2. Intelligence Gathering Organization

The Late Roman intelligence gathering organization consisted of five basic branches: 1) military intelligence (both in the interior and in foreign territory) gathered by regular military units (most of the military intelligence was gathered by units posted along the frontiers) which served under the *duces*, *comites* and *magistri militum*; 2) intelligence operations conducted by the imperial bodyguard units (mainly internal security but could also involve operations abroad, particularly true for the *protectores* and *excubitores*) that served under the *comes/comites domesticorum,comes scholariorum* and *comes excubitorum* (after Leo I created these), with the *magister officiorum* serving as their official commander; 3) intelligence operations conducted by the *agentes in rebus* (General Agents) and by other 'civilian' professions (e.g. the *notarii*), usually for internal security purposes with the *magister officiorum* serving as their overall commander; 4) the elusive *kataskopoi* (spies) who conducted spying missions abroad and who probably served under the *magister officiorum*; and 5) the informers who acted in this fashion in the hopes of rewards. The narrative sources also prove that the emperors could employ anyone who they trusted for internal and external security duties, and it is because of this that we see for example eunuchs of the Imperial Bedchamber wielding vast powers.

For the Romans there existed no clear divisions between internal and external security, and civilian and military intelligence gathering, so that it was possible for the civilian and military administrators to collect civilian, military, domestic and foreign intelligence. The *magister officiorum* served as a sort of spy master for most of the Late Roman period so that he not only served as a sort of prime minister/foreign minister/controller of the public post, but also as head of all intelligence gathering organizations both internal and external. In the eastern half of the Empire, the *magister officiorum* also acquired the duty of conducting yearly inspections of the *limitanei* in 443. His powers as spy master were limited only by the emperor's will, because the emperors could also use whoever they wished as their eyes and ears

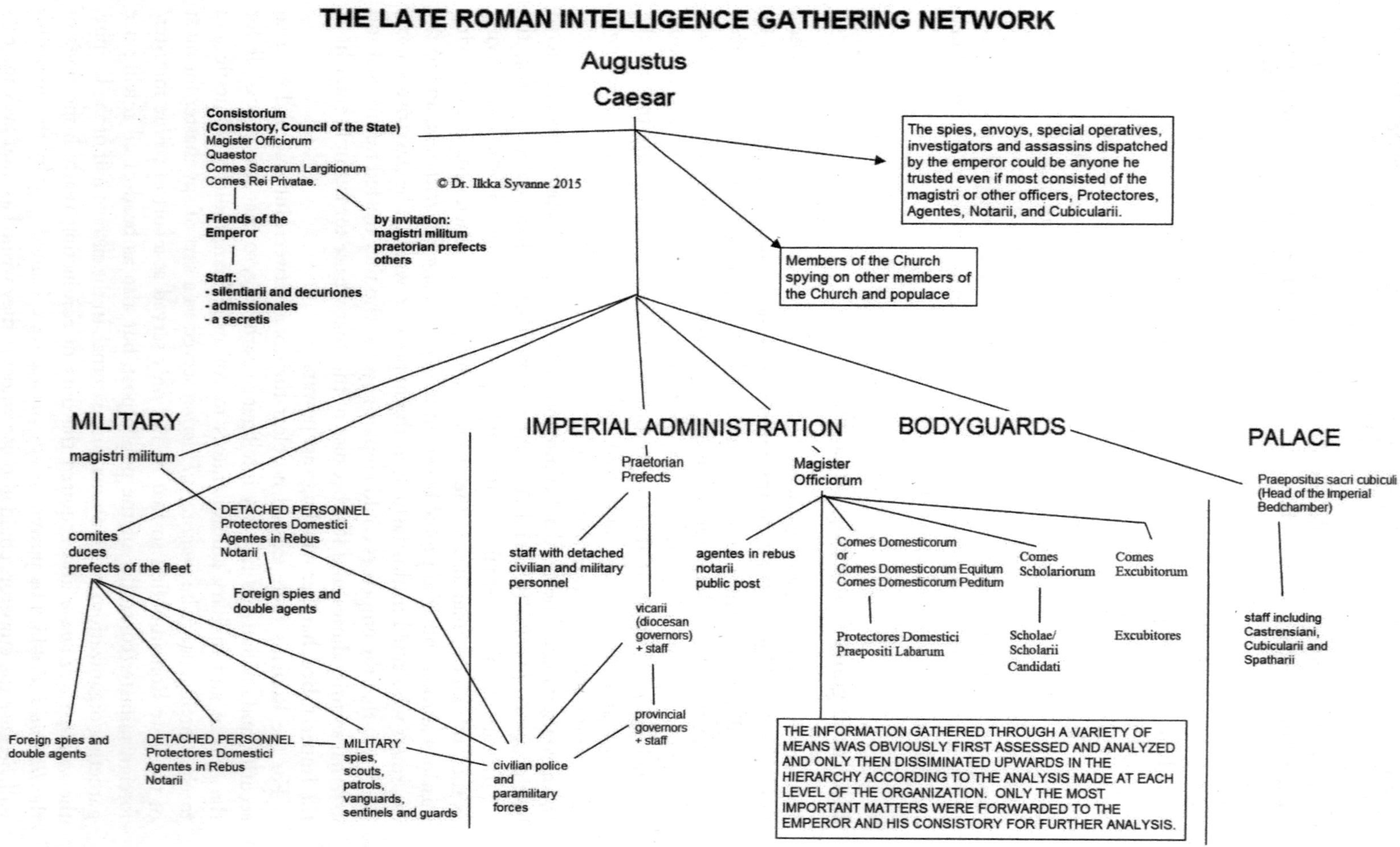
THE LATE ROMAN INTELLIGENCE GATHERING NETWORK
Augustus
Caesar
Consistorium
(Consistory, Council of the State)
Magister Officiorum
Quaestor
Comes Sacrarum Largitionum
Comes Rei Privatae.
Friends of the Emperor
by invitation:
magistri militum
praetorian prefects
others
Staff:
- silentiarii and decuriones
- admissionales
- a secretis
© Dr. Ilkka Syvanne 2015
The spies, envoys, special operatives, investigators and assassins dispatched by the emperor could be anyone he trusted even if most consisted of the magistri or other officers, Protectores, Agentes, Notarii, and Cubicularii.
Members of the Church spying on other members of the Church and populace
MILITARY
magistri militum
comites
duces
prefects of the fleet
DETACHED PERSONNEL
Protectores Domestici
Agentes in Rebus
Notarii
Foreign spies and double agents
Foreign spies and double agents
DETACHED PERSONNEL
Protectores Domestici
Agentes in Rebus
Notarii
MILITARY
spies,
scouts,
patrols,
vanguards,
sentinels and guards
IMPERIAL ADMINISTRATION
Praetorian Prefects
staff with detached civilian and military personnel
vicarii
(diocesan governors)
+ staff
provincial governors
+ staff
civilian police and paramilitary forces
Magister Officiorum
agentes in rebus
notarii
public post
Comes Domesticorum
or
Comes Domesticorum Equitum
Comes Domesticorum Peditum
Protectores Domestici
Praepositi Labarum
BODYGUARDS
Comes Scholariorum
Scholae/
Scholarii
Candidati
Comes Excubitorum
Excubitores
PALACE
Praepositus sacri cubiculi
(Head of the Imperial Bedchamber)
staff including
Castrensiani,
Cubicularii and
Spatharii
THE INFORMATION GATHERED THROUGH A VARIETY OF MEANS WAS OBVIOUSLY FIRST ASSESSED AND ANALYZED AND ONLY THEN DISSIMINATED UPWARDS IN THE HIERARCHY ACCORDING TO THE ANALYSIS MADE AT EACH LEVEL OF THE ORGANIZATION. ONLY THE MOST IMPORTANT MATTERS WERE FORWARDED TO THE EMPEROR AND HIS CONSISTORY FOR FURTHER ANALYSIS.

– during the period from the late-fifth until the early-seventh century this usually meant the use of the *comes excubitorum* as an alternative source of information.

As dictators the emperors considered internal security to be more important than external security, but this does not mean that they would have neglected external security and foreign intelligence. On the contrary, the emperors usually possessed good intelligence about enemy activities, while they lacked the same quality information about potential internal threats – as the many usurpations from this period nicely demonstrate. It is also not surprising to learn that the quality of intelligence gathering operations varied greatly during this long period, but that it reached its lowest levels in the late-fourth century and early-fifth century. This was the result of the weakening of the position of *magister officiorum* and emperor vis-à-vis the *magistri militum*, which was reflected in the simultaneous weakening of the position of the *agentes in rebus* vis-à-vis the *magistri militum*. When the *magistri militum* did not have to fear for consequences, the end result was the massive corruption of the military and the weakening of the availability of accurate intelligence on foreign and domestic foes during the latter half of the fourth century. In the east, the emperors were able to reinstate their own position vis-à-vis their *magistri militum* in the course of the fourth century, with the result that the level of corruption diminished in the east. In the west this problem was never solved and the emperors remained vulnerable to being exploited by their generals, with the result that the West Roman emperors were always vulnerable both against usurpers and foreign foes until the very end.

2.3. Unit Structures 1: Land forces

The Romans divided their armed forces into land forces and navy. The land forces consisted of the cavalry (light, heavy, multipurpose, and special types), infantry (heavy infantry 'shield-bearers', and light infantry) and paramilitary forces. The two basic types of land units were those created prior to the third century (*legions, alae, turmae, auxilia, numeri*) and the new types of units created in the third and fourth centuries and after (*vexillationes, cunei, auxilia palatina*, new legions without cavalry after the reign of Constantine the Great, *protectores* and *protectores domestici, scholae, excubitores, optimates, bucellarii, foederati*). The civilian paramilitary forces concentrated mainly on policing functions.

The unit structures, size and hierarchy varied according to the type of unit, but all were based on the tent group called *contubernium* (*dekarchia*), which was determined by the size of the tent. The ten-man infantry tent groups consisted of eight soldiers (one of which was the *decanus/dekarchês*, commander of ten men), a recruit and a servant. In the heavy infantry this eight-man group was deployed as a single file of eight men, which could be divided into a file of four or doubled with other files into files of 16 or 32. The files of the light infantry were organized in like manner. The recruits and servants were usually left behind in the marching camp during combat, but these could also be used to bolster numbers when considered necessary. The internal structure of the cavalry tent groups varied according to the quality of the unit, so that the best elite unit could consist of five horsemen (a single file in combat) and five servants/squires while the regular units had eight horsemen (a single line in combat) and two servants/squires. The servants/squires could be used to bolster the size of the unit when necessary.

The tent groups were then united to form larger units. In the case of auxiliary units and legions, these consisted of units of approximately 80–100 footmen (*centuria/hekatontarchia*), 160–200 footmen (maniple/*manipulus*), 480–555 footmen (cohort, *speira*), and 800–1,100 footmen (milliary cohort). These units or any number of detachments could then be combined to make up the legions or their equivalents so that the divisions in question had ca. 5,120 footmen and 500–600 horsemen, or ca. 6,100 footmen with 726 cavalry. The supernumeraries (officers etc.) are not included in the figures. The other alternative was that the larger battlefield units were formed up from detachments from various units, but it was typical that these too were grouped together to form up units of above sizes. The new non-legionary and non-auxiliary units of the era after Constantine the Great appear to have followed the Hellenic unit structures so that the basic eight men file was united to form up units of 16, 32, 64, 128, 256, 512, 1,024, 2,048, 4,096, 8,192 and 16,384 men. It is probable that the new type of *auxilia palatina* infantry units had a paper strength of about 1,024 men and the new smaller legions a paper strength of 2,048 men, but it is possible that the new legions actually retained their older paper strength of approximately 6,100 footmen. In practice strength could fall well below that, even if in times of war the same units could have more men than their paper strengths. For example, in reality the new legions could have as few as 800–1,200 men. It is therefore unsurprising that in practice these units and detachments of varying sizes were distributed and brigaded into actual combat units as the generals though appropriate. This was possible, because the organizing principles for all types of units were still the same. Confusingly the largest of these ad hoc units/divisions could also be called legions (ca. 6,000 footmen with attached 1,600 horsemen, or separate divisions of 6,000 footmen and 6,000 horsemen) even when these consisted of several titular legions. The division and brigading of the units to form combat units of various sizes is described in the *Strategikon*. It is probable that the *Strategikon* merely codified and unified the practices that had thereto been common practice, even if the ultimate decision had been left up to the commanding general to decide. The *Strategikon* instructed commanders to distribute their men into units of 200–400-man *arithmoi/banda/tagmata*, 2,000–3,000-man *moirai*, 6,000–7,000-man *mere*, and into *drouggoi* (irregular throngs of men) of varying sizes (*Strategikon* 2.2.1–3).

As noted, the resulting command structure in the *Strategikon* is as follows: 1) the whole army was commanded by the *strategos* (general), whose deputy/second-in-command was the *hypostrategos* (lieutenant general). The commander of the *meros* (division) was called *merarches* or *stratelates* and later also *tourmarches*. The commander of the *moira* ('regiment') was called *moirarches, taxiarches* (*moirarches* of the *optimates*), *doux/dux*, or *chiliarches*. The commander of the *tagmata/banda/arithmoi* was either the *komes/comes*, or *tribounos/tribunus*. The *ilarches* (the first of the *hekatontarchai/centuriones*) was the second-in-command (*vicarius*) to the *komes* or *tribounos*. The commander of 100 men was called the *hekatontarches/centurio* (centurion). The *dekarches* was in charge of ten men, and *pentarches* in charge of five men. The *tetrarches*

Late Roman Unit Organization

post third century units	Scholae	Strategikon
tribunus/comes	tribunus/comes	komes/tribounos
(protector detached into units)	domesticus	
vicarius	vicarius	bikarios
primicerius	primicerius	
senator	senator	
ducenarius	ducenarius	ilarches
centenarius	centenarius	hekatontarches
campiductor		kampidouktor
draconarii, signiferi,		drakonarioi, bandoforoi, ornitoforoi (inf.)
primus (inf.)		lochagos (inf.)
circitor (secundus inf.)s		dekarches
tubatores		boukinator, toubator
biarchus		pentarches
semissalis		tetrarches
miles (pes, eques)		stratiotes, (kaballarios,
		hoplites, aspidoforos, psilos)

was the last man in a file. The soldiers (*stratiotai*) consisted of the cavalry *kaballarioi*, heavy infantry *skoutatoi* (shield-bearers), and light infantry *psiloi*.

The command structure among the auxiliary and legionary forces was traditional, consisting of the recruits (*tirones*), soldiers (*milites*) of various pay grades (cavalry *equites*, and infantry *pedites*), infantry *decani* (commanders of ten) and cavalry decurions, centurions, tribunes, and prefects. The temporary units and armies consisting of units detached (originally known as *vexillationes*) from their mother units of varying sizes (from 50 men up to actual armies) had temporary commanders who could be centurions, tribunes, *duces* etc. These temporary units could be called *numeri* or *arithmoi* (or in the case of cavalry also *vexillatio*), but from the third century onwards these ad hoc groupings were also called in military slang 'throngs of men' (Latin sing. *drungus* pl. *drungi*; in Greek sing. *drouggos/droungos*, pl. *drouggoi/droungoi*). The *drungus/drouggos* could also mean irregular combat order. The commanders of the larger *drouggoi* of ca. 1,000–5,000 men or more came to be known as *drungarius* (in Greek *drouggarios/droungarios*) by the seventh century. By the eighth century the *drouggarioi* were subordinates of the *tourmarchai*, but at this time there appears not to have been any distinction in grading. The post-third century units had the command structure depicted in the above diagram, which was then further simplified by Maurice in the *Strategikon* so that both the old (auxiliary and legionary) and new units followed the same command structure (nominal titles being replaced by generic titles).

The antiquarian list of John Lydus[4] of the types of units and soldiers shows how many different types of older troops and grades there existed at the beginning of the late Roman period. The following list of terms is based on the edition and translation of Bundy (Lydus, pp.69–75) and Syvänne (2015) with some changes.

Lydus' Legions:

alai apo ch hippeôn	alae of 600 horsemen (former auxiliary cavalry)
vexillatiōnes apo f hippeôn	vexillationes of 500 horsemen (former legionary cavalry)
tourmai apo toxotôn hippeôn	tourmae of 500 mounted archers
legiōnes, legiones apo hexakischiliôn pezôn	legions of 6,000 infantrymen
tribounoi, dēmarchoi,	tribuni, tribunes
ordinarioi, taxiarchoi, ordinarii	ducenarii and centurions?
signiferai, sēmeioforoi,	signiferi, standard-bearers (during Vegetius day called draconarii)
optiōnes, optiones	chosen men (centurion's deputies/vicars) or registrars
vēxillarioi, doryforoi,	vexillarii, spear-bearing men belonging to vexillationes, i.e. legionary cavalry
mēnsōres	mensores, camp-surveyors
toubikines, salpistai pezōn	tubicines, infantry buglers
boukinatōres, salpistai hippeōn	bucinatores, cavalry buglers
kornikines, keraulai	cornicines, horn-blowers
andabatai, katafraktoi	andabatae, cataphracts; note that the andabatai/andabatae (sing. andabata) were gladiators who fought blind with such helmets that had no eyeholes. This is clearly a referral to the katafraktoi helmets with very small eyeholes. It is of note that both Modestus and Vegetius (see Syvänne, Aurelian and Probus, 242–3) include the cataphracted footmen as a counter measure against elephants. It is therefore possible that the andabatai/katafraktoi mean both the specialist heavily-armoured and spiked footmen used against the elephants and the extra heavy cavalry cataphracts.
mētatōres, chōrometrai	metatores, land-surveyors
archytēs kai sagittarioi, toxotai kai beloforoi	arquites and sagittarii, archers and arrow-bearers
praitōrianoi, stratēgikoi	praetoriani, praetorians, general's men
lagchiarioi/lanchiaroi, akontoboloi	lanciarii, lance-throwers
dekemprimoi, dekaprōtoi	decemprimi, heads of 10 horsemen, decani
benefikialioi, hoi epi therapeia tōn beteranōn tetagmenoi	beneficiales, those giving medical aid to the veterani/veterans.
torkouatoi, streptoforoi, hoi tous maniakas foreuntes	torquati, torc-wearers who wear necklaces (rewarded for bravery), and those who wear maniaces arm-guards.

brachiatoi, ē toi armilligeroi, pselioforoi	brachiati or armilligeri, bracelet-wearers (rewarded for bravery)
armigeroi, hoploforoi	armigeri (armour-bearers), arms-bearers (hoplon-bearers)
mounerarioi, leitourgoi	munerarii, servants or soldiers (munifices) doing fatigues and services
dēputatoi, afōrismenoi	deputati, deputies appointed for a specific task
auxiliarioi, hypaspistai	auxiliarii, auxiliaries (note the use of hypaspistai/shield-bearers for foreign troops which is suggestive for their later use as a term for bucellarii)
kouspatōres, fylakistai	cuspatores, gaolers
imaginiferai, eichonoforoi	imaginiferi, imaginarii, image-bearers, i.e. bearers of the emperor's image
okreatoi, pezoi sidērōi tas knēmas peripefrakmenoi	ocreati, infantry with iron greaves to protect the calves
armatoura prima, hoplomeletē prōtē	armature prima, first arms service
armatoura sēmissalia, hoplomeletē meizōn	armature semissalis, advanced arms practic
hastatoi, doryforoi	hastati, spearmen
tessarioi, hoi ta symbola en tōi kairōi tēs symbolēs tōi plēthei perifēmizontes	tesserarii, who announce the watchword to the soldiers at the time of encounter
dracōnarioi, drakontoforoi	draconarii, the bearers of the dragon standard
adioutōres, hypoboēthoi	adiutores, adjutants
samiarioi, hoi tōn hoplōn stilpnōtai	semiarii, the polishers of arms
baginarioi/vaginarioi/thēkopoio	vaginarii, scabbard-makers
arkouarioi, toxopoioi	arcuarii, bow-makers
pilarioi, akontistai	pilarii, javelin throwers
beroutarioi, veroutarioi, diskoboloi	verutarii, throwers of verutum/spiculum javelin (Veg: shaft 3.5 ft, iron tip 5 in.)
founditōres, sfendonētai	funditores, slingers
ballistarioi, katapeltistai (katapeltēs de estin eidos helepoleōs, kaleitai de tōi plēthei onagros)	ballistarii, catapult-men. A catapult is a kind of city taker/siege engine; it is called by the soldiers/multitude onager (wild ass)
binearioi, vinearioi, teichomachoi	vinearii, wall-fighters or men who fought with the siege sheds
primoskoutarioi, hyperaspistai, hoi legomenoi protēktōres	primoscutarii, shield-bearers who are now called protectores
primosagittarioi, toxotai prōtoi	primosagittarii, first archers (i.e. mounted bodyg. or commanders of LI?)
klibanarioi, holosidēroi. kēlibana gar hoi Rhōmaioi ta	clibanarii, the horsemen who wear iron armour, for the Romans call iron coverings

sidēra kalummata kalousin, anti tou kēlamina	*celibana*, that is to say *celamina*
flammoularioi, hōn epi tēs akras tou doratos foinika rhakē exērtēnto	flammularii, who bear at the end of their spears scarlet banners
expeditoi, euzōnoi, gymnoi, hetoimoi pros machēn	expediti, well-girt, lightly clad and mobile, ready for battle (i.e. non-encumbered with baggage train and lightly equipped for ease of movement)
ferentarioi, akrobolistai	ferentarii, skirmishers
kirkitōres, hoi peri tous machomenous periiontes kai chorēgountes hopla mētō epistamenoi machesthai	circitores, who go about the fighters and give them arms
adōratōres, beteranoi, teirōnes	adoratores, honourably retired soldiers; veterani, those who had grown old while in service; tirones, recruits not yet permitted to fight.

When reading the list of Lydus one should keep in mind that it was possible to combine several of these specific types in a single soldier so that he could belong simultaneously for example to the *ocreati, brachiati* and *armigeri*. The same is true for the diagram on p.56 depicting the various different types of troops in the Hellenistic military theory, which is still useful for its categorization of the men.

The Roman infantry can be divided into the following general categories:

1) The heavy infantry (*hoplitai, scutati, skoutatoi, skoutarioi*, which consisted mainly of the regulars (legionaries, auxiliaries, *numeri/arithmoi*). The typical heavy infantryman was equipped with a shield (oval, rectangular, round), helmet (ridge, segmented, single-bowl, pseudo-Attic, pseudo-Corinthian), metal or leather armour (segmented, scale, lamellar, muscular), protective padded garment worn under armour (*subarmalis, thoracomachus*), spear (*hasta, lancea, contus, kontarion*) and/or javelin (*pilum, spiculum, akontion*), darts (*plumbata, martiobarbulus*), long sword (*spatha, spathion*), short sword (*gladius, semi-spatha*), and dagger/knife. In addition, the heavy-armed could be equipped with a battle-axe (*bipennis* or *securis*) or pickaxe.

The type of shield used by the troops determined both the classification of the troops into different categories and their use in combat. In the early-tenth century, Leo VI the Wise (*Taktika* 5.2) classified shields into four basic categories, which appear to be representative of the shields of our era too. These were: the *skoutaria* (*scuta*); *skoutaria megala* also called *thureoi* (sing. *thureos*); *skoutariskia* (small shields) of the infantry known as *peltai* (sing. *peltê*); and round and polished *skoutaria sidera* (iron *scuta*). The *skoutaria megala / thureoi* were presumably the large oblong *scuta* which Syrianus Magister described to be about 163.8cm in length (which had the 7.8cm spike in the boss) and which the *Sylloge Tacticorum* (39) stated to be ca. 140.4cm long (described to be almost the height of the man). The width of the shield is not known, but I would suggest that it was ca. 93.6cm – ca. 94cm, the width of the cavalry and infantry file in the close order *pyknosis* (shields ca. rim-to-rim). This was clearly an infantry shield. The *skoutaria* appear to have been smaller *scuta*, which appear to have been used by both cavalry and infantry. I would suggest that the width of this shield was also ca. 93.6cm = ca. 94cm, the width of the cavalry and infantry file in the close order *pyknosis*, while the length was probably about 105.3cm, the size of the large cavalry shield in the *Sylloge Tacticorum* (39.1). Both of these variants were obviously employed by the *skoutatoi/hoplitai* heavy infantry. The *Sylloge Tacticorum* (39.8) also offers a smaller version of the oblong shield for the cavalry which was only 93.6cm long, with the implication that its width was also correspondingly smaller. It is probable that this version was also used by infantry. The *skoutariskia/peltai* were obviously small round infantry shields. It is probable that some of the cavalrymen were equipped with similar small round shields called *parma equestris* or *peltê*. The probable diameter for the *peltê* shield is about 64cm (the width of the shield in Macedonian monuments) or 70.2cm (this diameter of shield used by the cavalry javelineers in the *ST* 39.8). The round iron *skoutaria* were obviously the equivalent of the round bronze hoplite *aspis* and were therefore primarily employed by footmen. One may assume that the iron *skoutarion* could also be constructed from bronze, just as previously. The *Sylloge Tacticorum* (38.1) also includes a round infantry shield that had a diameter of 81.9cm, and one wonders if this was the intended size for the iron *skoutarion*. For cavalry shields, see the analysis in Chapter 6.8.[5]

2) Multipurpose heavy-armed footmen. Ever since the second century BC the Romans had trained at least a quarter to a third of their legionaries to use the bow in combat. This training scheme was retained, so that there were heavily-equipped footmen (e.g. in Modestus, Vegetius and *Peri strategikes*) that carried bows and arrows in addition to their spears and swords. However, when the commander decided that he needed more foot archers, it was still more typical for the heavy-armed dual purpose footmen to be used as real light infantry without the heavier gear (armour, helmet, large shield, spear).

3) 'Medium infantry', the *peltastai* (peltasts). This category consisted of those units (e.g. *auxilia palatina*) that carried lighter gear (i.e. no armour), but which still usually fought in close order or tortoise formation in pitched battles. In other words, medium infantry which fought like heavy infantry. In addition to this, the 'medium infantry' also included the heavy-armed *hoplitai/scutati/skoutatoi* (shield-bearers e.g. legionaries) who had been equipped in light gear without armour, helmet and spear to fight in difficult terrain. The instruction concerning the helmet and shield varied. In the *De rebus bellicis* (*DRB*) the soldier was expected to wear a helmet, while in the *Strategikon* the lightly-equipped *skoutatos* was not expected to use a helmet. The *DRB* preferred the small round *clipeus*, while the *Strategikon* preferred the 'larger shield' (either medium sized or large depending on how one defines this). This means that the preferences had changed slightly by the time the *Strategikon* was written. However, on the basis of the *DRB* it is clear that the medium infantry was not only to fight as heavy infantry in difficult terrain, as implied by the information in the

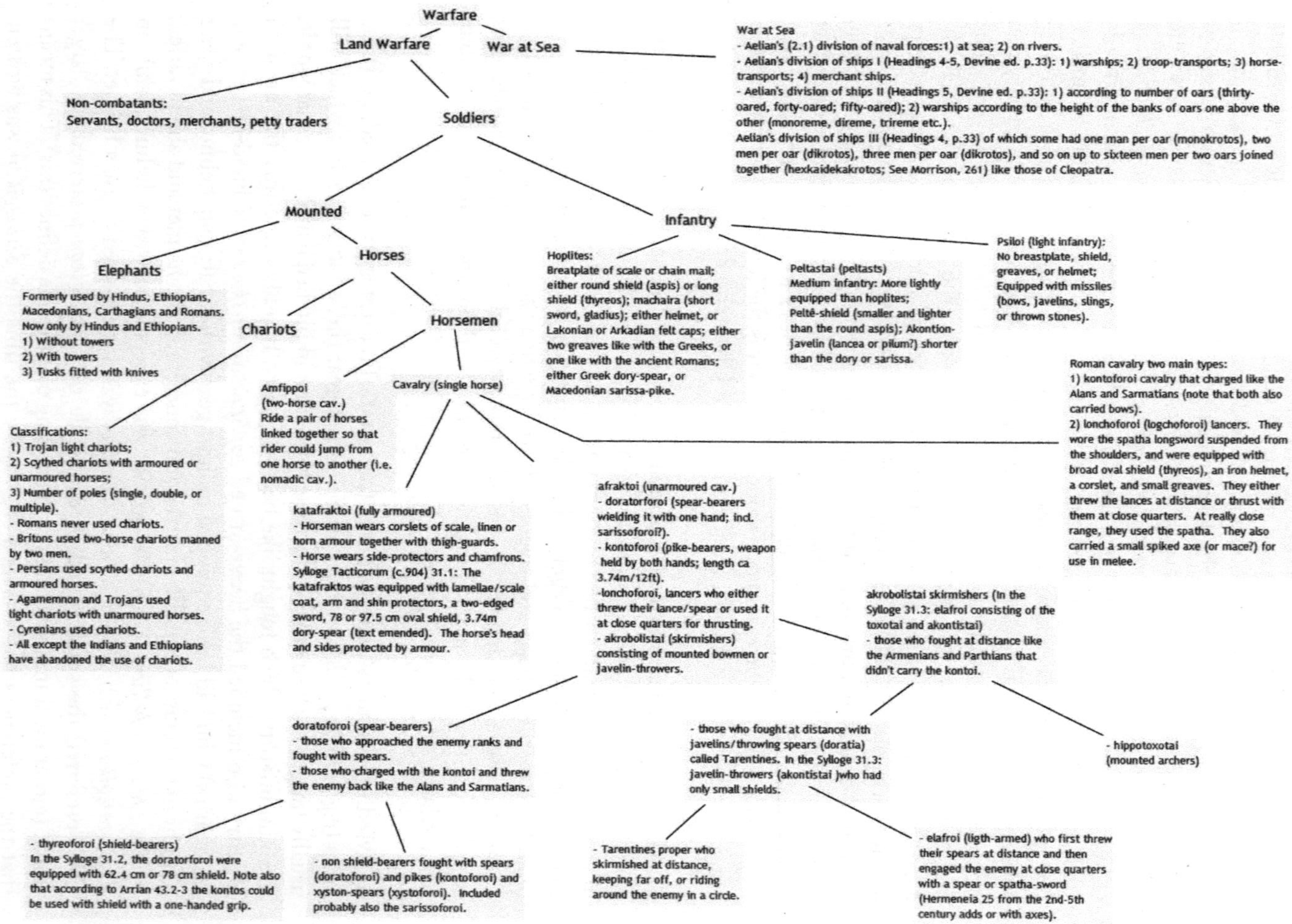

The division of the different arms of service according to the Hellenistic military theory as found in various treatises following this theoretical model.

Strategikon. The principal reason for the discarding of armour in favour of the lighter gear was that it enabled the commander to use the medium infantry as skirmishers in the irregular *drouggos* (throng) formation if there were not enough light infantry *psiloi*. When the *Strategikon* was written such *drouggoi* consisted of light infantry *psiloi*, so that there were three to four javelineers and one archer, but it is likely that such irregularly-organized fighting groups had been used well before Maurice's day because the Roman legions had always possessed men equipped as skirmishers. In addition to this, there also existed a special category of medium infantry which consisted of clubmen/mace-bearers, which on the basis of the existing works of art were usually equipped like a German clubman with only trousers, shoes, wooden club and shield. These clubmen were a specialist force meant specifically for use against cavalry *clibanarii/catafraktarii*.

4) Light infantry psiloi. The light infantry proper consisted of foot archers, slingers and javelineers who were usually lightly equipped. The lightly-armed could consist of specialized light units or of the heavy-armed who had been equipped in light gear either as javelineers (all could be used as such), slingers (all could be used as such), or archers (all could use the torsion crossbow and at least up to a third could use the composite bow, but there may have been time periods during which there were more dual purpose footmen).

Categories of Roman cavalry in 284–583:

The Roman cavalry consisted of several different general categories (stirrups were known throughout the era, but became more common only in the sixth century) during the period 284–583. It is because of this that Vegetius (3.16) stated that the Roman *dux* was to know which type of cavalry *globus/drungos/drungus* he was to employ against which type of enemy cavalry *globus/drungus*. The different types of troops also used different kinds of combat formations (rectangular formations; rhomboids, wedges, irregular). According to Vegetius some troops fought better against some, while other types of troops fought better against others. This means that when Vegetius wrote his treatise the Roman cavalry was still divided into different types of cavalry categories. Procopius's description of the Roman cavalry (*scholarii* using javelins in the same manner as cavalry in Arrian's texts, Heruls with their native equipment, regular cavalry from Thrace, *bucellarii*, Huns, Persian auxiliaries and so forth) during the reign of Justinian I confirms this (see *MHLR* Vol.6 for details).

1) Clibanarii or cataphract super heavy cavalry. The Romans had used cataphracts modelled after the Alans since the second century, but the real *clibanarii* cavalry were actually copied from the Parthians/Persians in the third century. In the fourth century the emperor Constantius II considered this type of cavalry as his elite forces. The typical distinguishing elements were the following: a) both man and horse fully armoured (trooper equipped variously with lamellar, scale, segmented and chain mail; horse armour consisting of: chamfron, crinet, peytral, flancard, crupper) so that only the eyes were visible; and b) the men used the two-handed *contus/kontos* spear/pike at close quarters combat (i.e. the true *clibanarii* did not use a shield, but this category still included also troopers who used the shield while mounted). The *clibanarii* could also be equipped with bows so that they could soften up the enemy with arrows before charging. The Romans appear to have used this type of cavalry until the sixth century, so this type of cavalry was abandoned by the time the *Strategikon* was written. Its principal weakness was the burden that the horse had to carry, which slowed it down in combat and made it less durable over long distances. This category also includes foreign units (e.g. Persians, Armenians, and Arabs) in Roman service.

2) Heavy cavalry/shock troops. This refers to the use of cavalry as shock troops and not to their equipment. The heavy cavalry consisted mainly of the versatile regular Roman cavalry (trained to fight as shock troops, skirmishers and mounted archers), which was equipped for shock combat and close quarters melee. In addition to this certain Federate (*foederati/foideratoi*) and allied cavalries (e.g. Goths, Heruls, Lombards, Gepids, Taifali, Franks, Alamanni, Armenians, Parthians/Persians and Arabs) could fight in this manner. The tactics of the heavy cavalry varied. The only unifying factor was the readiness to charge into contact with the enemy without any prolonged skirmishing, but if it was deemed necessary the regular Roman shock cavalry could also be used for skirmishing as well. Sometimes the heavy cavalry unit contained a missile element within it (could be either bows or javelins or both) so that these peppered the enemy during the charge (e.g. this was the expectation when the *Strategikon* was written) while at other times the unit lacked these (e.g. when the unit consisted of recently recruited Germanic tribesmen) and just charged without any missile fire in its support.

The type of equipment worn by the shock troops varied. The horses could be armoured (e.g. chamfron, crinet and peytral) or unarmoured. The amount and type of equipment worn by the troopers also varied. In some cases, the trooper was fully armoured with shield while in other cases they could discard both shield and armour. The type of martial equipment also varied. Typically, the troopers carried various spears or javelins used as missiles and for shock effect – a) Gallic *contus* (*kontarion*) usable with a one-handed grip; b) *xyston*-spear; and c) *lancea/akontion* or *lanceae/akontia* (if the intention was to throw lances/spears/javelins before engaging the enemy with one spear/lance/javelin retained for melee) – after which they used swords, spears (*lancea pugnatoria*) or axes. It was also possible that the heavily armed troopers carried two to three throwing spears or even more throwing javelins/darts in a quiver/holster that could be used as missiles in combat and pursuit.[6] However, there were also 'heavy cavalry' that were very lightly-equipped (no armour and in some cases no shields either) and which used only swords or spears/javelins, but which still fought like regular shock troops (e.g. the lightly-equipped troopers of Constantine the Great and the Herulian cavalry).[7]

3) Skirmishers, mounted archers and scouting units. The skirmishers were a heterogeneous group consisting of the regular Roman cavalry, Federates and ethnic units (e.g. the *akrobolistai* in Arrian's *Taktika* and the *koursores* of the *Strategikon*).

These could be either javelineers who skirmished in front of their unit (Arrian, *Taktika* 4.3, 4.5, 32.3–42.5 with Syvänne, *MHLR 6*, 355) or mounted archers (Arrian, *Taktika* 4.3, 4.5; see *koursores* in the *Strategikon*, Gammillscheg ed. Index). The tactics employed by the skirmishers also varied. In some cases, they just engaged the enemy with missiles (either javelins or arrows) and came to hand-to-hand combat with the enemy only if it fled, while in other cases they engaged the enemy with a prolonged barrage of missiles and then charged against the enemy in the same manner as the heavy cavalry. The scouting units or selected scouts from the regular units were obviously lightly equipped for this purpose. The typical gear appears to have consisted only of clothes, sword, spear/javelin and shield.

The Roman cavalry in ca. 583–641

As already noted, the cavalry that we find in the *Strategikon* differs from earlier Roman cavalry. The emperor Maurice introduced uniform training, equipment and tactics. All horsemen were trained as mounted archers and lancers and all units used the oblong rank-and-file unit formations the depth of which varied according to the quality of the unit from five ranks to ten ranks. Only young foreigners were allowed to use javelins and spears instead of bows before they learnt the use of the bow well enough.

The equipment of the troopers in cavalry combat formation depended on their position and rank. The officers, the *boukellarioi* (*bucellarii*) and *foideratoi* (*foederati*) had better equipment than the rest. All of the horsemen were to be equipped with a hooded coat of mail reaching to the ankles (lamellar, scale and plate armours were also used), plumed helmet (usually segmented type, but ridge, pseudo-Attican, Pseudo-Corinthian and other types were also used), round neck piece of the Avar type made with linen fringes on the outside and wool inside, bow(s) suited to the strength of each man, bow cases, spare bow strings, quivers (30–40 arrows), two ca. 3.74m lances of the Avar type with a leather thong in the middle of the shaft, shield, sword (*spathion* ca.85–115cm in length), lasso, two iron stirrups, hobble, saddle bag and other relevant equipment, such as an extra-large cloak or hooded mantle of felt, and an Avar-style tunic made from linen, wool or goat's hair etc. The size of the shield probably varied from the small *pelte* to the larger infantry model. Maurice fails to give specific measurements for the cavalry shields, but from the tenth century treatises we learn that the length of the large cavalry shield usually varied from 93.6–117cm, the diameter of the round shields varied from 70.3 to 93.6cm, the small *pelte* being 70.3cm in diameter. The widest versions of the cavalry shields enabled the horsemen to form up a cavalry *testudo* (see p.152). The *Strategikon* further specified that the horses of the officers, two front ranks and the rear rank in a cavalry unit were required to have protective pieces of armour about their heads and breasts either of iron or quilted/padded leather or such as the Avars used. In other words, the position of the horseman in the battle array determined his armament. For example, (see *Strategikon* 3.1–2) in the ten-ranks-deep formation (i.e. in the worst units), the first (or the first and second) and the last ranks were equipped as lancers; the ranks two to four (or three to five) were equipped with bows and no shields; and the ranks five to nine or six to nine with whatever weapon they could handle. The elite forces such as the *bucellarii* were expected to possess fancier looking equipment suited to their elite status. The general was required to ensure that the baggage train carried extra arms, especially bows and arrows, to replace the lost weapons while the horsemen were required to have servants to see to their needs (at least one per three or four soldiers) and pack animals to carry the coats of mail, tents and other equipment.

The accompanying images drawn after works of art depict most of the types of land forces that the Romans fielded during the late Roman period.

A Late Roman Murmillo?

'Byzantine' ivory figurine in the Bargello Museum, Florence, usually dated to the 10-13th centuries. David Nicolle (1992/1994, 24) suggests a 4th-7th century date on the basis of the helmet (crest and face-covering visor), the tall rectangular shield and gaiter-like protection on the left leg.

The figurine does indeed resemble the *murmillo*-gladiator (a chess-piece?) so it is quite possible that Nicolle is correct in his re-dating. However, the fact that the rectangular shield remained in use throughout the late Roman era until High Middle Ages makes the dating uncertain at best. The figurine could even depict Persian *murmillo*! Furthermore, it is always possible that later artists used some older work as their model. However, I have here tentatively accepted Nicolle's suggestion because there is definitely enough evidence for the use of this type of gear for the late Roman period.

Drawn after Nicolle's photo.

Roman elite horseman equipped with scale armour and segmented helmet (Arch of Galerius)

Roman footman from the Arch of Galerius. Note the muscle armour and shoulder pieces.

Infantry archer from Meroe (Nubia). Drawing based on the Arch of Constantine the Great. The legs, bow-string, and mission portions of the arrows emended to the illustration.

© Dr. Ilkka Syvanne 2013

1

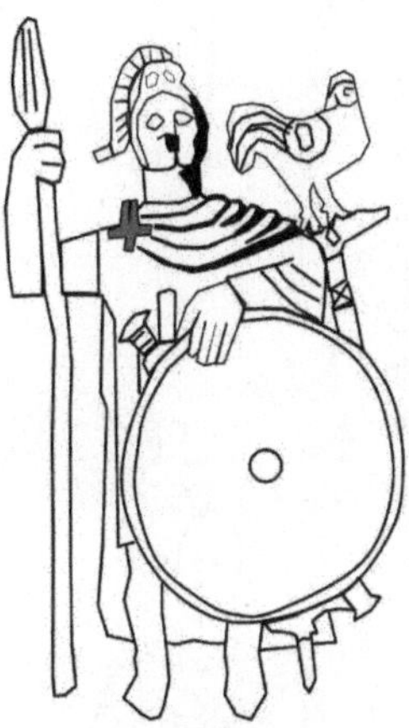

2

Fourth Century Tombstones
1. Lepontius (Strasbourg)
2. unknown protector (Aquileia)
3. unknown (Gamzigrad)
Note the great variety of equipment.
Drawn after Bishop&Coulston with some changes

4 Soldiers wearing coifs in the Vatican Manuscript (drawn after Bishop & Coulston)

3

4

Aurelius Sudecentius, legionary of the *Legio XI Claudia* (first half of the fourth century AD)

Funerary stele of Valerius Ianuarius Ursariensis (legionary 4th century)

Coin of Constantine II
(source: Cohen)
Note the use of the aegis and the victory image of the mounted emperor destroying enemy infantry.

Unknown *miles*,
4th century (Aquileia)

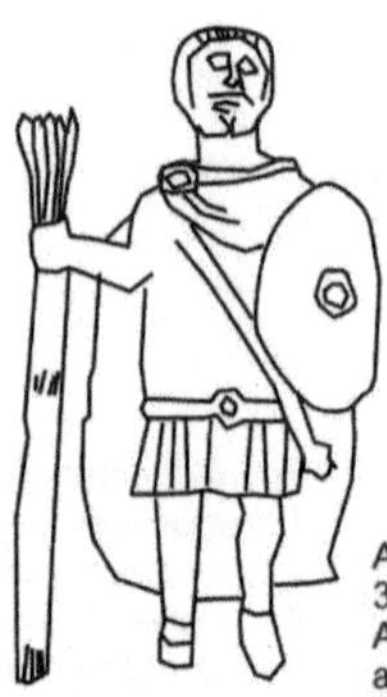

Aurelius Mucianus
3rd century tombstone from Apamea. Equipment: a shield, a sword, no armour and at least five lancea-spears.

Top left: Germanic club-bearers in the Column of Trajan (18th century drawing)
Top centre: Germanic club-bearer in the Column of Theodosius I (no longer extant, drawing by Menestrier).
Top right: Gravestone of Aurelius Alexianus of Sparta wearing *pilos* hat or helmet (typical for ancient Sparta). A member of Caracalla's Spartan cohort ca. 212-217. (drawing by Syvanne after Cowan/Richardson)

Below:
Left: Valerius Maxantius, *Numerus Cataftactariorum*, 4th century AD (Worms)
On the basis of the presence of the chiselled large holes in the armour I have made the educated guess that Valerius was equipped with chain mail armour, which I have 'emended' to the illustration with dots. This same style of armour was later recommended by the 6th century *Strategikon*. Note also the way in which Valerius holds the *contus*-spear. His grip clearly resembles the couched lance technique, and in fact, I would suggest that the Roman heavy cavalry did use this technique with the horned saddle.
Right: A member of the elite scouting unit, *explorator* Respectus (tombstone Heidelberg, drawn after Cascarino/ Flavius). Note the light equipment that resembles the equipment worn by the Moorish cavalry units.

Roman infantry above and Germanic prisoners below in the Column of Theodosius I (no longer extant). Drawing by Menestrier. Note the use of ridge, pseudo-Corinthian, and pseudo-Attican helmets, and also the use of the leather muscle armour.

Above: Roman infantry above and baggage train (*impedimenta*) below carrying Roman spears and shields in the Column of Theodosius I (no longer extant). Drawing by Menestrier. See also Chapter 9.6.
Below: Roman cavalry in the Column of Theodosius I (no longer extant). Drawing by Menestrier.

Top left: emperor in the Column of Theodosius I the Great (no longer extant, drawing by Menestrier). It is probable that the horseman represents the hero of the Column, the emperor Theodosius I himself.

Centre: Three Gothic commanders identifiable from their fur-cloaks in the Column of Theodosius I (drawing by Menestrier).

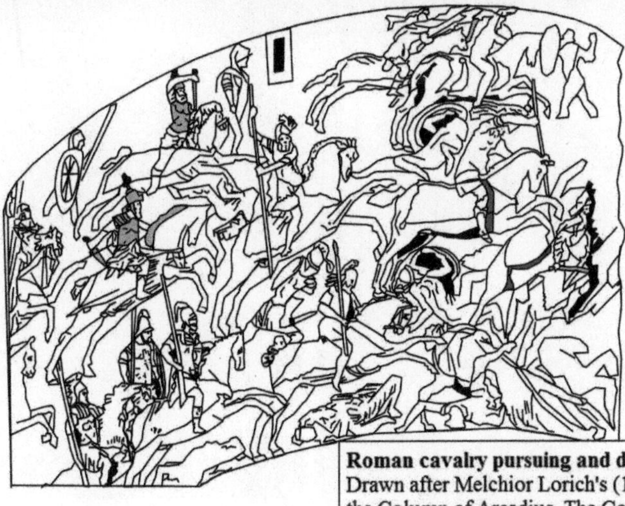

Roman cavalry pursuing and destroying Gainas' forces
Drawn after Melchior Lorich's (1527-1583/1588) drawing of the Column of Arcadius. The Column is no longer extant.

Ilias Ambrosiana (late fifth or early sixth century)
Source: Angelo Mai (Milan 1819), Scene 32.

Note that the Greeks and Trojans wear late Roman equipment.

Note the defensive fighting technique employed by the man on the centre-right, which can be found in most of the scenes of the Ilias Ambrosiana! This does suggest the likelihood that the Romans quite often grasped their cloaks with their left hands so that the lower part of the cloak protected the area below the shield. Needless to say, this was a very good defensive technique.

Isola Rizza Dish (6th cent.) depicting either a Roman or Lombard lancer fighting against Germanic foes. Note the use of the lamellar armour and the two-handed grip of the lance.

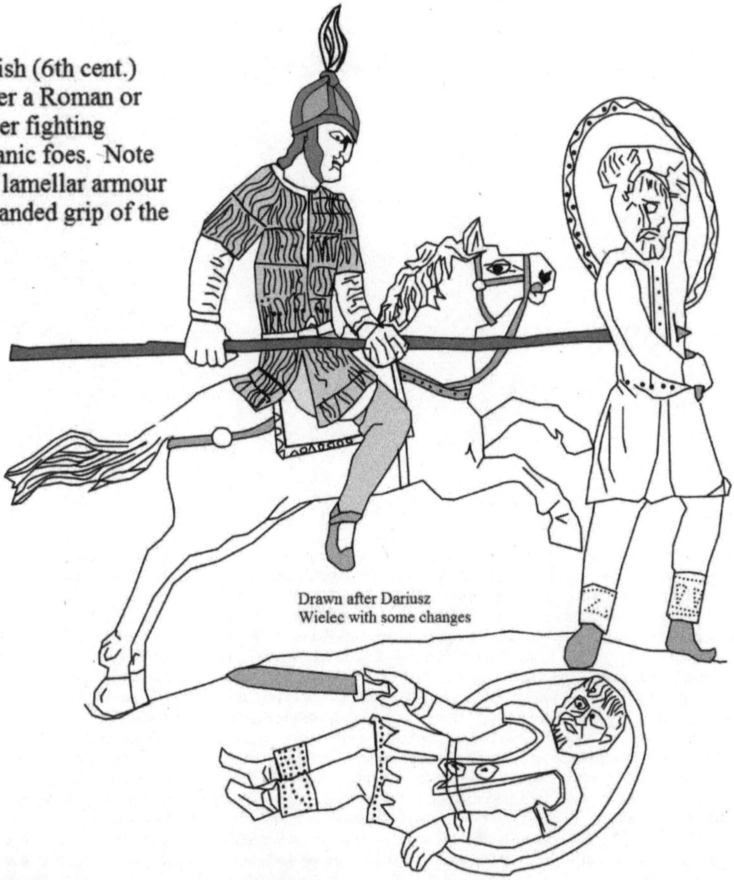

Left: A late Roman mounted archer and legionary.

Drawn after a 6th century Egyptian ivory (slightly simplified), Rheinisches Landesmuseum, Trier).

The shield emblem identifies the legionary as a member of the *legio V Macedonica* (identification of the unit by McDowall)

Top left: A soldier in the "Siege of the Citadel of Faith" (wooden carving, Ashmunayin, Coptic Egypt), usually thought to date from the fifth century. Drawn after Nicolle, 1997, fig.83 A. Nicolle suggests a later dating on the basis of the presence of the stirrups in the carving (he dates it 5-8th cent.), but on the basis of the shield emblem of the *legio V Macedonica* (identification by McDowall) I suggest 5-7th cent. In other words, I am inclined to accept the traditional fifth century dating. Note the use of the rectangular shield. The identification of the armour is problematic. It can represent lamellar or chain mail or scale. Note the use of the rectangular shield.
Top centre: Carved ivory panel in the Pulpit of Henry II, Coptic Egypt (Cathedral, Aachen, Germany) dated to the 7th century, which Nicolle suspects as too early (suggests 7-9 cents A.D). Drawn after Nicolle, 1997, fig.85. Once again on the basis of the *legio V Macedonica* shield emblem I would suggest a date 5-7th cents A.D. It is probable that at this date the trooper did not belong to the legion proper but to a cavalry unit detached from it. However, it is worth noting that this cavalry *vexillation* was still posted alongside with its old mother unit in Egypt, which suggests that the separation of the legionary cavalries from their mother units was not quite as drastic as usually suggested – the other option is that these units had been reattached back to the legions after early fifth century Note the use of the metal *pteruges* for the leg and the use of similar metal protection for the shoulder and upper arm – both widely used in the later centuries. Note also the way how the trooper grasps the spear which suggests the use of the couched lance technique in combat.
Top right: Bronze statuette, Coptic Egypt 6-8th cents (Ch. Ratton. Coll., Paris, France). Drawn after Nicolle, 1997, fig.87. I would suggest on the basis of the shield emblem that it is again possible (but less certain than in the other instances depicted on this page) that the rider depicts a member of the *legio V Macedonica.* If the statuette dates from the same period as the rest, it depicts a trooper who does not use stirrups.

Left: A wall painting of Muslim mounted archer from Qasr al-Hayr al-Gharbī, dated ca. 728/9 AD (Syrian National Museum, Damascus) according to David Nicolle (1992, 17-9). The archer is a hunter who pursues a deer. The military equipment and horse furniture are almost entirely Turco-Iranian, but the costume is Middle Eastern, Syrian or Arab. He uses the iron stirrups, but it should be noted that most of the Muslim troopers did not use those before the turn of the eight century. He carries two bows. The one he uses is an angled Sassanian style composite bow used with the thumb-draw. The unstrung second bow is in a bowcase on the left side of the horse. Nicolle suggests that the spare bow would be double curved Turkish bow. In other words, the rider used two different bows and shooting techniques just like the Romans after the reign of Maurice. The quiver is unusual because its opening points towards the rear. The belt with decorative pendants shows steppe influence because such belts were typical for Turkish and Avar warriors. Author's drawing.

Military saints in sixth century Roman military gear depicted in a weight located in the British Museum.

Mounted archer depicted in a textile. Egyptian sixth century. Source: Diehl.

Cataphracts and clibanarii

a) A graffiti of a cataphract. Dura Europos, 3rd cent. Drawn after von Gall.

b) A graffiti of a mounted archer and armoured fully armoured horse. Dura Europos 3rd cent. Drawn after fig. 17 D James.

c) Seal impression of Persian Wahram (6th cent.).. Drawn after Patryk Skupniewicz.

d) Seal impression of Persian Wixtaxm (6th cent.). Drawn after Patryk Skupniewicz.

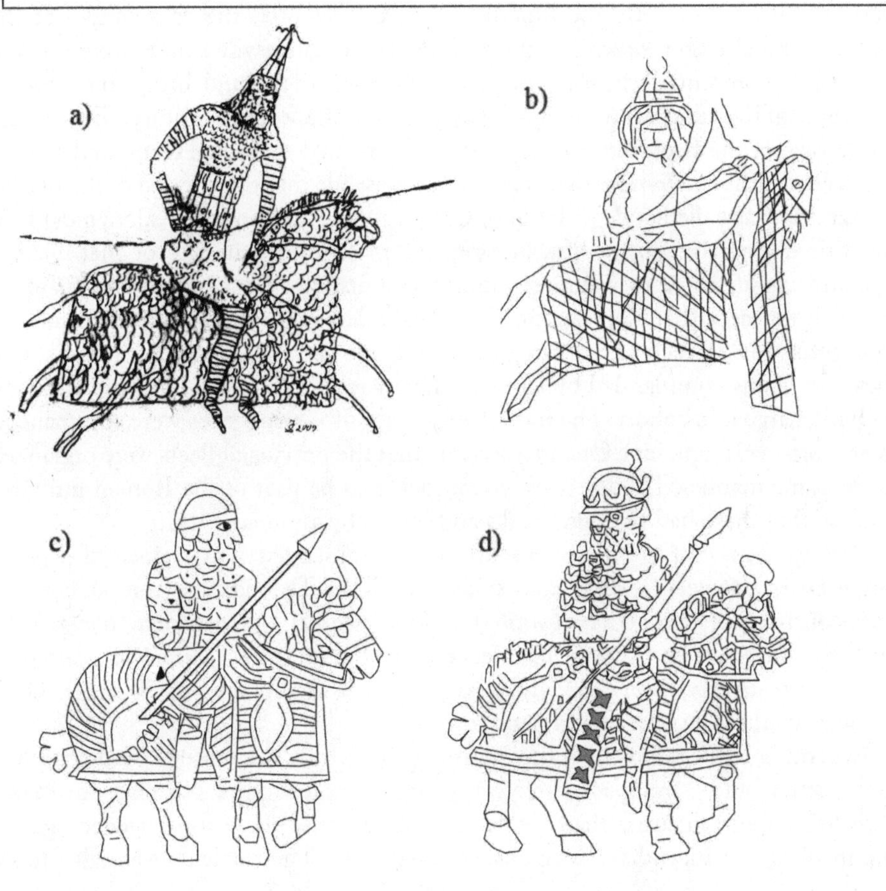

2.4. The navy in land warfare

The Romans also expected their naval forces to be able to fight on land. The legions attached to the praetorian fleets obviously fought like other legions, but in addition to this the marines, mariners, sailors and rowers could be expected to fight on land. This was particularly true of the members of the praetorian fleets accompanying the emperors as Pseudo-Hyginus's treatise on marching camps shows for the earlier period. The imperial fleets (in other words the praetorian fleets) with the legions originally consisted of the Fleets of Ravenna and Misenum in Italy but later each new emperor added his own to the numbers so that each emperor had his own imperial fleet. In the end, however, there was only one imperial fleet, the Fleet of Constantinople. For these developments, see the *MHLR* series.

Each of the fleets was commanded by a Fleet Prefect (*praefectus classis*) who was assisted by a *subpraefectus* and by an administrative staff. The hierarchy of the temporary fleet commands was as follows: the detachments which included ships from the praetorian fleets were commanded by a *praepositus vexillationis* or a *dux* and the detachments from frontier fleets were commanded by a *praepositus classis/classibus.* The fleets were always placed under land commanders when these were present so that during most of the Late Roman period the overall commanders of the fleets were *magistri militum.* By the time of Maurice (582–602) the commander of the fleet was called either *nauarchos* (the admiral; presumably when there were no land commanders present) or the *strategos* (general in charge of the land forces). It is possible that some of the admirals were called *drouggarioi* in the seventh century.[8] Depending on the size of the fleet, the *nauarchos/strategos* also had divisional commanders who were called either *moirarchai* or *merarchai.* It is possible that the admiral of the fleet of Alexandria was called *archegos* because Calonymus of Alexandria (Kalonymos) held this title during the North African campaign of Belisarius in 533, or that *archegos* was just one of the titles meaning admiral. Confusingly the *navarchi* and *trierarchi* could also mean either ship captains or commanders of groups of ships.[9] According to Vegetius (4.32), each of the two praetorian fleets had a legion of marines. Each of these fleets was commanded by a prefect, below each of whom served ten tribunes, each in charge of a cohort. The individual galleys of varying sizes were commanded by the *navarchi* (captains). One may assume that the provincial fleets were organized in the same manner. The fleets were considered to be part of the Roman military/army so that these had the same ranks and hierarchy as on land.

The *navarchus* or *triararches* in his turn commanded and trained the pilots (*gubernatores*), rowers (*remiges*) and soldiers (*milites*). The fighting component of a fleet consisted of the soldiers (*milites*), sailors (*nautae*) and oarsmen (*remiges*) led by their officers. The officers included e.g. fleet centurions (*centuriones classiarii*), *optiones navaliorum*, *suboptiones* and *nonegenarius* (a commander of 90 men). Their numbers could be bolstered with land forces.

According to Vegetius 4.44, the marines proper (not the rowers) were equipped with better and stronger protective equipment (cataphract-armour or *lorica* with helmet and greaves) than the land forces because these were needed against the missiles and incendiary weapons used on ships. Their shields were also to be stronger and larger than on land to withstand the stones. The sailors and marines on board included specialist elite *propugnatores* (front-rank fighters), *balistarii* (users of mounted field artillery, crossbows and torsion *manuballistae*), *sagittarii* (archers), and *urinatores* (divers). The sailors obviously included specialist classes of their own, just like there were other specialists of all sorts in the fleets including junior officers plus secretarial and headquarters staff. The basic building block of the seagoing ships appears to have been a crew of one *ousia* (108 or 110 men), which were multiplied for the largest vessels. This number included the rowers and their spares. The smaller vessels presumably had divisions of these (54–5, 27–8, 14–5), but there is no concrete evidence for this.

2.5. Recruitment

During the Late Roman period the recruitment of regular soldiers was for the most part based on acquiring volunteers whose primary motivation for being in the service was the money. However, conscription (based on taxable land) and hereditary military service were also still in use. The *foideratoi/foederati* (Federates) consisted mostly of foreigners, but also contained Romans. Some of the units were named after their place of recruitment or on the basis of their station (e.g. Thracians, Armenians, Isaurians etc.). As one can guess the conditions of service were variable during so long a period as is under discussion here. There were periods when the salaries were paid regularly, just as there were times when those were in arrears. There were periods of military mutinies and other troubles, just as there were periods without these. There were also periods when corruption was rampant and causing local or wider problems, just as there were periods when the corruption was under control or not detrimental. However, on average the Late Roman military hierarchy operated at least adequately, so that the soldiers were happy with their position in society and ready to perform their duties. For a detailed analysis of the varying conditions of service, see the *Military History of Late Rome* series.

2.6. Peace and Wartime Organization

Late Roman military structures were extremely well organized for the standards of the day. The laws and bureaucracies regulated everything from tax collection down to the actual organization of the armed forces, which included for example supply structures, recruitment, conditions of service, hierarchies, payment of the salaries, production and distribution of equipment. The military doctrine expected that the peacetime operations and the wartime campaigns were well coordinated so that the Romans always possessed accurate information on enemy activities, which then enabled the Romans to plan accordingly. Military doctrine also expected that the logistical network functioned well so that the Romans could choose when and where to engage the enemy if they so decided. In practice this ideal was not always achieved.[10]

2.7. The Training in General and Winter Quarters[11]

We possess several military treatises which depict the Roman training system. These treatises include Arrian's *Techne Taktike*, the *Epitome* of Vegetius, *De scientia politica dialogus* (*On Political Science*), *Peri Strategikes*, and the *Strategikon*. These prove, and which is only natural, that the soldiers were first taught the elementary skills and only when these had been mastered did they proceed to teach more complicated skills and military manoeuvres in various kinds of formations. All soldiers were taught how to march, carry heavy loads for long distances, swim, run, jump over obstacles, build camps and other basic skills required from the men while the horses and other animals were trained for their specific tasks. The training was usually performed during the winter break in hostilities. Military doctrine recognized the need to keep the soldiers busy during the winters because too much leisure could result in weakness and mutinies.

The recruits were first assigned to a unit (e.g. century, *tagma, arithmos*) which belonged to a larger unit and then to a squad/tent group called a *Contubernium* which formed one file in the battle formation. The soldiers swore the military vows soon after entering the service. The tent group consisted of ten men, and in the cavalry the fighting component consisted of five to eight horsemen plus their squires while in the infantry the fighting component consisted of eight men, one recruit (*tiro*) and one servant. The training of the soldiers was therefore conducted in tent groups and units. The officers were expected to check the equipment of each soldier during the winter break (the soldiers received an allowance for this and the officers checked that the soldiers were properly equipped) and arm the new recruits. Each soldier was expected to be equipped according to his pay grade. The soldiers were accustomed to keeping silence, and to obeying spoken and visual commands in various kinds of situations. The treatises prove that the Romans recognized the importance of repetition, punishments and rewards, camaraderie of small groups, and the simulation of reality. In short, all of the soldiers received training in basic skills either as footmen or as horsemen, but the horsemen were also trained to fight on foot. In addition to this, there were specialist units such as the imperial guard units with their own distinctive equipment and the men belonging to the baggage train, or siege train (who operated e.g. the wagon/cart-mounted arrow-shooting ballistae and stone throwers), or supporting units like the artisans, bowyers, carpenters, engineers, architects, doctors and so forth. These specialists were obviously trained separately for their specific duties.

The period when the army was at winter quarters was not only devoted to furloughs and training and equipping of soldiers. The *strategos* and officers under him also had other duties than the training of the soldiers, units and army during the winter quarters. For example, the *strategos* was to make certain that the baggage train carried extra arms, especially bows and arrows, to replace lost weapons since arms were essential for fighting. In addition to this, the horsemen and footmen were required to hire servants to help them to carry out their duties. On the basis of Maurice's requirement for each group of three to four horsemen to possess at least one servant/squire and pack animals to carry their coats of mail, tents and other equipment, it is probable that the soldiers were in the habit of saving money by not hiring servants. It was the duty of the officers to ensure that the soldiers had servants to perform these duties. Since Maurice's text was based on Roman military traditions it is clear that similar demands had always been made.

Roman military doctrine also required the *strategos* assemble his forces prior to a campaign so that he could evaluate their skills, weaknesses and overall strength. This was necessary for two reasons: to see if the army was ready for combat; and to familiarize the *strategos* with the soldiers and officers serving under him so that he could use them in the most advantageous way. It was also considered a good idea to familiarize the soldiers with the enemy and combat by skirmishing with the enemy before engaging them in pitched battle.

2.8. Cavalry Training

The regular Roman cavalry had been taught to fight in all possible combat roles (close order forces, skirmishers, dismounted cavalry) ever since the days of the emperor Hadrian and military theorist Arrian, and this can also be detected in the treatise of Vegetius. Each individual was required to be proficient in the use of javelins (*lanceae*), various types of spears (e.g. *lancea, xyston*, Gallic *contus* used with one hand, Sarmatian *contus* requiring the use of both hands), sword, shield, composite bow, throwing of rocks, slings, and crossbows. The troopers also varied the amount and type of armour and helmets according to the situation, just like they varied the amount and type of armour worn by the horses from unarmoured to frontal half-armoured (chamfron, crinet, peytral). The size of the cavalry shield appears to have varied from one unit to another, from the small shield (*pelte, parma, clipeus*) to the larger infantry model, but it was more typical for Late Roman cavalry to use smaller shields than the heavy infantry.[12] The units composed of foreigners (depending on the time period, e.g. Moors, Armenians, Parthians/Persians, Heruls, Arabs etc.) used their own native gear and tactics even when included as part of the Roman combat formation. The Romans added the *clibanarii* type of cavalry (trooper without a shield and using *contus* with two hands; horse and man fully armoured so that only eyes were visible; horse protected by: chamfron, crinet, peytral, flancard, crupper) to this basic system in the third century (the favourite type of cavalry for Constantius II) and it remained in use until at least the sixth century. The cavalry and their horses were also taught how to swim and how to operate in different kinds of terrains and weathers and combat situations, but the success of this training scheme appears to have been limited because there were times when some of the cavalry units were clearly unable to cross water obstacles.[13] The Romans had always required their horsemen to be able to fight on foot, and this requirement appears to have been true for most of the regular cavalry during the Late Roman period. Even though the military treatises do not specifically mention all of the following exercises, it is easy to see that the training of cavalry and specialists within it included the following elements: physical exercises, hand-to-hand combat (wrestling, boxing, *pankration*), swimming, marching, foraging, scouting, how to fight in a siege, the building of fortified camps, and so forth. This situation persisted at least until the emperor Maurice streamlined the combat doctrine and system in his *Strategikon*.

Maurice required his Roman troopers to be similar multipurpose troops (both mounted and dismounted[14]) so that only the young foreigners were exempted from the need to be able to use the bow on horseback. The troopers were taught how to shoot to the front, the rear, the right and the left while their horses galloped and they were taught how to use the bow and lance (and obviously also the shield and other types of weapons) consecutively in different types of terrains and weather circumstances and battle roles. However, one can detect one major difference in the stress that Arrian, Syrianus and Maurice put on the use of mounted archery in training and combat. Arrian's *Taktika* (32.3–44.2) clearly emphasize the use of spears and javelins as missiles over the other techniques as a sign of a good horseman, while both Syrianus (*PST* 44–7) and Maurice (e.g. 1.1, 1.2.28–34, 1.2.83–5) clearly considered mounted archery to be the key element in cavalry training and combat, even if Maurice also stressed the use of the spear alongside the bow. Vegetius's text (2.14) appears to stress both the melee skills and skill with the bow equally, with the implication that it represents the middle stage in the development of the importance of mounted archery in Roman military thinking. The second major change in the *Strategikon* is that the soldiers were taught to shoot rapidly either in the Roman or Persian manner. Before this, Roman horsemen were expected to be proficient only in Roman style archery even if there obviously had always been archery specialists (e.g. Persians and Armenians) in Roman service who could use both archery styles.[15] We also find separate archery training techniques for accuracy, power and speed of fire in the *Peri strategikes* (44–7) for situations in which the archer stood in place or was moving. The fast shooting style was the Persian style. It has been recognized for long that this text has been lifted out from some earlier text, which means that some Roman units had also trained to use the Persian technique well before this. The Roman manner with stiffer bows allowed more powerful shots, while the Persian manner allowed more rapid firing. The troopers could also switch between the methods of release so that their fingers would not become exhausted. When Procopius wrote his *Wars* (e.g. 1.28.30–4) the Romans used only the more powerful Roman/Hunnic release, which means that either Maurice or Tiberius II (much likelier than Justin II) added the Persian release to the repertoire. It should be stressed that Maurice did not introduce the Persian archery technique to the Roman armed forces. He only made it a compulsory element in the training of the archers. The third major change was that Maurice does not mention the *clibanarii* type of cavalry at all, with the implication that its use was probably phased out at the latest during his reign. Maurice also required that the Roman troopers varied their equipment according to their position in the combat line (front and rear of the cavalry unit equipped for melee and the horsemen in the middle armed with bows but not using shields) or their rank (the officers, *bucellarii* and *foederati* had better equipment than the rest).

Maurice's *Strategikon* provides us with a good glimpse into the variety of cavalry tactics during Maurice's reign. According to the *Strategikon*, cavalry units were trained to perform the varying roles of the different parts of the battle line, so that they could do anything that was asked of them: dismounting to fight as infantry, various circular movements, to close and open ranks, the open order, the close order, the irregular order (*drouggos*), to wheel right and left, to about face, to halt, to charge, to move in close, open or irregular order in all terrains, counter-march, to divide or double the units, ambushes and other particular roles including the drills for the *defensores*, *koursores*, ambushers, outflankers and flank guards. In addition to this, the troopers were also taught the Scythian Drill, Alan Drill, African Drill, Illyrikian Drill and the Italian Drill. As the names imply, the reformed tactics of Maurice were based on earlier Roman combat methods that he just simplified and streamlined so that earlier variations in equipment, tactics and unit formations were abandoned. It is of note that even if Maurice required all of his troopers to possess two iron stirrups, just like Vegetius (2.14) he also demanded them to be able to mount their horses by leaping on their back. This was a necessary skill even when the troopers had stirrups because it enabled them to mount their horses faster when necessary.

In sum, the ideal horseman from the second- to the seventh-century was a multipurpose trooper able to use both the bow and lance with equal skill, and at the same time the horseman was also required to be able to fight as a footman. Regardless of this, the Romans continued to employ also other types of cavalry with different tactics until the reign of Maurice, who finally instigated a military reform that unified the practices. This required a fair amount of training and one can justifiably call the Roman cavalry an elite force.

2.9. Infantry Training

Infantry training differed from cavalry training in that the Romans did not expect all of their regular footmen to be proficient in archery – only those who showed some natural aptitude for this were trained as such. Most of the footmen were trained as spearmen/javelineers and swordsmen. According to Vegetius (1.15), only a quarter to a third of the force was trained as archers, but we know on the basis of the *Peri strategikes* and Procopius that there were also multipurpose footmen, all of whom were trained to be equally adept as heavy infantry and as foot archers. It is probable that this concerned only selected elite units, because during the late-sixth century the *Strategikon* (12.B.9) specified that in a large infantry army of over 24,000 footmen only a half of the force should consist of foot archers, with the implication that this was the approximate achievable maximum number of foot archers. In fact, for a regular army of less than 24,000 infantry, the *Strategikon* suggested that only a third of the force was to consist of foot archers. Regardless, this still suggests that the Romans increased the number of foot archers by the sixth century so that at least half of their infantry force consisted of men able to use the bow. This, however, does not mean that the rest of the men would have lacked ability for long distance missile attacks – all men were trained to throw stones and use the sling.

The heavy-armed infantrymen were trained to use spears and javelins for throwing and the use of these in melee with a shield. The men were also trained to use both the short and long sword (*semi-spatha/machaira* and *spatha/spathion*), together with a shield for close quarters combat. Some of the units also received training in the use of lead-pointed darts. The foot archers were trained to shoot with stiff composite bows in a manner that was known as the Roman/Hunnish manner until the late-sixth century, but at least from the reign of Maurice onwards the foot archers were required to be equally adept in the Persian archery technique which specialized in the shower archery (speed of delivery preferred over power of the archery shot). During the reign of Maurice (*STR* 12.B.3) the light-infantry were also trained to use the bow while carrying a shield. It is probable that some of the foot archers were trained in like manner even before this, but we know that before his reign there were also infantry archers who did not carry a shield at all. The archery technique with the shield probably resembled the corresponding Muslim practice as depicted in the wall painting shown below and dated to the eighth century.

The footmen were trained according to their task in combat. The heavy infantry was trained to fight either in heavy gear (various types of shields, spears, armours and helmets) in the open, or in light gear (no metal armour, javelins instead of spears) in difficult terrain. The preferred equipment for difficult terrain appears to have changed by the reign of Maurice. As already discussed, the fourth-century *DRB* equipped the footman for difficult terrain so that he carried a javelin/spear, small round *clipeus*-shield, sword, *subarmalis/thoracomachus*, and helmet, whereas in the *Strategikon* the shield-bearer carried javelin/javelins, a sword, no armour, no helmet and a 'larger-type' of shield. The heavy infantry was trained to fight both as line troops in combat formations and as lightly-equipped forces outside the phalanx formation in irregular formations (*drouggoi*). The multipurpose, bow-armed heavy infantry was obviously trained to fight as heavy and light infantry, while the light infantry specialists were trained to fight as required by their role in various different combat formations or as irregular *drouggos* troops outside it. In short the infantry learned how to march or halt, reduce or divide the depth of the files, march steadily in close order for a good distance over various kinds of terrains, close or tighten their ranks in depth and width, march in a *foulkon* formation (a tortoise), engage in a mock battle, divide into double phalanx and back, face to the right and left, march to the flank and back to their original position, countermarching, change their front to the right and left, broaden and thin their formation, deepen or double the depth of the files, and about turn and back. These manoeuvres enabled the soldiers to adopt the various combat formations mentioned in

Left: An archer shooting at a fortified place.

- A fragment of a wall painting preserved in Khirbat al-Mafjar, ca. 743/4 A.D. (Palestine Archaeological Museum, East Jerusalem). Drawn after Nicolle, 1997 fig. 15.
- According to Nicolle, the style of painting indicates strong Byzantine or Coptic influence while possibly reflecting current reality. The archer shoots under his round shield as described in Islamic archery treatises. *The Strategikon* also required the *psiloi* to be able to use the bow with the shield.

the military treatises, such as for example the lateral phalanx, double phalanx, a square, a half-square, a wedge, a circle, a pincer, outflanking, an oblique attack, etc. On the basis of Urbicius's *Epitedeuma* (see *MHLR* 5, 256–7) we know that even if the Romans always trained their infantry to use all of the various ancient phalangial formations, they were not always able to use these in practice in situations where the Roman infantry forces consisted of recent green recruits. It was because of this that Urbicius favoured the use of the hollow square with additional safety measures.

The infantry forces obviously also included dismounted cavalry and disembarked naval forces. These were obviously not expected to be as good as specialist infantry in infantry warfare and were not expected to be able to do all of the same things as the specialist footmen. It should be noted though, that the dismounted cavalry and naval forces included all of the same specialist elements within the infantry proper, with the difference that practically all of the dismounted cavalry forces were expected to be able to fight as multipurpose archers and spearmen, often with a smaller shield, while the marines could actually be more heavily equipped, with armour and larger shields than the typical heavy infantry forces.

2.10. Training the Officers[16]

The Romans did not possess any military academy in the modern sense of the word. The closest thing to such was service in the imperial bodyguard units. The vast majority of the officers received their training by practice in the course of their career, meaning that the members of military families had already learnt the basic skills from their parents before they entered the army. The military traditions, drill manuals and drillmasters (*campiductor/campidoctor*) ensured that ordinary native officers knew the basic skills required of them.

The officers were expected to be literate so that they understood the written orders that they received from their superiors, but this was not always achieved when they had humble origins or were foreigners. In fact, there were even foreign born officers who did not speak either Latin or Greek. The officers hailing from wealthy native families were better off in this respect because they were often well-educated. They could therefore complement their military training with their own study of written histories and military manuals. The foreigners who were admitted into the Roman armed forces as officers had obviously learnt their trade earlier, so they had to be familiarized with Roman combat methods through some sort of quick crash course, but we do not know any details of how this was done in practice.

Thanks to the lack of a military academy, the level of military competence varied greatly during this era. Some of the commanders were very well-educated and very learned, for example Urbicius and Maurice, while others were illiterate, like Theoderic the Great and Justin I. Education did not secure success and neither did the lack of education mean that the commander would have been automatically bad, as the case of Theodoric the Great demonstrates. Regardless, it is still clear that the study of history and military treatises was very advantageous. The examples of the *strategos* Belisarius and emperor Heraclius demonstrate this nicely. Both had initially very limited knowledge of military history and theory, but both were able to learn from the past to achieve success. According to Procopius, at the siege of Auximum in 539 Belisarius and his soldiers did not know that the trumpet (*salpigx*) could be used to order the footmen either to attack or retreat by using different strains. This means that at some point in time in the past the military traditions had broken, so that the well-educated Procopius had to point out to Belisarius that the ancient Romans had used the trumpet in this manner. It was thanks to this that Belisarius was able to save the situation. In the case of the emperor Heraclius, in 622, after a series of defeats, he finally started reading and studying the ancient military treatises. It was thanks to this study that Heraclius was then able to drill his army to perfection and then inflict a series of defeats on the Persians. In sum, the usefulness of military manuals, military drill manuals, and military histories is clear.

On paper, advancement through the ranks was based on the principles of seniority, ability, and performance, but in practice birth, family background and personal relationships mattered just as much. The emperor appointed all the senior officers, generals and members of the bodyguard units, while the generals divided the officers and units according to the personal assessment they made. The highest ranking commanders could therefore be men without any military background whatsoever, as the appointment of eunuchs and courtiers demonstrate nicely. It should be noted that the lack of military background did not necessarily mean that the commander would be bad. In fact, the eunuch Narses and emperor Maurice are the best examples of this, because both can be considered to have been among the best commanders in human history. Unsurprisingly there were many mediocre, good and bad commanders during the long period that is called the Late Roman era.

Despite the lack of a formal education, the Romans expected their generals (*magistri, strategoi, duces*) to be very skilled and versatile. The generals were expected to understand espionage, spying, scouting, reconnaissance, logistics, use of ruses and stratagems, guerrilla warfare, siege warfare, infantry warfare, cavalry warfare, and naval warfare. Fortunately for them, the Romans obviously had specialists in each of these fields to advise them.

2.11. The Training of Militia/Paramilitary Forces and Police

The urban and rural militias/paramilitary forces appear to have been trained to fight just like the regulars in various kinds of phalangial and irregular formations. The training of these paramilitary forces was the responsibility of the local authorities and/or guilds and circus factions. The civilian paramilitary forces were usually used only in the immediate neighbourhood of their homes, so that they typically fought against enemies in defence of their own cities, just in front of their city walls or from the walls. Otherwise, these paramilitary forces, together with professional police forces, were used for the policing of their cities and the surrounding areas. The best-trained militias were obviously located in places that were threatened by invaders and bandits. In rural communities, the local landlords could possess large forces of their own, consisting of their dependants/tenants. The equipment worn by the paramilitary forces reflected their personal wealth and/or the locally available arsenal of weapons provided by the city, guild or faction.

Chapter Three

Strategy and Pitched Battles

3.1. Grand Strategy and Campaign Strategy[1]

Even if it can be said that the Late Romans possessed a single long term grand strategy because they always sought to reconquer land lost, it would still be wrong to say that this was true all of the time. In practice, the grand strategy depended on the situation and the wishes of the ruler (depending on the time period either the emperor or generalissimos). Furthermore, in some cases the rulers were not free to implement any long term grand strategy that they would have wished, either because the empire faced a civil war, or many enemies simultaneously, or both. In other words, the different emperors/men behind the throne had different grand strategic and strategic goals depending on the situation and personality. Some of the emperors, usually the better ones, or generals like Aetius, had aggressive foreign policies, but such aggressive goals could also result in massive disasters when employed in the wrong circumstances or in the wrong manner. The best example of this is the reign of the incapable and incompetent Honorius – the reason for this was that Honorius's only grand strategic goals were to regain power from the hands of his *magister peditum* Stilicho while seeking to implement a racist policy towards the barbarians within the Roman armed forces. Other very good examples of the use of the aggressive policies in the wrong circumstances and in the wrong manner are the reigns of Julian and Justin II. However, there were also those rulers like Constantine I, Marcian, Justinian and Maurice who used aggressive policies in the right circumstances and were therefore able to reconquer land lost or actually managed to enlarge the empire on certain parts of the border. These rulers realized the relative strengths and weaknesses of their forces vis-à-vis their external and internal enemies and were able to formulate the right kinds of policies for the circumstances. Some of the rulers adopted purely defensive policies in the right circumstances. Good examples of the successful employment of defensive strategies were the reigns of Constantius II and Tiberius II. Both lacked the means for the adoption of a more aggressive stance and both realized this. The key for the successful implementation of the grand strategic and strategic goals was the understanding of the situation.

The grand- and campaign-strategies were formulated in the imperial councils where the emperor could listen to the advice of his administrators and generals. These men were expected to know their trade so that the emperor knew what the strategic situation facing the empire was. On paper, the emperors therefore had all the necessary tools at their disposal for the making of right decisions as regards the grand strategy and strategy, but in practice the ability to arrive at the right decisions also depended on the abilities of the advisors and on the personal ability of the emperor – and both could leave a lot to be desired. However, it is still clear that the energetic, able emperors could form a relatively accurate picture of the objective and what were the resources, capabilities, and needs of the Roman army. The able rulers usually sought to avoid simultaneous wars on multiple fronts, but bad emperors such as Justin II did not. The able rulers sought alliances or bought peace so that they could concentrate a numerically overwhelming force with adequate supplies and salaries to the theatre of operations of their own choice. This was obviously not always possible, because in some cases enemies could not be bought off or alliances formed, meaning that there were wars on multiple fronts. Alternatively, possibly due to the limitations of transport or supplies or because the soldiers had not received their salaries. In some cases, a numerical superiority was not even considered necessary because the enemy was considered tactically inferior. See the Chapter *When and Where to Fight.*

The other important variables for the success of the adopted strategy were: 1) the availability of well-motivated and trained soldiers, which was obviously not always possible after defeats or because of arrears of payment, logistical difficulties, or because the emperor or his predecessor had neglected the upkeep of the armed forces; 2) the availability of able military commanders (e.g. Justinian and Maurice had plenty of able commanders to choose from, while Heraclius did not have a single really able commander); 3) the emperor chose the wrong commanders or was forced to seek commanders outside the regular military establishment because the existing commanders refused to fight (this happened under Anastasius); 4) the danger of giving large field armies to able commanders, which could influence their decisions, as happened for example in the case of Germanus during the reign of Justinian or when Phocas could not place the generals of Maurice's day in command of large field armies; 5) there were also several instances in which officers failed to cooperate, or even sabotaged each other because of personal ambitions (see especially the *MHLR* vols. 6–7); and 6) the ability of the enemy to respond to the Roman plans.

It should not be forgotten that the military structures, diplomatic efforts, terrain, type of enemy and the distribution pattern of forces and fortifications were a reflection of Roman grand strategy and that these also influenced the operations. The most important of these was the distribution of military resources, as this had a direct importance to the availability of forces for each theatre of operations, as well as the quality of the forces available. The presence of strong armies also deterred enemy invasions. One of the greatest Roman military assets was the Roman navy. It was only when the enemy could challenge Roman naval dominance by using their own resources against them that the Romans lost complete control of the Mediterranean. This happened when the Vandals gained control of Carthage and when the Persians and Muslims gained control of Syria and Egypt. And it was the naval supremacy that gave the Romans the ability to support military operations in all places that could be accessed through water routes (seas, rivers and lakes). The possession of naval mastery in the Mediterranean was the key asset in this respect, because it gave the Romans the ability to use the inner lines of communication. In fact, the Roman Empire was based on its naval superiority over all of its enemies.

On the basis of Vegetius, *Peri strategikes* and *Strategikon*, it is clear that the Late Romans took into account in their plans the perceived moral implications of the defensive and offensive warfare upon the soldiers. Maurice's view was that Roman soldiers were less inclined to fight in their own territory because there were many ways they could use to save themselves, which meant that they did not want to take any unnecessary risks. The reason for this was the presence of forts and cities. The soldiers could always seek a place of refuge in such places rather than take risks. The same is implied by both Vegetius and Syrianus. Therefore, the recommendation was to avoid the risk of an open battle unless one had a clear superiority in numbers against the invader, especially in the early stages of an enemy invasion. The military treatises recommended that commanders delayed battle, denied access to provisions by gathering the population, cattle and wheat inside fortified places, and the army ambushed and destroyed enemy foraging parties. The stated aim of the treatises was to weaken the enemy by ambushes, skirmishes and by shadowing warfare, and then finally destroy the already weakened enemy as he was on his way back home. Excluding the walls of Chersonesus, the Long Walls of Anastasius, and the Wall of Thermopylae, the fortifications did not block invasion routes physically, but if invaders marched past them they endangered their route of retreat. Additionally, the presence of the fortifications with their garrisons made it difficult for the enemy to collect provisions and forage. By allowing the enemy to advance deeper into Roman territory, the Romans could destroy the enemy piecemeal if they spread to forage, even when outnumbered by the invading force.[2] And if the Romans managed to collect a sizable field army, its presence ensured that the enemy would have to concentrate its forces together, with the consequence that they consumed faster those provisions and fodder they possessed, causing them to have to leave the area. This tactic minimized both the damage caused by the enemy invasion and also the risks the Romans had to take in defence of their territory. Roman combat doctrine quite correctly gauged that it was especially important to avoid defeat in defensive warfare, because that would have allowed the enemy to spread out and/or besiege cities and cause more destruction without incurring any danger. The Romans realized that it was far easier to defeat the enemy warriors when they were about to leave Roman soil, possibly encumbered with plunder, tired out, and getting closer to their homes. The last mentioned made the enemy more prone to flee. The Late Roman defensive strategy was based on the defence in depth concept. The frontier forces engaged small invasion forces already on the border, but if the enemy launched a large-scale invasion, they shadowed and harassed them until the field army could be brought in to handle the situation. Therefore, one should not see all of the instances of deep penetration of Roman territory by enemy forces as instances of military failure, because from the point of view of military doctrine the system was only working as expected. Defensive warfare could also include the use of a diversionary attack against enemy territory to draw the invaders back to defend their homes. In addition to this, the Romans naturally varied their strategy and tactics according to the type of enemy they faced.[3]

In real life, Roman strategy varied greatly according to the circumstances. The capable emperors and military leaders always aimed to match their resources to the situation. Sometimes there were not adequate numbers available even for the above kind of guerrilla warfare against the invading enemy. In such situations, the Romans could either buy the enemy off, or they could try to induce another tribe to invade enemy territory in order to create a diversion, or they could play for time while gathering reinforcements from every possible corner. Furthermore, in real life there are instances in which the Romans engaged the enemy immediately close to the border because they had both adequate forces and prior knowledge of the enemy plans. Additionally, there are also instances in which the Romans conducted pre-emptive strikes across the border. In addition, the Romans did fight pitched battles even when outnumbered, but this usually took place just outside their own cities and forts or field fortifications which gave them the advantage of position over the enemy. Procopius adds yet another variety, because according to him Justinian actually sometimes forbade his generals of Thrace and Illyria to attack the invaders (Huns) so that the Huns could be used as allies against the Goths or other enemies. Stratagems and foreign alliances were also employed when possible. Furthermore, sometimes successful defensive warfare then resulted in the invasion of enemy territory, while unsuccessful invasion of enemy territory resulted in the invasion of Roman territory.[4] In short, defensive strategies were varied and flexible. The following diagram,

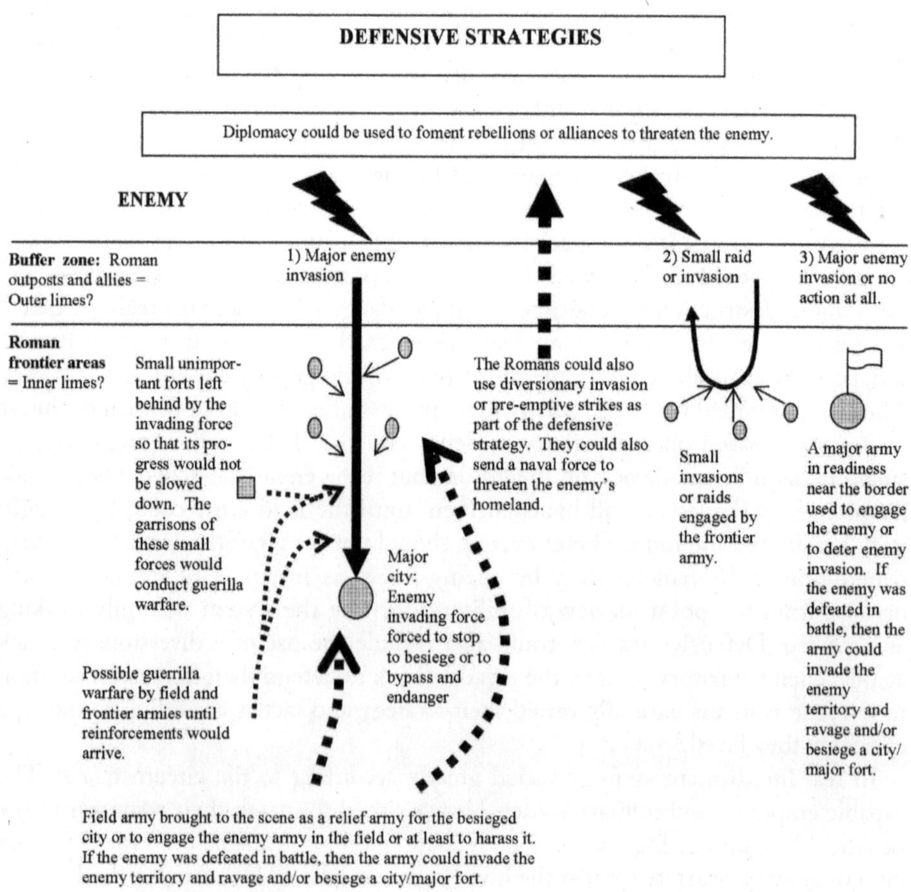

borrowed from my doctoral dissertation, summarises the main features of Roman defensive strategies.

On the basis of Vegetius and Syrianus Magister, plus the long list of precautionary steps to be taken before resorting to battle given in the *Strategikon*, the usual mistake made by modern historians concerning Late Roman and 'Byzantine' period warfare is that it would have been purely defensive and passive in nature. Nothing could be further from the truth. The Romans resumed the offensive every time they had enough resources for it and/or the situation was favourable. In fact, none other than Maurice recommended that the general always had to be on the lookout for emerging opportunities and pretexts, because it was more beneficial to conduct wars in the enemy's territory. The idea was to surprise the unprepared enemy, especially an enemy who was regarded as being strong, like the Persians. It was because of this that an invasion was to be carefully planned and then swiftly carried out before the enemy had the opportunity to react. The *Strategikon* also considered it safe to invade hostile territory if the enemy had been defeated in battle or when it was known that the enemy was unprepared or unfit for action – the Slavs and Antae were considered so weak that their territories could be invaded whenever there were enough forces available for this. The *Strategikon* also considered it safe to invade enemy territory if the Romans could seize and hold a strong position, such as a riverbank or mountain pass, from which it was possible to hurt the enemy without being touched by them.

The Romans also invaded enemy territory for limited purposes, such as the building of border fortresses by stealth or to besiege enemy fortresses. In the former case, the *Strategikon* instructed the commander to use a feint to distract the enemy's attention elsewhere by spreading rumours and by posting troops there, after which he was to move the men suddenly to the chosen place and build a fortress. The Romans and their enemies also tried to surprise the enemy by using unexpected invasion routes or an unexpected time of the year (i.e. winter). The Romans could also use two (or more) armies so that one army acted as a shielding/covering/ diversionary force for the other that invaded or besieged a city. The narrative sources prove that all of these methods were used during the Late Roman period, just like they had been

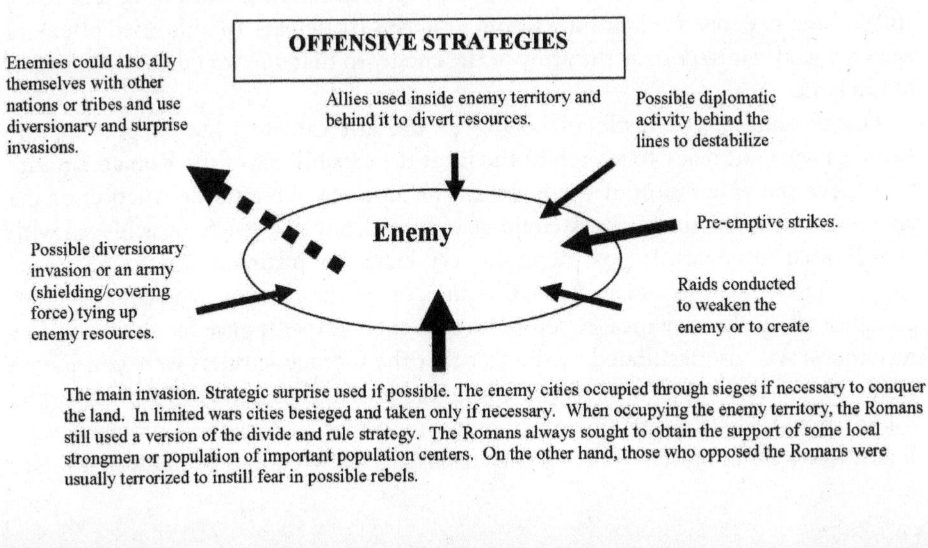

used before and were after.[5] The accompanying diagram, borrowed from my doctoral dissertation, shows the main variants of Roman offensive strategies.

The importance of the surprise in warfare has also found favour in our own times. The quote below from Erfurth's *Surprise* is a good example of this thinking. It was important to achieve surprise both on the strategic and tactical levels and the Romans recognized this.

> Every military plan and its execution should be conceived in view of the necessity of surprise. Surprise thus appears as the primary objective of military planning. … Inversely, success in war depends upon the commander's ability to prevent the enemy from accomplishing his own surprise.
>
> *Erfurth's Surprise*, tr. by Dr. Stefan T. Possony and Daniel Vilfroy 1943.

The Late Romans fought wars over unlimited and limited objectives and they employed several unorthodox methods in warfare. These included the use of hostages, bribery, marriage contracts and assassinations. In defence, limited wars were fought to defend some territorially-limited area, or over monetary payments, or other concessions (e.g. commercial), while in offence the wars were fought over limited territorial acquisitions, or over payments, or other concessions. In contrast, unlimited wars were fought over possession of an entire country. In offence, this meant the total conquest of some country, while in defence the very existence of the Roman Empire was at stake, like it was for example in 626. In limited wars, the Romans usually sought to force the enemy to accept their limited goals through a war of attrition or by other means that would bring about the desired result. Excluding the reigns of Constantine the Great and Julian, the wars against Persia belong to this category, but with the difference that, despite the limited objectives (quarrels over tribute payments and possession of border regions, cities and forts), the Romans were always ready to prioritize the Persian front because a serious defeat in that theatre of war could endanger the very existence of the Roman Empire. The Persian wars demanded the simultaneously use of large armies and their conservation. In other words, the Romans fought for limited goals but with the maximum resources while still seeking to conserve their forces from unnecessary defeats. In unlimited offensive wars the goal was to destroy the army of the enemy so that one would gain possession of the land.

The reconquests of Justinian belong to the last category, but in practice the Romans were not ready to stretch to the limit the capabilities of the Roman Empire to achieve the annexation of these areas. The aim was the total destruction of the enemy army so that the territory could be occupied, but this was to be achieved with very limited resources. It was therefore very lucky for Justinian that he possessed two outstanding commanders, Belisarius and Narses the Eunuch, who could achieve conquests with the very meagre resources that their emperor gave for their use. The reconquest was also facilitated by the fact that the Germanic rulers were considered aliens by the local populations who continued to see themselves as Romans. The role of pitched battles was obviously more important in unlimited offensive wars than these were in limited wars. As this discussion shows, unlike in modern military theory, in the context of the Roman wars the limited and unlimited wars sometimes overlapped in their execution.[6]

The Romans also fought wars which are difficult to categorize in terms of unlimited and limited goals because the nature of the war was actually determined by the type of enemy. For example, during the reign of Maurice the wars against the Slavs assumed the nature of wars of annihilation, in which the aim was the total destruction of all Slavic settlements, villages, fields and tribes bordering the Danube. The aim was to create a no-man's land, while subjecting the rest of the Slavic tribes under the Roman yoke. This was possible because the Slavs possessed permanent civilian settlements that could be targeted. Similar tendencies can be detected in the wars against the Germanic tribes and some Sarmatian tribes close to the Danube border during the third and fourth centuries, so that one can say the idea was not new. Such total destruction also served to create healthy respect for Roman military might further away from the border regions so that it was doubly beneficial.

The same strategy did not work against nomadic (or semi-nomadic) peoples like the Huns, Avars, Moors, and Bedouins, because there were usually no permanent villages or settlements or fields to destroy, and even when there were, the people were usually able to flee to somewhere else. Neither could the use of economic sanctions, namely the denial of commerce, bring these nations to their knees: it only induced the nomads to start raiding Roman territory. The only effective way to counter the nomads was to defeat them successively in combat, so that the prestige of the leading tribe would falter, while paying some chosen tribes to join the Romans. This means that pitched battles assumed greater importance in the wars against the nomads, even if the Romans could also defeat the nomads with guerrilla warfare as the example of Comentiolus's campaign in 586 demonstrates nicely.[7] It was next to impossible for the Romans to pursue the nomads deep into the steppes or deserts (due to logistical limitations), which meant that the Romans had to engage the enemy inside their own land or in the immediate vicinity of their borders. Since the wars were usually fought on Roman territory, the wars against the nomads could also become fights for survival, as happened when Attila led the Huns and when the Avars and Persians cooperated in 626. In sum, the migrant lifestyle exercised by the nomads limited Roman ability to respond to this threat. The Romans could only aim to limit the damage done by the nomads by instilling fear (i.e. respect of Roman arms) amongst them by inflicting a series of military defeats, while attempting to divide them through diplomacy and bribes.

Regardless of which type of war we are dealing, the treatment of the civilians was purely utilitarian and pragmatic so that it depended on the Roman aims. If it was in their interest, the Romans treated the civilians well, but when the aim was solely to hurt the enemy the civilians and civilian property were treated as legitimate military targets. However, the civilians could also suffer severely as a result of poor discipline, even when the official policy was to treat the civilians well. This was true for Roman-held territory (when the soldiers were billeted on civilian property or en route to some locale), allied territory (e.g. in Armenia) and in occupied lands (e.g. in reconquered Italy). The use of terror was a legitimate tool in warfare both for the Romans and their enemies. The fear of terror kept the subject peoples in control. The conquest of territory could also involve the mass transferral of population to another place to secure the territory for the conquerors, or the transferral of population from outside the borders to within the empire to inhabit territory either vacated through the transferral of local population to some other place or uninhabited thanks to the ravages of war. The capturing of civilians and their property were considered as military objectives. The prisoners could be sold as slaves or be ransomed by the enemy. The Romans had strict rules for ravaging operations. When done by the book these followed strict safety procedures so that the force sent to pillage was divided into two divisions, one that pillaged and the other defending them. In practice, however, the soldiers often disregarded these sound precautions and pillaged without any order whatsoever, which sometimes resulted in disasters. The total disregard towards civilians was also in evidence in the tactical sphere, because all prisoners regardless of age, sex or occupation were used as human shields if the enemy approached the Romans. Similarly, the Romans were in the habit of using hostages to secure the loyalty of their allies and subjects, but this was the universal practice at the time.

Excluding the Sasanian Persians, the Romans had two major advantages over all their enemies: 1) the Romans possessed the necessary skills to build fortifications; and 2) the Romans possessed unsurpassed siege skills both in defence and offence. It is therefore not surprising to find out that the Romans secured their own territory and new conquests always by building, rebuilding, and repairing fortifications. This gave the Romans the ability to control vast areas with relatively few professional soldiers.

In sum, pitched battles were just one of the components in the strategy that the Romans could adopt, even if it is still clear that the strategy revolved around them. In all cases, it was the relative advantages and disadvantages of the situation that determined whether the Romans would decide to fight a battle. The battle was often the culmination point of a campaign and it was not to be undertaken lightly. The Romans used a great variety of other methods to overcome the enemy that did not necessarily demand the use of full-scale decisive pitched battles. Roman military doctrine aimed at the most effective use of one's own resources in relation to the enemy's resources.

3.2. The Strategic Situation and the Sizes of the Field Armies[8]

It is dangerous to make generalizations for the army sizes for the whole Late Roman period because these varied greatly according to the availability of forces, needs, time period and location. However, if one wants to make generalization regarding the whole period, it is that the Late Romans usually possessed sizable professional forces which were adequate for their intended use. However, when there were several conflicts going on simultaneously the armies were sometimes stretched to breaking point. Some of the emperors, for example Justinian, also intentionally lowered the availability of soldiers by not filling up the ranks or by not paying the salaries. However, the most important limiting factor for the size of the armies was the availability of supplies. The larger the army the more difficult it was to provision.[9]

After the reforms of Diocletian and Constantine the Great, the defensive strategy was based on the concept of defence in depth (each front having frontier armies and field armies in addition to which came the imperial reserve forces), while the offensive strategy was based on the grouping of detachments from the field and frontier armies to the front of the intended offensive. Major operations could be conducted only in such areas where the Romans or their enemies could acquire adequate supplies. In practice, this meant that major operations were conducted only in well-inhabited areas or areas close to the supply dumps, major roads, rivers or the sea. The Romans also needed to garrison the cities and forts, which in its turn limited the availability of men for combat.

The overall size of the Late Roman military establishment was truly impressive. According to John Lydus (*De Mensibus* 1.27), at the very beginning of the Late Roman period Diocletian had 389,704 soldiers and a navy of 45,562 men. This would have represented the army of the eastern portion of the Roman Empire. One may therefore assume that Maximian had a roughly similar number of men in the western portion of the Empire. Diocletian had quadrupled the size of the military establishment because it had dwindled seriously in the course of the Third-Century-Crisis, while their barbarian enemies had bolstered the sizes of their own armies by forming tribal confederacies. On the basis of Agathias and Procopius, the Romans appear to have increased the size of their military establishment even more because both claim that before the reign of Justinian the Romans had over 500,000 soldiers in arms in the eastern portion of the Empire.[10] The probable reason for the increased size of the armed forces is that the Romans needed to bolster their numbers in the course of the fifth century because they faced ever increasing numbers of enemies, the most notable of which was the massive tribal confederation of Huns under Attila. The quality of these new Roman forces, however, was no longer the same as it had been previously and ever increasing number of these new recruits consisted of hired barbarians, especially in the western portion of the Empire, with the result that West Rome was eventually overtaken by these barbarians by the end of the fifth century. Agathias (5.13.7) claims that the size of the Roman army had dwindled from 645,000 men to a mere 150,000 men under Justinian. The likeliest explanations for Agathias's figures are: Firstly, it is probable that the figure of 645,000 men dates from the reign of Anastasius, who increased the size of the military establishment while refortifying Roman territory, the last of which required the creation of new garrison forces to man the new fortifications; Secondly, the lower total for the reign of Justinian probably refers to the fact that (Procopius, *Secret* 24.12–14) Justinian had not paid the *limitanei* regularly and had taken away from them the name of regular soldiers (*stratiotai*); thirdly, it is probable that the 150,000 men means only the field armies, the *comitatenses* and the praesental armies. In short, it is clear that the total number of soldiers did not fall as low as 150,000 men: it was only that the *limitanei* were no longer called soldiers under Justinian. In practice, despite Justinian's stingy policies concerning the armed forces, the Roman armed forces still consisted of over 500,000 men, even if their numbers were no longer 645,000 strong. This is proven by the statement that Procopius puts into the mouth of a Frank envoy. According to Procopius (*Wars* 6.28.10, 6.28.17), the Frankish envoy claimed that not less than 500,000 Frankish fighting men (an exaggerated figure) had crossed the Alps, which the Roman envoys countered by stating that the emperor's army surpassed all others in the numbers of soldiers. In other words, Procopius implied that the Romans had more than 500,000 men at arms at a time when the Romans were reconquering Italy.

The number of soldiers used for actual combat duties varied greatly according to the task at hand and situation. At the lower end of the spectrum were the small scale policing operations, commando strikes and guerrilla campaigns, and at the upper end of the spectrum were the massive concentrations of forces for major wars. The typical medium- to large-sized mixed armies of infantry and cavalry consisted of anything from ca. 10,000 soldiers up to ca. 40,000 soldiers, while truly large scale concentrations of forces consisted usually of roughly from about 40,000 soldiers to about 120,000 soldiers. The servants and other supporting elements should be added to these figures. The typical land armies of the period from 284 until the early sixth century were combined forces of infantry and cavalry, while from the early sixth century onwards it was more typical for the army to consist either wholly of cavalry or so that the main striking force of the field army was its cavalry – the sole exception to this rule being the late-sixth century battles fought against the Avars, because in these encounters the infantry formed the main component of the combined force field army. The cavalry armies or armies in which the cavalry formed the main striking force had different size classifications, which will be detailed below in the context of the sixth-century developments.

The two largest concentrations of troops for the Late Roman period were the civil war between Constantine the Great and Licinius in 324 and the battle of the nations on the Catalaunian Fields in 451. In the former case, the two emperors massed together all of their resources. According to Zosimus, Constantine the Great had 200 thirties (thirty oars), more than 2,000 transport ships, 120,000 infantry, 10,000 seamen, and 10,000 cavalry for his war against Licinius. Constantine's fleet consisted mostly of Greeks, which is a good indication of the method he had used to gather together so large a force. He levied civilians both into his army and fleet just like the Romans had always done during civil wars[11] and just as always these units were disbanded once the crisis was over. On the basis of the *Strategikon* we can estimate that the Roman infantry had one servant per nine soldiers and one servant per four horsemen, so that there were altogether about 16,000 servants and at least the same number of other supporting elements (medics, doctors, artisans etc.) so that the supernumeraries of the land army consisted of about 32,000 men. If we estimate that the transport ships had crews that averaged about 50 sailors and guards (all large merchant ships had a contingent of armed guards and it is difficult to imagine that merchantmen used as warships would have lacked such) because there were great variety in the tonnage of merchant ships at the time, this gives us 100,000 sailors which with the 10,000 seamen of the fleet gives us a total of 110,000 sailors. In sum, Constantine therefore had 120,000 infantry (with servants ca. 136,000), 10,000 cavalry (with servants ca. 12,500), 110,000 sailors and about 16,000 other supporting elements. The corresponding numbers for Licinius were 350 triremes, 150,000 infantry, and 15,000 cavalry. Licinius would have similarly levied additional forces to match the numbers of Constantine. If we assume that each of the 350 triremes of Licinius had on average about 170 men, the fleet consisted of ca. 60,000 men in addition to which would have come the transport ships which Zosimus fails to mention for Licinius. This means that Licinius had 150,000 infantry (with the servants ca. 167,000), 15,000 cavalry (with the servants ca. 19,000), 21,000 men in the supporting units and (excluding the likely crews of the transport ships) ca.60,000 men in the navy. If we exclude the likely transport ships of Licinius, the total concentration of Roman forces would therefore have been approximately 542,000 men.[12]

At the battle of the Catalaunian Fields in 451 both the Roman side and the Hunnic side had about 500,000 soldiers. As noted in the previous analysis of the civil war between Constantine the Great and Licinius, the figures are by no means impossible. However, in this case the Roman field army under Aetius did not consist solely of the Romans but also of a conglomeration of barbarian nations (Franks, Alans, Sarmatians, Armoricans, Liticians, Burgundians, Saxons, Ripuarian Franks, Alans, Visigoths and other barbarian tribes) most of which served in the capacity either as *foederati* or as allies, because these tribes had taken possession of large tracts of the West Roman territory. The same was true of the Huns of Attila. He brought to the scene of combat most of the tribes from the region of Rhine up to the steppes of modern day Ukraine and Russia. This was the largest battle of antiquity and a decisive one for the fate of humanity because it saved western civilization from the Hunnic nightmare.[13] Those who think such figures impossible should consider the question of why such figures were possible during the Napoleonic wars, and in particular during Napoleon's ill-fated Russian campaign which was fought in far less densely populated areas than Gaul. Napoleon led no less than 600,000 men to Russia and fought another battle of nations, the Battle of Leipzig in 1813, in which the French had about 225,000 soldiers and the allies about 380,000 soldiers for a total of about 605,000 soldiers. The diminished French numbers in this battle were the result of the losses the French had suffered in Russia and Spain. The logistics of the Napoleonic era were still based on wagons, horses, mules and oxen. Most importantly, Napoleon as a practicing commander well versed in logistics saw no problems in accepting the figures running into hundreds of thousands in the context of the barbarian nations in his commentary of the wars of Julius Caesar.[14] The audiences of Zosimus and Jordanes did not doubt their figures, why should we when even Napoleon considered similar figures plausible?

The size of the field armies diminished after the battle of the Catalaunian Fields and even more so after the destruction of the massive East Roman navy by the Vandals in 468. Under Anastasius the Romans once again fielded large armies, but under Justinian the numbers were far more modest thanks to his stingy financial policies, so that under Justinian the largest field armies had only about 60,000 soldiers. The field armies grew larger once again under Tiberius II and Maurice, but in most cases these field armies consisted only of cavalry. The *Strategikon* classed cavalry armies so that a small cavalry army consisted of less than 5,000/6,000 horsemen (Maurice gives two alternative figures), the medium sized force 5,000–10,000/12,000/15,000 horsemen (Maurice gives several alternatives) and a large cavalry army of over 15,000 horsemen. Diagram 3.8 of the *Strategikon* describing the large cavalry army (see the chapter on cavalry formations) has 103 to 113 *banda* ('flags' of 200–400 horsemen with the average in the text being 310) which means that the diagram has a minimum of 31,930 men with 103 *banda*,[15] and a maximum of 35,030 men with 113 *banda*.[16] If we add to this figure the missing numbers of ambushers (max. 8 *banda* or more), the large cavalry army in the exemplary diagram had at least 37,510 men and with the allowable maximum of 400 per *bandon*, the army of 113 *banda* had 45,200 men plus the ambushers.[17] The maximum size of a normal large cavalry army (calculated on the basis of the maximum numbers for the *mere* 6,000–7,000 men) would have been ca. 42,000–48,000 horsemen, but this was not the maximum envisaged by the *Strategikon*. Maurice recommended the posting of extra troops that could not be included in the *meros* structure either as reserves, or as ambushers, or on the flanks.[18] See Chapter 5 on cavalry formations. In addition to these, the cavalry armies obviously had servants/squires (a minimum of one per three or four horsemen)[19] and other supporting services (artisans, engineers, doctors etc.). Unsurprisingly the existence of these larger cavalry armies is confirmed by the narrative sources.

The *Strategikon* envisaged mixed infantry and cavalry armies in excess of 24,000 footmen (extra men posted outside the cavalry wings) and 12,000 horsemen, but in this case we are in the dark concerning the envisaged maximum figures. The sources fail to give us any specific figures for the combined armies that the Romans operated against the Avars in the 590s but on the basis of the typical massive Late Roman armies massed on one front one may guess that these may have consisted of figures of 80,000 to 120,000 men, so there could even have been two massive armies operating simultaneously against the Avars. It is unlikely to be a coincidence that we find Heraclius using armies of 90,000 up to 120,000 men against the Persians.

In summary, the Roman state was militarily effective on a strategic level when ruled by able emperors. The Romans possessed professional armies that were usually better trained and equipped than any armed force that their enemies possessed. The military operations and campaigns were usually planned with clearly defined goals. The intelligence gathering systems were also usually up to their task, so that able emperors could form an accurate picture of the capabilities of the Roman military and of the enemy. The existence of the permanent army and navy, and the sophisticated Roman fortifications, usually ensured that the Roman emperors had adequate numbers for the tasks at hand, but there were periods when this was not true. This was the case under incompetent emperors or when the efforts were hampered by massive corruption or by civil wars. The continuing existence of the civilian paramilitary forces, emergency conscription and levying of civilians, and the availability of foreign mercenaries could be used to alleviate these problems. In short, the Romans were usually able to raise adequate numbers of men to meet the military emergencies, even if the quality of the newly recruited men was not the same as it would have been with professional soldiers. Therefore, it is clear that the principal strategic long-term weakness of the professional Roman army was precisely the lack of trained professional reserves – the availability of civilian levies and foreign mercenaries did not entirely remove the problem of not possessing vast numbers of trained reserves for use in situations in which the Romans had lost significant numbers of their professional soldiers in military defeats – a case in point is the reign of Phocas.

3.3. Marching to Combat[20]

All military plans started with the evaluation of the situation which was based on the availability of information about the enemy and its plans and movements. The single most important item for the success of a military campaign was the quality of intelligence on both strategic and tactical levels. Roman combat doctrine expected that Roman generals would always possess accurate and timely information of the enemy and its activities so that they could make calculated decisions on the basis of this. The availability of strategic intelligence obviously varied greatly during so long a period as is under discussion here, but it can still be stated that the Romans were usually in a position to know what their enemies were up to so that the Romans usually knew if their enemies were planning major operations against them and they also usually knew the intended invasion routes that the enemy would take. Similarly, the Romans usually knew if the situation was opportune for them to conduct major invasions of enemy territory. It was relatively easy for both the Romans and their enemies to detect any major military operations well in advance because it was impossible for them to hide such major preparations from prying eyes. It was thanks to this that the Romans were able to formulate their operational plans in the best possible way.

The gathering of strategic intelligence, obtained primarily by spies and secondarily from deserters and merchants, was obviously the single most important element in the formulation of defensive and offensive plans, but in order to succeed one also needed to possess accurate tactical intelligence of the enemy activities. The other major concerns were the safety of the marching route, and the availability of water, food, fodder and so forth. In order to be able to choose when and where to fight, the commander needed to possess adequate supplies and such terrain that made it possible for him to avoid having to fight in unfavourable circumstances. The gathering of tactical intelligence included several different layers of spying, reconnoitring and assessment. In addition to this, the Romans also needed to know the tactically-important details of the terrain where the armies operated. The Romans could obtain such information from spies, scouts, patrols, former intelligence reports, maps/itineraries of the area, locals and from deserters.

The first layer of tactical intelligence consisted of the information brought by the spies and scouts, and possibly also by deserters and merchants, whose accounts the former needed to confirm. The second layer of tactical intelligence took place when the Roman armed forces were closer to the enemy. It consisted of the use of patrols, sentinels, strong vanguard- and rear-guard detachments, and finally of the intelligence gathering conducted by the officers, heralds and generals in person. The several layers of security measures indicate that the Romans recognized the dangers present during the marching. Vegetius summarized the situation as follows:

> Those who have studied military affairs more studiously have asserted that more dangers are accustomed to be present on the march than in the battle line itself. For in battle all are armed and they both see the enemy close at hand and come to the fight prepared in mind; on the journey the soldier is less armed and less alert and is unexpectedly disturbed by the force of a surprise attack of the enemy or by the trickery of the enemy sitting in ambush. ... the soldiers must be warned beforehand to be prepared in spirit and to have their weapons in their hands; for in sudden danger they might be terrified, but in foreseen danger they are not usually in fear.
>
> Vegetius 3.6. tr. by Stelten.

According to the *Strategikon*, the scouts and patrols were chosen from the ranks of the army and this appears to have been the standard method in the Roman army throughout its existence, even if there is also evidence for the existence of specialized scouting units too. The specialized scouting units came mainly from the ranks of the frontier armies, but the evidence also suggests that the *protectores* and other imperial bodyguard units, plus members of the *bucellarii* units, could be expected to perform spying and scouting missions. The evidence suggests that in most cases the scouts were indeed drawn from the ranks of the frontier armies or from the imperial bodyguard units. The Romans also exploited the special skills of ethnic units (e.g. Moors, Arabs, Britons, Slavs, Germans etc.) for scouting duties and for the capturing of prisoners in their native areas or in areas where their native skills could be used effectively. The *Strategikon* recommended the choosing of a whole *tagma* or one of their subdivisions for scouting duties and it is probable that this was the standard method used by all field armies. The mission was dangerous, so it was preferable that the men knew and trusted each other. Only when this was not possible, for example when the majority of men were reluctant or horses were in poor condition, was it acceptable to pick men from different units. The scouts were to be lightly equipped, well above average, reliable and intelligent, and mounted on fast horses. When deemed necessary the general and the high ranking staff officers could also conduct a so-called 'officer's reconnaissance' in order to get a better picture of the terrain, and of the strength and dispositions of the enemy force in order to make judged decisions.

The operating procedure for the approach towards the enemy was as follows: 1) when the enemy was still at distance only scouts and spies were sent to detect the troops strengths and their movements; 2) when further information was needed the commander dispatched a patrol under the guidance of the scouts to capture prisoners for interrogation; 3) when the Roman army was nearing the enemy, the spies and scouts formed the outer layer of an intelligence gathering network all around the army, while the scouting patrols formed the inner layer of security; 4) if the situation was considered dangerous, an additional layer or two of scouting patrols (i.e. a maximum of three layers) were posted in all four directions behind the preceding ones in such a manner that the numbers of scouts per patrol increased the nearer they were to the Roman army; and 5) the commander conducted a personal inspection of the enemy force if he considered it necessary. The effectiveness of the patrols was ensured by the officers who were required to conduct surprise inspections of them. In fact, Roman tactical intelligence gathering was usually successful, but as always there are some examples of failures resulting from the negligence of duties or treachery, but this is only to be expected when humans are involved. It is probable that the some of the patrols had dogs with them because it is known that the Roman field armies were always accompanied by guard dogs. The scouts, patrols and the men of the vanguard left signs on the road for the main army to show which road to take. When the Romans marched towards the enemy they could also post a separate lightly equipped rear guard at a distance of ca 15–20 miles (22.2–29.6km)[21] behind the main army. The rear guard was expected to send its own patrols to guard against the enemy and also to catch any stragglers. If the Romans faced difficult terrain, the *Strategikon* recommended the use of patrols of light infantry and cavalry at a distance of a mile away from the main army and since this text is based on older Roman traditions it is very likely that similar procedures had also been followed in the past.[22]

Roman combat doctrine expected for security reasons that scouts moved during the night, and observed and rested during the day, but obviously there were exceptions to the rule. The spies obviously did not need similar precautions, because they were expected to intermingle with the enemy populations and armies in disguise. The narrative sources suggest that the typical scouting parties consisted of about 20 horsemen and their leader, or 30 to 70 men when the intention was to capture a prisoner for interrogation.[23]

Unfortunately, Late Roman sources fail to describe the exact deployment pattern of the scouts during marching, but fortunately we possess more detailed instructions in the tenth century military treatise *De velitatione* – the use of approximately the same numbers suggests that the tenth-century Romans continued to use the same deployment pattern as their ancestors, hence the inclusion of this tenth century array. In the example given by the *De velitatione* the Romans were shadowing the enemy so that each of the groups of scouts was in visual contact with each other, so that if necessary the last two could relay messages to the *tourmarches* in charge of the vanguard. The vanguard in its turn relayed messages of the enemy movements to the general. The role of the vanguard was to attack and test the quality of the enemy troops if the situation was opportune. For the tenth-century deployment of scouts and army see the attached diagram, which has also appeared in my doctoral dissertation *The Age of Hippotoxotai*.[24]

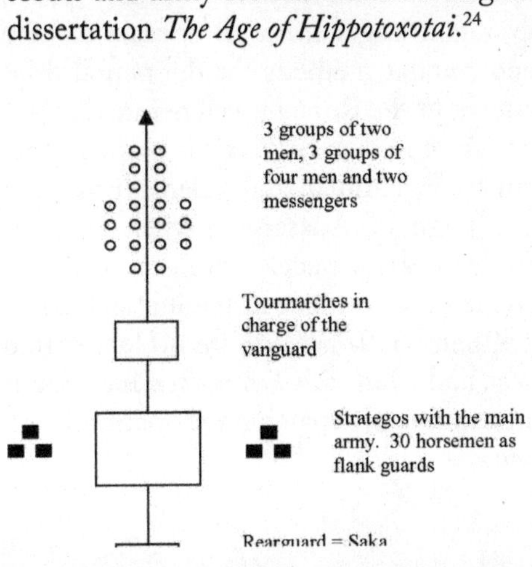

Marching Formation with Scouts

10th century cavalry deployment pattern of 20 scouts: one possible method also for the 6th century.

Source of diagram: *De velitatione*/ Dagron/Mihăescu, Schéma no.III, p. 201. The vanguard was used to flush out the enemy forces from their places of ambush as well as for skirmishing.

The Romans recognized the need for secrecy. They used several different methods to prevent the enemy from making observations of the Roman army: The war plans were kept secret, false rumours were spread, military manoeuvres were intended to fool any observers, men of the same race as the enemy were to be sent away before the battle (not always followed), special camp routines were used to capture spies, various stratagems were used to make the size of the army appear numerous when small and small when numerous, before the battle they could use cavalry screens, the army could be marched forward with lightning speed, the military patrols acted as lookouts against enemy scouts and patrols, the vanguard prevented the making of observations about the main army, and so forth.[25] The success of these operations varied. Sometimes they succeeded and sometimes not. Nevertheless, it is still clear that when the Romans had secured the availability of provisions and operated by the book and posted several layers of spies, scouts and patrols around the army, and the men in charge of these operations did not make errors, the Romans were in a position to exploit any mistakes the enemy made while they themselves could avoid being surprised by the enemy.

The basic marching formations for the mixed armies of cavalry and infantry consisted of the following:

1) a single column in extremely difficult terrain or when there did not exist any immediate threat of enemy attack;
2) double column (a variant of the 'hollow square');
3) triple column,
4) four columns;
5) *epikampios emprosthia* (forward-angled formation);
6) *epikampios opisthia* (rearward-angled formation);
7) the infantry divisions deployed as wedges (with cavalry deployed where needed, but usually behind or on the flanks of the infantry) during marching through forests and other difficult terrain;[26]
8) hollow square/oblong;
9) wagon laager (*carrago*; a variant of the hollow square/oblong) which was a moving fortress with the wagons and cart/wagon mounted artillery pieces placed as a defensive outer layer. The wagon laager (*carrago*) appears to have been one of the standard marching and combat methods for the period from ca. 366 until the 530s. Even if it was known to the Romans well before this (see e.g. my biography of Gallienus) on the basis of *De rebus bellicis* it is probable that it was introduced into the Roman army by Valentinian and Valens. It received an additional modification during the reign of Anastasius, when Urbicius introduced the new anti-cavalry 'Spanish riders' to make it even more secure. The use of the wagon laager as a marching and combat formation appears to have fallen out of favour during the 540s and 550s because we no longer find this method in Syrianus's *Peri strategikes*, and Maurice's *Strategikon* included it only as a form of fortified encampment and not as a marching formation;[27]
10) marching formations for difficult terrain.

Column Formation

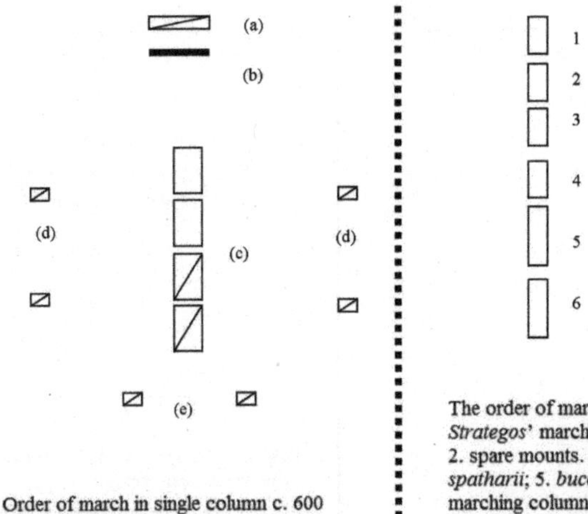

Order of march in single column c. 600 after Haldon, 1999, 157: a. advance guard; b. quartering parties; c. main column; d. flank guards; e. rear guard.

The order of march in column formation: i) *Strategos'* marching column: 1. bodyguard retinue; 2. spare mounts. 3. *strategos* and officers; 4. *spatharii*; 5. *bucellarii*; 6. baggage. ii) Regular marching column: 1. retinue; 2. spare mounts; 3. commander and guards; 4. *spatharii* (sword-bearers, aides); 5. regular troops; 6. baggage. SOURCE: Haldon, 1999, 158.

The posting of the front towards the potential sides of enemy attacks in these marching formations was important because that placed the best men against the enemy.

The baggage train was usually posted so that the baggage train had soldiers between them and the side where it was feared that the enemy could attack, meaning that if the Romans feared attacks from all directions, they posted the baggage train in the middle of a hollow square/oblong, and if they feared enemy attack from behind they posted the baggage train in front and so forth. The only exception to this rule of thumb was when the Romans formed up their wagons as a wagon laager for marching, combat and camping. All of these marching formations were expected to be surrounded by scouts, patrols, rear guard and vanguard. The vanguard consisted usually of cavalry which could be used independently when considered advantageous. If the enemy force proved too strong the cavalry vanguard retreated either to the flanks of the infantry or behind it, either via the flanks or through the infantry formation if it opened up its formation for this. The basic structure of the column formation is given in the attached diagrams borrowed from my doctoral dissertation drawn after Haldon. For descriptions of the other arrays, see the infantry battle formations.

The basic marching formations of the cavalry consisted of the following:

1) the column formation in friendly territory or in wooded terrain;
2) vanguard with the main army and rear guard;
3) the regular cavalry battle formations when near the enemy.

Cavalry marching and combat formations

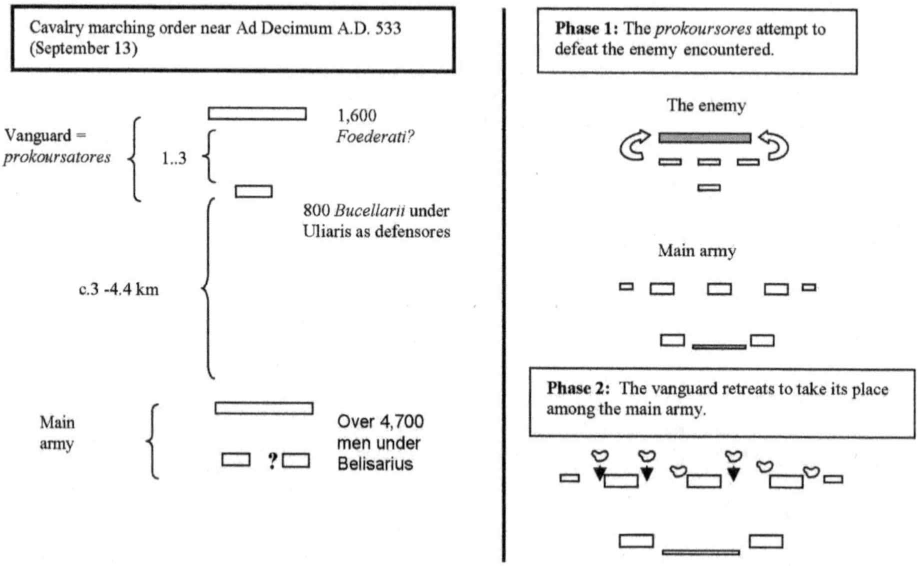

All of these cavalry marching formations had scouts and patrols around them. If the cavalry had a smaller force of infantry following, then this was posted behind the cavalry army but ensuring that it had a separate cavalry force following as its rear guard. For the uses of regular cavalry formations, see the relevant chapter. The variant which had a separate vanguard and rear guard was used in the following manner if it encountered the enemy; 1) the vanguard (*prokoursores*) could engage the enemy on its own if this was considered advantageous; 2) if the vanguard considered it impossible to attack the enemy on its own or had been defeated when doing that it retreated to the main army which became its defenders (*defensores*) so that the vanguard units (now acting as *koursores*, i.e. as runners/skirmishers) retreated to the flanks of each division (*meros*) in the first line. The attached diagrams from my doctoral dissertation *The Age of Hippotoxotai* explain the procedure.

The Roman marching formations were preceded by surveyors, quartering parties and by parties of men assigned to clear and level the route when needed, but in such a manner that if the enemy was near these were to maintain eye contact with the main marching formation. If the terrain was rough, wooded or mountainous, the military doctrine expected that these would be secured first by sending the lightly equipped infantry to occupy them. This enabled the Romans to travel safely even through defiles and mountain passes. If the Romans faced a river which could not be forded, they usually used either the navy to ship them across or they used pontoon bridges consisting of boats/ships/rafts. The pontoon bridges in their turn usually consisted of boats or ships and other parts that could be taken apart and assembled as needed. These were usually carried in wagons and carts, or transported along water routes. It is also probable that the Romans used inflatable calf-skins as described by the *De rebus bellicis* as an alternative lighter version of a pontoon bridge which did not require the use of wagons and carts to carry them. There are also some examples of the Romans swimming across the river in front of them, but this option was not always feasible because there were often troops who could not swim.[28] The standard combat doctrine called for the use of ship-mounted ballistae, stone throwers and archers if the opposite bank was occupied by the enemy and the crossing could not be done in some other place. Once the ship artillery and their crews cleared the opposite bank, it was secured by an infantry detachment and by building a fortified camp to protect the crossing of the main force. This is the manner which the *Strategikon* and *Peri strategikes* instructed the general to use and it was based on long standing Roman principles used when crossing sizable rivers should the enemy opposed the crossing. The following quote from Dio describes how the Romans used basically the same method during the reign of Marcus Aurelius, the only difference between Dio's text and the *Strategikon* being that in the latter the preferred method was to use the artillery-carrying warships to protect the building process. However, it is clear that the procedure was the same as described by Dio when the warships proper were not accompanying the army.

> By the Romans the streams and rivers are bridged with the greatest ease, since the soldiers are always practicing at it, and it is carried on like any other warlike exercise on the Ister [Danube] and the Rhine and the Euphrates. The manner of doing it (which I think not everybody knows) is as follows. The ships, by means of which the river is bridged, are flat. They are anchored upstream a little above the spot where the bridge is to be constructed. When the signal is given, they first let one ship drift down stream close to the bank that they are holding. When it has come opposite the spot to be bridged, they throw into the water a basket filled with stones and fastened with a cord, which serves as an anchor. Made fast in this way the ship is joined to the bank by planks and bridgework, which the vessel carries in large quantities, and immediately a floor is laid to the farther edge. Then they release another ship as a little distance from this one and another one after that until they run the bridge to the opposite bank. The ship which is near the hostile side carries also towers upon it and a gate and archers and catapults.
>
> Dio 71.3, tr. by Foster p.249 with my corrections (*naus* consistently translated as ship instead of the boat used by Foster) and addition inside the square brackets.

Unsurprisingly, the *Peri strategikes* (19) includes somewhat similar instructions for the crossing, namely the use of boats for the bridging and the use of ships equipped with a parapet and artillery to shoot missiles and stones. However, Syrianus also includes several other variants. One of the alternative ways of crossing the river comes from Apollodorus's military treatise (Trajan's military engineer and architect) while the other examples come from his personal experience or from works of history that were also used as sources for collections of stratagems. Apollodorus suggested the use of a single long raft with the fortified raft being placed at the opposite end so

Lipsius' (16th cent.) reconstruction of the bridging of a river.

that it was built along the bank of the river and then shoved off downstream so that it swung to the opposite bank. Syrianus criticizes Apollodorus's version heavily and for good reason. In his opinion this was impossible to do along small narrow rivers (quite true) and any changes in the depth of the river or in the swiftness of the current in different parts of the river could lead to troubles. All of his points are valid, but thanks to the incompleteness of the source material it is impossible to know if it was still used because, after all, Apollodorus's treatise continued to be read by generations of commanders – and Syrianus felt it necessary to criticize the method. The rest of the examples provided by Apollodorus were definitely used. These include the fording of shallow rivers, the distraction of the enemy with a stratagem so that the army could get across the river without opposition, the use of ramps and levelling of the river banks to enable safe fording/crossing, and the canalling of the river to make it fordable.

The *Strategikon* included also the alternative in which the Romans had to clear the river of enemy vessels first before the bridging could be done. This operation naturally required the presence of the navy. The commander of the fleet was either the admiral (*nauarchos*) or *strategos* and under him were *merarchai*, *moirarchai* and other officers. The ship sizes were medium or small. If the Romans had a large fleet of dromons (galleys) it was divided into three divisions. In such instances the warships, the dromons, lead the convoy of ships; the cargo ships were in the middle; and some warships were left as rear-guards. For the actual naval battle, the warships were arrayed in a single line abreast for the entire width of the river and if there were extra ships these were to be posted in a second or third line. The lines were separated from each other by intervals of about one bowshot. It is clear that this approach had been followed throughout the Roman era when the enemy opposed the Romans with a fleet of their own and the Romans had brought their own fleet to the scene to clear the route. However, there are also instances in which the Romans did not bring their navy to the scene or when they employed stratagems instead of the procedures described here. For good examples of such, the reader is advised to read in particular the *MHLR* volumes two, six and seven.

The infantry based Roman armies were expected to be able to march ca. 30km in five hours in good weather and at a faster pace, covering about 35.5km, along good roads or tracks, after which they built a fortified marching camp, but even faster rates are recorded for instances in which it was necessary to surprise the enemy by marching on the double. It is of note that these distances were considered such that these did not tire the soldiers too much. Roman soldiers were real athletes. Pure cavalry detachments/armies could easily achieve up to 80km per day, provided that the horses were regularly rested and well-nourished and the army was unencumbered, but once again there are examples of faster pace too. The marching speed obviously dropped in difficult terrain, poor weather or when enemy harassed the marching formation. However, the principal limitation for the marching speed of the major armies was the baggage train and its beasts of burden. If the intention was to campaign for long periods of time, or there was a need to besiege a location, the army was usually accompanied by a sizable baggage train that included ox-driven wagons, which in turn slowed down the speed of marching. However, when the commander decided that greater mobility was needed to surprise the enemy, he had the option of leaving the wagons behind so that the soldiers carried their provisions (could be three to four days' provisions for a short distance, or eight to ten days' worth as instructed by the *Strategikon*, or 17–20 days' rations as the traditional Roman system expected for longer distance) on their persons (horsemen carried theirs on their mounts) and pack animals (camels, mules, horses). If the army marched along water routes, then the transport ships of the fleet carried the provisions. Roman soldiers were expected to respect the rights of local inhabitants while in friendly territory, but there are numerous examples of the soldiers pillaging and looting while on friendly territory which means that discipline was often not as good as expected by military doctrine. In contrast, in enemy territory the civilians and their property could be the actual target of the invasion so that the enemy territory was pillaged mercilessly, unless of course the intention was to return through it.

In friendly territory, the Romans usually marched their forces as several widely separated marching columns in order to avoid damage to civilian property while also making it difficult for the enemy to estimate the size of the Roman army. The existence of roads and supply dumps/depots on friendly territory facilitated the marching speed of the army. When the enemy was about six to ten days away the Romans united their different columns and set up fortified camps. Roman combat doctrine called for the use of fortified camps in safe locales with access to water and fodder because the availability of these eased the logistical burden. The Romans also recognized the need for maintaining good hygiene/sanitation. It was because of this that the cavalry was brought inside the marching camp only when the enemy was already near. The marching camps were always built with protective barriers of ditches and earth walls (usually with a wooden palisade) if the terrain permitted. The Romans posted a protective layer of scouts and patrols around their fortified marching camp so that there were patrols in every direction. The *Strategikon* instructed the general to change encampments frequently so that the soldiers would not start suffering from illnesses, but it is clear that this was not always possible, for example when the Romans besieged a location.

The general was obviously required to gather intelligence of enemy activities constantly, but if the enemy still managed to surprise the Roman army on the march and the terrain and the situation was unsuitable for battle, the general was instructed to assemble his forces together and occupy a position suitable for a camp and delay until a better opportunity presented itself. If the Romans were forced to retreat, the military doctrine called for the use of prisoners/hostages as human shields.[29]

In the cavalry section of the *Strategikon*, Maurice stated that when the general intended to engage the enemy or when he could not avoid battle that he should pitch a fortified base camp about 48 to 80 kilometres away from the enemy. The general was to leave all the extra baggage, possible family members of the soldiers, horses and men not needed for the battle in this camp.[30] If the enemy consisted of large numbers of cavalry, the army did not change campsites before the battle, which means that they did not use both base camps and intermediate camps when advancing into battle. In such cases the soldiers' families were apparently left further behind in some city near the scene of operations. The idea behind this precaution was to prevent the enemy from targeting the families and personal property of the soldiers.

When the main fighting component of the army consisted of cavalry and wagons were accompanying the army, these were always left behind in the camp and never taken to the battlefield. The same was true of the spare horses in most cases. However, if the enemy surprised the Romans then these were posted behind the cavalry forces in the manner described on page 168.

After having left the superfluous behind, the general then moved closer to the enemy (1–2 days' march) and pitched a fortified intermediary marching camp where he then rested his men and horses before the battle. If infantry accompanied the army, the foot soldiers could be left behind in the base camp, but if these accompanied the cavalry force to the intermediary marching camp, then the footmen could be posted in front of the marching camp to protect the cavalry if it needed to retreat.

Each cavalry trooper was required to carry with him about 20–30lbs of hardtack, flour and provisions worth 7–10 days' rations (3lbs/day) or 3–4 days' worth if the trooper used some of this to feed his horse (3lbs per man, 5lbs per horse + grazing). The margin of operations before being forced to return to the base camp was therefore about 3 to 8 days. The general was also required to post couriers at intervals between the two camps so that the men left in the base camp could react to the outcome of the fighting. These could be ordered to remain in place, or they could be ordered to move to another place, or they could be ordered to join the main army.

The instance in which Comentiolus engaged the invading Avars in 587 demonstrates that the same principles could also be used in a modified form when one intended to engage the enemy in skirmishes by separate roving cavalry divisions. In this case the Roman army divided itself into separate divisions so that the unfit were left behind in a base from which the cavalry divisions then advanced to engage the enemy. The wing divisions then advanced in front of the centre division to skirmish while the third division in the middle formed a mobile reserve force for both. If the enemy passed between the front divisions, these could in turn attack its flanks and rear if aware of such an action. See the attached illustration of the marching formation of Comentiolus on page 102.

The combined armies used several different methods. The baggage train and wagons usually followed the army to the scene of combat where this was possible and were then deployed behind the infantry for its protection. However, it is still probable that the instruction included in the cavalry section of the *Strategikon* not to take the families and pricier personal property of the soldiers close to the battlefield held true also for the infantry, which means that these were also usually deposited in the last fort, city, or a base camp, and that only the essential parts of the baggage train with their wagons were taken to the battlefield. For reasons of hygiene, the cavalry was to stay outside the camp until the enemy was close by. The real difference between the cavalry and the infantry army was that the latter could not cover the same distances as the pure cavalry army, so it could advance a shorter distance from the base camp. However, there were also instances in which the infantry could not take the wagons with them. This happened when the terrain was impassable for the wagons or when the army needed to conduct a forced march to reach some strategic location fast. In such cases, the infantry took 8–10 days' worth of provisions (*Strategikon*) or 17–20 days' rations (the traditional amount carried by legionaries), which were carried by the packhorses, camels or mules. During sieges or long campaigns, the Romans organized separate supply trains with protective escorts of soldiers or had ships carry the provisions where needed.

The Romans tried to ensure that most troops reached the site of battle in fighting condition, but in practice the number of men fit for combat varied, so that the faster the marching speed and the longer the campaign, the fewer the numbers of men and horses were fit for combat. This resulted from accidents (broken bones, twisted ankles, broken ligaments etc.), exhaustion, disease, 'cowardice', repetitive stress injuries, etc. These men were left in the base camp when possible.

The attached diagrams from my doctoral dissertation *The Age of Hippotoxotai* detail the principal facets of campaign logistics and their effect upon tactics. The greater the distance from a major supply dump/depot, the fewer chances there were for the general to avoid being forced to fight a pitched battle in an unsuitable time and place and vice versa. The greater the amount of provisions, the more there was freedom of movement and options available for the general in command. The logistical side of Roman campaigns was left in the hands of specialists, so that generals could concentrate on the purely military side of the campaign. Typically, special praetorian prefects were appointed in charge of the logistical side of large campaigning forces, but during the sixth- and seventh-centuries the man in charge could also be a *sacellarius* (usually a eunuch in charge of the privy purse of the emperor) or a *logothete* (financial officer serving under the praetorian prefect).

The principal sources for Roman marching camps are Hyginus/Pseudo-Hyginus's *de munitionibus castrorum* (variously dated between ca.70–220, but I would date it to the period between ca. 160–220s[31]), Vegetius's *Epitoma rei militaris*, Syrianus Magister's *Peri Strategikes*, Maurice's *Strategikon* and archaeological finds.[32] It is

Cavalry Marching Methods and Logistics. Left: Recommended method of advancing into a battle. Alternatively, the place of troops concentration/mounting area, base camp and refuge could also be the same place (i.e. fort/city close to the border).
Right: Comentiolus' variation of the same method, but with the intention of only skirmishing with the enemy. After skirmishes the troops were withdrawn back to the base camp for resupplying.

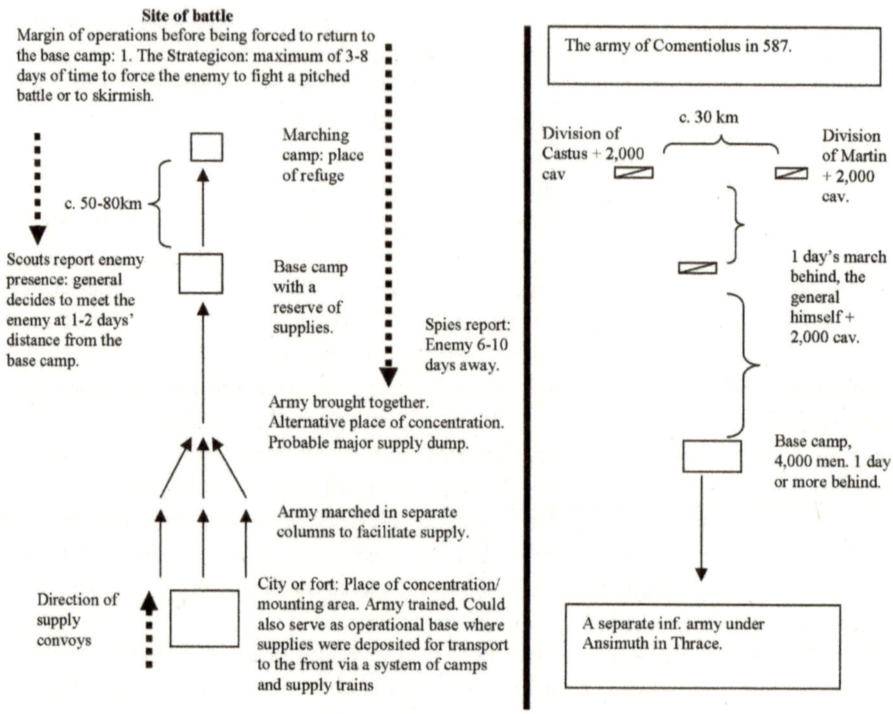

Methods of marching the infantry

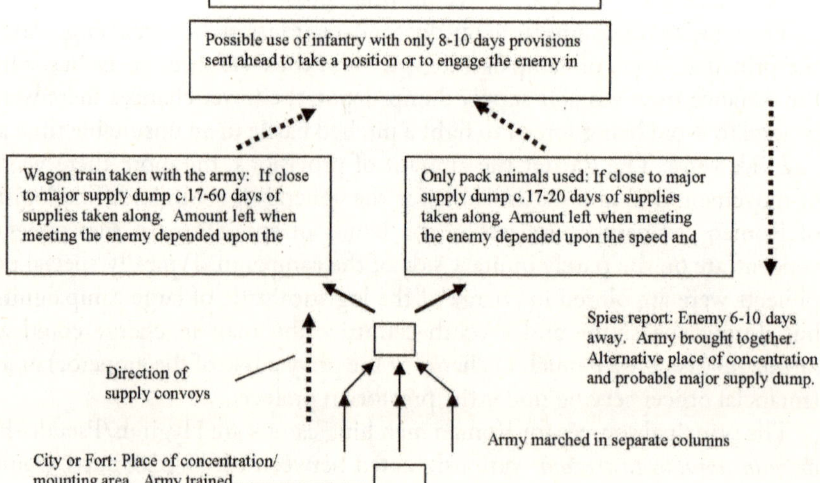

usually assumed that Pseudo-Hyginus's treatise describes the rectangular Roman marching camp as it appeared during the Imperial period. See the attached diagram overleaf.[33] For example, Giuseppe Cascarino (55ff.) does so in his monograph *Castra*. It is not known how and when this imperial-era fortified oblong marching camp became the square/oblong with the two crossing main streets as depicted in the *Peri strategikes* (27–9) and *Strategikon* (esp. 9.5, 12.B.20.91–8, 12.B.22, 12.C; Maurice includes also the variant with wagons posted along the edges). Cascarino (*Castra*, 191–5) does not attempt to date the change. He only states that a new type of marching camp which appears in the sixth-century *Strategikon* is not the same as we find in Hyginus (i.e. Pseudo-Hyginus). On the basis of archaeological data (see e.g. Syvänne, *Caracalla*, 107), and the use of the wagon laager as marching, combat and encampment formation during the period ca. 366–530s, and the use of earlier treatises as sources by both Syrianus and Maurice, I would suggest that the square/oblong shaped marching camp dates from the fourth century or before. One could actually suggest that this type of encampment was adopted during the early-third century, because we find the Romans using the hollow square as their principal marching and fighting formation during this era with the implication that the Romans marched and encamped in the same formation.[34] This, however, is by no means conclusive because the variant described by Pseudo-Hyginus also has infantry on the outside and cavalry inside.

One could also use the instructions of Vegetius (ca.390–450) as counter arguments against the early adoption of the hollow square/oblong formation with two crossing main streets, because he (3.8) states that the camp was to be either square, or round, or triangular, or oblong according to the lay of the land. This is by no means conclusive, because Vegetius was using very ancient material as his sources. Maurice (12.B.52–8) also noted that the ancient sources described many different kinds of shapes but noted that only rectangular formations were in his opinion practical. On the basis of his sources Vegetius also added that the best camps had a width that was a third longer than their depth. Maurice (12.B.52–8) agreed with this assessment, because he recommended the use of rectangular oblong marching camp with a broad front located on high ground (if the army had enough provisions) because this was bound to impress the enemy scouts, but the equally sided square is the formation which the *Strategikon* actually describes in greater detail both in the text and as diagram (12.C). Obviously, the Roman marching camp was always fortified according to the impedimenta available and terrain, but it is clear that the preferred shape was the square or four-sided oblong during the Late Roman period, and probably also before this because this was the shape that we find in Pseudo-Hyginus's treatise and in most of the permanent fortresses.

Constantin Zuckerman (1994, 368–9, 373–85) has demonstrated that Maurice used at least two earlier sources for his description of the Roman marching camp, the longer version (esp. 9.5, 12.B.20.91–9, 12.B.22) which can also be found in the later military treatises *Apparatus bellicus* (75/6 directly utilized the longer version from some other source than Maurice), *Sylloge Tacticorum* (22.1–12 based on Maurice) and Leo's *Taktika* (11.1–41 based on Maurice), and the shorter version (12.C) which found its way to the military compilation of Nikephoros Ouranos (in Zuckerman

MARCHING CAMP OF PSEUDO-HYGINUS

drawn after Lenoir

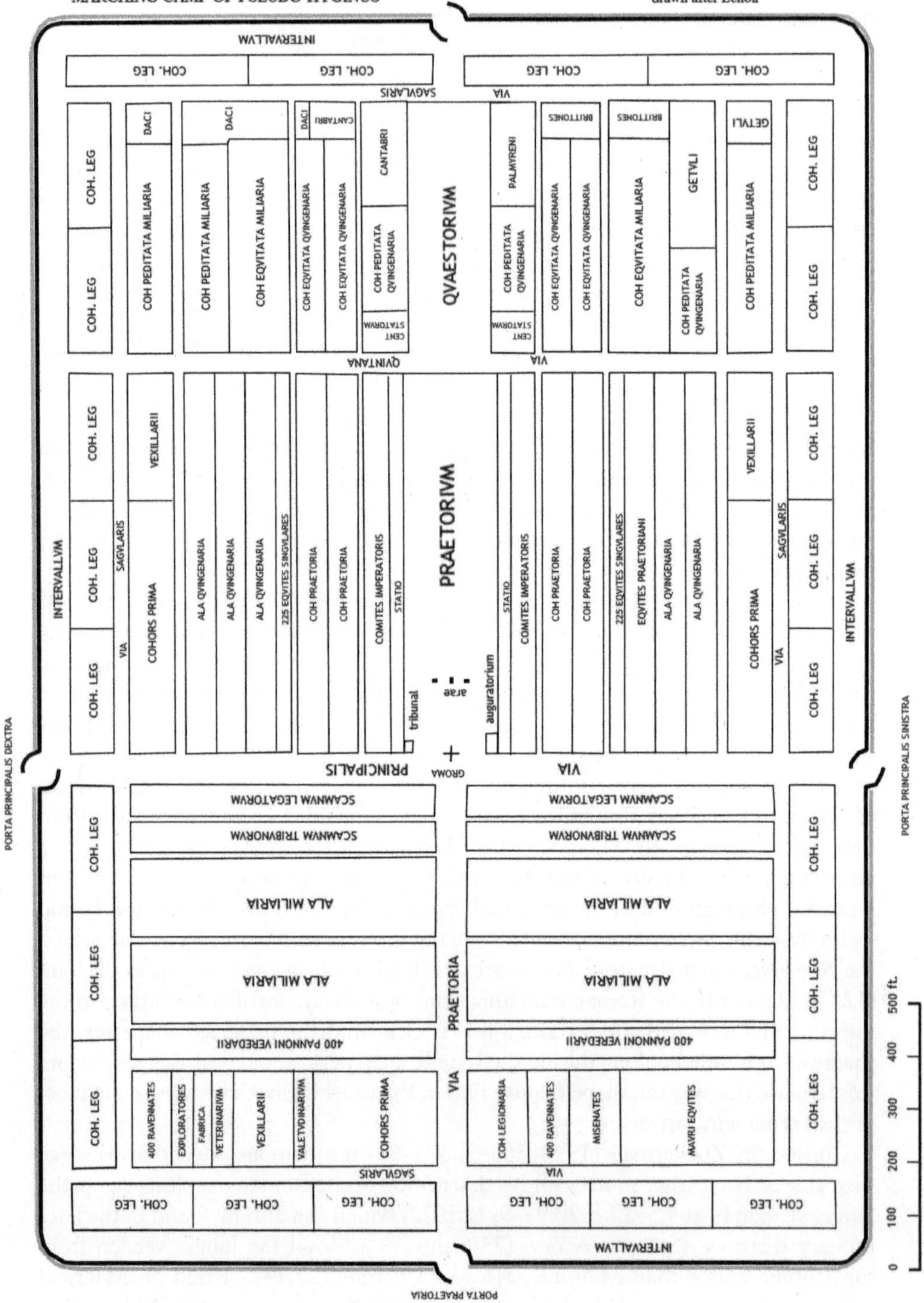

p.381) with greater number of details. I have here analyzed the material on the basis of Zuckerman's conclusion that there existed two different literary traditions by the late-sixth century, but I caution the readers to keep in mind that it is possible that there was only one large treatise on encampments on which all of the Late-Roman and Byzantine Roman treatises were based, and that all of the treatises mentioned just picked and chose their instructions as they preferred. The principal differences between the longer treatise on encampments and shorter version are: the longer version (12.B.22) has an interval of 300 to 400 ft between the infantry posted next to the wagon fortress/laager (*carrago, karagos*) and the rest of the tents in the middle and it has only one trench/ditch (*fossa*); the shorter version (12.C) has an interval of 200 ft between the infantry and cavalry and it has one extra ditch/trench outside the caltrops. The *Strategikon* therefore has two slightly different versions of the marching camp which were presumably in use at the time. The principal differences between the camp of Pseudo-Hyginus and Maurice were: the camp structure was simplified in the Late Roman period so that there were always only two main roads and four main gates and when the wagons followed up the army these could be used as wagon fortresses (*carrago, karagos*); and the interval between the Roman infantry and cavalry was increased from the 30 to 40 feet-wide *via sagularis* in Pseudo-Hyginus (20) either to 200 feet or 300–400 feet in the *Strategikon*. The increase in the interval between infantry and cavalry improved the safety of the marching camp because enemy arrows were not expected to reach the tents of the cavalry and their horses when the caltrops and other protective measures surrounding the camp kept them at a distance.

The *Strategikon* provides us with the best description of how the Romans formed their marching camps in practice. Since Maurice based his treatise on actual Roman practices, it is clear that his procedures had been followed also by earlier commanders. The existence of the other military treatises that have drawn from the same sources as Maurice helps to explain some of the problems present in the *Strategikon*. If the enemy force was in the vicinity and the cavalry army was accompanied by wagons and included infantry, the wagons were parked first as a wagon fortress (*carrago, karagos*), the oxen were hobbled or tied, and a trench/ditch (*fossa*) was dug around it with the earth thrown inside to form a wall of earth while the cavalry patrols, cavalry and 'hoplites' protected the building of the camp.[35] In emergencies, the wagons could be dug in if the terrain allowed this. The ballistae wagons and carts were posted among wagons. In the *Peri strategikes* the encampment with the ditch was not protected by wagons, but by foot soldiers posted along the outer edges so that these placed their shields and spears in such a manner that it formed a spiked obstacle towards the enemy. The longer version of the marching camp in the *Strategikon* (12.B.22) instructed the general to use caltrops and small pits with stakes (*Peri strategikes* has only caltrops) outside the ditch as an additional form of protection. Urbicius obviously included also his own version of the *chevaux de frise* at such distance outside the wagon fortress that the enemy archers could not harm the Romans, while the Roman ballistae wagons would be able to decimate them. It is probable that this device or something similar continued to be used when the commanding general was well-versed in the art of war.

Maurice places the tents of the *psiloi* (light-armed) just inside the wagons with the best tribunes posted to guard the gates, after which followed an empty space of 300 to 400 feet and after which were posted the rest of the tents. Maurice fails to mention where the heavy infantry (*hoplitai, skoutatoi*), were but he does mention that the cavalry was posted in the middle of the camp, which implies that the heavy infantry was stationed together with the *psiloi*. We can clarify this further from the *Apparatus bellicus* because both have used the same source. The *Apparatus bellicus* (75/6.95–9) places behind the trench/ditch (*fossa*) either the wagons, or a palisade, or some other construction and along the *fossa* the *hoplitai* (hoplites) and behind them the *psiloi* (light-armed). This gives us the following alternatives: 1) it is possible that the hoplites were placed between the wagons and trench and the light-armed behind the wagons (least likely); 2) that the *hoplitai* and behind them the *psiloi* were placed behind the trench when the wagons were not accompanying the army; or 3) when the wagons accompanied the army, the *psiloi* were placed behind the wagons and the *hoplitai* were placed behind the *psiloi*, because the wagons provided adequate protection for the light-armed so that the heavy-armed could act as reserves. I would suggest that alternatives two and three are the likeliest, but this is not conclusive because Nikephoros Ouranos (p.381 in Zuckerman, 1994) gives us a fourth alternative by placing the wagons between the cavalry and infantry so that the *psiloi* were closest to the *fossa* (with the hoplites behind them). We should remember that this treatise has used the same source as Maurice in the *Strategikon* 12.C. It is therefore quite possible that all four alternatives were used in different circumstances, even if Maurice fails to state this, but I would still consider the options two, three and four as the likeliest. The *Peri strategikes* places the infantry next to the trench, but so that the outer edges had men armed with spears and shields – these were clearly the multipurpose foot soldiers which placed their shields and spears as anti-cavalry devices towards the enemy while they employed bows from behind this obstacle. In other words, there is no real difference between Maurice and Syrianus in this case and it pays to remember that it is certain that this method was used when the Romans had multipurpose footmen, as they did at least until the reign of Justinian I. Even if this is stated nowhere in the *Strategikon*, one may imagine that, when the cavalry army was not accompanied by infantry, some of the horsemen dismounted to become footmen armed with both bows and spears along the edges of the encampment while their horses were taken to the middle of the camp just like happened when a cavalry army retreated and needed infantry forces as a double phalanx/column for its protection (*Strategikon* 9.4).

The shorter treatise on encampments presented by the *Strategikon* (12.C) and Nikephoros Ouranos (p.381 in Zuckerman, 1994) has two ditches/trenches, one outside the wagons and another outside the caltrops. It was presumably thanks to the existence of the second *fossa* that it was possible to shorten the distance between the infantry and cavalry to 200 feet, so that it was still wide enough to protect the cavalry and its horses from enemy arrows. The text quoted by Nikephoros Ouranos places the horse-breakers (pits) outside the second ditch so that the protective zone before the camp was even wider than with the second *fossa*. I would suggest that the Romans did indeed use this variant in situations where there was not enough infantry

6th Century Roman Marching Camp.
Source: *Strategikon* 12.C after the so-called "short treatise on marching camps".

According to Corippus (*Iohannidos* 8.324-9), the general's tents as well as a large altar for sermons were placed in the centre of the camp. This implies that the placing of the general's tent to one of the quarters and not in the crossroads was probably added to the genre by the Emperor Maurice who did not like being bothered by outside noises.

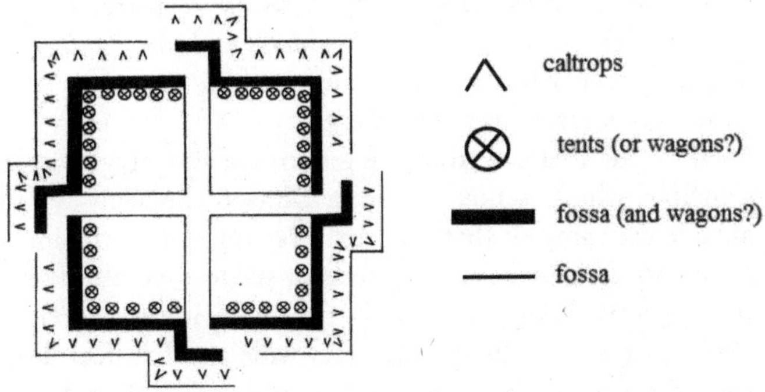

to form as large a rectangle as the wider interval of 300 to 400 feet between infantry and cavalry would have required. The widening of the protective zone in front of the marching camp would have compensated for the lack of infantry. Considering this, it is entirely possible that there was only a single prototype source for the building of fortified marching camps which all sources used selectively.

According to Maurice and other sources in the same tradition, the encampment had two major roads 40- to 50-feet wide intersecting in the middle, four major gates and many smaller gates. The *Peri strategikes* (28.11ff.) has two main roads and four gates, and states that for reasons of security the other minor roads did not go through the trench. It is possible that this derives from the shorter treatise on encampments or from a larger treatise that included all of the different variants, and that this system was used by commanders that preferred the extra security it provided. In the *Strategikon*, the *strategos* and his *meros* did not pitch their tents in the crossroads of the main roads like the Romans had done in the past (see e.g. the diagram depicting Pseudo-Hyginus's camp).[36] The general's tent was placed off to one side so that it would not interfere with traffic. The second of the reasons was that the noise made by the traffic could disturb the general. It was because of this that Maurice forbade dancing and hand-clapping and anything else that he considered a frivolous waste of time. It is quite probable that these were Maurice's own personal additions to the genre, because Theophylact's text proves that Maurice was personally bothered by such noise.[37]

This receives further support from the statement of Corippus (*Iohannidos* 8.324–9). Corippus states that the general John Troglita placed his tents in the middle of the marching camp in 548.[38] This implies that the general's tents were still in their traditional place in the middle of the marching camp during the reign of Justinian I the Great. The building of fortified camps was neglected during the periods when discipline was lax. It was because of this that the Romans suffered a major defeat in Armenia in 577. It was then that Maurice was appointed as supreme commander of the eastern armies. He reintroduced discipline into the army and the army started building fortified camps again (Menander, Blockley ed. frg. 23.3).[39] It is therefore quite plausible that he changed the practice of placing the commander's tent centrally at this stage because he disliked loud noises – this is not surprising because he had started his career as notary.

In contrast to the general, the *Strategikon* instructed each *merarches* to place his tent precisely in the middle of his *meros* and place a messenger in attendance at the *strategos*'s tent, while each tribune placed a messenger at the *merarches*'s tent. This ensured the fast delivery of orders.

All of the treatises from the text of Pseudo-Hyginus onwards have one unifying principle, which is that if terrain allowed the army was instructed to dig a ditch outside the camp so that the earth was thrown on the inner side to form a rampart. In the *Strategikon* the ditch is five- or six-feet wide and seven or eight deep, while in the *Apparatus bellicus* the ditch is five-feet wide and seven- or eight-feet deep. In the *Peri strategikes* the ditch is five-feet wide and five-feet deep. Once the fortifications were finished the baggage train proper and soldiers then entered the camp in an orderly manner.

If wagons did not accompany the army, then the soldiers dug a trench, scattered caltrops outside the trench, and built a wall or palisade inside the trench, as the *Peri strategikes* instructed the army to do always. If the enemy was not near, then it sufficed to leave a *drouggos* or *meros* as guards while the rest of the army pitched the camp. The last-mentioned instruction implies that the irregular-sized *drouggos* was a large detachment comparable to a *moira* (max. 3,000 men) and *meros* (max. 6,000–7,000 men), which perhaps means that it was an irregular-sized unit between those which had around 4,000–5,000 men on average. This indeed seems to have been its typical size during the early-seventh century.[40]

The *Strategikon* (9.5.43ff.; 12.B.22.48–51) also included the option that the marching camp was not fortified. This could happen when the ground was rocky or it was very late. In such instances the soldiers spread caltrops around the encampment and placed outposts of guards around according to the requirements of the terrain. If the enemy had only one narrow approach to the site, then few outposts were enough, but if the terrain was wide and open, then several layers of patrols and outposts were needed.

Roman camping procedures also included safety features against enemy spies. For example, they used periodical trumpet calls for the troops to get inside their tents so that possible enemy spies could be caught. The generals were always accompanied by their own personal retinues of bodyguards who protected them against assassination attempts and military mutinies. These safety precautions were wise indeed, and made futile all possible assassination attempts by enemy spies and assassins. None of the commanding generals was assassinated in his tent, which proves that the precautions were effective. Even when the Roman army revolted, the commanders were always able to flee from their quarters inside the marching camps, even if they were later killed by the mutineers.[41]

It is clear that when the regulations concerning the use of the spies, scouts and patrols were followed to the letter, just like the instructions concerning the fortification process of the marching camp, while adequate provisions were secured for campaigns, then the Roman field forces were very difficult to defeat. However, there were periods in which discipline in the army was so lax that the regulations for building a camp were not followed. A good example of this is the period just before Maurice was appointed as supreme commander of the Roman armies in the east.[42] In that case the Romans were surprised and defeated in their encampment, but it should be remembered that these instances were rare during the Late Roman period. The Romans followed their military doctrine to the letter for most of the time. The Romans also did not build fortified camps in situations in which the cavalry army stopped to eat. In such cases they only sent scouts and patrols around.

Once the army had been encamped securely, the *Strategikon* advised the sending of a *bandon* or two closer to the enemy to observe its movements and the terrain, and to prevent the enemy from making similar observations. When the general then decided to engage the enemy in battle, the heralds, officers and other chosen men were sent ahead to reconnoitre the battlefield. If the army consisted of cavalry, then the Roman ambushers also performed a reconnaissance in force when they advanced towards the enemy as their mission was either to ambush the enemy or to prevent the ambushing of the Roman army. The general was expected to ensure that the men and horses had a sufficient amount of energy available in the form of food, water and fodder to continue the battle or to flee if needed.[43] The food and water were carried in the wagons or on packhorses and served by the servants when needed. The marching camp served as a place of last refuge unless the retreat had to be continued. If the cavalry army included infantry, then some of these were posted in front of the caltrops as a phalanx for two purposes: the cavalry could rally behind it and repeat the attack; and if the cavalry was forced to retreat to the marching camp, then the infantry protected its flight.

3.4. When, Where and How to Fight[44]

> The general is wise who before entering into war carefully studies the enemy, and can guard against his strong points and take advantage of his weaknesses ... Wild animals are taken by scouting, by nets, by lying in wait, by stalking, by circling around, and by other such stratagems rather than by sheer force. ... To try simply to overpower the enemy in the open, hand-to-hand and face-to-face, even though you might appear to win, is an enterprise, which is very risky and can result in serious harm. Apart from extreme emergency, it is ridiculous to try to gain a victory which is so costly and brings only empty glory.
>
> *Strategikon* 7. Preface, tr. by G.T. Dennis, 1984, 65.

The above quotation describes the Graeco-Roman approach to warfare concisely. It is a mistake to think that this would have been only the Late Roman or 'Byzantine' approach. Skilled Greek and Roman commanders always took into account the prevailing conditions and adjusted their combat methods to the situation in such manner that took into account the forces at his disposal and their strengths and weaknesses vis-à-vis the enemy he faced. The idea was to try to seek out the weak points of the enemy that could then be used against him. As I stated in my doctoral dissertation, the skilled commander used his intellect to overcome the enemy. The wise among the Romans idolized these better commanders so that their exploits were recorded in the works of history and in the military treatises such as those that collected the stratagems used (e.g. Frontinus and Polyaenus). It is therefore unsurprising that we find the same approach to war in such Late Roman works as Vegetius, the *Peri strategikes* and the *Strategikon*. However, as always, the actual implementation of these sound rules to warfare was a question of personality. In the past, there were sensible commanders like Fabius Maximus, Scipio Africanus Sr. and Aemilius Paulus (Varro's cautious colleague at Cannae in 216 BC), and there were rash and bold commanders like Varro (the commander who lost at Cannae). However, it should be kept in mind that bold rash action did not always result in disaster and, vice versa, the cautious approach did not always ensure success. The best example of this is the civil war between the cautious Pompey and bold Caesar. Furthermore, it should be kept in mind that bold decisive action has always had its supporters. In the ancient period the principal reason for this was the emulation of Alexander the Great. Regardless, it is still clear that a bold, incautious attitude in warfare was always considered risky in ancient military theory. Therefore, it is not surprising that the Late Roman period treatises extolled the intellectual approach to warfare over the use of bold brute force. This, however, did not ensure that there would not have been those who acted incautiously. There were rash commanders like Julian, John Troglita, and Tiberius II, and there were cautious commanders like Constantius II, Aetius, Belisarius, Narses the Eunuch and Maurice.[45]

Roman military doctrine required the commanding general to take into account many different variables before he committed his forces either to guerrilla war or pitched battle. The Roman general was required to devise his plans according to the forces at his disposal, while keeping in mind the nature of the enemy force, the terrain, the time of day (direction of the sun in relation to the army, night attacks etc.), the weather (wind or rain could prevent effective archery fire), time of the year (affected the availability of food and fodder), and the personality of the enemy general (i.e. was he rash, devious, careful, cowardly etc.). The general was required to assess the strengths and weaknesses of each side before deciding what kind of approach to the situation he would adopt. As is obvious, the availability of accurate intelligence of the situation was the decisive factor in the making of these assessments. If the enemy was unfamiliar, the general first tested the enemy with skirmishes to learn the strengths and weaknesses of the enemy force. This also familiarized the soldiers with the methods used by the enemy and removed the fear of the unknown.

The role of good intelligence was obviously paramount. Without it the army would have been blind and practically helpless. However, as already noted, it was generally good. Against unfamiliar peoples it was not recommended to engage them right away in battle. One was first to engage them in skirmishes and ambushes to familiarize the men with the enemy and to improve morale. The Roman rules of engagement were very practical. Consequently, the general was required to adjust his methods and mix of soldiers to meet the type of enemy in question. Roman military doctrine also advised the general not to use the same methods against the same enemy even when successful, because the enemy could learn from their mistakes.[46] The same instruction can be found in the *Strategikon*.

The general was also required to choose his own battle site, and most importantly he was required to avoid being forced to fight at an inopportune place or time. If the enemy's strength lay in archers, he was to array his army on higher ground or on level open ground, but never on the lower slopes or in difficult terrain. He was also required to have the sun, wind, and dust behind his own line of battle and in the face of the enemy. Roman military theory paid particular attention to the rear of the battle formation. It was because of this that the Romans had reserves. In addition to this, it was recommended to possess some protective terrain feature (river, lake etc.) or some artificial feature (camp, wagons) to protect the rear of the battle formation. If the Romans used a cavalry army, they could hide the presence of a second line by posting it just behind a hill or in a valley or in forested terrain. For obvious reasons, it was advisable for the general to avoid battling against superior enemy forces, especially when these fought in 'organized' fashion like the Persians. If the general was forced to fight against a numerically superior enemy force, he was to resort to guerrilla war, or to the using of restrictive terrain (hill, valley, pass, river etc.), or field fortifications, or a fortress to negate the enemy advantages, or to the use of outflanking operations against the flanks and rear of the enemy by using outflankers and ambushers. The narrative sources also prove that very skilled generals, such as Philippicus, could manoeuvre in such a way that the enemy army would be forced to march through difficult terrain that debilitated their horses.[47]

The type of conflict also had a bearing on the way the war was fought. In defensive warfare, Late Roman combat doctrine recommended the initial avoidance of pitched battles so that the invaders were subjected to a period of guerrilla warfare which consisted of scorched earth tactics with skirmishing, while the population and their property were taken inside walls. This approach could also include the use of diversionary campaigns against the enemy. The idea was to force the enemy to leave with minimum risk, because the presence of fortifications on Roman territory diminished the readiness of the soldiers to fight bravely – it was far easier for the soldiers simply to avoid danger by fleeing inside the cities. It was of paramount importance to keep the field army in existence, because its presence prevented the enemy from foraging and besieging cities. If the Romans lost a pitched battle at the beginning of an enemy invasion, the enemy could then freely forage, pillage and besiege places. Therefore, it was recommended to engage the enemy only when it was already retreating from Roman territory, exhausted and close to their own homes. If the Romans suffered defeat in such conditions the enemy would still continue its journey home to deposit the possible plunder and to rest. This was military doctrine, but in practice when the Romans possessed adequate forces on the frontier area, they usually tried to block the advance of the enemy rather than follow the abovementioned instruction. This is what Constantius II did against the Persians in 361.[48]

Contrary to popular image regarding Late Roman and 'Byzantine' warfare among both the academics and general public, the Romans never stopped waging offensive wars in an effort to occupy new territory if the circumstances for this were opportune. Military victories and territory added to the Roman Empire brought military glory to the general, army and emperor. It was because of this that Constantine the Great enlarged the Empire, and it was because of this that emperors such as Maurice and Heraclius did the same.[49] Even the *Strategikon* advised the general to look for suitable opportunities and excuses to invade enemy territory whenever the enemy had suffered a defeat, or was known to be weak and unprepared for some reason. Offensive war called for the use of surprise. On the strategic level the surprise could be achieved by using some unexpected invasion route, or time of year, or the exploitation of the enemy's troubles; on a tactical level by surprise attacks against careless enemy, night attacks and by ambushes. Obviously, in practice surprise was not always achieved. In offensive war it was preferable to engage the enemy in pitched battle in the initial stages of the offensive when their army was still unprepared, small and/or demoralized. The crushing of the enemy field force at the beginning of the invasion gave the invaders the chance to ravage enemy territory and/or to besiege a city unhindered. Belisarius's first campaign against the Ostrogoths was a special case, because he was essentially fighting a defensive war even though he had invaded enemy territory. The reason for this was that most of the local inhabitants received the Romans as liberators, so they opened the gates of the city of Rome to the invader who thereby became the defender against the Ostrogoths. Belisarius then exploited this to the hilt by destroying his enemy with guerrilla warfare without resorting to any single decisive pitched battle. It was Belisarius's cautious approach that became the model to be followed in warfare against the light-haired peoples, in other words against the Germanic peoples like the Lombards, Visigoths and Franks. It should also be noted that in practice there were also instances in which the Romans and their enemies could agree to limit the scope of the conflict, or when they agreed to fight a pitched battle in the open, or when one or the other managed to fool the other side with a stratagem.[50] In sum, the true quality of generalship was shown before the battle.

The maxims of the *Strategikon* also preserve some other pre-battle precautions which date both from the imperial and the Late Roman periods, and the following discussion details some of the more important of these precautions. For example, the maxims advised that if there was an allied force present, it was preferable that it consist of various different nationalities so that these would not unite against the Romans. It was also inadvisable to have an allied force in Roman territory if it was larger than the Roman army.[51] The *Strategikon* noted that the allies of today could be tomorrow's enemies. They were not to learn the methods of the Roman army. In order to achieve this, foreigners were to be kept separate from the Roman main battle formation and were to set up their own camp and march separately.[52]

In Book 11 the *Strategikon* introduced a new element to the genre of military treatises, which was the adjustment of military methods according to the generic type of enemy faced, which were classified into four groupings; the Persians, Scythians (Avars, Turks, and other Hunnic peoples), light-haired peoples (Franks, Lombards and other Germanic peoples), and Slavic peoples (Slavs, Antes and the like). The Romans had obviously always adapted their combat methods according to the enemy faced, because the above-mentioned instructions already take this into account, and even the *Strategikon* includes references to these in its maxims. For example, the maxim 8.2.88 refers to the specific qualities of the Parthians, Gauls,

Spanish, Ligurians, Britons and Germans. It is therefore clear that it is only an accident of history that the *Strategikon*'s version is the only one still extant. However, the description is still very valuable as it preserves for us the specific instructions of Maurice when dealing with the various different categories of enemies, and because these categories then became military doctrine for the future generations of Roman commanders, as is demonstrated by their preservation in Leo's *Taktika.*

Against Persians[53]

The *Strategikon* (11.1) unsurprisingly classed the Persians among dangerous well-organized enemies because they were undoubtedly the most powerful enemy the Romans faced. According to Maurice, the Persians were patriotic and obedient to their rulers out of fear and therefore not easily bribed. The wars between the two superpowers show this to be true. The Persian approach to warfare was orderly, so that they preferred to win their wars through generalship. They were also very skilled in diplomatic negotiations. They could bear all hardships (heat, thirst, lack of food) easily. The Persians were excellent besiegers and even more formidable when besieged. It was this that made them more dangerous than any of the other foes. They were the only enemy the Romans faced that were equal to them in this field. Maurice's view was that the principal elements of the Persian soldier were: the Persians wore armour and mail; were equipped with bows and *spathion*-swords; and their archery fire was more rapid, even if not powerful as, the archery practiced by any other nation. From other sources we know that this is a good summary of the typical Persian knight, but with the caveat that Maurice has not included all of the pieces of equipment worn by the Persian troopers. It is well-known that the elite Persian cavalry consisted of dual purpose cataphracts who acted as mounted archers and lancers (in other words, the Persian troopers were not only equipped with *spathion*-swords for close range combat), while their allies provided most of the lightly-equipped mounted archers, the unifying element being a concentration on the use of archery to defeat the enemy before advancing into close combat. The lightly-equipped mounted archers were usually posted in the front line, but there were also instances in which both lines consisted of cataphracts, or when the light cavalry was posted beside the cataphracts. The omission of the lance/spear from Maurice's list may indicate that when the Persians used the bow, they typically grasped their swords after they had put away their bows. It also suggests that the Persians were skilled with the sword. The required martial equipment for the regular Persian knights consisted of the following pieces from the reign of Chosroes I onwards, and it is clear that in practice this was also the panoply worn during the reign of Maurice even if he fails to list all of the pieces of equipment: helmet, hauberk, breastplate, mail, gauntlet, girdle, thigh-guards, *contus*-spear, sword, battle-axe, mace, bow case, two bows and bowstrings, a quiver with 30 arrows, two extra bowstrings, a lasso, a sling and stones, a shield, and horse armour (either metal or leather).

According to the *Strategikon*, the Persians deployed their cavalry battle array in three divisions (*mere*), centre, right, and left, in which each *tagma* was drawn up in the first and second line or phalanx. The depth of the formation was not uniform, but the front of each phalanx was even and dense: in other words, the Persian units were

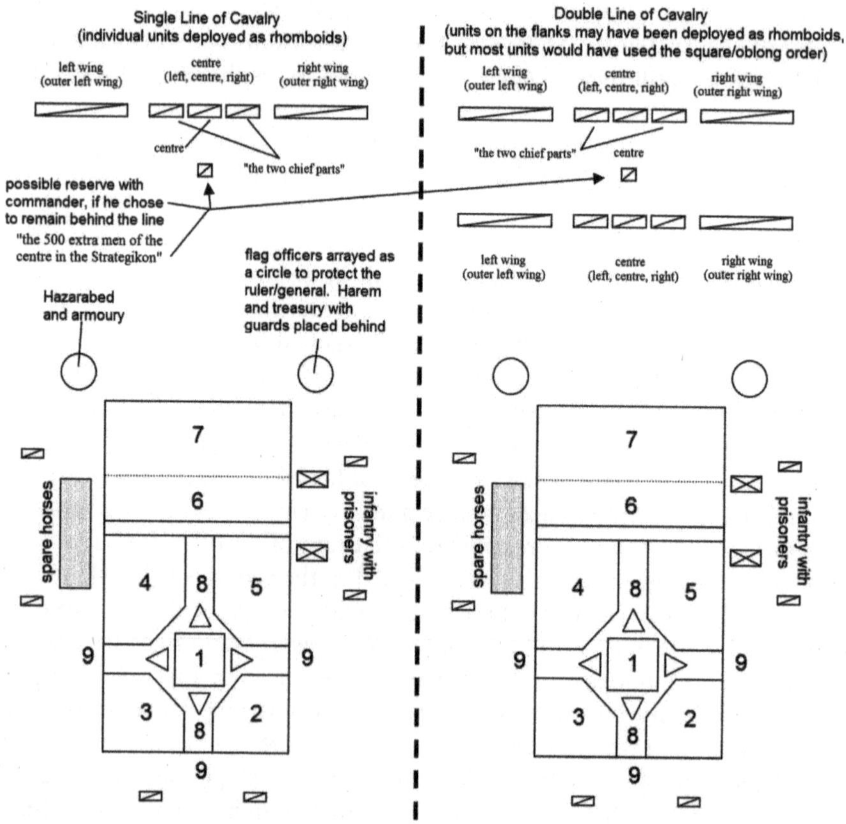

The two standard Persian cavalry formations with infantry and camp placed behind (not in scale)

- the probable structure of a Persian marching camp based on the supposition that later Muslim practices mirrored those of old Persia: 1) King or general, his entourage, guards, war chest etc. 2-5) Officers, officials, doctors, elephant keeper, entourages, guards, servants etc. 6) Cavalry. 7) Infantry. 8) Roads. 9) Gates. The Persians also posted a strong guard unit on one side of the camp to act as ambushing and guarding forces against any army trying to surprise them in their camp.

- the rear half with the ruler's entourage (1-5, 8-9) was the camp proper; and the front half (7) with infantry and wagons, hospital (could also include a trench and caltrops) was the portion facing the enemy (with no gates) that served as a bulwark against attackers. The cavalry advanced from the camp proper (6) to form the cavalry battle array. When the camp was built behind the battle line to protect the army in battle, it could include separate cavalry detachments to protect the flanks and rear, which I have added to the illustration on the basis of the later practices. The reason for this reconstruction is that the Tafrij, which includes these detachments, includes several borrowings from the earlier Persian treatises as a result of which it is possible that these were also used by the Sasanians.

- After Syvänne (2004, 2009, 2015) based on the *Strategikon* (11.1), *Tarfij* (The 5-lines formation), *Gotha* (Square formation), and illustration of the ancient Persian battle array by Fakhr-i Mudabbir.

typically deployed as rank-and-file oblongs. In fact, the *Strategikon* warned against the use of the feigned flight tactic against the Persians because during the sixth century the disciplined Persians did not break their formation during pursuit. The Persians had stopped using disorderly gallop in pursuit when they had been defeated as a result of this by the Hephthalites in 484. If the general still decided to use the feigned flight against the Persians, Maurice advised him to direct the attacks against the vulnerable flanks and rear of the Persian formation because the Persians did not use adequate numbers of flank guards. The lack of adequate flank guards made the Persian flanks vulnerable also for such attack in regular battles.

The sources that preserve Persian military treatises suggest that the Persians divided their formation into two phalanxes consisting of the outer left, left, centre, right, and outer wings. The left and right wings of the Persian tradition appear to have belonged to the division that the *Strategikon* considered the centre division, and not as part of the insufficient flank guards because the latter would not have constituted actual divisions in the battle line. According to Maurice, the Persian centre had four or five hundred additional picked troops, which I interpret as the commander's personal reserve. It is quite possible that these four- to five-hundred men were arrayed as a massive wedge, as the 'Byzantine' Romans did with their elite forces in the tenth century. The Persian mode of fighting was to advance methodologically forward in perfect order. The Persians placed their spare horses and baggage train a short distance behind the main line. The baggage train was placed behind a ditch as a last place of refuge. The Persians usually posted their infantry and servants behind this ditch together with the baggage train. This means that the typical Persian battle formation consisted of two long lines, one behind the other, each divided into three divisions, but the *Strategikon* also notes that the Persians also continued to use the single line with three divisions for the purpose of outflanking the enemy.[54] The attached diagram shows the typical Persian cavalry formations.

Maurice noted that the favourite tactic of the Persians was to shower the enemy with arrows, which is confirmed by the narrative sources. It was because of this that Maurice recommended that the general choose an open, smooth, and level terrain, because these were suited to the charge of the Roman cavalry lancers. Such charges were to be performed in an even and dense order so that the entire force would be forced to participate in the charge behind the protective cover of the shields and also because this looked impressive to the enemy. The swift charge to contact was beneficial, because the Persians were prone to rapid flight if charged to contact and did not usually employ the nomadic feigned flight in battles. The Persians attempted to counter this tactic by posting their forces on difficult terrain or behind trenches, ditches, pits, or caltrops, from which they could then bombard their enemies with barrages of arrows. The terrain or obstacles in conjunction with the archery could disrupt the Roman cavalry battle array so that they would then fail to reach the Persian lines. When faced with this, the *Strategikon* expected the general to dismount some of his men so that the Romans would then use these dismounted men as infantry to overcome the obstacles. This instruction was necessary for the East Roman field armies of the fifth- and sixth-centuries, as it was typical for these to consist mostly or solely of cavalry. During the late-third and fourth centuries, typical Roman armies had an infantry majority, so this instruction was not as relevant for those centuries. The *Strategikon* noted that the Persians were vulnerable to compact infantry formations, weather conditions (south wind and rain) harming their archery, to cavalry lancers in unobstructed terrain, and to hand-to-hand combat in general.[55]

The advantages and disadvantages between the Romans and Persians/Parthians had remained basically the same throughout their mutual history, even if the Romans had increased the numbers of horsemen and archers in their armies thanks to these contacts.

The narrative sources prove that the Persians sometimes included their infantry also in the battle formation proper and not only as rear guards for their cavalry. In such instances the infantry was usually posted in the middle. In emergencies, the infantry formed up a circle around the rest of the army. Typically, the Sasanian infantry consisted of spearmen equipped with large shields and infantry archers both of which were usually of dubious quality usable only as 'cannon fodder'. Excluding elite units such as the allied Daylami or the *murmillones*-type infantry, the Persian infantry was therefore usually no match for the Romans, but there are instances in which the footmen posted behind the cavalry still managed to save the Persian army from complete defeat. A good example of this is the Battle of Nineveh in 627.[56]

Against the Scythian nations (Avars, Turks and other Hunnic nomads)[57]

Maurice classified the Huns, Avars, Turks and other nomadic Scythian tribes as a single category on the basis of their mode of living and style of fighting. In his opinion, only the Turks and Avars were interested in organizing their military in an orderly manner, and in his opinion both were stronger than any other Scythian nation thanks to this. This is a sound estimation. All Scythian nations had a monarchical form of government, and endured heat, cold and want of necessities. They were considered superstitious and very treacherous and greedy. Their preferred methods of waging war were the use of deceit, surprise attacks and denial of supplies to the enemy – the large numbers of cavalry made the last one easy for them to achieve.

According to the *Strategikon*, the Scythians were equipped with mail, swords, bows and lances, meaning that in combat they attacked so that the lance was slung over their shoulders while they shot with their bows and then used other weapons as the situation required. In other words, the nomads were always ready to charge into contact and this made them formidable in pitched battles. In addition to this, the horses of their illustrious men were always armoured either with iron or quilted/padded armour. The Scythians were also always accompanied by vast herds of horses, which made them vulnerable to the denial of fodder (for example in campaigns in such times that fodder was not available). The Scythians did not fortify their encampments, but posted so large a number of sentries at distance that it was difficult to surprise them. However, as Comentiolus's campaign in 587 demonstrates, this was still possible. The Scythians did not divide their battle line into three divisions like the Romans and Persians, but into several irregularly sized *moirai* ('*moirais drouggisti*'), which were closely joined together to present one long battle line. They also posted a separate reserve force which they could use to ambush the enemy or to support their main line. The spare horses were placed close to the battle line to enable the riders to switch those fast or to protect the rear of their formation. The baggage train was placed to the left or right, a mile or two behind the battle line with a sizable unit of guards. The depth of their battle formations (*acies*) was not uniform but they preferred to make them as deep as possible while they made their front even and dense. This refers to the use of two or even three battle lines, as mentioned by Maurice earlier in the treatise (*Strategikon* 2.1). In combat the nomads preferred to fight in scattered wedges ('*kounas taxesi*') while using bows at long distance, ambushes, encircling, and feigned flights. One wonders if the *drouggos* (throng, irregular throng) mentioned above in the context of the *moirai* also implied the use of the wedge-formation (*cuneus, cunei*) by the units of the *moirai*. The pursuit of the Scythian nations was always relentless, aimed at the total destruction of the enemy.

Maurice's description of the so-called Scythian drill and hunting together with other extant sources enable one to reconstruct both the basic battle array and the various forms of Scythian outflanking operations. The attached diagrams have been taken from my *The Age of Hippotoxotai* and from the *Military History of Late Rome* series.

The *Strategikon* recommended the following counter measures. The general was to deny fodder to the nomads. The recommended mode of fighting was to use infantry placed in front, preferably in the *epikampios opisthia* formation (see pp.332–45), while the cavalry was placed behind. The ethnic units belonging to the infantry were always deployed in the customary manner of their nation within the *epikampios opisthia* array – this presumably referred to the fact that the Romans also possessed soldiers recruited from subject nations and from allies. The use of infantry worked well because the nomads did not dismount to fight as infantry. For the battle the general was to choose unobstructed level terrain, preferably with some defensive feature like a river, marsh or lake behind the army. The general was to post scouts on all four sides of the formation. If the Romans only had cavalry, the flanks were strengthened with extra forces of good quality. The rear guards were adequate for the task. In pursuit the *koursores* (runners, skirmishers) were not to advance more than

The Basic Scythian Battle Formation

This has only a single battle line with a reserve but the variants had two to three battle lines.

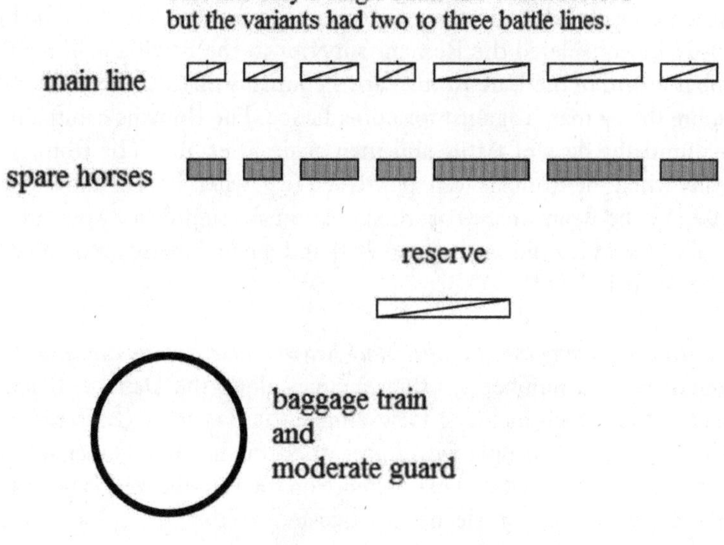

Scythian Manoeuvres

According to the Scythian Drill of the *Strategikon* (6.1), the Scythian units were all formed in the same manner (no *koursores* or *defensores*) in one line so that it was divided into two *moirai* (instead of three). In combat these units were used so that the two flanks advanced as if to encircle and then moved towards each other so that they continued in a circle so that the right wing was on the outside and the left on the inside until their reached the opposite part of the line previously occupied by the other left. Since it would have been very foolish to place the right-handed men inside the circle it is probable that the left wing galloped in front of the enemy line to the right while the right wing galloped around the enemy to the left. This interpretation has been inspired by Roy Boss' recent analysis of Hun tactics. This makes more sense than placing the left wing inside a circle where they would not have been able to shoot towards their enemy on the right. The encircling of the enemy would have been done by using scattered wedge formations.

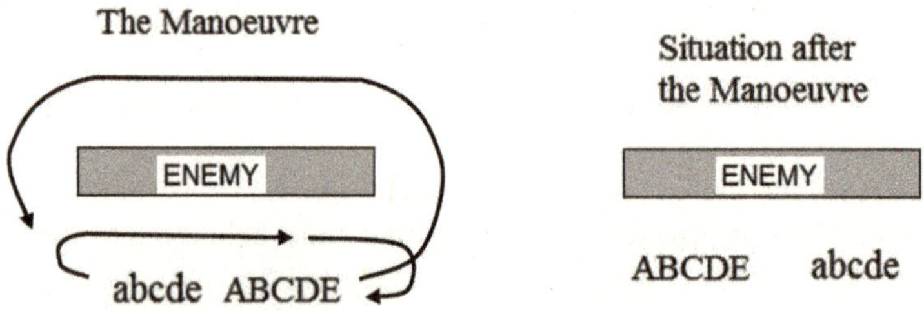

Maurice includes a variant of the Scythian drill in his discussion of the hunting methods (12.D.73-4/104-5) in which the right flank was inside the encircling formation. This is plausible as a battle formation because it places the right-handed mounted archers in such position that it was possible to shoot at the enemy. Note the use of wedges by the Scythian nomads (in practice there would been a lot more of those), which the Romans did not use with this training system. The intention has been only to show the principle.

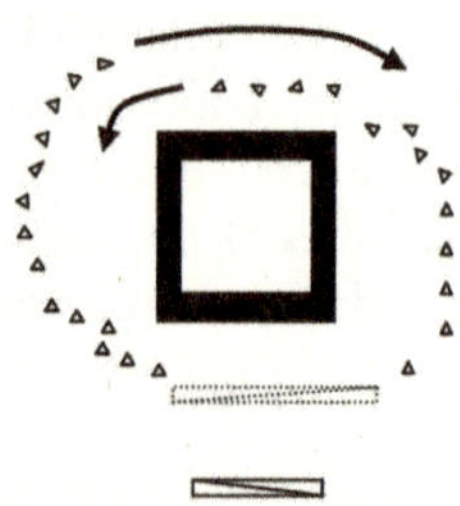

three or four bowshots away from their *defensores* (defenders). It is therefore clear that Maurice considered it advantageous to engage the nomads in pitched battle. In other words, he considered the Romans superior to the Scythians in regular battle. The combat record of the Late Roman army confirms this and shows that they were usually more than a match against for nomadic foe. The Romans habitually defeated the Huns until the days of Attila and then again after him. The Romans suffered defeats only when the Romans were poorly led (e.g. when Priscus allowed himself to be surprised by the Avars in 588) or when the nomads significantly outnumbered the Romans, like the Huns did in the late-440s and when Tiberius was forced to fight against the Avars in 574.[58]

The Lesser Enemies along the Danube, Black Sea and North of the Caucasus Range[59]

The Romans faced a number of other enemies along the Danube, Black Sea and Caucasus frontier, which included for example: the Bastarnae (Bastards, an ancient mixed Celtic-Germanic people with traces of Sarmatian and Thracian bloodlines); the Bastarnae who lived on the island of Peuce on the Danube were called the Peucini; the Sciri (the pure-ones or clean-ones boasted of their pure Germanic blood);

Dacian Carpi; Sarmatians (Argaragantes, the Free Sarmatians, and Limigantes, their 'slaves'); and remnants of the former Alan realm which were assimilated into the Tervingi Goths and Bosporan Kingdom.

The Sarmatians of the Danube frontier were the archetypical nomadic 'heavy' shock cavalry which consisted of cataphracts (*katafraktoi*) and *contus/kontos*-bearers (*contarii/kontoforoi*). These were equipped with the *contus Sauromatus* (Sarmatian *contus/kontos*) that was wielded with a two-handed grip. The cataphracts wore scale or mail armour for both the man and the horse, a helmet, a *contus*, a long-composite bow and arrows/quiver, longsword on the left, short-*akinakes* sword on the right, and dagger. The *contus*-bearers were equipped with the *contus*, longsword, short sword, bow/quiver/arrows, and dagger, but their horses were unarmoured and the amount of armour worn by the rider depended on the wealth of the man. Some of the mounted Sarmatians along the Danube had also adopted the use of the shield. Some of the less wealthy Sarmatians and Alans were equipped only as light cavalry mounted archers (*hippotoxotai*), but the poorest (presumably consisting of the subject peoples and slaves) fought as infantry.

The Sarmato-Alan battle tactics were based on the use of the *koursores* (runners/skirmishers/assault troops) and *defensores* (defenders) to harass the enemy with advances and retreats. The Romans had copied it as early as the second century and it was known as the Alan Drill Formation. In this combat system the *moirai* (at most 2,000–3,000 men) were arrayed two to four-hundred feet (ca. 62–125m) apart. The attack began with the *koursores* charging out of the formation at a gallop, after which they turned and retreated into the intervals (i.e. performed feigned flight) between the *defensores* and reformed themselves. Then the *drouggoi* of *koursores* together with the *defensores* charged the enemy. It was also possible for the *koursores* to attack the flanks of the pursuing enemy unit if its flanks were not protected.

The *kontos*-equipped Sarmatians were no match for the well-drilled combined forces which the Romans possessed during the third and fourth centuries. The typical hostile encounter between the Sarmatians and Romans was that the Sarmatian cavalry forces invaded across the border and raided and pillaged while attempting to avoid contact with the Roman army. The Roman combat doctrine was to seek contact with them, which was not always possible thanks to the speed of the Sarmatian invading forces. It was because of this that in the fourth century the Romans launched punitive expeditions across the border against the Sarmatian settlements. The existence of Sarmatian villages, towns and cities made them vulnerable. In contrast, in the late-fifth century the Goths achieved far greater successes against the Romans with a similar tactical system, the use of the *kontoforoi* charge, the difference between the two periods being the poor quality of the Roman infantry from about 468 until 530s. The Romans needed high-quality infantry against the *kontos*-bearing *katafraktoi* and *kontoforoi*. It should be noted, however, that the Sarmatians and Alans who lived along the Danube in the fourth century were eventually assimilated into the Vandal and Gothic forces, or were enrolled into the Roman armed forces as ethnic *foederati*. It was because of this that the Romans also possessed these types of units in their armed forces.

The Alans who lived north of the Caucasus posed a different kind of threat. They had retained their nomadic lifestyle and the Romans usually faced them only when they invaded across the Caucasus Range against the Georgians or Armenians, or when they attacked the 'Cimmerian Bosporus'. These Alans fought like the 'Scythian' nomads and the Roman response to this followed the already described tactics.

The Bastarnae and Sciri were also heavily influenced by their neighbours the Sarmatians, and therefore fought mainly as cavalry. The Bastarnae, Peucini and Sciri were small tribes that did not pose any major threat to the Romans. The Carpi were a sedentary people whose military forces appear to have consisted mainly of lightly-equipped infantry and cavalry of the Sarmatian type. The high quality combined forces that the Romans possessed in the third and fourth centuries defeated all of these lesser foes with great ease. Roman combat doctrine was to seek a battle when possible.

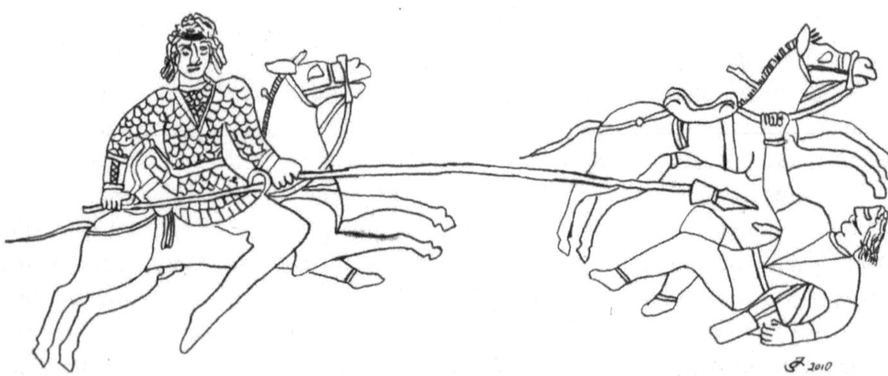

The so-called Kossika vase (Russia) 1-3rd Century AD (drawn after Brzezinski & Mielczarek, 15). It is usually thought that the men represent duelling Sarmato-Alans, but I would suggest the possibility that the *contus*-bearer on the left would represent a Goth while the man on the right would represent a Sarmatian or Alan mounted archer. However, it is impossible to be certain for both used the same tactical systems. Of note is the fact that the *contus*-bearer had shot at least two arrows before he charged.

Against Light-haired peoples (Franks, Lombards, Goths, Gepids, Alamanni, Burgundians and others)[60]

According to Maurice (11.3), all light-haired races were bold and undaunted in battle so that they considered even a short retreat a disgrace and they also often sought to avenge their dead comrades regardless of what it took. The Germanic peoples excelled at hand-to-hand combat either on horseback or on foot. Their martial armament consisted of shields, lances, and short swords slung from their shoulders. The narrative sources prove that there were great variations from one tribe to another in the equipment worn, but the general idea behind Maurice's description is still true. All Germanic tribes concentrated mainly on the use of short range weaponry (javelins, throwing axes, spears, pikes, swords, hand axes, battle axes). Their battle line was formed according to tribe, kinship, and common interest. The front of their battle line was even and dense, which means that the units were deployed as in rank and file oblongs. Their attacks on horseback or on foot were impetuous and undisciplined and they were disobedient to their leaders. They frowned upon good order, and even more so on horseback. The Germanic peoples were not interested in anything that was at all complicated. The narrative sources confirm the accuracy of these views.

Maurice recommended the approach that had been previously used by both Belisarius and Narses the Eunuch to overcome the Ostrogoths. The general was instructed to avoid engaging the Germans in pitched battles in particular during the early stages of fighting. The general was to use ambushes, surprise attacks, and stratagems. The aim was to reduce Germanic boldness and bravery by a shortage of provisions, especially of wine, and by the discomforts of heat or cold, or by corrupting the leaders with money. Maurice advised the general to pitch the marching camp on rugged and difficult ground because the Germanic cavalry lancers were hindered by uneven and wooded terrain. In other words, the *Strategikon* advised the general to use the Persian tactics against the Germanic lancers. Maurice also recommended the use of ambushes against the flanks and rear, because the Germans did not use adequate numbers of scouts and other security measures. The impetuous lancer attack made the Germans vulnerable to the nomadic tactic of feigned flight. The Germanic marching camps were also disorganized and therefore very vulnerable to night attacks with archers. In spite of these precautions, Maurice was still ready to accept that generals engage the Germans in pitched battles if the circumstances were favourable for this. In other words, even if Maurice was more cautious in his approach to fighting against the Germanic peoples than were the third- or fourth-century Romans who usually sought to engage the Germanic peoples in pitched battles rather than resort to the use of prolonged guerrilla warfare, he was not totally opposed to such. In fact, it is clear that the Romans did indeed fight pitched battles against the Lombards and Visigoths during his reign, even if this can usually only be guessed from the territorial gains of the Roman armies. In sum, the *Strategikon*'s recommended methods against the light-haired peoples were: 1) the use of nomadic tactics of feigned flights, ambushes, surprise attacks and archery; 2) the use of Persian archery tactics on difficult terrain; and 3) the use of standard battle tactics when the circumstances were favourable. Roman tactics were very versatile, but it should still be kept in mind that the third- and fourth-century Roman approach to fighting the Germanic peoples was far more aggressive than the approach recommended by Maurice, which was presumably as a result of the great successes of Belisarius. The likeliest reason for this is that the composition of the typical Roman field army changed in the course of the Late Roman period. When Roman field armies consisted mainly of infantry, as it usually did from the late-third century until the early-fifth century, the Romans were readier to engage the Germanic peoples in pitched battles. When their field armies consisted mainly of cavalry they preferred to use nomadic or Persian style cavalry tactics against the foe.

The Slavs and Antes[61]

The way the Slavs and Antes fought was based on their place of abode (forests, rivers, lakes, and marshes) and on their culture, which stressed freedom. According to the *Strategikon*, the Slavs were typically armed with two short javelins and shield and fought on foot, and used archers equipped with wooden bows and arrows smeared in poison (therefore Maurice instructed the soldiers to prepare for combat by taking an antidote), but from other sources we know that the Slavs also possessed some excellent cavalry. The typical forms of combat for the Slavs were ambushes, sudden attacks, and raids. They excelled at crossing rivers. The Slavs were not effectively organized as combat forces because they did not possess a unified government and their kings/chieftains usually quarrelled with each other. Therefore, the Slavs were usually reluctant to fight a close order battle on open and level ground. According to Maurice, the favourite tactic of the Slavs was to post their disorderly infantry in front of woods/forest after which they first shouted all together (possibly howled like wolves) and moved forward a short distance to see if their opponents would lose their nerve. If this happened, the Slavs attacked. If not, the Slavs retreated to the woods. As light infantry, the Slavs were very adept at fighting in such difficult terrain. The idea was to lure the enemy to follow them into the woods. See the attached diagram.

The counter measures against the Slavic tactics in the *Strategikon* consisted of several different alternatives. As is obvious from this, the Slavs as light infantry were inferior in the open against the better equipped combined forces of the Romans and this is confirmed by the recommendations of the *Strategikon*. The disorganized manner of conducting wars, and the lack of ability to fight on open and unobstructed ground made the Slavs vulnerable to such. Therefore, Maurice recommended the use of archery, sudden outflanking manoeuvres, hand-to-hand fighting, and fights on the open and unobstructed ground. If the Slavs occupied a strong position, which secured their rear, Maurice advised the general to use feigned flight to draw them out to the chosen place of ambush. Since the Slavs inhabited difficult territory, Maurice recommended the use of combined arms consisting of both cavalry and infantry. In fact, in such conditions the use of light-armed infantry with javelins was necessary for success and Maurice recognized this. The *Strategikon* also recommended the use of dromons, pontoon bridges, and inflatable skins for the crossing of the rivers, or alternatively fighting in winter when it was possible to cross the rivers on ice and then fight against unprepared enemies when the foliage did not help them and the tracks on snow made them easy to find. The last-mentioned was to lead to the downfall of Maurice, when he required his men to fight in the winter 602 – Maurice was undoubtedly correct about the best strategy, but he failed to take into account the wishes of his own soldiers. Slavic forces were so weak in pitched battles that it was not necessary to make the Roman battle line deep. It was to be shallow. However,

The Slavic tactics in the open consisted of posting a horde or a disorderly line in the woods or edge of the woods and then charging forward. They probably used ranks and files but the lack of regular drill ensured that the formations were disorderly. The place of the cavalry in the battle array is conjectural. The sources do not mention their place or role. They may also have been dismounted.

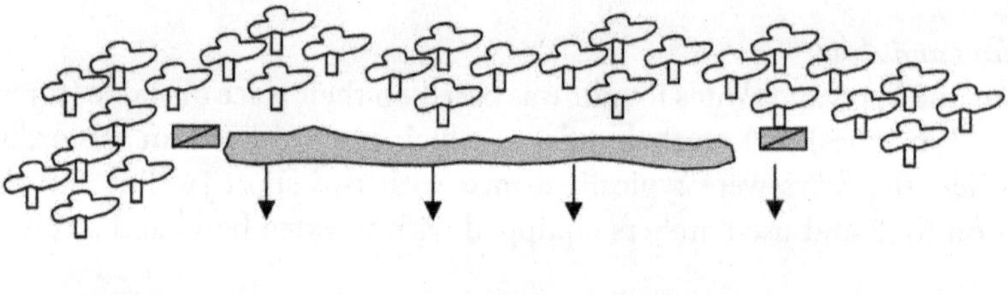

the Slavs also had some advantages over the Romans. They were very dangerous foes in forests and woods, so the general was advised to avoid wooded terrain as much as possible in summer and operate in the open areas, but, as the instructions for fighting in difficult terrain make clear, it was sometimes necessary to engage the Slavs in such conditions. For additional details see the use of the lateral phalanx formation in difficult terrains in Chapter 9.4.

Maurice also recommended the use of surprise attacks so that the Slavs would be unable to unite and hide in the woods and the relentless ravaging of the Slavic lands during summer. Such attacks followed the standard recommendations regarding offensive wars and surprise invasions (see above), but Maurice added other material which was specific to the Slavs. Firstly, when the Romans invaded Slavic lands across the river, the general was to leave the superfluous baggage and a *moira* of cavalry behind. The *moira* both guarded the crossing and acted as a threat against the Slavs, so that they would be unable to unite. The Roman force was to be equipped with lighter equipment for the troops, and with materials for the building of bridges. The navy was also to assist if present. The local kings could also be bribed. If possible, the invasion was to be made simultaneously from two directions so that the Slavs would be unable to flee and unite (Tactic 1). In such instances, the *hypostrategos* was to advance c.22–30km through unsettled land behind the Slavic settlements and then began to pillage, while the *strategos* advanced from the other direction. Maurice noted that the Slavs could be dangerous if the attack was done only from one direction and the army then got stuck in conquering the first Slavic settlement. In such instances the other Slavs had time to assemble their forces, which they could then bring to the assistance of those still fighting the Romans. The soldiers were instructed to kill everyone who resisted so that the progress of the attack would not be slowed.

If there was only one suitable road, the army was still divided (Tactic 2), but this time the *hypostrategos* took half or more of the army, while the *strategos* followed. The *hypostrategos* would detach about one to two *banda* per Slavic settlement during his advance and continued his march as long as there were enough *tagmata* in his command, but not further than c.22–30km. He was required to retain three to four *banda* (c.1,000 men) for emergencies. The *hypostrategos* then turned back, and gathered up the pillagers along the route of his march while the *strategos* did the same from the other direction. The logic behind this strategy was impeccable. The Slavs were considered weak and disorganized as enemies so it was preferable to target them at their homes rather than on Roman territory when they were raiding Roman lands.

Unsurprisingly the Romans were usually very successful in their endeavours against the Slavs, but there were periods in which this was not true. Good examples of this are the time periods when the Romans lacked adequate forces to engage the Slavs or when the commanders were not up to their task. This was true for a part of Justinian's reign, and also during the reigns of Justin II and Tiberius II. The situation was restored under Maurice, but once the Romans again lacked the necessary numbers under Heraclius, Heraclius was forced to ask other Slavic nations (Croats and Serbs) to assist him against the invaders. This was done by settling

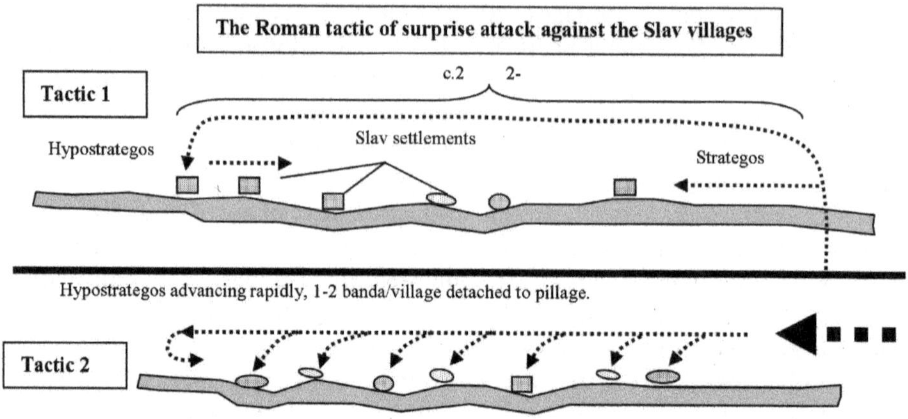

these tribes on Roman territory. Therefore, even if the Romans could always expect to prevail over the Slavs in regular pitched battles, they did not always possess the necessary numbers for this and were therefore overrun by them.

Against Other enemies[62]

None of the extant Late Roman military treatises, the *Strategikon* included, has any detailed instructions on how to deal with the Picts, Irish tribes, mountain tribes of the Caucasus range, Georgians (Iberians and Colchians/Lazicans), Armenians, Moors, Nubians, Blemmyes or Arabs.[63] However, we know that such texts did exist on the basis of the *Strategikon's* maxim at 8.2.88. Fortunately, we can reconstruct the methods used on the basis of the narrative sources. It is because of this that we know what types of enemies these constituted at this time and how the Romans dealt with these.

The Picts of modern day Scotland posed a double threat to the Romans, consisting of both land and naval forces, while the Irish were primarily pirates who raided Roman territory. The same was true for some of the Germanic peoples, such as the Franks, Saxons and Vandals during the Late Roman period. For a fuller discussion of naval matters, see Syvänne's *The Age of Hippotoxotai and the Military History of Late Rome* volumes 1–8 and the forthcoming separate analysis of the Late Roman navy. Excluding the fifth-century Vandals, none of the enemies of Rome could challenge them at sea. The evidence for the Pictish military is scarce, but as far as can be gauged from the evidence the principal fighting force of the Picts consisted of a spear-armed infantry phalanx, bow- or crossbow-armed men, and small numbers of cavalry. They also appear to have used war-chariots. The Irish appear to have used similar equipment. Both concentrated on raiding activity and both were poorly organized by Roman standards, so that when one adds to the equation the fact that neither the Picts nor Irish possessed truly large armies, it is clear that they posed no serious threat to Roman Britain. It was only when these united with the other tribes, as happened for example when the Saxons and Franks cooperated with them during the barbarian conspiracy of 367, that they posed serious threat thanks to their numbers.

The mountain tribes of the Caucasus range consisted mostly of lightly-equipped infantry, but these did not usually pose any serious problems to the versatile Romans. Rather the principal difficulty for the Romans consisted of the mountainous territory. However, when the Romans were determined to engage the mountain tribes, they always won in the end. The crème-de-la-crème of the Georgian armed forces consisted of the heavily armoured cataphract type of cavalry that they had copied from the Parthians and Persians, but they also possessed infantry able to fight in a phalanx formation. After the adoption of Christianity, the Georgians usually sided with the Romans or were disloyal allies of the Persians. The Armenians also possessed well-organized infantry and cavalry forces, but the crème-de-la-crème of their forces also consisted of the cavalry cataphracts copied from the Parthians and Persians. Unlike the Persians, the Armenians usually preferred to use these forces for lancer charge, which made them dangerous to face in pitched battles in the open, which they did independently and as allies of Persia. It was therefore a good thing for the Romans that the Armenians became Christians in the early-fourth century. The conversion to Christianity made the Armenians perfect partners for the Romans in their wars against Persia.

The Moors, Nubians and Arabs can be categorized as desert peoples whose forces consisted of lightly-equipped infantry and cavalry. The Moors were brave but very poorly equipped, so that their typical martial equipment consisted only of a shield and two to three javelins both for the infantry and cavalry. The Moors possessed some foot and cavalry archers but their numbers were insufficient against the Romans. In addition, the Moors also employed camels as cavalry, but Vegetius (3.18) quite correctly considered that the camels were useless against anyone except those who were unfamiliar with them – and the Romans were not. The Moors were poorly organized and could cause problems only when large numbers of tribes united as a confederacy. The Romans usually sought to engage these in pitched battles because the Moors were poorly organized and the Romans could expect to win the battles even when heavily outnumbered. The bravery of the Moors made them extremely well-suited as auxiliary forces for the Romans, so the emperors even possessed Moorish bodyguards until the reign of Theodosius I.

By Roman standards the Blemmyes can be considered to have been very lightly-equipped, both as infantry and cavalry. Roman combat tactics were always to seek a pitched battle if at all possible. Whenever the Romans had adequate numbers close to the scene of operations, they did not have any serious problems with the Blemmyes.

The Nubians were usually better organized than the Moors or Blemmyes, and were particularly famous for their archery skills, but once again they were too lightly-equipped (i.e. most of their men did not wear armour) to face the Romans in pitched battles during this era, meaning that the Romans always sought to engage them in regular battles to limit the ravages that their raiding could cause. The fact that the Muslims had significant problems when fighting against the Nubians proves that the Roman successes against the Nubians during the Late Roman period were considerable achievements.

Before the arrival of the Muslims, the principal and feared component of the Arabic forces was their cavalry. The Arabs employed camels for transport and horses for combat, and were particularly well-known for their skills as lancers and swordsmen and hence very dangerous in close quarters combat, plus the Arabs were also known as ferocious fighters. In addition to this, the Arabs possessed fair numbers of mounted archers and the Lakhmids also cataphracts. The Arabs usually sought to avoid battles and concentrated on hit and run raids before the Romans could mount an effective counter attack and the Arabs flee to the deserts before the Roman army reached the scene. Unsurprisingly, the Roman tactic was to seek battle with the Arabs and they employed Arabs both as their allies and as a camel corps to enable them to achieve this. For most of the time the Romans were successful in this, but under skilled leadership, such as were provided by queen Mavia in the fourth century and Alamoundarus in the sixth century, the Arabs could cause major problems for the Romans. It is therefore not surprising that the Muslims turned out to be the most feared of the enemies the Romans ever faced. The Muslims added to the superb Arabic cavalry the Macedonian style pike phalanx brought from Yemen, the numbers achieved from the uniting of the tribes of Arabia and the united will provided by the new faith Islam and Caliph. When the Romans then faced the Muslims under Heraclius (610–41), the Roman army was no longer the army that Roman combat doctrine expected it to be. It had become a cavalry army of the same type as had been fielded by the Ostrogoths against Belisarius but with the difference that the Roman cavalry consisted of multipurpose cavalry equipped with lances and bows. The tactics that Heraclius adopted against the Muslims were actually an adaptation of the methods advised by Maurice in his instructions on how to deal with the light-haired peoples. In other words, Heraclius recognized the inherent dangers present when engaging a foe that fielded a pike phalanx and which was motivated by Jihad. This approach failed because the Roman forces were poorly led and poorly motivated when Heraclius was actually persecuting them for religious reasons, meaning that many of the Armenian and Christian Arab commanders defected to the Muslim side.

Unsurprisingly, we find no discussion of the methods to be used against usurpers. The military treatises were concerned only with foreign foes. Small troublesome ethnic groupings, such as the lightly-equipped Isaurians, Samaritans and Jews, caused temporary localized troubles, but these cannot be considered to have been major problems for the Roman armies. The revolts by these and other ethnic groupings were always crushed, but when the Germanic tribes were settled on Roman territory after 382, these formed a far greater threat than any of the above-mentioned and these eventually overtook the West Roman Empire. The native domestic enemies such as usurpers were obviously usually the most dangerous ones to the emperor, because these used the very same methods and forces as the loyal armies and because the division then present within the military community made the loyalty of everyone suspect. The only counter measure against this was timely information of such plans, but as the success of Julian and Phocas shows, even when the security apparatus informed the emperor of the problem the usurpation could still succeed if the circumstances for it were right. In such instances the decision to fight a pitched battle obviously depended on the general characteristics of the situation and on the analysis of the strengths and weaknesses of each side, but the principal means used by the usurpers and emperors usually consisted of attempts to undermine the support the other one had through bribes and negotiations.

The Roman Approach to Fighting

As we have seen, Late Roman combat doctrine possessed very detailed and effective instructions for generals to use in situations in which they faced the decision when, where and how to engage the enemy. When followed to the letter, these very sound practical instructions gave the generals the necessary tools for success. The reality was of course varied, the key component usually being the personality of the general and his personal qualities and skills.

As I noted in my doctoral dissertation, the true advantage of the Romans over their enemies arose from their greater tactical versatility. This resulted from the conquests of large tracts of lands which had brought the Roman Empire into contact with different cultures, geographical areas, and a wide variety of enemies. This in its turn had forced them to adapt their military methods to deal with the new challenges, and the readiness to copy and adapt had given the Romans an advantage over any single one of the enemies, because they were able to draw upon a greater variety of methods than any of their enemies. Since the versatility and better organization of the Romans usually gave them a tactical advantage over their enemies, the decisive factors in deciding who would actually prevail depended upon the individual circumstances, availability of troops and generalship. Consequently, the ability of the Romans to defeat their enemies depended primarily upon the amount of enemies they had to face simultaneously and the generalship.[64]

Chapter Four

Cavalry Unit Orders and Formations

4.1. Roman Cavalry Unit Formations and Orders

The principal unit orders for the cavalry units that were organized as rectangular rank-and-file units consisted of the following: 1) the open order (*araiôsis*), which was used mainly for marching and fast attacks at the gallop; 2) the close-order (*pyknôsis/puknôsis*), which was used for the main charge and always by the *defensores*; 3) and the irregular order (*drouggos*), which was used for ambushes, pursuit, support etc. (i.e. in situations in which the rectangular rank-and-file orders were too slow). These three unit orders and the crescent (*menoeides*) unit order (means the snaking column order) by the flank units for outflanking the enemy wings, were the only ones that Maurice included in the *Strategikon* for a very good reason. The cavalry did not really need any more than those to perform everything that was expected of it and it is quite probable that this represented the standard set of unit orders after the reign of Maurice.[1]

However, we learn from other treatises and narrative sources that the Romans also used other unit orders. For the regular rectangular rank-and-file formations it consisted of the following. Arrian, the copies of whose treatise on tactics (*Ars Tactica* 16.13–4) were also reproduced during this era (e.g. by Urbicius), and the *Peri strategikes* (17.8–11) include a variant in which the heads of the horses of the second rank were lined up with the shoulders or flanks of the horses in the first rank. These two variants in which the ranks were interjected into each other were therefore probably used at least from the early imperial period until the reign of Tiberius II, and possibly also later by commanders who preferred this. In addition to this, Arrian (*Technê Taktike* 36.1) and the tenth-century *Sylloge Tacticorum* (43.6–7) include a cavalry formation that Arrian calls *xynaspismos* (= *synaspismos*) and *chelone* (tortoise, i.e. *testudo*). There is no definite period referral for this, but the circumstantial evidence suggests that it was used. See the discussion below.

Besides the above formations that were based on the use of rectangular unit orders, there were also specific unit order variants that were based on the cavalry unit formation itself. These included the different versions of the wedge (*cuneus, kounion, embolos*) and rhombus, both of which appear to have continued in use at least until Maurice streamlined the practices in about 583, unless the *drouggos* meant these – the case for non-rank-and-file versions as one of the meanings for the *drouggos* is actually strong, as we shall see. After Maurice, most of the other variants were abandoned by the regular Roman cavalry so that the Roman cavalry proper continued to use only the open order, close order, *drouggos* and crescent. What happened to the use of the cavalry *testudo* is not known, because we find this once again in the *Sylloge*

Tacticorum. It is therefore possible that it survived Maurice's reforms. It should also be remembered that the Romans employed various ethnic units as mercenaries and allies, which means that it is quite clear that at least some of the units in Roman service continued to use some or all of the versions of the wedge and rhombus.

4.2. The Rectangular Rank-and-File Unit Formations before Maurice's *Strategikon*

The principal Roman cavalry combat formations throughout the existence of the Roman Empire were the different variants of the rectangular cavalry formations consisting of files and ranks. In Hellenistic military theory, copied by the Romans, it had two main variants, the square or oblong, which in practice meant several different unit variants based on the use of ranks and files.

A rectangular cavalry array three deep and nine wide for a total of 27 men is mentioned by both Aelian (18.7) and Arrian (*Taktike* 16.12). This is obviously five men short of the strength of the Roman cavalry *turma* of 32 horsemen (which could also be 33 horsemen strong when the decurion is added to the figure). The evidence suggests[2] that the Romans used this shallower three men deep array when the *turma* was below strength or when the decurion or his superior decided to do so. In the latter case, the surplus men were posted somewhere else or united with some understrength unit. The other cavalry unit formations mentioned by the military theorists were based on multiples of four or five, but these could also be modified according to the situation. The extant evidence backs up the use of these variants too.

The rectangular arrays which had files divisible by two and four are the rectangular arrays of Aelian (18.5–6), Arrian (16.9), and Asclepiodotus (7.4). Aelian and naturally the Byzantine interpolations of his treatise mention the array which was eight wide and four deep, which was the same as the 32-horseman *turma*. Asclepiodotus adds an array used by the Persians, Sicilians and Greeks which was sixteen files wide and eight ranks deep for a total of 128 horsemen. It is clear that this variant was also used by the Romans because a 128 horsemen rectangle was divisible with the size of a *turma* and it is also the precisely same number of horsemen that we find in the rank-and-file rhomboid used by the Macedonians, Romans, Armenians, Parthians, Sasanians and Muslims.[3]

Aelian (18.6) and Arrian (*Taktike*, 16.11) both mention cavalry 'square unit formations', which were clear precursors of the cavalry formations that we find in the *Strategikon*. These variants are the ten-wide and five-deep (50 horsemen), and the twenty-wide and ten-deep (200 horsemen). Arrian (16.10–11) judged the square array to be the easiest to control and organize. Most importantly, its shape enabled its best men, the officers, to fall in simultaneously as a single united body against the enemy. Its weakness was that it was slow to manoeuvre and wheel, a weakness which was recognized by both Arrian (*Taktike* 16.8) and Maurice (*Strategikon* 4.5). The resulting organizing principle when using the figures based on fives would probably have consisted: a) 3 x 32 plus 3 decurions and one centurion for a total of 100 men; and b) 6 x 32 horsemen (192) plus 6 decurions and 2 centurions for a total of 200 men. It is therefore no wonder that this system became the standard method by the time the *Strategikon* was written in the 590s.

In the chapter devoted to the depth of the cavalry formation, the *Strategikon* (2.6) specifically noted that up to his own times the universal practice among the Romans was to use cavalry formations ten deep. He considered this a bad practice because it made it easier for the enemy to estimate the size of the Roman army. It is possible that the Avar habit of forming their units with irregular depths served as an inspiration for Maurice in this case. Maurice noted that the ancients had deemed the depth of four men to be enough – and noted that this was also true because horsemen beyond the fourth could not contribute to the combat effectively with their spears or bows – but also noted that in his own day the depth had to be deeper because the number of excellent fighters in each *tagma* was limited. His solution was to organize the depth according to the quality of the unit, so that the best units could have the depth of five ranks and the worst ten ranks. As examples, Maurice mentioned that the formations of the *foideratoi* (these could have one rank of squires following) and *vexillationes* could be seven deep while the array of the *Illyrikianoi* was to be eight ranks deep.[4] The elite *optimates* (the best) were to be arrayed five men deep with two ranks of squires following. The regular units were to be arrayed eight- to ten-deep and the foreign contingents according to their own customs, but noting that foreigners were considered particularly valuable as *koursores* (skirmishers, runners, pursuers) and mounted archers. Maurice stated that it had been a universal practice to array the cavalry *tagmata* ten deep. This is probably an exaggeration, because it is unlikely that all of the cavalry units would always have been up to their paper strengths and we have to remember that the commanders also needed to adjust the length of their battle lines according to the situation. Regardless, it is still clear that Maurice must have had a good reason for his comment, which means that at some point in time in the past the Romans adopted the ten-ranks-deep formation as their standard combat formation. On the basis of the *Peri strategikes*, it is probable that this change had taken place before the reign of Justinian, but it is impossible to pinpoint the exact date.

According to the *Peri strategikes* (17), the file leaders, the four ranks behind them, the rear guards, and the troops on the edges of the formation and the men next to them were to be more experienced and braver than the rest (i.e. there were additional ranks between the rank five and last rank) and they were also to carry spears and shields. The hooves were to be protected by iron plates (horse-slippers), and the horses of the front rank were to be equipped with iron armour on their heads, breasts and necks. The horses of the front rank were to be well trained and used to the confusion and noise of the battlefield. Syrianus also likened the armament and structure of the cavalry phalanx to the infantry phalanx. This means that the horsemen in the middle of the formation carried javelins or other weapons usable as missiles. I would suggest that this includes bows and arrows. In sum, it is clear that Syrianus envisaged a cavalry array that consisted at least of seven or eight ranks, but in light of Maurice's referral to the imprudent universal use of the ten-ranks-deep formations until his own time, it is likely that Syrianus meant the ten-ranks-deep cavalry formation. In other words, it is probable that the Romans had adopted the ten-ranks-deep formation as their preferred cavalry array well before the reign of Justinian, as this was already recommended by Arrian. However, as stated, I would still suggest that in practice able commanders always varied the depth according to the quality of the unit and situation. Maurice was originally a notary who learnt his military trade mainly from military treatises, so it is probable that his comments refer mainly to standard combat doctrine as he found it in the military drill books and treatises.

Arrian (*Techne Taktike* 17.3–5) includes two additional versions of the oblong cavalry formation (*heteromekes taxis*), which could be deployed either in width or in depth (*bathos*). The deployment in width was considered the better of the two in combat. The deployment of the oblong formation in depth, the cavalry column, was judged to be better only when the intention was to cut through the enemy formation or when one wanted to conceal the numbers of the cavalry from the enemy. The concealment of numbers from the enemy was one of the ancient stratagems used to fool the enemy and undoubtedly therefore used also in practice, but in such cases the intention was not to use the column array for the actual battle – the idea was to spread the formation out before engaging the unsuspecting enemy. The actual column array appears not to have been purposefully used in pitched battles except perhaps when the enemy appeared in front of the Roman formation when it was still in marching order. However, the column array was not without its advantages, as Arrian's comment shows. In the sixteenth century the so-called *Reiter*-cavalry found it very useful precisely for the purpose of breaking through the enemy array. At the other extreme of the rectangular oblong arrays was the depthless formation consisting of a single front/row (*metōpon*). Arrian believed it to be useful for raids against unsuspecting enemies or when one wanted to trample something underfoot, but he judged it unsuited for long-lasting pitched battles. This array was later used to great effect by medieval knights, but even they recognized its weakness in prolonged combat and usually deployed their squires behind them as a separate line. There is no definite evidence for the use of this array for the Late Roman period, but one cannot entirely rule out the possibility that it was still employed by some cavalry forces in desperate situations when the intention was just to charge through the enemy. On the basis of the earlier works of art, Principate era Roman cavalry had probably used this method to trample enemy footmen underfoot. On the basis of the *Strategikon's* instructions for the use of the so-called fill-up *banda/tagmata* between the divisions of the second support cavalry line (see below) it is clear that even Maurice expected that the cavalry units that had only two or four ranks of horsemen could be used effectively in combat in the right circumstances. The *Strategikon* (12.D.55) also included the variant of one rank in the context of hunting as a form of military training, which implies that there were occasions in which the cavalry could also charge in a single rank. One may therefore expect that the shallower oblong formations were also used throughout the Late Roman era when the situation for this required it.

The Romans had also always arrayed their cavalry in larger units than described above. Traditionally these units had been called *equites* and *alae* and then according to John Lydus (*De Magistri.* 1.46) with the new names *alae* of 600 horsemen (former auxiliary cavalry), *vexillationes* of 500 horsemen (former legionary cavalry) and *turmae* of 500 mounted archers. One wonders if these should be identified with the *arithmoi, tagmata* and *banda* of Maurice's day. In other words, the larger Roman cavalry units had paper strengths of about 500 or 600 horsemen, which in their turn consisted of the smaller units, the centuries, 32-horse *turmae* and tent groups. It is unfortunate that Arrian, Aelian and Asclepiodotus fail to mention how the square and oblong formations were then grouped together to form up these larger combat units. Nevertheless, one may make the educated guess that this was done according to the principles that we find in the *Strategikon* so that the commanders combined the smaller units into larger units later called *arithmoi, tagmata* and *banda*; or alternatively that these larger units (*alae, vexillationes, turmae* or the *cunei, numeri*, etc.) consisted of units 200, 100 or 50 strong, all of which were used in the building of larger units that could be 500 or 600 strong, or stronger; or that the smaller units were just combined together with very small intervals between them to form up whatever sized units were thought preferable.

The grouping of different cavalry units as separate larger or smaller ad hoc combat units is also confirmed by Aelian (Matthew ed. 21) and Arrian's *Ektaxis kata Alanôn*. In Arrian's text, the eight combat divisions consisting of the different auxiliary cavalry units were called *lochoi* so that there were six *lochoi* in the middle, two *lochoi* of Armenian cavalry on the wings, and behind the wings two units of legionary cavalry (*logchoforoi, kontoforoi, machairoforoi, pelekoforoi* = legionary cavalry), and between them the *epilektoi hippoi* (picked cavalry) under Arrian himself. The eight *lochoi* were divided into *koursores* (runners) and *defensores* (defenders) in the Alan manner (see later) so that every other *lochos* pursued the enemy as *koursores* while the remaining *lochoi* behind were their *defensores*. It is clear that Arrian's eight *lochoi* were not uniform in size because these had been formed from various different-sized units, and on the basis of the list of units[5] it is probable that some of the *lochoi* had more than 500 horsemen while others had less. In contrast, Maurice favoured the use of smaller, irregular sized *arithmoi, tagmata* and *banda* (unit strength varied between 200 and 400 horsemen) so that the existing regular units were either divided or united to form these irregular sized units, which were then combined to form the even larger units of *moirai* and *mere* with their *koursores* and *defensores*. Maurice's idea was to make the units irregular in size so that it was impossible for the enemy to make accurate estimations of the size of the Roman army. Maurice preferred smaller sized units for the units, called *arithmoi, tagmata* and *banda*. To sum up, the different-sized regular cavalry units accompanying Arrian's army were for combat purposes arrayed as new combat entities called *lochoi* so that some of these were designated as pursuers (*koursores*) in irregular or open order (required by the speed of pursuit) while the rest served as their defenders (*defensores*) in close order. Notably, this larger scale battle formation of Arrian's day (three divisions in front, two reserve units on the flanks and commander's reserve in the middle of the second line) already followed the basic structure adopted by the medium-sized cavalry formations of Maurice's day (see Chapter 5.3). As we have seen above, when there were significant numbers of cavalry present, the *alae/lochoi* sized larger units were further united into even larger entities that during the reign of Maurice were given the names *moirai* (sing. *moira*) and *mere* (sing. *meros*). During Maurice's day the *moirai* consisted of a maximum of

2,000 to 3,000 men while the *mere* consisted of a maximum of 6,000 to 7,000 men. It is probable that earlier maximums for these larger 'regiments' and 'divisions' were roughly the same because the division of the cavalry battle line into the left, centre and right divisions (called *mere* by Maurice) remained the same. It is unlikely to be coincidence that the cavalry 'legions' of John Lydus (1.46) were 6,000 strong, just like the *mere* of Maurice.

The rectangular cavalry units which were armed with javelins or spears could be employed in three different manners. Firstly, it was possible to send individual files forward from each of the small cavalry units (32 horsemen *turma*, 50 horsemen unit, 64 horsemen unit) to skirmish with the enemy while their 'mother' units remained behind as reserves. The purpose of this was to skirmish with the enemy. This system enabled the units to rotate the files between those skirmishing and those left behind if the enemy did not attack, and if the enemy attacked then the men left behind could charge forward to protect the files skirmishing. In the latter case, the men posted behind acted as sort of *defensores* (defenders) to use Late Roman terminology. Secondly, it was possible for the entire units to advance forward, throw their javelins or spears and then engage the enemy in melee with swords or axes. Thirdly, it was possible for the units that carried *conti*-spears (or other spears) to charge into contact with the enemy unit if it held its ground. The mounted archers obviously either skirmished by using bows, or charged while shooting arrows after which they engaged the enemy with spears and swords.

4.3. The Basic Unit in the *Strategikon*: The *Tagma* or *Bandon* or *Arithmos*

The basic building block of the cavalry formation in the *Strategikon* was called the *tagma*, or *bandon*, or *arithmos*. Maurice's recommended number of soldiers for this unit ranged from 200 to 400, the only exception being the *banda* of the *optimates*,[6] where additional soldiers were allowed. In short, excluding the units of the *optimates* where extra men were allowed, Maurice wanted most of his cavalry units to be smaller than the paper strengths of the former *alae, vexillationes, turmae, cunei* and *numeri*. This implies a change in the preferences, but as noted above not conclusively, because it is quite likely that the commanders had always varied the strengths of combat units according to the availability of soldiers and the situation – a good example of this being the above-mentioned *lochoi* of Arrian in the *Ektaxis kata Alanôn*. In the *Strategikon* the depth of the cavalry formation reflected the fighting quality of the unit, the shallower the *bandon* the better the division, but it is likely that the same principles had always been followed when necessary. As already discussed, the depths of the units varied between five and ten so that the best units were allowed to have only five ranks of horsemen while the regular units had eight to ten ranks. For example, the elite *banda* of the *optimates* that were usually stationed in the second line were to have five regular soldiers followed by two 'squires' or servants.[7]

Maurice expected that the two front ranks in each of the *banda/tagmata/arithmoi* were to be fully protected by armour and were to be equipped with lances and shields for close combat. Their horses were also expected to be armoured. The last man in a file, the file closer, or in some cases the last two ranks in a file, were to be equipped with a lance and shield to make the unit double-fronted against possible threats from behind. The flank files could also be formed of the better-equipped men. The commander (*archôn* who held the title of tribune or count) and the flag-bearer in the middle of the unit were also to be protected by 'better' chosen men. The *hekatontarches* and *eilarches/ilarches* (second-in-command to the *archôn*) were posted on the edges of the formation to control the width of the array. The *dekarchês* commanded ten men and the *pentarchês* five men, while the *tetrarchês/tetrarchos* (commander of four, also known as *fylax*/guard and *ouragos*, the rear guard) commanded the rear half of the file. The depth of the unit provided security for the formation because only the best qualified men for hand-to-hand combat formed the front ranks and because the presence of the well-armoured men and horses in front protected those behind. Additionally, the presence of the horses and the tightness of the unit order during actual combat forced the entire block of men to participate in the charge. The commander of each unit and the flag were guarded by chosen men. The mounted archers were posted between the second and last rank, but if there were no archers present then these were to use whatever weapon they could handle. Nonetheless, the expectation was that at least the third, fourth and the last in a file were to carry bows, but the last one also with a shield. Procopius confirms the variation of equipment within the cavalry array with his description of Belisarius's *bucellarii*: some of these were equipped only with armour, bows and swords, while others also had spears and small *aspis*-shields attached to their shoulders. It is probable that shields were still carried by the other archers on some other occasions, placed behind their back. The above implies that there were not always enough mounted archers to fill all of the ranks in the middle of the formation and to act as *koursores* on the flanks of the *meros*. This was undoubtedly true in those cases in which the army included large numbers of recently recruited Germanic *foideratoi*. According to the text portion of the *Strategikon* (2.8), the file closers were to carry the lance, but in the diagrams the last men carry bows with shields. I take this to mean that the last men in the files, the file-closers, were always expected to be dual purpose soldiers armed with bows and spears just like their training scheme (*Strategikon* 1.1) expected.[8]

There exists one problem with the evidence, because the *Strategikon* includes conflicting information regarding the armament. In the section devoted to cavalry training (*STR 1.1*) the troopers were expected to be able to switch between bows and spears of the *kontarion*-type in the course of the combat. In the section devoted to the equipment carried (1.2.28–32), the *Strategikon* states that all Romans below the age of forty were to carry bows and two *kontaria* – in case the first missed its mark, one had a spare *kontarion* for use. The verb αστοχέω (to miss the mark, to miss, fail) suggests that the *kontarion* was thrown. This instruction is in direct contradiction with the section devoted to the internal organization of the *tagma* and *meros* in the *Strategikon* (3.1–6) in which the two front ranks carried the *kontos* (*contus*-spear) with the *skouton* (*scutum*-shield). The use of the singular (3.1.) implies that they carried only one spear. The text (3.5.32) also refers to the use of *kontos* and not *kontarion*. Therefore, one possible explanation for the difference would be to think that the *kontarion* was the shorter spear and the *kontos* the longer spear, so that the two *kontaria* were carried when the spear was used for throwing and the single *kontos*-spear when the intention was to lance/thrust with the spear. The key problem with this is that Maurice (e.g. *STR* 2.8 with 2.10) uses *kontos* as a synonym for *kontarion* and describes the cavalry *kontarion* as a long spear (*STR* 12.B.20.8).

There are three possible ways to explain this discrepancy: 1) it is possible that Maurice was careless in his instructions and actually meant that the troopers needed to possess two *kontaria*, not because two were needed if the first missed its mark when thrown, but because the second was needed if the first was lost in the course of the campaign; 2) it is possible that the troopers were sometimes required to carry two *kontaria* because the requirement to remove the lance pennons so that these would not interfere with archery (*STR* 2.10) imply that there were at least sometimes extra upright spears in front of the archers hanging against the back; and 3) it is possible that Maurice just omitted to include the spare spears of the front rankers and mounted archers. However, on the basis of combat logic, I would suggest that the diagrams regarding the formations of the *tagma* and *meros* accurately depict the usual armament used in the main charge, with the exception that the last rank also had spears in addition to the bow and shield as stated by the *Strategikon* in the text (2.8). As noted above, this is also confirmed by Procopius's description of Belisarius' *bucellarii/boukellarioi*. The second reason for this conclusion is that if the two front ranks had a spare *kontos/kontarion* placed behind their backs this could have interfered with the archery of the rear ranks, even when the lance pennons were removed. The same is true for all of the ranks between the second and last rank. The last rank was also required to use the bow during the attack, which means that any upright spears in front of the last rank could have interfered with their archery. In other words, it is very likely that none of the ranks in front of the last rank (which carried a *kontos* behind their back) had spare spears placed on their backs during the charge when it was done by the book because this could have interfered with archery. However, on the basis of the requirement to remove the lance pennons prior to combat (*STR* 2.10), it is still clear that the soldiers were at least occasionally equipped with the two *kontaria* for combat. In such situations the *kontarion* could obviously have been used as a thrown weapon. In fact, it is obvious the Romans used the *kontos/kontarion* in both manners, as thrown weapons and for thrusts/stabs/lancing (*STR* 1.1, 2.10).

The method of posting the best fighters with the best equipment in the front and rear ranks was beneficial to the overall combat efficiency of Roman cavalry. The formation as a whole was always in readiness for close combat, regardless of the direction the enemy came. The heavy armour of the front-rank men and horses improved the confidence of all of the troopers, because the heavily armoured veterans in the front led the rest of their men forward by their own example and shielded the weaker elements from the worst of the battle stress. The positioning of both heavy and light elements within the same combat formation also removed the unfortunate tendency of the dual-purpose troops to concentrate on ranged fire at the expense of melee.[9] The front ranks with their armour and shields also shielded the rear ranks from direct archery fire, and during the approach stage of combat the front ranks also shielded the rear ranks with their upright lances from arrows shot at a high trajectory. The lances/spears were of course lowered at the beginning of the charge,

tagma, bandon, arithmos

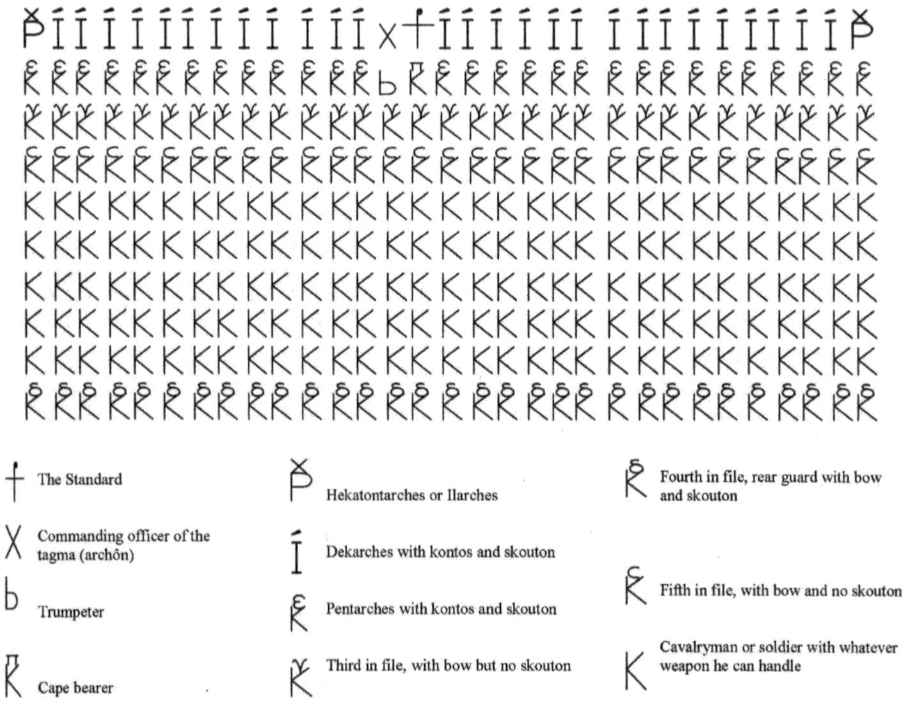

but in this case it actually facilitated Roman archery. The lowering of the spears would have signalled to the Roman archers that it was the time to start firing. The order to charge and begin archery was obviously done by the officers posted in the front ranks. The fact that this was always done at a specific distance from the enemy gave these troopers also the correct range and angle to the enemy, which increased their chances of hitting the target. The fact that the commanders and soldiers were instructed and trained to do all of their manoeuvres according to the distance of bowshots enabled the officers and soldiers to perform their actions in a similar manner every time, which in its turn improved the efficiency of the archery. It should also be remembered that the volleys of arrows were directed against groups of people and horses, which increased their chances of hitting the enemy.[10]

And when had the Romans adopted the system of placing both lancers and mounted archers in the same cavalry unit and not on different specialized units? It is practically certain that the Romans had possessed these types of units among their Parthian and Armenian auxiliary units from the first century onwards and among their regular cavalry units from the reign of Hadrian onwards, because Hadrian (Arrian, *Tactica* 44.1) required the Roman cavalry to be proficient with the tactics employed by the Parthian and Armenian mounted archers, and with the feigned flights and about turns of the Sarmatian and Celtic *contus*-bearers. The variety of protective equipment used during cavalry training (Arrian, *Tactica* 32.3.1ff.) demonstrates that the regular Roman cavalry also had the necessary pieces of armour to equip the frontal parts of the horses (chamfron, peytral, crinet) posted in the front ranks in the same manner as the Parthians, Armenians and Sarmatians did when they had enough fully armoured men and horses only for the front ranks. The importance of the use of armoured horses and riders had already been recognized during the fifth and fourth centuries BC, for example by Xenophon (*Hipparchos* 1.6–7; *Peri hippikes* 12.1–10), so it should not be surprising to find these posted at the front for the protection of those horses and riders who did not possess these. According to the *Sylloge tacticorum* (31, 33), the Macedonians had actually adopted for their four rank deep formations the practice of posting two ranks of heavy armed in front and two ranks of light armed behind them. The Macedonians appear to have also invented (or copied from the Persians?) the *koursores* and *defensores* system (see the next chapter) because when the cavalry was deployed on its own as four divisions the two centre divisions consisted of heavy cavalry and the flanks of light cavalry (these consisted of the javelineers and mounted archers, so the javelineers were presumably in front). It would be easy to dismiss this on the basis of the fact that the source is late, but this would be a mistake because the author had access to sources which we no longer possess. It is therefore quite possible that the Romans adopted both the *koursores/defensores* system and the posting of mounted archers behind the heavy armed from the Greeks. However, the Macedonian *koursores/defensores* system had three major differences, which were that it consisted of four *mere* instead of three *mere* (see the next chapter) so that the flank *mere* were also posted slightly in advance of their two centre *mere* and the units were only four ranks deep.

However, it is still clear that the Romans continued to possess units that carried only the equipment meant for some specific type of combat tactic, the best examples of this being the ethnic units (e.g. Moors, Heruls, Arabs), and that regular Roman cavalry units were sometimes designated for some particular type of cavalry combat because their training scheme and great variety of different types of equipment (javelins, spears, bows, various different types of protective gear) enabled this. As Vegetius (3.16) noted, the cavalry commander was expected to know which type of cavalry *drouggos* to place against which *globus*. For the meaning of these words, see the discussion later in this chapter. In other words, before Maurice unified the cavalry combat methods and the internal organization of the cavalry units, there was an expectation that the cavalry commander would vary the type of equipment and tactics according to the type of enemy. However, as the previous discussion of the use of ethnic troops and the adaptation of the strategy and tactics to the type of enemy has already implied and the following discussion will clarify, the Romans never stopped using different types of equipment and tactics that were geared according to the type of enemy and situation. Regardless, it is still clear that, when the commander decided to use the cavalry for a frontal charge, that the unit composition shown here and which we find in both the *Peri strategikes* (see below) and *Strategikon* was the one they usually used, at least from the second century onwards. In other words, in such situations the Romans preferred to place their lancers in the front ranks and archers behind.

It is therefore obvious that the pattern of having both elements, melee and archery, simultaneously within the same unit formation as depicted by the *Strategikon* was not the only one used by the Late Romans, even after the reign of Maurice, because we know that the Romans continued to use ethnic units, that they always adapted their tactics to the situation, and we know from the narrative sources that the Romans used dual purpose units for archery at distance, meaning that all horsemen, the front ranks included, used bows, which is actually also included in the *Strategikon* as a tactical alternative when facing Germanic peoples. In such cases the readiness of the unit for melee obviously depended on its quality and situation. We also know that the Romans did not always possess units that could be used as dual purpose cavalry, so that they had to settle on using cavalry units that specialized either in melee as spearmen/lancers (some Germanic cavalries), or as swordsmen (e.g. Heruls), or as skirmishers with bows (e.g. some Armenian, Persian and Hunnish units), or as javelineers (e.g. some Moorish, Roman and Germanic units). The units assigned to melee roles were used as *defensores* (defenders) and the units assigned to skirmishing roles were used as *koursores* (runners, skirmishers). For the meaning see the next chapter dealing with *moira* and *meros*. This obviously diminished the effectiveness of the lancer charge because the entire combat line was not composed of melee-ready forces, but if there was enough time for prolonged skirmishing before the Romans launched their charge, in ideal circumstances the skirmishers had disordered the enemy formation sufficiently for this to be meaningless. In fact, as we shall see, the use of the *koursores*-skirmishing was one of the main tactics used by Roman cavalry, the downside being that it was usually effective only against enemies that lacked adequate numbers of mounted archers. If the enemy consisted of mounted archers such as the nomadic peoples or Persians, the preferred tactic was to charge into contact immediately. In this case the skirmishing units were less effective than the melee-ready units.

4.4. The *Moira* (max. 2,000–3,000 horsemen) and *Meros* (max. 6,000–7,000 horsemen) in the *Strategikon* and their uses in drilling and combat

During the reign of Maurice the largest battlefield units were called *mere* (divisions), which were divided into *moirai* and *tagmata*. The *meros* with its *moirai* in its turn was divided into specialist components, with the horsemen of the centre *moira* being designated *defensores* (defenders) and the horsemen of the *moirai* of the flanks as *koursores* (runners, skirmishers). The ideal proportions for the *moirai* were for the *moira* of the *defensores* in the centre was to consist of two-thirds of the *meros*, while the flanking *moirai* of the *koursores* on the flanks were to consist of one-third of the *meros*. However, in practice the proportions of each varied. The entire *meros* was commanded by a *merarches* (or *hypostrategos* in the case of the centre *meros* in the first battle line) who was positioned in the centre of the formation. The flanking *moirai* were commanded by the *moirarchai* who were posted in the centre of their own units. The *moirai* of the *koursores* were expected to consist of mounted archers. In other

Meros of Federates

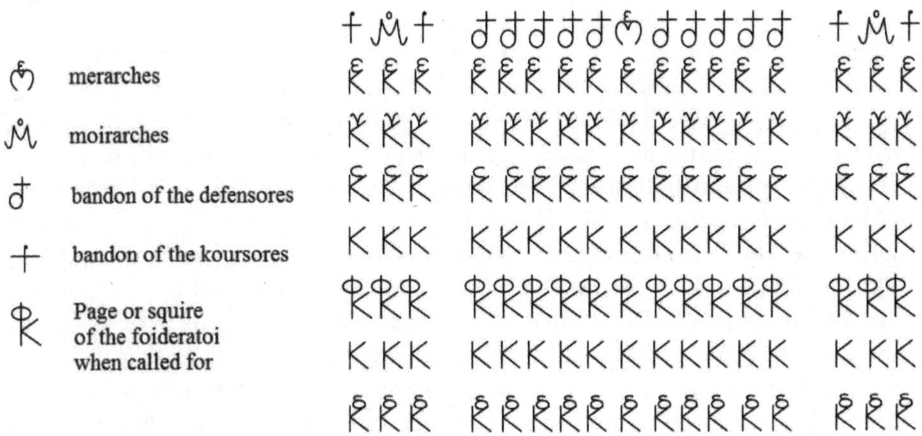

words, all men were trained to this standard. This means that they were expected to belong to the better units. The *De militari scientia* (4) stated the same by requiring the best *tagmata* to be placed in the middle with the *merarchês* and on the extremities of the *meros*. This followed the so-called 'Homeric Principle'.[11] This was the *meros*-structure used in the African and Italian Drill Formations – the latter, with reserves, wing units, ambushers etc. was the preferred combat formation during the reign of Maurice. It is probable that the cavalry divisions of the earlier era could also be called 'legions', as is implied by John Lydus' (*De Mag.* 1.46) referral to 6,000-strong cavalry legions.

The *defensores* in the centre of the *meros* were expected to use the close compact order at all times during the attack, but depending on the situation the *koursores* used a variety of formations. When the entire *meros* consisted of regulars, the standard combat tactic was to charge against the enemy in close order by using the canter. In other words, the entire *meros* was compacted towards the centre and all *moirai* (including the *koursores*) assumed close order and charged at the enemy. However, when the *koursores* consisted of foreigners, they used whatever formation they were used to (*Strategikon* 2.6.15–7). The *banda* of the *meros* were organized in such manner that better *banda* were posted in the centre and on the flanks of the *meros*, which gave psychological support for those in between. The *koursores* pursued the enemy at the gallop (command: *cursu mina*) and the *defensores* at the canter by retaining close order (command: *cum ordine seque*). The use of the gallop in pursuit and the different manoeuvres associated with it disordered the rank-and-file structure of the units assigned as *koursores* so that their unit order became an irregular *drouggos* (*Strategikon* 3.5.41–50, 4.5). In instances in which the *koursores* were used to perform skirmishing missions before the battle or were forced to retreat after pursuit, they took their place in the intervals in the irregular *drouggos* formation. The *moirai* of the *koursores* and *defensores* were separated by small intervals in order to facilitate their differing roles. This separation of the *moirai* meant that if one of those panicked it did not necessarily result in the flight of all three. The successful *moirai* could support the less successful *moira* and the *meros* as a whole could support the less successful neighbouring *meros* by sending the *koursores* or flank *banda* against the enemy flank.[12]

The above describes the standard way of employing the *meros* in the Italian Drill Formation, which resembles the African Drill Formation (see below). The Italian Drill Formation proper included three *mere* (each organized into three *moirai* with *defensores* in the centre and *koursores* on the flanks) in the front, the two wing units, the support line and ambushers. For a fuller discussion, see Chapter 5. On the basis of Maurice's description of the earlier Roman combat drills (*Strategikon* 6) and other sources like Arrian, the earlier versions of the *meros* structure included other variants.

In the Scythian Drill Formation, the *meros* was not organized as *defensores* and *koursores* but as two *moirai* so that all *tagmata* were equipped in like manner. In training and in combat these performed an encircling movement so that the *moirai* advanced towards each other in an encircling motion with the right wing on the outside and left wing on the inside. The literal interpretation of this would have placed the left wing with its archers towards their right side – the shieldless right side where they could not shoot – which means that we should interpret the text as shown in the accompanying diagram overleaf. Since this was the Scythian Drill, it is probable that the *tagmata* of the *moirai* became *drouggoi* (units in irregular order), which in their turn usually meant wedges (*kounai, cunei*). In the hunting exercise (*STR* 12.D.73–4) the right column of horsemen was on the inside and the left on the outside so it is probable that this version was actually the other variant used in encircling enemy formations. According to Maurice, the Roman cavalry had been in the habit of using this exercise regularly as a part of their cavalry games in March. This obviously means that it was no longer a part of the regular training scheme during his reign, even if he recommended its inclusion as an advanced form of drilling. It is probable that this Drill with its method of arraying the *tagmata* of the *meros* in like manner was copied from the Huns, but obviously this is just an educated guess. On the basis of the description of cavalry combat in the texts of Procopius, it is probable that the *bucellarii* of Belisarius, the Huns in Roman service, and the cavalry in general employed this *meros* structure when it was deemed suited to the situation. The best examples of the use of this system are the descriptions of the cavalry skirmishes of Belisarius's cavalry during the first siege of Rome.

The *meros* structure could also be varied in other ways. In the Alan Drill the *meros* was divided into moirai of *koursores* and *defensores* so that these had wide intervals, 200- to 300- feet wide. As already discussed (pp.4–6 and 119) the Alans usually had every other unit designated as *koursores* and *defensores*, but since in the Roman system described by Maurice the training was to be performed by a single *meros*, the *meros* structure was actually the same, namely that the *defensores* were in the centre. The only real difference was the wider-than-usual intervals. In the training scheme the *koursores* galloped in advance, with *defensores* following in close order, after which the *koursores* fled to the intervals, regrouped and then together with the *defensores* advanced against the imaginary foe. In the alternative scheme, after having returned to the intervals, the *koursores* then turned and attacked the flanks of the *defensores*. The latter variation trained the *koursores* to attack enemy flanks. The fact that the

The Scythian Drill of the *Strategikon* (6.1)
This diagram offers a solution to the problem of the left wing being on the inside in the Scythian Drill Formation. It is very likely that the encircling was performed by using the *drouggoi* unit order which in the case of the Scythians usually meant the use of scattered wedge formations. Note that I have changed the interpretation here slightly from my earlier versions in the *MHLR* series.

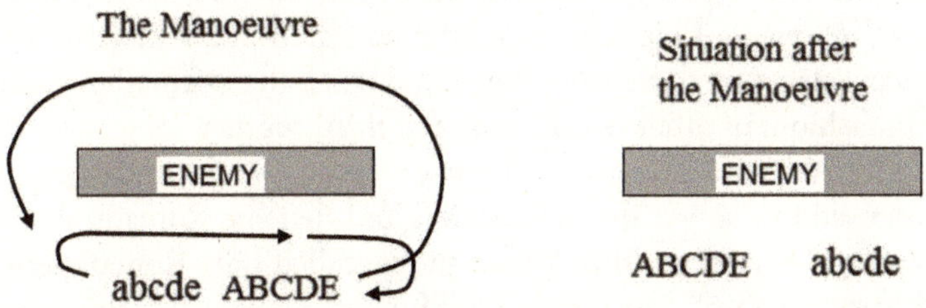

The hunting version
- Note that this version does not describe the *meros* but an actual army.
- In this version the wing *mere* surrounded the wild game in a column unit formation so that the right wing was on the inside when the circle was tightened to make it smaller. Note that the *mere* in question were not divided into *koursores* and *defensores*. This method is plausible because it placed the archers in the right place.
-The outflankers and flank guards were posted slightly behind the wing *mere* and charged forward to encircle the wild animals if these attempted to flee before the edges of the columns met.
-The centre *meros* formed the bottom of the array.

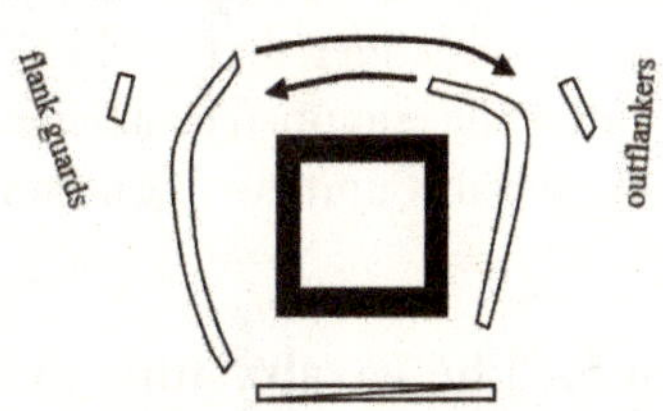

Alan system was retained in the advanced form of training suggests that the system remained in use even after the reign of Maurice. As we have seen, it was already in use during Arrian's day.

In the African Drill array the *meros* was deployed in the regular manner, but so that it consisted only of a single line, which Maurice claimed to have been the usual practice until his own day. This appears to be an exaggeration, because the Romans had certainly used cavalry reserves ever since Republican times. We should therefore understand Maurice's statement to refer to some very recent cavalry practices that date from the reigns of Justin II and Tiberius II, or that Maurice meant that the cavalry had usually performed its training only by using a single line, or that Maurice simply referred to the fact that there had been several commanders who preferred the single line array before he reformed the system, or that the referral to the single line meant the bad practice of sending the reserves to the flanks to outflank the enemy (this is the likeliest alternative).[13] Regardless, one may make the educated guess that the original African Drill had indeed been copied from the Moors, so that it really had only a single combat line. The likeliest period when the Romans could have sometimes used this single line version is the era when Moorish cavalry was fashionable, namely the period from the second- to the fourth-centuries. The fashion is likely to have ended with the arrival of the Huns and the elimination of the Moorish Guards by Theodosius I the Great.

In the African Drill, the *meros* pursued the imaginary foe in regular manner, but so that when the *moirai* of *koursores* turned back, one of these remained behind (this simulated differing combat success) while the other galloped back to their own interval, passed behind the centre *meros* of *defensores*, and emerged from the opposite side after which they advanced towards the other *moira* of *koursores*. The *moira* that had remained behind turned and advanced towards the rear. Both units of *koursores* then charged towards each other and halted just before colliding.

In the variant drill called the Illyrikian Drill, the centre *moira* consisted of *koursores* and the wings of *defensores*. One may make the educated guess that the Illyrikian version of the *meros* structure dates from the period when the Dalmatian and Illyrian cavalries were in fashion; in other words, from the third century.

In sum, it is clear that the Roman cavalry *meros* and its *moirai* and *tagmata* could be deployed and used in many different manners and that the Romans already used the Alan, African and Illyrikian Drills before the so-called Late Roman period. The Scythian Drill was probably copied from the Huns in the late-fourth century. The Italian Drill with its cavalry reserves appears to have been developed in stages prior to the Late Roman period, reaching its final form during the reign of Gallienus, but in such manner that during Gallienus's reign the actual combat units were not as uniformly equipped as during the sixth century – these still appear to have consisted of specialist units with different armaments.

4.5. The cavalry unit orders for the regular rank-and-file formations: Open, close, *drouggos*, crescent, interjection of ranks, tortoise

The rectangular cavalry formations had six different unit orders: Open, close, *drouggos*, crescent, interjection of ranks in the *Peri strategikes*, and tortoise. The only period source which offers measurements for any of these is the *Strategikon*, which states that in the close order *pyknôsis* the width of the cavalry files was three feet (93.69cm = ca. 94cm) and the depth of the rank eight feet (249.84cm = ca. 2.5m). This means that the open order was roughly the double of that. We find confirmation for these figures from the tenth-century *Sylloge Tacticorum* (43.6–7), where each footman and horseman occupy in width an *orguia* during march (= 4 *pecheis* = 187.38cm = ca.187.4 – which is double the close order figure given by the *Strategikon*); half an *orguia* when battle was expected (= 3 feet = 3 x 31.23cm = 93.69 – which is the figure in the *Strategikon* for the *pyknôsis*); and a third of an *orguia* during battle (= 2 feet = 2 x 31.23cm = 62.46 = *synaspismos/chelonê/syskouton/tortoise*). The *Strategikon* does not include the cavalry tortoise, but it is very likely that it was used during the Late Roman era because we find it in both Arrian and later in the *Sylloge Tacticorum*. The *Peri strategikes* also gives us a variant of close order in which the two front ranks were partially intermingled and which appears to have been used by some commanders. The *drouggos* was an irregular non-rank-and-file formation that appears to have had many different variants, while the crescent was a column formation used by the flank units.

***The extended or open order* (araios, araiôsis)[14]**

The open or extended order (*araiôsis*) was a cavalry unit order consisting of ranks and files in which the men were far enough apart for each one to be able to about turn their horses easily. The evidence suggests that the cavalry files in this formation had a width of 187.4cm and the ranks a depth of 5 metres. The open order was used during marching, during the initial advance to the battle, and probably also when the *koursores* advanced forward at gallop to skirmish before the entire *meros* assumed close order. It is actually clear that the open order was used for attacks in such situations in which the cavalry units had not yet assumed close order because there were situations that called for immediate charges at a gallop. In fact, the cavalry attack at gallop in loose order was a frightening sight to those who were not used to seeing such. The command to march in open order was '*Equaliter ambula!*' ('*ekoualiter amboula!*'). The command to charge at gallop was '*Cursu mina!*' (*koursou mina!*'). The principal ideas behind using the open order were: 1) to avoid exhausting the men and horses while marching; 2) to make the battle line more impressive to the enemy with extra width; and 3) to enable the cavalry to manoeuvre at greater ease to the desired position before the actual charge in close order, the latter of which limited its ability to manoeuvre.[15]

The close/compact order **(pyknôsis)**[16]

Close or compact order (*pyknôsis*) meant a cavalry formation formed out of the ranks and files in such a manner that each trooper assumed the closest possible position in relation to each other for the final charge into contact, which the *Strategikon* (9.5.9–12) estimated to mean 3 feet (ca. 94cm) in width for each file and 8 feet (ca. 2.5m) for each rank in depth.[17] In other words, Maurice expected the horses to be so close to each other that the knees of the riders almost touched while the horses were almost nose to tail (average length of the horse is ca. 2.4m). However, I would suggest that in practice the space occupied by each horse/rider in the formation was sometimes slightly greater, and I would suggest that this was definitely true in the case of the depth of the ranks in the array, which may easily have been ca. three metres in practice. The figures given by Maurice should only be seen as rough estimations.[18]

The cavalry assumed close order just prior to the main charge. The tightening of the formation began by closing the files at a canter towards the centre (could be the centre of the *bandon, moira, meros*, first line, or second line) and standard bearer (*bandoforos*) with the command '*Ad latus stringe, ad decarchas, ad pentarchas!*' ('*ad latous strigge, ad dekarchas, ad pentarchas!*'). When this had been performed, the command '*Iunge!*' ('*Iugge!*') was given to close up the ranks from the rear towards the front and when the entire formation had assumed close order, the order to charge at a canter was given with the command '*Percute!*' ('*perkoute!*'). It was then that the *dekarchai* and *pentarchai* leaned forward, and covered their heads and part of their horses' necks with their shields while holding their spears as high as the shoulders. The levelling of the spears by the two front ranks cleared the skyline for archery and helped the archers in the estimation of the distance to the enemy. The rear ranks timed the start of their archery to take place at the same time. If the enemy fled, the *defensores* retained their close order and pursued by using the canter with the order '*Cum ordine seque!*' ('*koum ordine sekoue!*') while the *koursores* pursued at gallop with the order '*Cursu mina!*' ('*koursou mina!*'). As already discussed, as a result the *koursores* lost their close order and cohesion of the formation, so that their formation became irregular (*drouggos, drouggisti*).[19]

The Cavalry Close-Order from Front and Side

Above: The cavalry close order (*pyknosis/puknosis*) from the front. The example shows four files from the fill-up *bandon* with two ranks charging towards the enemy approaching from behind. The width of each file was approximately 94 cm. The *skouton*-shield (*scutum*) shield that the *Strategikon* instructed the men to use was also ca. 94 cm in width, which means that it was easy to find the correct width for the files on the basis of the shields. When deployed as fill-up *banda* it is possible that the troopers used their bows and spears alternatively as needed in the same manner as is described in the training of the cavalryman by Maurice (*STR* 1.1). However, I have here depicted them with the spears levelled for the charge.

Below: The cavalry close-order (*pyknosis/puknosis*) from the side. The example shows an elite unit with five ranks. The two front ranks are here equipped by the book with a single *kontos/kontarion*-spear, bow, *skouton*-shield and *spathio*-sword. The file-closer, the last man in a file, is also equipped with the full panoply. In addition to this, I have assumed that the soldiers carried extra javelins in a holster which we know was still in use during the sixth century (see e.g. the battle scenes describing the combat of John Troglita in Syvänne, *MHLR 6*, 272-96). According to Maurice, the depth of each cavalry rank was ca. 2.5 m, but since the average length of a horse is ca. 2.4 m, I have assumed that in practise the depth of the rank could also be slightly longer so that the horses would not collide accidentally for example when the horse archers concentrated on delivering their volleys of arrows.

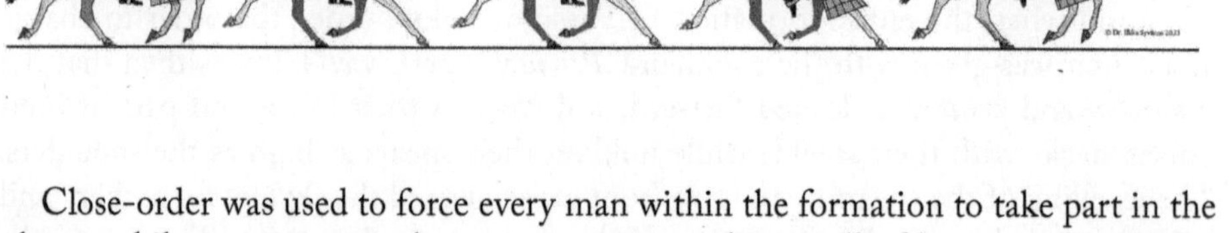

Close-order was used to force every man within the formation to take part in the charge while presenting to the enemy an approaching wall of horses and men. On the basis of the usual success of the Roman cavalry charge, it is clear that the sight of the approaching tight Roman cavalry formation usually made the enemy cavalry hesitant, which in its turn usually caused them to flee in order to avoid collision with this wall of men and horses. This use of the mounted archers within this array contributed to this. The enemy was not only disrupted by the sight of the approaching wall of men and horses in readiness for melee, but also by arrows hitting them from above. The reason for the readiness of the enemy to flee before such approaching force is that humans and men instinctively tried to avoid having to make contact with a solid object with suicidal speeds. If neither side flinched and fled, and both sides used close order, the likeliest occurrence was that both sides halted just in front of each other and began engaging each other in melee.

The irregular order globus/ drouggos[20]

The Roman cavalry also used the so-called *drouggos* unit order, which has usually been translated as irregular order. The name was probably Celtic in origin, but the Germanic version of the same word may also have influenced the Romans in the adoption of the word. Its Latin name was *globus,* which signified an independent throng of men operating separately from the main body of troops.[21]

Philip Rance and John Haldon have established that the words *drouggos/drungus/ globus* and *drouggisti* had several different meanings depending on the context, so that it meant a unit arrayed in an irregular non-linear compact mass and also units of irregular sizes (could be anything from four to five footmen up to 5–6,000 horsemen), so that *drouggos* (originally an irregularly sized unit composed of several smaller units) eventually came to mean a *moira.* Rance notes that the *Strategikon* refers to Avars forming their combat formations in various *moirai* as *drungi* all linked together, and that the Avars favoured the use of wedge-shaped formations. This does indeed connect the use of wedges with the *drouggos,* as I also noted in my doctoral dissertation. Rance is also correct to note that both the wedge in particular and the rhombus of the Hellenistic military theory fit the description in Maurice of how the concealed *drouggos* was to be used in combat, but he suggests that the principal meaning for the *drouggos* as combat unit order was probably the wedge unit order.[22] I suggested the same in my doctoral dissertation. However, even if it is clear that the *kouna/kounion*-wedge (pl. *kounai/kounia,* Latin *cuneus, cunei*) of the *Strategikon* (11.2.54, 12.A.7.23) can be equated with the *drouggos* (*Strategikon* 11.2.40–3, 11.2.54–5) it is still probable that *drouggos/drouggisti* meant also the non-rank-and-file rhombus and any other non-rank-and-file cavalry formation in which the troopers followed their standard and commander in any way they could. This means that the standard size for the regular *drouggos* was the *tagma/bandon,* because it was the smallest combat unit that carried the standard. As a result, it is clear that the troopers of such *banda* that had lost their rank-and-file organization as a result of the use of the gallop or because these had originally been deployed irregularly simply followed the standard during combat as best they could. The resulting formation would have in most cases been a wedge or a round lump of horsemen. The last of the alternatives actually represents a return to the interpretation that I and everyone else appears to have done concerning the meaning, namely that the *drouggos* simply meant an irregular formation. The new element here concerns the non-rank-and-file rhombus.

The best evidence for the identification of the rhombus as one of the meanings for the word *drouggos* comes from the analysis of the context in which the Latin word *globus* is used in Roman military texts and from the new interpretation of the diagrams in the *Strategikon* 3.10 in combination with the use of the word *drouggos* in Leo's *Taktika* and *Sylloge Tacticorum*.[23]

Let us begin with Leo. Leo includes the cavalry versions *drouggisti êgoun ôs maza* (*Taktika* 7.34) and *drouggisti êgoun pyknôs* (*Taktika* 12.64) and the infantry version *drouggisti kai ôs maza* (*ST* 45.10).[24] The *Sylloge Tacticorum* (45.10) also specifies that the infantry *tagmata* often attacked without ranks and files, which sometimes had a wide front, while at other times it contracted as a mass of men (*drouggisti kai ôs maza*) which fits perfectly with how the non-rank-and-file cavalry rhomboids changed their shapes. Both versions clearly mean a tight formation and imply that it did not have ranks and files.

We can take the meaning a bit further by analysing the meaning of the Greek word *maza*, which means a mass, lump, and a round barley cake, while its Latin equivalent *massa* similarly means a mass, lump and something that adheres together like dough (i.e. like barley cake). In other words, *maza/massa* implies a round-shaped mass. The Latin word *globus* means a round body, a ball, sphere, globe, globular mass and throng. In other words, the *globus* also implies something which is massed together as round-shaped.

This brings us to the use of the word *globus* in Roman military texts. Cato in his *De re militari* (frg. 11 in Jordan 1860, 82) lists *cuneus (*wedge), *globus* (globe), *forfex* (pincer), *turris* (tower, column), and *serra* (saw). Aulus Gellius in his *Attic Nights* (10.9.1) lists *frons* (front), *subsidia* (reserves), *cuneus* (wedge), *orbis* (orb, ring, circle), *globus* (globe, mass), *forfices* (shears), *serra* (saw), *alae* (wings) and *turris* (tower, column). Vegetius (3.17, 3.19) lists the following unit formations: *cuneus* (wedge), *forfex* (pincer), *serra* (saw), *caput porci* (swine head), and *globus* (globe) which is the same as *drungus*. The important point in all of these is that every one of the authors lists *cuneus* and *globus* separately, with the implication that these were separate unit orders. The rounded shape (*maza, globus*) implies that it faced all directions simultaneously. The only ancient cavalry formation which matches this is the *rhombus* formation and it had an irregular non-rank-and-file version too. I would suggest that this was indeed one of its meanings and I would also suggest that the infantry employed its own version of the *globus* which was a hollow rhombus with the name *caput porci(num)* (pig's head) which was known to the Vikings with the name *Svinfylking*. We know that the Late Romans used the rank-and-file rhombus during this era because it was used by cavalry in one of the versions of the infantry *epikampios opisthia* (see pp.341–3), and because it was also used by Armenian and Persian cavalries in Roman service. We also know that the Romans continued to use both rank-and-file and non-rank-and-file versions of the wedge, which implies that they also continued to use the non-rank-and-file version of the rhombus. Unsurprisingly, this is the only version included by Arrian in his book on tactics.

Further support for this comes from the analysis of the diagrams in the *Strategikon* 3.10. The *Strategikon* (3.4–4.5) stated that one or two *banda* were to be detailed as ambushers behind the wing units (flank guards and outflankers). The following two diagrams (outflanking on both wings; outflanking by right wing) are taken from the Ambrosian version of the *Strategikon* and these show hidden *drouggoi* deployed as ten separate circles behind the right wing unit, each group deployed as a wedge. In other words, the *drouggoi* were simultaneously rhomboids and wedges. The ten balls/circles correspond with the unit sizes for ten rhomboids if these were irregularly grouped as non-rank-and-file units. If all were deployed as 25-horseman rhomboids, there were 250 horsemen in the hidden *drouggos*; if there were 36 horsemen per rhombus, then the corresponding figure is 360 horsemen. Both of these figures are within the 200 to 400 men allowed for a *bandon*. If there were for some reason two small *banda* behind the wings as allowed by the *Strategikon*, it was possible to form 49-horseman rhomboids so that the total would have been 490 horsemen. The largest variant allowed by the *Strategikon's* limits would have been 64 horsemen per rhombus for a total of 640 horsemen (two *banda* of 320 horsemen). This was also possible if the commander deployed the ten units as non-rank-and-file wedges (10 x 21 = 210; 10 x 28 = 280; 10 x 360 = 360; 10 x 45 = 450; 10 x 55 = 550; 10 x 57 = 570; 10 x 69 = 690). However, it should also be remembered that the sizes of the rhomboids and wedges could vary from one unit to another so that the totals were within the limits allowed by the system having between 200 and 400 men per *bandon*. However, the standard version of the *drouggos* would still have been the single *bandon/tagma* in irregular order (usually a wedge or a round mass).

The key to understanding the meaning of *drouggos* is therefore that it was a non-rank-and-file unit order used by smaller units than the *bandon/tagma* or by the *bandon*, and that each *bandon/tagma* consisted of *drouggoi* probably in such a manner that there were usually about ten *drouggoi* in each *bandon/tagma*.

Double outflanking (hyperfalaggesis) and outflanking (hyperkerasis)

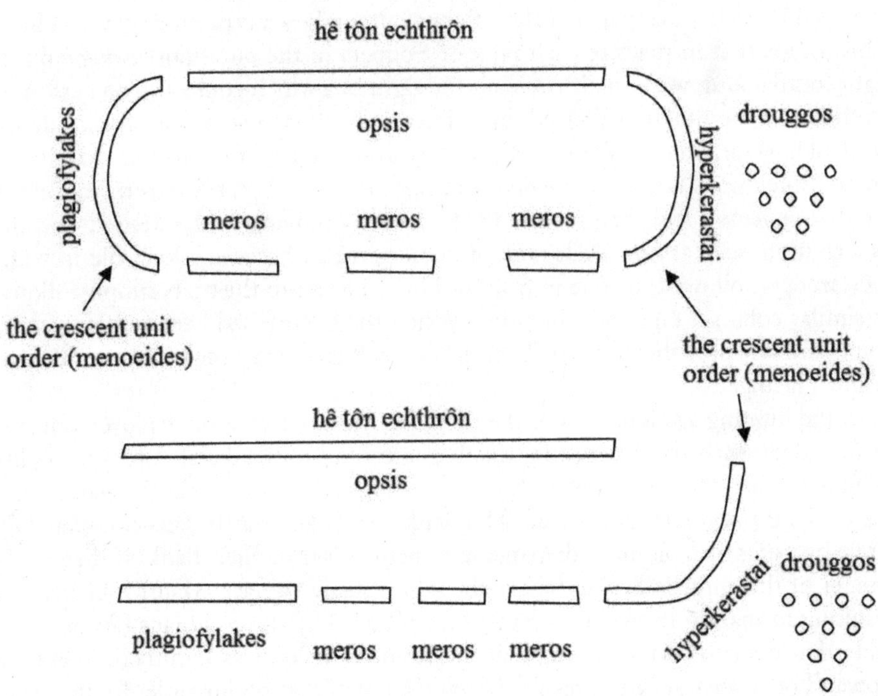

The best evidence for the actual use of the *drouggos/globus* in combat comes from the pen of Maurice. The *Strategikon* considered the *drouggos* particularly useful for ambushes, for attacks against the enemy flank or rear or baggage train, for reconnoitring, for pursuing, and for the bringing of support to units in distress. In short, this unit order was used by: 1) the scouts and vanguards for reconnoitring and skirmishing; 2) the *koursores* for the pursuit of the enemy; 3) the ambushers for the ambushes; 4) the concealed wing units for the outflanking of the enemy; and 5) the reserves for the bringing of help to units in distress. This was the most mobile and manoeuvrable of the cavalry orders. It was ideal for rapid movement and for concealing a unit in ambush – absolutely necessary for the kind of warfare that the Romans excelled at, in other words for skirmishing, ambushing and other guerrilla tactics that were designed to exhaust the enemy. Its use was highly recommended by the *Strategikon*.[25]

The Crescent order (menoeides)[26]

The unit orders in the *Strategikon* include also the crescent formation (*menoeides*) used in the encirclement of the enemy. In combat situations it was used by the flank units (flank guards/*plagiofylakes* and outflankers/*hyperkerastai*) to outflank the enemy line, while in the section dealing with hunting as a form of military training it was used by the flank units and left and right wing divisions (*mere*) to encircle the game. The best clues to the meaning of the crescent order in cavalry combat are the diagrams that depict it (see the drawings dealing with the *drouggos* order) and the description of how it was used in hunting.

The diagrams clearly depict the wing units (flank guards and outflankers) deployed in such a manner that their flanks/outer edges advanced first as a column. This means that in practice each rank of troopers in the outermost *bandon* on the right (outflanking was usually done on the right but which could also be performed on the left if the situation allowed) or in the outermost *banda* on both flanks (double outflanking) turned as individuals towards the outer edge of the formation with the orders 'Shieldward turn, move!' ('*Ad scuto clina, move*') or 'Spearward turn, move! ('*Ad conto clina, move*') and then initiated the attack by moving in the desired direction as a column, so that now each rank of horseman had become a long file in which each trooper followed the one in front of him. The rest of the units simply followed as similar columns one after the other. When the enemy had been outflanked, the command was probably '*Return*' ('*Redi*'), because that turned the units towards their original facing.

In the hunting section, each of the three *mere* (divisions) were deployed one, two or four deep with the outer units (flank guards and outflankers) deployed slightly behind so that the flank units advanced by the flank to encircle the game using the crescent (*menoeides*) formation. The flank guards and outflankers completed the circle by using the column order (men advancing towards their flank) if there was a danger of the animals fleeing before the wings made contact. Maurice likened the resulting formation to the encirclement tactics used by the Scythians (Avars, Huns and other nomads), while noting that the Roman version was slightly different and slower. The relative slowness resulted from the use of the column order for the attack (note the depth of one, two or four ranks which makes this interpretation certain), while the Skythians used the *drouggos* order for its units which was faster. It was because of this that the Romans posted the hidden ambushing *bandon* behind the wing unit as a *drouggos*, and it was because of this that the Romans trained their wing units in the use of the *drouggos* order too. The last mentioned detail means that the crescent encircling order could also be performed in the same manner as the Scythians, as irregular units (*drouggoi*) or wedges (*kounia/cunei*).

***The cavalry orders in the* Peri strategikes (17)**[27]

As already noted, the cavalry formation of the *Peri Strategikes* required that the file-leader and four men behind them, the file-closers, and the men at the outer edges, were to consist of the better horsemen. The probable depth of this formation was ten ranks, even if Syrianus states that the infantry and cavalry formations were to be the same in armament and formation. The comparison with the infantry formation means simply that the inner ranks were equipped with missile weapons and outer edges with spears. The horses of the front rank were expected to be equipped with iron armour on their heads, breasts and necks, and were expected to be well-trained and used to the confusion and noise of the battlefield. The hooves of the horses were also to be protected by iron plates, in other words by horse-slippers. According to the *Peri strategikes*, the cavalry phalanx still differed from the infantry phalanx. It was not as crowded as the infantry phalanx. In contrast, the cavalry phalanx was loose (*araios*, *araiôsis*) because it was used for cavalry charges at full speed. It is probable that the command to begin the charge was '*Cursu mina!*' ('Gallop, Threaten!') just like in the *Strategikon*. The intention of this cavalry tactic was to frighten the inexperienced enemy soldiers.

The problem here is how loose was the cavalry array of the *Peri strategikes* on the basis of this description? The width and depth of the infantry file and rank in the *Peri strategikes* (16.37) was a *pechus* (ca. 62.46cm), which means that the cavalry formation was wider than this. Furthermore, it was wide enough for the interjection of the horses from the second rank with the horses of the first rank, so that the heads of the horses reached either the shoulders or flanks of the horses of the first rank. This was possible also with the close order *pyknôsis* of the *Strategikon* which gave each cavalry file the width of ca. 94cm. It is also clear that the interjection of the horses does not mean open order, because there would not have been any need to place the heads of the horses only up to the shoulders or flanks if the array was an actual open order. The heads of the horses were brought up to the shoulders or flanks of the front rank horses because both alternatives left the knees of the rear rank horsemen behind the knees of the front rank horsemen. However, it is possible that the width of a cavalry file in the *Peri strategikes* was still wider than this, because the *Strategikon* expected that the charge at full speed resulted in the loss of cohesion, and in the end the *koursores* that used the gallop lost their order and became deployed as irregular units (*drouggisti*, *drouggos*). However, I would still suggest that the looser cavalry formation of the *Peri strategikes* is still the close order of the *Strategikon* because the *koursores* started their pursuit of the enemy when they were deployed in close order. They just lost the cohesion of the unit order as a result of the charge. This is how we find the

The interjection of the two front ranks by 200 horsemen

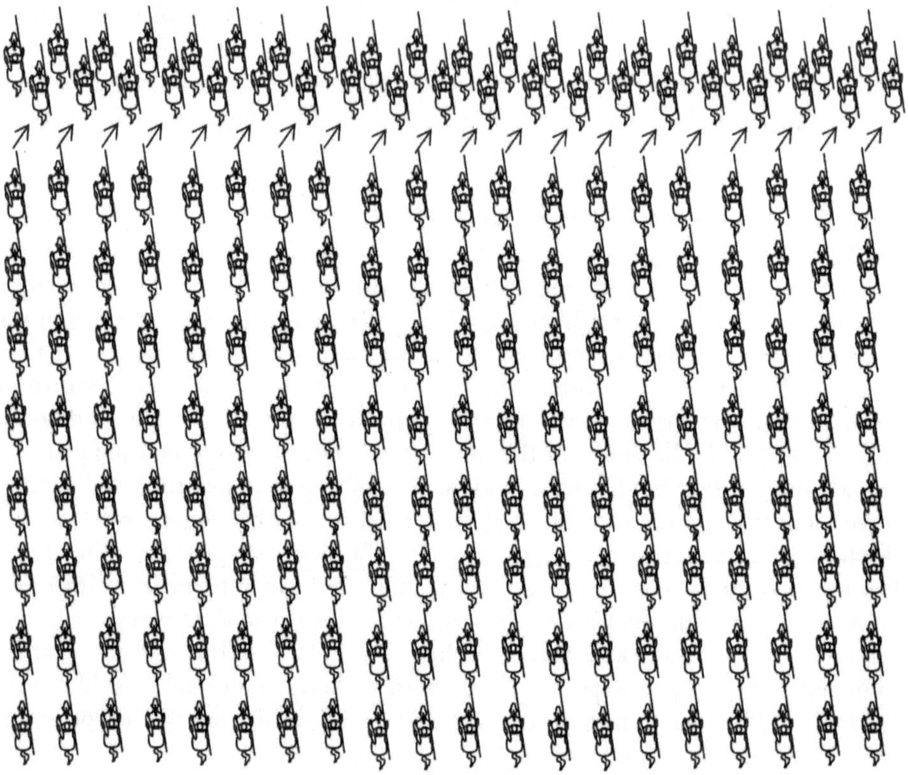

European cavalry of the eighteenth and nineteenth centuries operating. These could charge at full speed when in close order (in fact even in a closer knee-to-knee order than in the *Strategikon*) and with the same result as in the *Strategikon*: they lost the cohesion of their array and had to regroup. In short, it is clear that at the time when Syrianus wrote the *Peri strategikes* the Roman cavalry was used in slightly different manner than it was used during the reign of Maurice, and the narrative sources confirm this conclusion. We find the cavalry of Belisarius employing the full charge and shouting during the reign of Justinian I, both of which were disapproved by Maurice. See Syvänne, *MHLR 6*.

The second of the cavalry formations in the *Peri strategikes*, the interjection of the first and second rank horses, is the same as we find in Arrian (*Technê Taktika* 16.14). However, the instruction of how this was done in the *Peri strategikes* contains a problem. Syrianus merely states that some preferred to tighten the cavalry phalanx so that they lined up the heads of the horses of the second rank with the shoulders or flanks of the horses of the first rank. This leaves open whether Syrianus meant only the first two ranks or whether all of the ranks were intermingled in the same manner. However, I would suggest that he meant only the intermingling of the first two ranks, because this would have been sufficient for the appearance of a continuous line and the attached illustrations show this alternative. As noted, it is likely that the resulting

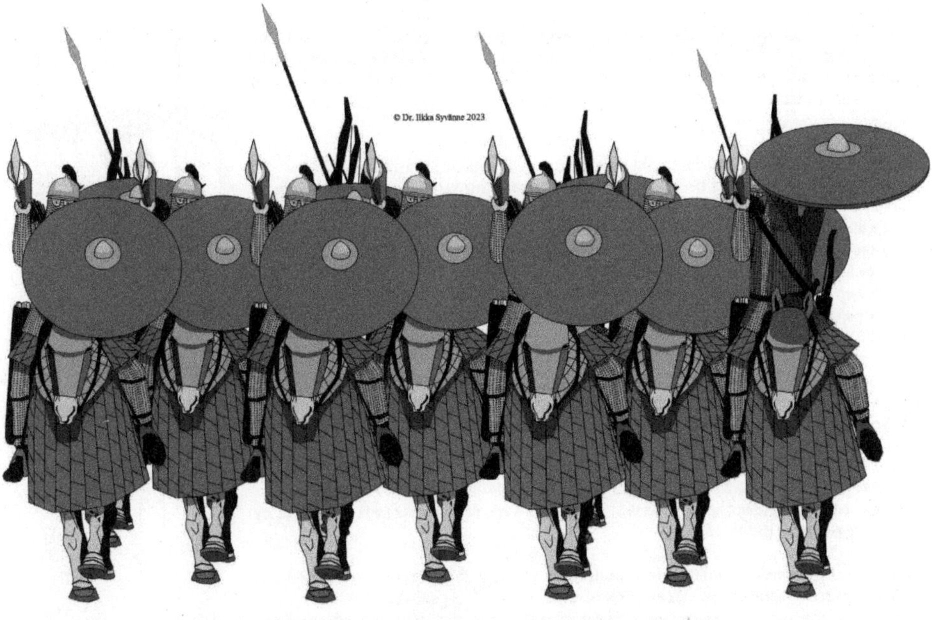

The super tight cavalry attack formation of Arrian and Syrianus Magister from the front. The troopers are depicted with the round shield (*clipeus, aspis, pelte, parma*).

formation was based on the same file widths as the close order in the *Strategikon* so that the heads of the horses were brought so far forward that the knees of the second rank riders were behind the knees of the first rank riders. Arrian disapproved of this method because in his opinion it only disturbed the horses, and it is actually very likely that this was the reason for Syrianus's comment that the horses of the first rank were to consist of well-trained horses used to the confusion and chaos of combat. In other words, Syrianus and the other military thinkers he was referring to knew the danger and took it into account in the instructions. I would therefore suggest that the Late Romans did use the interjection of the horses of the first two ranks to obtain an even tighter and more impressive cavalry formation for the charge at gallop. The placing of the knee behind a knee and charge at gallop was also later used by the Swedish cavalry wedges during the reign of Charles XII (Karl XII). It required well-trained and disciplined men and horses.

The Cavalry testudo

Arrian, (*Ars Tactica, Technê Taktika* 36.1) and the *Sylloge Tacticorum* (43.6–8) mention a tight cavalry formation called *synaspismos/chelone/testudo* (tortoise). The period sources provide one likely proof for its use for our era, and conjectural evidence suggests other instances. According to Procopius (*Wars* 5.18.13) the Roman horsemen formed a protective cover for both Belisarius and his horse using their shields at the battle of Milvian Bridge in 537. Additionally, it is quite probable that when Roman cavalry formations fought against, for example, the Persians, and both armies were stationary and exchanged volleys of arrows, that the shield-equipped Roman

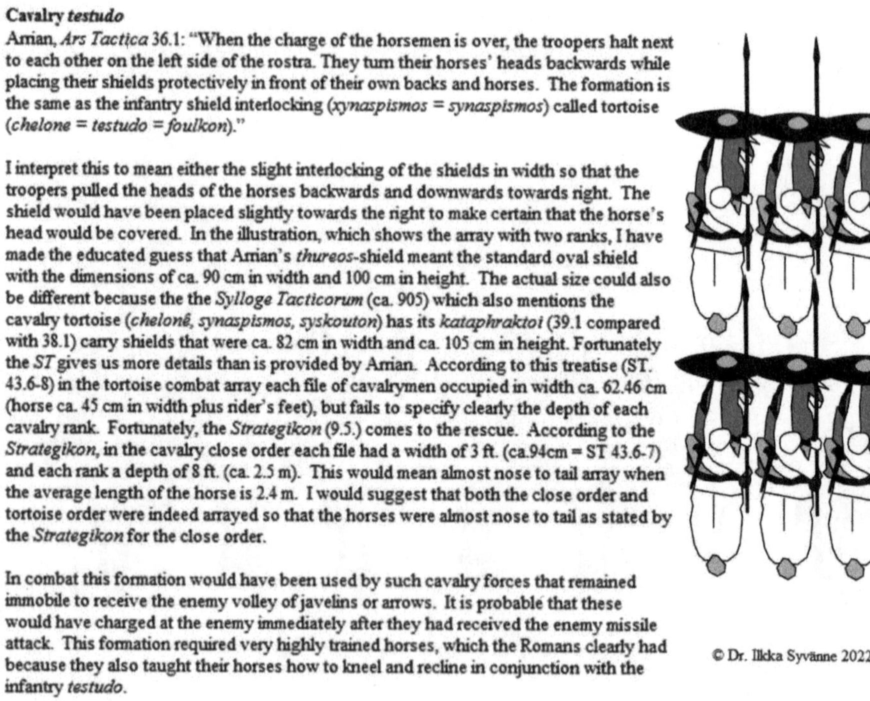

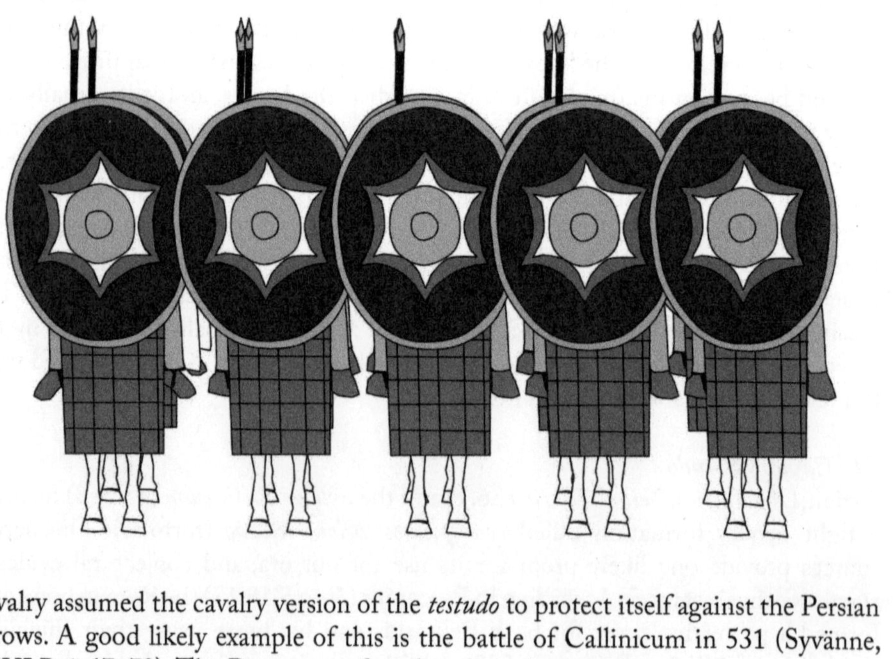

cavalry assumed the cavalry version of the *testudo* to protect itself against the Persian arrows. A good likely example of this is the battle of Callinicum in 531 (Syvänne, *MHLR 6*, 67–70). The Persians preferred such prolonged periods of archery before actual close-quarters combat, so the adoption of the cavalry *testudo* was probably a more typical phenomenon when fighting against them, but it is clear that similar instances could take place whenever the Roman cavalry was forced to stay in place under enemy fire. According to the *Sylloge Tacticorum*, each file of cavalrymen in the tortoise array occupied in width ca. 62.46cm and I would suggest that it is probable that the depth of each rank was the same as in close order (ca. 2.5m).

This formation required very highly trained horses, which the Romans clearly had because they also taught their horses how to kneel and recline in conjunction with the infantry *testudo*. Maurice in his *Strategikon* did not include the array. This was a sound decision, because cavalry's principal contribution to combat was its mobility and the cavalry's greatest psychological impact on the enemy cavalry and infantry could only be achieved through charging and movement. This was not possible with stationary cavalry forces: these did not impact the morale of the enemy.

4.6. The wedge and rhombus (see also Chapter 9.6)

As we have seen, it is probable that the non-rank-and-file versions of the wedge and rhombus unit orders were variants of the irregular *drouggos*-order. In addition to this, we know on the basis of the Roman use of Armenian and Persian units (*Byzantine Interpolation of Aelian*, Devine ed.45.2; Dain J2; with Syvänne 2010, 2014 and 2017b) and on the basis of the older version of the *epikampios opisthia* infantry formation (see Chapter 9.6), that the Romans also continued to use the traditional rank-and-file versions of the wedge and rhombus. Note that all of these variants also had their own versions of open and close order.

The ancient military treatises in the Hellenistic tradition in conjunction with the later 'Byzantine' adaptions of those (mainly the so-called *Byzantine Interpolation of Aelian*) and the instructions concerning the older version of the *epikampios opisthia* in the *Apparatus Bellicus* allow one to reconstruct the different variants used by the Romans and Late Romans. Let us begin with the analysis of the rhombus and wedge in the Hellenistic treatises.

Asclepiodotus (7) describes both the rank-and-file and non-rank-and-file versions of the rhombus and wedge. The wedge was always roughly one half of the rhombus. Asclepiodotus gives us two different examples of the rhombus. The first is the rank-and-file version with 61 horsemen, so that the first rank had one man, second three, third five and so forth (1, 3, 5, 7, 9, 11, 9, 7, 5, 3, 1). The second of the examples was the non-rank-and-file version with 49 horsemen (1, 2, 3, 4, 5, 6, 7, 6, 5, 4, 3, 2, 1). Neither of the examples conforms to the unit strengths of cavalry, so it is clear that Asclepiodotus's examples are just examples of how to array the cavalry. Regardless, it is clear that both of these versions could be used when the situation required and there were not enough men for the forming up of larger or smaller units.

Similarly, Aelian (Matthew ed. 18–9) and its Byzantine Interpolations describe both the rank-and-file and non-rank-and-file versions of the rhombus and wedge. See the attached images. Aelian's example of the rank-and-file rhombus in the text had 113 horsemen (ranks consisting of horsemen: 1, 3, 5, 7, 9, 11, 13, 15, 13, 11, 9, 7, 5, 3, 1). As I have noted before (Syvänne, 2010, 2014, 2017b) this is a mistake because we know on the basis of the Arabic version of Aelian and Arabic version of the rhombus (*kurdus*) that its paper strength contained 128 horsemen, which

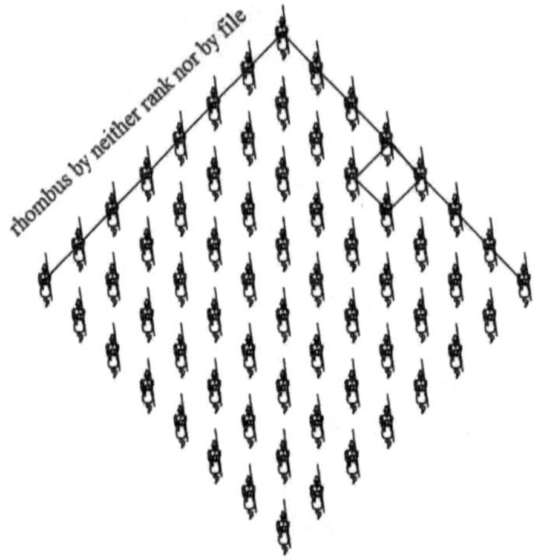

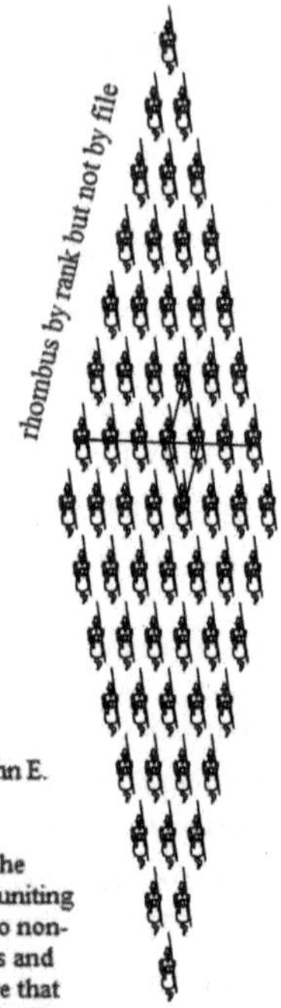

The non-rank-and-file cavalry rhombus/rhomboid
(adapted from J.E. Lendon's idea but with some changes)
Note that my reconstruction of the rhombus places the horsemen closer to each other than John E. Lendon's reconstruction (2011) and that my rhombus is 64 horsemen strong.

On the basis of the unit structure it is clear that the Romans used the 64 horsemen version of the non-rank-and-file rhombus. This means that the Romans could form up this version either by uniting two 32 horsemen *turmae* or by dividing the larger 128 horsemen rank-and-file rhombus into two non-rank-and-file rhomboids. It is also clear that the non-rank-and-file rhombus used by the Greeks and Romans was a close order formation because both Aelian (19) and Arrian (*Ars Tactica* 17) state that the heads of the horses were arrayed at the shoulders of the horse in front. The existence of the file but not rank version also implies this. This 64 horsemen non-rank-and-file rhombus could also be divided into three non-rank-and-file wedges (three 21 horsemen wedges and one commander) just like the 64 horsemen rank-and-file wedge.

John E. Lendon (2011) is likely to be correct in his statement that the rhombus by-rank-but-not-by-file and the rhombus by-file-but-not-by-rank are but variations of the non-rank-and-file rhombus so that these were formed from this array when the need for this arose. The reason for this is that if one draws the file-but-not-rank version in the manner explained by Aelian (the horses of the next file always adjacent to the interval beside) one ends up having the non-rank-and-file rhombus. This means that the intervals between the files in the rhombus with files-but-not-ranks must have been larger with the implication that the rhombus with files-but-not-ranks was the open order version of the rhombus. Therefore the likely campaign and battle usages for the variants were: 1) the non-rank-and-file version was the standard battle formation; 2) the rank-but-not-file version was used to get through intervals between units or through some terrain feature; 3) the file-but-not-rank version was used when the terrain or situation required the use of open order.

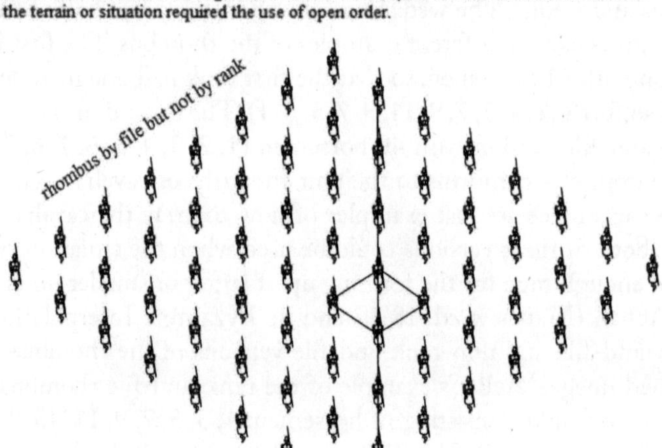

could be divided into two 64 horsemen wedges. The 64 horsemen rank-and-file wedge in its turn was also the wedge of Polybius, as the Greek versions of Aelian note. In other words, these two figures conform to the cavalry unit structure 32, 64 and 128. The size of rank-and-file rhombus (128 horsemen) is also confirmed by the extant narrative sources depicting the Parthian army which used the rhombus formation and also by the *Apparatus Bellicus*. However, the size was obviously smaller in instances in which the unit had suffered casualties as my analysis of the battle of al-Qadisiyyah demonstrates.[28]

The text version of the non-rank-and-file rhombus has two variant sizes: 36 horsemen (1, 2, 3, 4, 5, 6, 5, 4, 3, 2, 1) and 100 horsemen (1, 2, 3, 4, 5, 6, 7, 8, 9, 10, 9, 8, 7, 6, 5, 4, 3, 2, 1). The former is not compatible with the unit sizes, but the latter is (3 x 32 plus 3 decurions and one centurion for a total of 100 men) so one may make the educated guess that it was also used by the Romans. However, it is probable that the 36 horsemen and other variants were also used if there were not enough men to form larger rhomboids. It should be noted that the diagrams of the *Codex Burnley* do not correspond with the information provided by the text, so that the non-rank-and-file wedge has only 15 horsemen (1, 2, 3, 4, 5), and the rhomboid 25 horsemen, and rank-and-file rhombus 13 horsemen. It is therefore clear that these are meant to show only the principles and not the reality. The illustrations in the other manuscripts have other variants. See Matthew's edition for another extant variant. Matthew (159) notes quite correctly that the rhombus was adaptable for various different alternative unit strengths. Arrian (*Technê Taktika* 16.3–8, 17.1–3) describes the non-rank-and-file (first rank one man, second two, third three etc.) versions of both without giving any numbers for these units.

Let us now turn to the period evidence (*drouggos* being a non-rank-and-file formation; Persians, Armenians and Arabs using the 128 horsemen rhombus/*kurdus*) and information provided by *Apparatus Bellicus* (see pp.341–5). In the *Apparatus Bellicus* the 128-horseman *hippilarchia/epilarchia/epeilarchia/rhombus* consisted of two 64-horseman wedges deployed back-to-back. When deployed as a rectangle the 64 horsemen unit consisted of the 64-horseman *ilê* (wedge), which consisted of 32 *kontaratoi* (i.e. *kontoforoi, kontos*-bearers), 8 *spathatoi* (*spathaforoi, spatha*-bearers), 8 *rhiptaristai* (javelin-throwers), and 16 *hippotoxotai* (horse-archers). The terms *spathatoi* and *rhiptaristai* are tenth century terms, which means that the anonymous author has updated the terminology, which we find the Romans using already during Arrian's time. During Arrian's time, the Roman cavalry consisted of the *kontoforoi* and *logchoforoi* (lance-bearers who either threw the spears or used them in melee together with the *spatha* sword), which means that the *logchoforoi* of Arrian consisted of the *rhiptaristai* and *spathatoi* of the *Apparatus Bellicus*. In short, the 64-horseman unit when deployed as a rectangle consisted of 32 *kontaratoi*, 16 *logchoforoi* and 16 *hippotoxotai*. However, the armament changed when the same 64-horseman unit was deployed as a rank-and-file wedge. When the 64 horsemen wedge was ordered to advance through the interval in the infantry phalanx, it divided itself into three small 21-horseman wedges (*embolos, trigônikos;* the commander was outside this structure) to enable the horsemen to pass through the interval more easily, and each of these smaller 21 horsemen wedges had different armament. The three 21-man wedges were formed up so that the *hippotoxotai* were posted on the left flank, the

Late Roman rhomboid and wedge

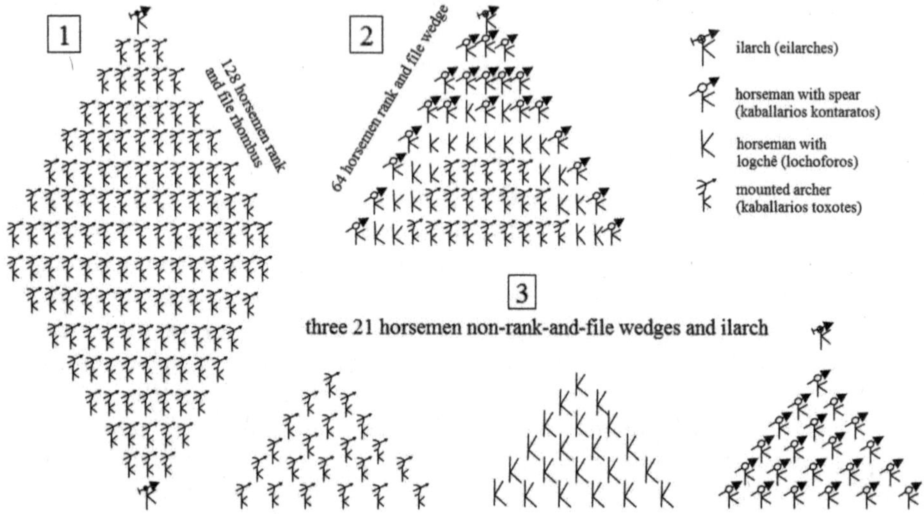

The rank-and-file 128 horsemen rhombus (*hippilarchia*) together with the 64 horsemen *eilê* arrayed as rank-and-file wedge and the three 21 horsemen non-rank-and-file wedges formed out of the 64 horsemen *eilê*.

Symbols borrowed from the tenth century "Byzantine" military treatises

1: Four *turmae* (128 horsemen) deployed as a rhombus. The mistake made by Asclepiodotus, Aelian and after them by most modern historians has been to believe that the military rhombus was an exact rhomboid in shape and would therefore have consisted of 113 horsemen. The practice of using units of 32, 64, 128, 256, 512 etc. units was the Macedonian practice at least from the reign of Alexander the Great onwards and if the rhombus consisted of two wedges then the obvious answer is that it meant 128 horsemen. The equivalent term for 128 horsemen rhombus in the Muslim military treatises was *kurdûs* (pl. *karadis*). It should also be noted that the basic Macedonian *ilê/eilê* was the 64-horseman wedge, two of which made up the rhombus. It is very likely that the regular Roman units employed the rank-and-file rhombus formation during this period because the Parthian/Persian and Armenian mounted archers (Aelian 45.1-2; Dain J2-3) and the Muslim cavalries also employed it (Syvänne, 2014, 2017b, *MHLR 8*) it, and it is also included in the *Apparatus Bellicus* – a text which is a collection of older treatises. The Romans appear to have adopted the rhombus (presumably both rank-and-file and non-rank-and-file) early on because we find it in Arrian (*Taktika* 16.1-2). Furthermore, the fact that Caligula (Suetonius, *Caligula* 55.2) placed Thracian gladiators in command of the German bodyguard cavalry (Thracians used the wedge), and Claudius (Suetonius, *Claudius* 21.3) had both the Praetorian cavalry and the superb Thessalian horsemen perform on the arena suggest that the Romans trained their elite horsemen in the use of both wedges and rhomboids. The rhombus was a highly maneuverable formation that allowed fast deployment on all directions of the compass.
2: The 64 horsemen *eilê* deployed as a rank-and-file wedge (*embolos, cuneus*). This was one half of the 128 horsemen rhombus. The inner structure of the wedge is an educated guess based on the types of equipment used by the three smaller non-rank-and-file wedges. Note that the 64 horsemen *eilê* could also be arrayed as non-rank-and-file rhombus.
3: The three 21 horsemen wedges and *eilarchês* formed out of the 64 horsemen rank-and-file wedge. The location of the *eilarchês/ilarch* shows just one of the alternatives. He could choose his place as required.

logchoforoi (*rhiptaristai/spathatoi*) in the middle and the *kontaratoi* on the right flank. The variation in the armament in the different unit orders demonstrate nicely how the troopers were trained as multipurpose forces able to change their equipment and tactics according to the needs of the moment.

The example given shows that the troopers could assume different combat formations on the spot or when moving forward so that the same unit could be deployed as a 128-horseman rectangle (e.g. 16 files, 8 ranks), a large rank-and-file rhombus of 128 horsemen consisting of two 64 horsemen rank-and-file wedges both of which could be further divided into three 21-horseman non-rank-and-file wedges and their commander.

The *Apparatus Bellicus* does not mention the 64-horseman non-rank-and-file rhombus (1, 2, 3, 4, 5, 6, 7, 8, 7, 6, 5, 4, 3, 2, 1), but on the basis of the versatility demonstrated one may assume that the 64 horseman rank-and-file wedge could also be deployed as such.

On the basis of the above evidence, one may assume on the basis of the paper strengths of the differing units that the standard deployment pattern for the rhomboids and wedges were the following: the 128-horseman rank-and-file rhombus consisted of two 64 horsemen rank-and-file wedges; and the 64-horseman rank-and-file wedge which in its turn could assume the 64-horseman non-rank-and-file rhombus formation, or it could divide itself into three 21 horsemen non-rank-and-file wedges and one commander. When the units were not up to their paper strengths, these could assume different sized variants of the rank-and-file and non-rank-and-file rhomboids and wedges.

4.7. The use of the unit formations and orders[29]

Each of the unit formations and orders were designed for specific purposes. The open order (*araiôsis*) was used during marching, approach to the battle site, and when there was no time to compact the formation. The close-order (*pyknôsis*) was the principal combat formation and used for the main charge at a canter. The *synaspismos* was used by stationary cavalry in defence. The irregular non-rank-and-file order *drouggos* was an all-purpose formation for use in situations that required mobility (pursuit, ambush, support). The division of the soldiers into *koursores* and *defensores* enabled the Romans to maintain reserves in close order when they could use the galloping *koursores* in the *drouggos* order for effective pursuit. The crescent order (*menoeides*), the snaking column, was used when outflanking the enemy with the main line. Before the reign of Maurice, the Roman cavalry also appears to have used the gallop when charging at close order. This was a particularly frightening sight to soldiers who were not used to seeing a fast approaching wall of horses and men. The further tightening of the cavalry array, which is also depicted in the text of Syrianus, made the approaching Roman formation even more threatening. The downside of the use of the gallop was that the formation lost its cohesion so that it became an irregular *drouggos*. However, for most of the period the standard Roman cavalry tactic appears to have been the well-coordinated close order cavalry charge at a canter and the existence of the other unit orders ensured that the Roman cavalry tactics were flexible enough in defence, offence and pursuit.

The following quotes from Arrian and Maurice give us a good understanding of the way how the ancient military thinkers saw the advantages and disadvantages of each unit formation and order in combat.

Arrian, *Technê Taktika* (16) free translation by author (resembles closely Aelian Matthew ed. 18–9)

16.1–2: The cavalry formations are various and diverse, some square, some oblong, some rhomboid, and some grouped together as a wedge. All these formations are good when used at the right time and one cannot choose to consider one of them superior to the others because in another situation against another enemy, one could find another formation more useful than one's favourite formation.

16.3–4: … And it [*rhombus*] is very suitable for all kinds of turning manoeuvres and is safe because it is the least likely to be attacked in the rear or flanks.

16.7–8: This formation [*wedge*] is considered useful because it places the leaders [*hêgemones*] as a circle and it can break through enemy formations with ease because its front ends in an apex [*Asclepiodotus 7.3 correctly notes that the rhombus had the same advantage*]; And it also facilitates the use of the quarter-turning/wheeling [*epistrofê*] and wheeling back to original position [*anastrofê*] because it is difficult to turn in the square formation [*this explains nicely why Maurice required the troopers to be trained in the use of the drouggos-order in Strategikon 4.5. The drouggos-order complemented the regular rectangular rank-and-file formations arranged in dekarchies and pentarchies.*].

16.10: … The square formation… is more easily organized than others because [*the troopers*] are drawn up in ranks and files so that it allows easy charging and retreating; and only in this formation the file-leaders [*hoi hêgemones, i.e. the leaders posted in the front rank*] attack the enemy together as a single mass. [*Arrian (16.11) considered the best versions of the square formation to be such that had double the front to the depth so that there were to be ten files to five ranks, or twenty files to ten ranks*]

16.13–4: … the horsemen deployed in depth do not bestow the same aid as deep infantry formations… because one horse cannot press another like footmen with their shoulders and sides; and nor do the horses become a single mass by adjoining themselves with those arrayed in front [*means the bringing up of the horses posted in the rear rank between the files of the front rank*]; On the contrary, if they press together and tighten [*the formation*], they will only panic the horses.

17.3–4: … The oblong formation [*heteromêkês*] has greater depth than front, or the front is greater than the depth; the latter is better in combat except when one seeks to get through the enemy's formation, because in that case the narrow deep formation would be very useful [*this tactic was employed for example by the cavalry of Henry/Henri IV against the shallower cavalry formations of the Catholics towards the end of the 16th century. The shallower formations of knights using the gallop led to the loss of cohesion so that those who wanted were able to flee leaving the deeper formations of Henri IV as victors*], or when one wants to conceal the numbers of horsemen to provoke the enemy into hasty action [*this was used as a stratagem to hide the size of the force so that the formation then suddenly widened to surprise the enemy*].

17.5: The depthless [*abathos*] single-faced formation is useful for surprise raids or when we want to destroy or trample something under hooves, but in battles it is disadvantageous [*This was the array used by the knights and by the Catholic cavalry in the above example.*].

Maurice, *Strategikon* 4.5.1–22, free translation by author
We consider it self-evident that the soldiers chosen for ambuscades, for attacks against the rear or flanks of the enemy line, for guarding our rear, flanks and baggage train, for quick support of a *meros* in distress, or for minor reconnaissance duty are better served by using the *drouggos*-order [*drouggisti*] than by using the large battle line arranged in dekarchies and pentarchies. This [*the large battle line*] is certainly impressive looking, is more powerful, better-ordered, and able to charge with greater security in battle, but it is slow and not easy to manoeuvre in emergencies. The *drouggos*-order has the opposite qualities. It can be hidden easily in ambush; it does not require much space; and it is easy to manoeuvre in emergencies… the difference between the two formations is that the first [*the large battle line*] is powerful and secure for major operations, while the latter [*drouggisti, drouggos*-order] is meant for quick support, pursuit, sudden surprise attacks and for causing disorder.

The composition of the units varied greatly during this period so that there were different types of formations in existence. The units which consisted solely of javelineers or lancers could attack the enemy in several different manners. In the case of the rectangular formations it was possible to send individual files from the flanks of each 32-horsemen *turma* forward to skirmish with the enemy so that they threw javelins at the enemy without engaging them in melee (Arrian, *Taktika* 32.3ff., e.g. *Petrinos*, *Kantabrikê epelasis* = Cantabrian Charge), or for the entire formation to charge straight at the enemy while throwing javelins at them, after which the formation engaged the enemy in melee. One variant of this appears to have been to concentrate the volleys of some of the units at specific points in the enemy line so that the Romans would be able to penetrate the enemy formation at this spot. The units which were equipped as *kontoforoi* (in the example of Arrian these were using the Gallic-*contus* that could be used with one or two hands and shields) were expected to charge straight at the enemy, and if the enemy unit in front of them fled they were to attack in flank the enemy unit standing next to the fleeing one while moving their shield above their head to their backs because there were now enemies behind them.[30] Procopius's description of the Herul cavalry (no armour and equipped only with shields, swords and sometimes with spears) shows that these always attempted to charge straight at the enemy using either swords or spears. The Herul cavalry appears to have used the gallop and a looser unit order than the regular Roman cavalry. Regardless of the unit formation (rectangular, wedge, rhombus, *drouggos*), the units that were equipped as mounted archers obviously either skirmished with the enemy or charged into contact if the situation required. The same was true of the units that were equipped in the same manner as described by Procopius, Syrianus and Maurice as dual-purpose forces. These could be used for a great variety of different tasks, regardless of the unit formation used (rectangular, wedge, rhombus, *drouggos*) because their equipment allowed this.

The European military thinkers of the sixteenth to nineteenth centuries hotly debated the relative advantages of close order vs. other unit orders, the importance of the depth of formation, trot/canter vs. gallop, and the importance of the size of the horse in cavalry combat.[31] It is thanks to their efforts that we are in a position to make some judgements about the relative importance of these in cavalry battles. These theorists noted that when the cavalry units faced each other frontally, it was typical for one of the opponents to flee before making contact (it was likelier that a unit in open order or in irregular order fled when opposed by a close order unit using the trot/canter or the gallop, which had retained the cohesion of the formation), or both forces stopped their advance just before making contact to begin the melee (likelier when both used the close order with trot/canter), or both intermingled with each other so that individual horsemen and their horses fought as individuals (the likelihood of this increased if the unit had lost its cohesion as a result of the use of gallop), or passed through the intervals between the files (this was not likely when the formation had assumed the close order and had more than three ranks). The chances of these happening depended upon the type of cavalry formation and the surprise factor. The best summary of the different variants actually come from the pen of the great 19th-century French military theorist Ardant du Pick. He provides us with the best synthesis of the facts as follows:

> ***Cavalry combat according to Ardant du Pick*, tr. by J.N. Greely and R.C. Cotton**
> [*author's comments inside square brackets in Italics*]
> p.213–4: Cohesion and unity give force to the charge. The trot permits that compactness which the gallop breaks up. … the effect is moral above all. A troop at the gallop sees a massed squadron coming towards it at a trot. It is surprised at first at such coolness. The material impulse of the gallop is superior; but there are no intervals, no gaps through which to penetrate the line in order to avoid the shock, the shock that overcomes men and horses. These men must be very resolute, as their close ranks do not permit them to escape by about facing. If they move at such a steady gait, it is because their resolution is also firm and they do not feel the need of running away, of diverting themselves by the unchecked speed of the unrestrained gallop, etc.
>
> p.214: Galloping men do not reason these things out, but they know them instinctively. They understand that they have before them a moral impulse superior to theirs. They become uneasy, hesitate. Their hands instinctively turn their horses aside. There is no longer freedom in the attack at a gallop. Some go to the end, but three-fourths have already tried to avoid the shock [*in the sixteenth century, according to de la Noue, typically only 25 out of 100 charged to the end*] There is complete disorder, demoralization, flight. Then begins the pursuit at a gallop by the men who attacked at the trot… [*in the* Strategikon *this pursuit was conducted by the* koursores *while the* defensores *maintained close order and followed by maintaining the trot; in Arrian's* Ektaxis kata Alanon *the*

pursuit was similarly performed by the koursores consisting of mounted archers while reserves followed.]

The charge at a trot exacts of leaders and men complete confidence and steadfastness. It is the experience of battle only that can give this temper to all [*i.e. the cool steadfast charge at a trot required veteran troops*]. But this charge, depending on a moral effect, will not always succeed. It is a question of surprise. Xenophon recommended, in his work on cavalry operations, the use of surprise, the use of the gallop when the trot is customary, and vice-versa. ... the less a thing is foreseen, the more pleasure or fright it cause...

As a general rule, the gallop is and should be necessary in the charge; it is the winning, intoxicating gait, for men and horses. [*Ardant du Pick's comment regarding the use of the gallop reflected the typical view of nineteenth-century European military theorists, but there were some exceptions, such as Jomini who favoured the use of the trot. However, as we have seen, even du Pick had to accept that the use of the trot had its advantages too.*]

p.218–9: There is this important element in the pursuit of cavalry by cavalry. The pursued cannot halt without delivering himself up to the pursuer. The pursuer can always see the pursued. If the latter halts and starts to face about the pursuer can fall upon him before he is faced... The pursuit is often (p.219) instigated by the fear that the enemy will turn. The material fact that once in flight all together cannot turn again without risking being surprised and overthrown, makes the flight continuous. Even the bravest flee, until sufficient distance between them and the enemy, or some other circumstances such as cover or supporting troops, permits of a rally and a return to the offensive. In this case the pursuit may turn into flight in its turn.

pp.219–20: It is characteristic of cavalry to advance further than infantry and consequently it exposes its flanks (p.220) more. It then needs more reserves to cover its flanks and rear than does infantry. It needs reserves to protect and support the pursuers [*the* defensores *of the* Strategikon *and the second half of the* lochoi *in Arrian, and the second line of both authors*] who are almost always pursued when they return. With cavalry even more than infantry victory belongs to the last reserves held intact. The one with the reserves is always the one who can take the offensive [*as we shall see, Maurice sought to preserve the cavalry reserves intact as long as possible*]. With room to manoeuvre cavalry rallies quickly. [*It was because of this that Maurice separated the two cavalry lines from each other by three to four bowshots (ca. 990–1,320m).*

It is quite clear that the Romans came to the same conclusions as the European military thinkers of the sixteenth to nineteenth centuries and Ardant du Pick. Depending on the time and place, the Romans used the trot/canter or gallop and the close order to break up the will of the enemy to come into contact with their cavalry formation. It is obvious that there were differences in equipment and depth of the cavalry formations between the Roman and modern era, but the cavalry still operated similarly because the results of such actions were governed by the human psyche and by the animal instincts of their mounts. When two cavalry units in close order faced each other head on, there were two possibilities: one side fled before coming into grips with the enemy; or both sides stopped in front of each other to avoid the deadly collision. The horses and men collided only accidentally with each other, to their own demise.

The deep cavalry formations of the Late Roman period usually ensured that it was practically impossible for two cavalry formations to pass through each other (files entering the intervals between the enemy files) like happened sometimes during the eighteenth and nineteenth centuries, when the cavalry were deployed in shallower two- to four-rank formations. This means that if the one or both formations had lost their cohesion so that there were intervals between the files, it was far likelier that the units just became intermingled so that the horsemen fought at close range with spears, swords or axes. Obviously the likelihood of the units charging through each other increased if the Romans were deployed in shallower formations, such as those described by Arrian which were accepted for some occasions, or when the Roman cavalry was deployed in shallower order or in irregular order. For example, the *Strategikon* expected that the fill-up *banda* (see later) could be deployed only two or four deep while it is also clear that the galloping formations always lost their cohesions eventually which increased the chances of units galloping through each other or the intermingling of units.

The evidence suggests that, when the two close order formations halted in front of each other, only some of the men and horses were killed or wounded in the initial contact by the spear thrusts or throws, and it was only then that the front ranks advanced closer to engage each other in hand-to-hand combat until one side gave up. The sixteenth-century military theorists noted that when two deep cavalry formations fought, the cutting edge of these formations consisted of the two leading ranks while the ranks behind those contributed very little to the outcome – they just fled if the two front ranks were unable to contain the enemy attack. One can assume that this was also true of the Late Roman battles, and it was because of this that both Syrianus and Maurice placed the best fighters in the two front ranks. In other words, when the two front ranks of one side penetrated into the enemy formation up to the depth of two ranks, it was probable that the defending side fled. Nevertheless, the key deciding factor during the approach, charge and melee stages was always the respective morale of the soldiers on each side. The unit formations, unit orders, pace of the attack (trot/canter vs. gallop) and cohesion of the unit only influenced these but did not decide the outcome.

The sixteenth- to eighteenth-century European military theorists debated hotly about the importance of tall, heavy horses with tall riders in melee so that these were divided into two different schools of thought: 1) those who favoured heavier horses; and 2) those who favoured the stamina and agility of the horse. It was recognized that the taller heavier horses with tall men had a psychological impact on the enemy, so that it was likelier that they fled before coming to grips, and it was also recognized that when two formations then engaged in melee, the heavier horses with heavier riders had the edge. The taller, heavier horses could overthrow the smaller horses while the taller men from the taller horses could hit enemies from above with greater effectiveness. Most of the cavalry officers of the early modern period appear to have belonged to this school of thought. However, others opposed this because the heavier horses were more expensive, required more fodder, lacked stamina for long campaigns and battles, and were less agile and fast than the smaller mounts. The usual compromise solution that the Europeans adopted was that the heavy cavalry used the tall heavy horses while the light cavalry used the smaller, faster, agile horses.[32]

We can detect the same two schools of thought also in the Roman Empire. The size of the horse and rider had been recognized by the second- and third-century Romans because the giant Maximinus Thrax was recruited into the cavalry and then into the bodyguards of the emperor Septimius Severus. The horse which carried Maximinus must have been gigantic. The preference appears to have changed by about mid-third century because we find Probus and his men preferring stamina over size. The soldiers gave Probus a horse captured from the Alans which could travel 100 miles per day for eight to ten days in succession. The cavalry battle between the Roman and Palmyran cavalry near Immae in 271 also shows the Romans tiring the heavy Palmyran cataphracts with a long feigned flight. However, this appears not have found favour in every corner of the Roman Empire because while we find Constantine the Great favouring lightly-equipped cavalry with the implication that he favoured the smaller, faster, agile horses with greater stamina, his enemies favoured the heavier, cataphract type of cavalry which required heavier, taller horses. But then we find once again Constantine's son Constantius II favouring the super-heavy *clibanarii* cavalry with the implication that they sought heavier horses for these. The heavy armour worn by both the horses and men required taller heavier horses, just as during the late-medieval and early-modern eras. The horses for the heavy cavalry were obviously obtained from the imperial stud farms. Therefore, under Constantius II, the Romans appear to have resorted to the same system as the early modern era Europeans so that they used large horses for heavy cavalry, and medium- to small-sized horses for the rest of the cavalry. The same system appears to have continued until the sixth century, even if it is clear that the arrival of the Huns increased the importance of light cavalry. Syrianus and Maurice do not specifically refer to the size of the mounts, but both mention that the front rank horses were to be armoured, which means that it is probable that these consisted of the heavier breeds able to push over smaller horses in melee. Regardless, it is clear that neither of these considered the size of the mount to be of great importance in cavalry combat. Both considered the equipment worn, the pace (canter vs. gallop) and the tightness of the formation to be of greater importance than the size of the mount. In other words, it is clear that the size of the horse was not considered to be of great importance in cavalry combat. This is in fact the view adopted by most of the late eighteenth- to nineteenth-century European military thinkers and the reason why we do not find it in the above quote from Ardant du Picq. However, even if the size and speed of the horse did not cause any particular comment among military authors, it is still clear that the sixth-century Romans recognized the importance of speed and stamina for the light cavalry, because, for example, Evagrius commented about the extraordinary speed of the Arabic horses. The Roman cavalry was usually unable to catch the Arab raiders unless they had their own Arab *foederati* present. This last-mentioned fact had military importance in the wars of the late sixth century and early seventh, but fortunately for the Romans they almost always had some Arabic sheiks in their service.[33] The *koursores* and *defensores* system combined the advantages of both light and heavy cavalry in the same *meros*, so the Romans were able to exploit both to the fullest extent in their wars during most of the Late Roman period.

The following quote from the pen of Louis Edward Nolan shows the influence of the type of cavalry on the outcome of battle quite well:

> In a march of ten German miles, supposing it to be commenced with equal numbers, the hussars [*light cavalry equipped with smaller, swifter horses and armed with a sword*] would certainly have the advantage over cuirassiers [*heavy cavalry mounted on a large heavy horse with the men being equipped with a breastplate and sword*]. In the open country they would very much harass and dishearten heavy cavalry by continual skirmishing and hanging upon their flanks and rear; and the effect would be very much increased if the heavy horse should be provoked to charge, even though they should be so fortunate as not be in disorder after charging.
>
> But in such a length of march there must be at last some defile, or other obstacle, which would oblige this heavy cavalry, already much fatigued, to break off; and this is the moment for the light troops to act with the greatest vigour, and by continual pressing upon them in such situations (in which they can neither prevent being attacked, nor take their revenge), they will at length lose confidence; and the instant they either charge or disperse, they are generally certain of being vanquished. [*This is what happened to the Palmyrene cataphracts at the battle of Immae in 271, and what happened to the East Roman cavalry at the Battle of Manzikert in 1071. On the last occasion the Romans had abandoned the balanced* koursores/defensores *system present in the Maurikian cavalry army. This must have been one of the reasons that contributed to the inability of the Roman cavalry line to fight effectively against the Turks. A fuller analysis of the battle of Manzikert will be provided in the* Military History of Byzantine Rome, volume 5.]
>
> To remedy this disadvantage, the King of Prussia directed all his cuirassiers to be practised in the hussar exercises, which was, certainly, so far useful, but their horses are not proper for such light and active service. [*The Romans achieved this with their* defensores/koursores *system which combined the heavy and light cavalry in the same division. During the reign of Maurice, the Romans appear not to have had trouble in assigning the right kinds of horses to the role of the* koursores *even if the sources do mention the fact that the pursuit was not as effective as it could have been at times when some of the Arabic federates with their superb horses were quarrelling with Maurice.*] …
>
> The Prussian hussars are equally capable of every nature of service. In regular battles they have rendered the service of cuirassiers; they never hesitated to attack in close squadron whatever they have met with. [*The Romans obviously required all of their regular forces to be able to fight in close order so the Romans followed this practice too, so that they always possessed both light and heavy elements within their cavalry formation, even in situations in which the horses and men were not fully armoured, as was the case with the hussars of Prussia.*]…
>
> Nolan quoting Warnery (mainly pp.93–4, but some quotes have been taken from other pages) in his *Cavalry*, the 2nd London edition/1st American ed. pp.43–4; 3rd London ed. pp.75–6 with my comments in italics inside square brackets.

We also should not forget the importance of mounted archery during the Late Roman period, because it had an impact on the outcome. Effective mounted archery could cause a significant amount of disorder and casualties before the units made contact, so that the side which had suffered more fled. The effectiveness of the mounted archery depended on the type of formation and the type of enemy the Romans faced. If they faced more lightly armoured enemies in open order, it was very likely that the Romans in close order would defeat them even before coming to close range because the enemy just fled. If the Romans faced an enemy that was culturally predisposed to using mounted archery in compact formations, such as the Persians, the situation stayed the same because the Romans were more predisposed to charge into contact with the enemy. The Romans also had an advantage over those foes that employed shallower or irregular formations, because both gave individuals better chances of avoiding contact with the Roman formation. The only mounted enemies that the Romans had trouble with were the Germanic peoples, because these employed close-order cavalry predisposed to charge into contact with the enemy. Fortunately for the Romans, the Germanic peoples were often poorly disciplined and their use of the impetuous attack with gallop broke the cohesion of their formation. Both of these tendencies worked to the Roman advantage if the Romans managed to maintain the cohesion of their own formation. At close quarters the heavier armour worn by the Roman troopers could also give them advantage over their less-well armoured Germanic foes. Regardless, the wild impetuous charge by the Germanic peoples was always dangerous for the Romans in situations in which the Roman force consisted of inexperienced soldiers. It was because of this that the Romans were in trouble when they faced the Ostrogoths at the turn of the sixth century, and later when they faced the Lombards. On the other hand, the wild impetuous charge at the gallop was unlikely to succeed against a steady veteran force of cavalry, which the Romans possessed for most of the time even when they lacked experienced steady infantry forces like they lacked for most of the time during the period after the year 468 until early 530s.[34]

Below: The cavalry close-order (*pyknosis/puknosis*) from the side. The example shows an elite unit with five ranks. In this version the spear-armed soldiers carry two *kontaria*-spears, bow, *skouton*-shield and *spathion*-sword while the rear ranker is equipped with the full panoply. As can be seen when this was the case, the lance pennons could have interfered with archery, hence the demand for their removal.

Chapter Five

Cavalry Battle Formations[1]

5.1. Cavalry vs. Cavalry: Battle Formations

We find in the sixth century *Strategikon* four different variations for the same cavalry battle array, depending upon the size of the army. The unifying elements in these were the use of two battle lines, rear guards, the *koursores-defensores* system, and other additional separate detachments. The key structural difference between these variants lay in the amount of divisions in the second support/reserve line. Excluding the largest variant, we have evidence for the use of all of these variants in the narrative and other sources that describe Roman cavalry formations before the fourth century.[2] However, as already discussed, the *koursores-defensores* systems used in these formations varied from one era and place to another, the key variants in this case being the Alan, African, Scythian and Illyrikian Drill systems.

The largest cavalry battle formation was meant for armies that had more than 10,000–15,000 men. This version had four divisions in the second battle line. If there were more than could be incorporated into the *meros* structure (i.e. more than ca. 42–47,000 horsemen), depending on the situation the extra horsemen were to be posted either on the flanks to outflank the enemy and/or behind the second line as a third line and/or ambushers. The second major variant, the well-proportioned force or medium-sized cavalry army, consisted of variable numbers of horsemen (Maurice gives the variants: 5,000/6,000–10,000/12,000/15,000 men). This version had two reserve divisions (*mere*). The third and smallest version also called by Maurice a well-proportioned army, had 2,000–5,000/6,000 or fewer men and had only one *meros* in its second line. The flexible numbers suggest that the commanders were expected to demonstrate personal initiative when the numbers reached certain levels. These types of cavalry forces were the ideal, but we know both from Maurice's own complaints and from the narrative sources that until his own day the Romans had also used a single line formation and improvised formations.

5.2. The Cavalry Battle Formations for Armies in Excess of 10,000–15,000 Horsemen[3]

Large cavalry army (ca. 10,000–50,000)

The battle array for armies that had more than 10,000–15,000 horsemen but less than about 50,000 horsemen consisted of two or three lines if one counts the rear guard as such. Note that this battle array with four reserve divisions (*mere*) could be used even by armies as small as 10,000 horsemen, while the second variant of the

'Italian Drill' cavalry formation with the two reserve divisions could be adopted by cavalry armies as large as 15,000 horsemen. It was clearly up to the commander to decide which of these variants was appropriate in each situation. See the attached diagrams. By the late-ninth century the version with four reserve divisions could be used by armies as small as 4,000. Whether that was true already in the seventh century is not known, but certainly not impossible.

The first cavalry line (*promachos taxis*) consisted of three divisions (one *mere* each with a maximum of 6,000–7,000 men). Each *meros* was divided into three *moirai* (max. 3,000 men[4]). Each of the *moirai* consisted of the 200–400 men *tagmata*, *arithmoi* or *banda*. The *tagmata*, *arithmoi* and *banda* consisted of approximately two to four *hekatontarchiai* (sing. *hekatontarchia*, a century/*centuria* of ca. 100 men), each of which was commanded by a *hekatontarchês/hekatontarchos* (*centurio*). The middle *meros* was led by the *hypostrategos* (second-in-command) and the flanking *mere* by the *merarchai* (divisional commanders, sing. *merarchês*). The flanks of each of the *mere* consisted of the *koursores* and each of these was commanded by a *moirachês* (pl. *moirarchai*). All of the officers were posted in the centre of their respective units.

The narrative sources prove that the cavalry battle formation had three superior commanders for the left, centre and right. In the case of the large cavalry army it is probable that the overall commander of the left flank was usually the *merarchês* of the support-line *meros* posted on the left because he was in the position to direct the reserves, while the overall commander of the right flank was the *merarchês* of the support-line *meros* posted on the right. In the case of smaller cavalry formations with only two or one reserve *meros*, the wing commanders were the *merarchai* in charge of the front line flank *mere*. The commander of the centre would usually have been the *strategos*, who acted not only as overall commander but also commanded the two reserve *mere* in the middle, or just the one *meros* if the cavalry army was smaller than ca. 5,000 to 6,000 horsemen (see below).

According to the *De militari scientia* (*DRM* 4), the middle *meros* was to possess 1,000 extra troopers, while the illustration in the *Strategikon* placed the *foideratoi/foederati* under the *hypostrategos* (usually the *comes foederatorum* or one of the *comites*) in the middle.[5] In other words, the idea was to strengthen the centre of the first line either with better fighters and/or with more men. The posting of the 1,000 extra horsemen to the centre *moira* was also consistent with Maurice's instruction that the maximum size for the *moirai* was to be 3,000 men and the maximum size for the *meros* 7,000 so that the middle *moira* had 3,000 men and the flanking ones 2,000. The *DRM* (4) also pointed that the better *tagmata* were to be posted in the centre and flanks of each *meros*. This followed the so-called Homeric principle.

The *tagmata*, *banda* and *arithmoi*, each commanded by an *archôn* (commander with the title of *tribunus* or *comes*), were the smallest units of tactical importance. The second-in-command to an *archon* was called *eilarches* (*hekatontarches/centurio* with a higher rank than the regular *hekatontarches/centurio*), while the officer just below *eilarch* held the title of *hekatontarchês* (*centurio*). As noted already, the depth of the divisions (five to ten ranks) reflected their status and ability, the better the unit the shallower the formation. The *mere* of the cavalry battle formation adopted the close order and canter for combat while the *koursores* used the gallop and *drouggos*-

STRATEGIKON 3.8

Battle formation for a large cavalry army (ca.10-50,000 horsemen)

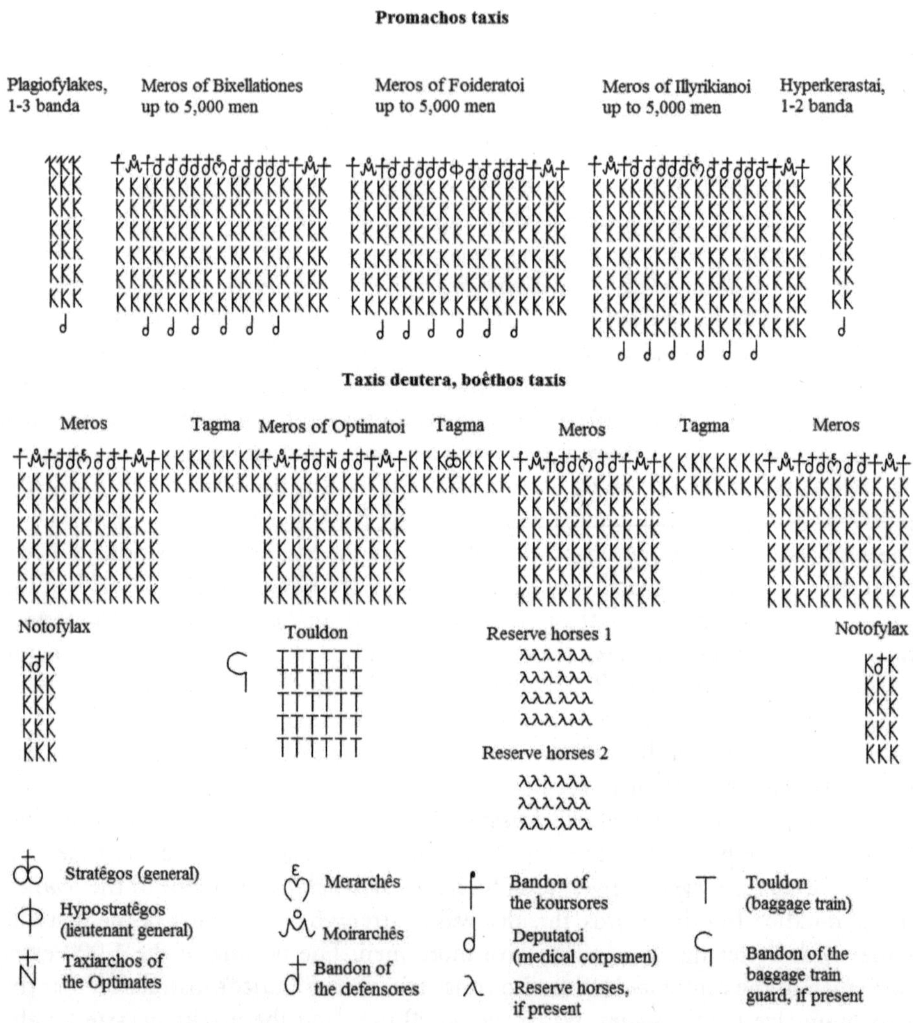

order in pursuit and skirmishing. When speed was needed the *drouggos*-order and gallop was also used by the reserves. The collection of similar units (e.g. *Bixellationes/Vexillationes, Foideratoi, Illyrikianoi, Optimatoi/Optimates*) into the same divisions facilitated the cooperation of the units belonging to those. It should be noted that the example given by the *Strategikon* shows only one of the possible ways how the different types of units were organized in the cavalry battle formation. For example, the *foederati* formed up the left *meros*, the *bucellarii* of Belisarius the centre *meros*, and the *stratiotai* (regular Roman cavalry) the right *meros* of the Roman first line in the battle of Tricamarum in 533. For the battle, see Syvänne (*MHLR 6*, 98–101).

However, there is still one common element to both battle formations, which is that the elite cavalry was posted in the centre *meros*.

The primary function of the flank guards (*plagiofylakes*) on the left flank was to protect that flank of the formation, but when the opportunity presented itself, the flank guards could also be used to outflank the enemy formation. The outflankers (*hyperkerastai*) were posted on the right flank to protect that flank and to outflank the enemy formation. Each *tagma* was also required to place eight to ten mounted medical corpsmen ca. 30m behind the first line to rescue the wounded, unhorsed and the ones out of action. This improved morale and enabled the unhorsed to obtain remounts so they could return to the battle.[6]

The second support line (*deutera taxis, boêthos*) consisted of four *mere*. Each of their intervals had one or more fill-up *banda* that eased the keeping of the correct distances between the *mere*. The fill-up *banda* were deployed two to four ranks deep. The *mere* of the support line were similarly divided into smaller units, the *moirai* and *tagmata*, and into the *moirai* of the *defensores* and *koursores*. On the basis of the diagram of the large cavalry army in the *Strategikon*, the proportion of the *defensores* in the first line (33 out of 51 *banda*) was greater in the first line than it was in the second line (20 out of 44), which reflected the greater importance of the first charge.

The large cavalry army had a small third line (*notofylakes* = rear guards) which consisted of one to two *banda* per flank, which were formed out of the flanking *mere* of the second line and placed just behind them. Each of these units was called a rear guard (*notofylax*). Their principal function was to protect the rear, which they did together with the fill-up *banda* of the second line when this was necessary. In addition to this, the array could have three to four *banda* per flank designated for ambushing duty.[7]

According to the *Strategikon*, the cavalry *mere* of the first line were to be close enough to one another, but still clearly separated. On the basis of the tenth-century *Sylloge Tacticorum*, which states that the divisions of the cavalry were to be separated by 13 *orgyia* from each other, I would suggest that the *Strategikon* envisaged similar distances, because the distances for the first line had to be smaller than the envisaged distances between the second line *mere* (one bowshot, i.e. ca. 330m intervals). The 13 *orgyia* equals ca. 24m, so I would assume that in practice the intervals varied between ca. 20–30m.[8] The obvious purpose was to present as continuous a line as possible to the enemy so that the enemy would be frightened by the approaching wall of cavalry. This is actually obvious also on the basis of the training scheme which called for the whole *mere* to train in closing their formation towards the standard. The end result was clearly a continuous line without large intervals. This, however, does not mean that the units would not have been separated from each other. On the basis of the diagram of the *meros* (see above), it is clear that even if the entire *meros* was compacted towards the middle *moira* for the charge in close order, that the *koursores* and *defensores* retained small intervals between them, and one may also assume that the same was true for the *banda*, meaning that there would have been a small interval between each *banda* to enable them to operate independently of each other. This is actually clear on the basis of the fact that the separate *banda* were given separate missions, for example when deployed in the second line as reserves. From the point of view of combat, the small intervals of a metre or two between each *bandon* still preserved the look of a continuous front which the enemy could not penetrate easily while allowing the units to act on their own. It is also clear that the separation of the *koursores* from the *defensores* helped the latter to conduct their charge unhindered if the enemy threatened the flanks of the *meros*.[9]

This means that the intervals between the cavalry units of the first line were so small that any attack through rough ground was impracticable. It is therefore unsurprising that Roman combat doctrine called for the use of open unobstructed ground in cavalry battles which allowed them to use their cavalry in orderly manner. The sources suggest that the Romans were usually able to achieve this and were able to use their cavalry to best effect. This could obviously be achieved only if the Romans possessed accurate information about the enemy and terrain they faced. Therefore it is unsurprising to find that the Roman combat doctrine expected each *tagma, meros* and unit of ambushers to post its own spies and scouts to reconnoitre and observe the enemy.[10] In addition, the Romans always left some soldiers together with squires to guard the camp and the baggage if these did not accompany the army, so that the enemy could not surprise them. When these instructions were followed to the letter it was not easy for the enemy to surprise the Romans.

The second support line (*taxis deutera, boēthos*) was initially posted about a mile (1,480m) or more behind the first line or immediately behind it so as not to advertise its presence. The idea was to induce the enemy to send its reserves to the flanks to outflank the Romans. It was only when the enemy was already so near that it could not change its tactics that the second line was to assume a position about three to four bowshots (ca. 990–1,320m) behind the first. Maurice considered the existence of the second line necessary, as its mere presence could be enough to prevent the men of the first line from deserting their posts since they knew that they fought under the gaze of their fellow soldiers. According to the text portion of the *Strategikon*, the second line was to consist of about a third of the whole force so that the four *mere* were about a bowshot apart (ca. 330m) from each other, but the accompanying diagrams show that in practice this varied greatly. It is also clear that the intervals of the second line were not always the same because the width of the interval had to be adjusted to the size of the army. The interval directly behind each of the first line *mere* had to be wide enough to provide a place of refuge. If any of the *mere* of the first line retreated, the fill-up *bandon* or *banda* in the interval directly behind it retreated to the level of the rear guards to provide the fleeing unit a place of refuge and a rallying point. The width of the second line was obviously slightly wider than the first so that each *meros* in the front line had an interval in the second directly behind. The extra width of the second line gave its flanks also improved chances to prevent flank attacks against the first line. The units of the second line were also double-fronted just in case the enemy attacked from the rear. It was expected that the third line, the rear guards, would be enough to protect the rear, but if this was not enough, then the commander dispatched the fill-up *banda* to their support, and if this was still not enough then the second line was expected to face the rear.

As the name support/reserve line (*boêthos taxis*) implies, the principal purpose of this line was to support and assist the first line. The commander was expected to keep the four *mere* structure as intact as possible for as long as possible so that before he committed any complete *meros* from the second line to combat he was first expected to employ the following measures: 1) to dispatch the flanking *tagmata/banda* of the second line to support the flanks; 2) to dispatch the fill-up *tagmata* from the intervals against enemies in the rear; 3) to use the ambushers as reserves for the flanks; and 4) to use the rear guards for reserve duties. This also means that the *strategos* was not expected to use the second line for lengthening the first line with the purpose of outflanking the enemy. The second line was to serve as the bedrock of the first line. The hiding of the presence of the second line from the enemy was actually meant to lure the enemy into an attempted outflanking of the Roman cavalry formation, sending its reserves (when it had those) to the flanks to outflank the Romans. In short, the aim was to entice the enemy into the adoption of a single combat line so that its outflanking sections could be targeted by the flank *banda* and *mere* of the Roman second line. Even if it is clear that the Germans in Roman service, such as those employed by Theodoric I the Great, used the single line cavalry formation and we find the Romans also using the single line formation in the Battle of the Nymphius River in 582, I would suggest that the real criticism of Maurice towards the use of the single line in cavalry combat was directed against its use in such a manner that the general dispatched the reserves to the flanks to outflank the enemy meaning that the formation which initially had reserves became a line without reserves.[11] The reason for this suspicion is that we know that the Romans had used cavalry reserves ever since Republican times, while Maurice instructed the generals to hide the presence of the second line so that the enemy could be induced into the adoption of a single combat line.

One can perhaps criticize the restrictions concerning the use of the second line because it prevented the general from using the second line for outflanking enemy, but the intention was clearly to avoid being ambushed by the enemy, and on balance this was clearly the safest course to take. Maurice, and probably also the military manuals that he used, expected that the wing units of the first line together with the ambushers were sufficient for outflanking purposes, so the second line could be devoted solely to the task of protecting the first line. Furthermore, it is clear that the second line was still used for the outflanking of the enemy wings, but only in situations in which the enemy had first outflanked the Roman first line and the wing units of the Roman second line had been sent against them: once the outflanking enemy forces were forced into flight, the wing units of the second line pursued and outflanked the enemy in its turn. The cavalry battle that took place close to Nisibis in 422 shows this nicely.[12]

The third line, the rear guards, was posted a bowshot (ca. 330m) behind the second line. It consisted of a *bandon* or two taken from the flanks of the second line. Its principal purpose was to protect the rear and flanks, but it is still clear that it could also be used as an emergency reserve. If needed, the rear guards were to be supported by the fill-up *banda* of the second line and in real emergencies by the units of the second line.

If the situation and terrain allowed, combat doctrine also expected that the commander posted three to four *banda* as ambushers (*enedrai, enedroi*) on each side.

The ambushers were not only expected to ambush the enemy but also to prevent the enemy from ambushing them on either side. This means that it was expected that the *strategos* could use as his emergency reserves, not only the second line with its fill-up *banda*, but also the ambushers and rear guards.

The numbers in the Strategikon, *Diagram 3.8.*

When one counts the symbols in large battle formation in the *Strategikon* which was meant as an example for cavalry armies in excess of 10,000 horsemen, one arrives at the following figures. The first figure inside square brackets represents 200 men per *bandon*, the second figure 310 men per *bandon* (the figure in the example of a generic *bandon* in the *Strategikon*), and the third figure 400 men per *bandon*:

> **First line:** *plagiofylakes* 1–3 *banda* (200–600; 620–930; 400–1,200); *meros* of the *vexillationes* 17 *banda* (3,400, 5270, 6,800); *meros* of the *foideratoi* 17 *banda* (3,400, 5270, 6,800); *meros* of the *Illyrikianoi* 17 *banda* (3,400, 5270, 6,800); *hyperkerastai* 1–2 *banda* (200–400; 310–620; 400–800). The total number of soldiers: 10,600–22,400.

The large cavalry array with all detachments included

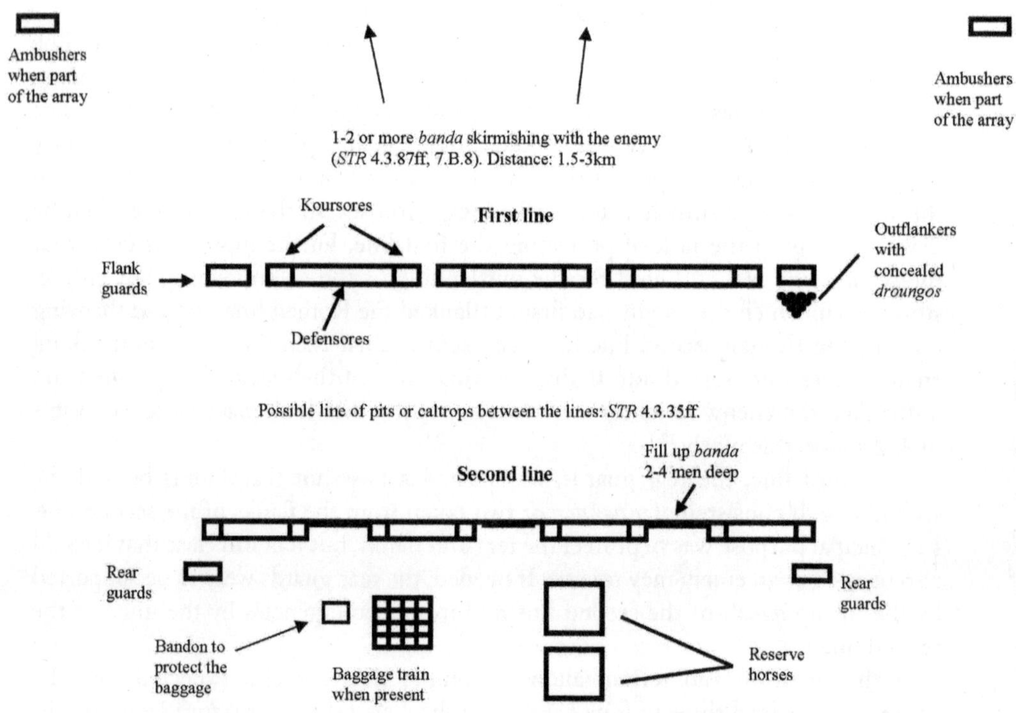

Second line: four *mere* each 11 *banda* (2,200; 3,410; 4,400), fill-up *banda* in the intervals one or more *banda* (here counted as one *bandon* per interval for the totals: 600, 930, 1,200). The total number of soldiers: 9,400–18,800.
Third line/rear guards: either a *bandon* or two from each flank for a total of two to four *banda* (400–800; 620–1,240; 800–1,600), or three *banda* per flank (as in the diagram) for a total of six *banda* (1,200; 1860; 2,400). The total number of soldiers: 400–2,400.

Baggage guard: 1 *bandon* (200, 310, 400). The total number of soldiers: 200–400.

The total number of soldiers in the diagram with the above assumptions is ca. 20,600–44,000 men. It should be noted that the figures could be even greater if there were more fill-up *banda* in the intervals of the second line and if the array had three to four *banda* per side as ambushers. With the assumption that there were two *banda* per interval in the second line it had six *banda* for a total of 1,200–2,400 men in the intervals, while the ambushers consisted of a total of six to eight *banda* for a total of 1,200–3,200 men. With these additions, the army in the diagram had 23,000–48,600 men. However, the diagram also contains another way of calculating the numbers which is to base it on the maximum number given by the diagram for the three first-line divisions, which is 5,000 men. If the three first line *mere* had each 5,000 men, then the total for the frontline is 15,000 horsemen plus the flanking units – on the basis of this one wonders if the cavalry army sizes in the *Strategikon* actually refer to the strength of the first line, because this conforms with the army sizes given in the text! By dividing the 5,000 by 17 *banda* one has ca. 294 horsemen per *bandon*, so that the four *mere* of the second line would have ca. 12,936 horsemen. This would give the seven *mere* the strength of 27,936 horsemen, in addition to which came the wing units, fill-up *banda*, rear guards and possible ambushers. With the average of 310 men per *bandon* given by the *Strategikon*, the array has a minimum of 31,930 men with 103 *banda*, and a maximum of 35,030 men with 113 *banda*. It is therefore clear that the diagram represents a cavalry army of approximately 30,000–35,000 horsemen.

Regular Cavalry Formation with Caltrops or Pits[13]

Maurice included in the *Strategikon* a variant of the regular cavalry array in which horse-breakers (pits) or caltrops were placed between the battle lines. In this case, the first line was placed a mile (c.1,480m) in front of the obstacles and the second, two to three bowshots (ca. 660–990m) behind the obstacles, or alternatively, the whole battle array was deployed behind the trenches. The trenches were to have four to five clearly marked passageways which the Romans could use for getting through the obstacles. The tactic as such was ages old, because it can be found in Polyaenus, but it is probable that the stratagem found its way to the *Strategikon* thanks to two instances. The Hephthalites defeated the Persians with this stratagem in 484 and the Persians defeated the Romans with the same stratagem in 528. It is not known if the Romans managed to overcome their enemies with this same system, because the sources do not mention such an instance – but this is not conclusive as the sources are not detailed enough for most of our period. For example, it is possible that the defeat of the Persians between the Roman cavalry lines in the battle of Constantia/Monocarton in 581 involved the use of trenches and pits, but the extant sources simply fail to mention this.

Foreign contingents[14]
If foreign contingents were present, they were drawn up according to their own customs, so that they did not disrupt the main battle formation. See for example, Syvänne (*MHLR 7,* 271–81). If included in the array, they were considered most useful as *koursores* or as ambushers (*enedra, enedrai ; enedros, enedroi*). This suggests that the foreign contingents most likely to join the Romans were drawn from such nations that could provide horse archers.

The super-large cavalry army of over ca. 50,000 horsemen[15]
If the army possessed extra men that could not be included in the *meros*-structure (max. 6,000–7,000 per *meros*), these were to be deployed outside the formation either

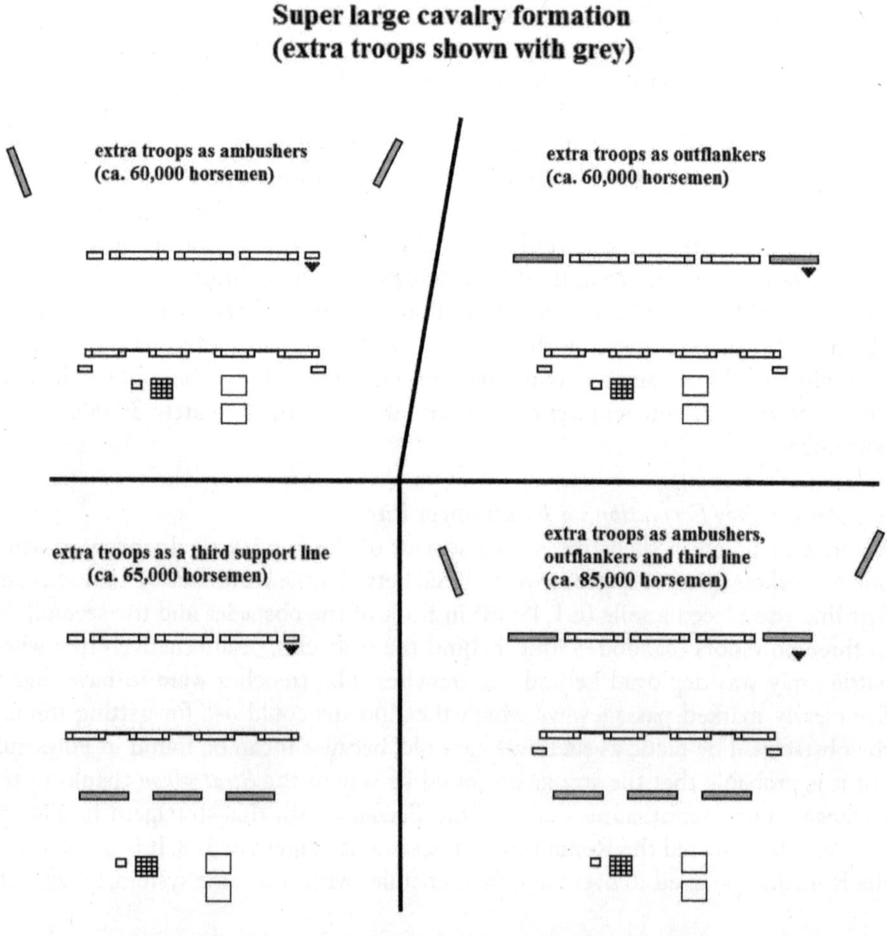

to support the second line, or to guard the flanks and rear of the *meros*, or to ambush the enemy, or to outflank the enemy. The following illustrations show some of the ways how the extra men could be posted in the cavalry array. I have here made the educated guess that when the extra men were posted to the flanks of the first line to outflank the enemy, the army no longer had separate flank guards and outflankers to protect them. This, however, is a mere guess. It is possible that the army still had these for their protection and I have in fact reconstructed the extra-large array in this manner in the *Military History of Late Rome, Volume 7* (page 214).

The instances in which the Late Roman army had more than 50,000 horsemen appear to have been scarce. The only certain instances are the years 577 and 593 (Syvänne, *MHLR* 7, 113–20, 271–80), but it is possible or even probable that there were more than 50,000 horsemen in some of the other major armies of the latter-half of the sixth century (see Syvänne, *MHLR* 7, 148ff.), and it is similarly probable that Heraclius had more horsemen during the years 622–9 (see Syvänne, *MHLR 8*, 150ff.), but the evidence for that is less certain because we do not possess a detailed rundown of the different components of these armies.

It is of note that, on the basis of the description of the joint campaign by the Romans and loyalist Persians of Chosroes II against the Persian usurper Bahram Chobin, the Romans used a variation of this system when they posted the extra troops to the right flank. The narrative sources of the battle show that the Persians were deployed as an extra-large right wing in the first line so that the right wing Romans of the second line were behind them. The only thing which is not known with certainty is the way in which the Romans deployed their reserves behind the

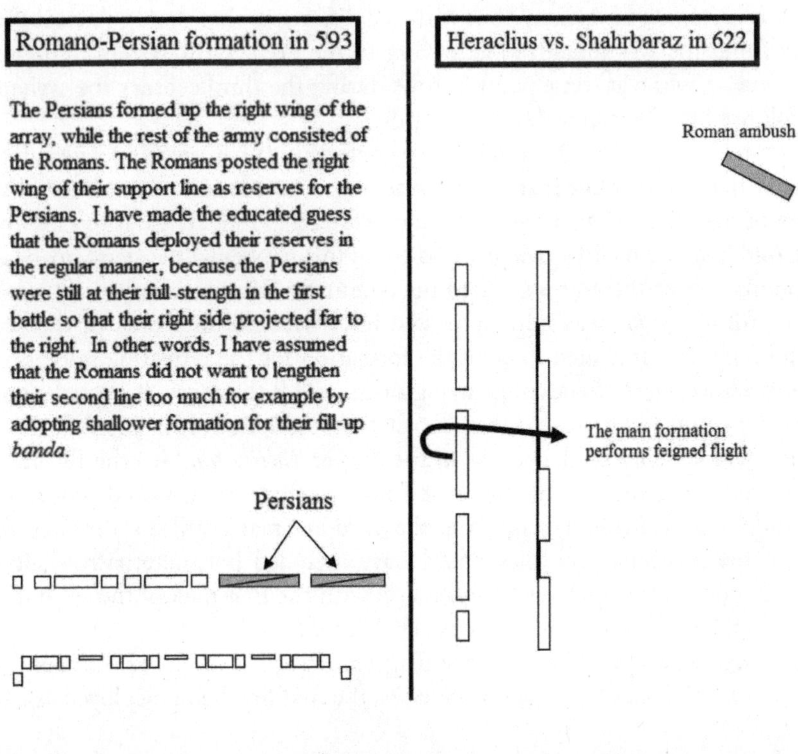

Persians. In other words, did they place the reserve in the regular place so that the extra Persians projected outside the right wing or did they spread the second line towards the right so that it covered it better? I have actually included both variants in the *Military History of Late Rome, Volume 7* (276–80), the reason being that in the first battle the Persians were at their full strength and in the latter battle they had suffered casualties, so it was possible that the Roman second line spread out further towards the right without endangering the *meros* structure. However, in the accompanying diagrams I show only the former alternative. The narratives of George of Pisidia and Theophanes suggest that in 622 Heraclius probably employed his extra troops as ambushers, but in that instance these were behind his main battle formation that performed a feigned flight in its entirety. See the attached images (593, 622). The extra troops are shown with grey colour.

5.3. The Cavalry Formation for Armies of 5,000–15,000 Men[16]

The medium sized cavalry army in the *Strategikon* consisted of more than 5,000/6,000 but less than 10,000/12,000/15,000 horsemen. The flexibility of the numbers at the both ends shows that the general was expected to decide which of the alternative versions of the cavalry formation he would use. The moderately sized army had only two *mere* in the second support line. These had a clear space in between proportionate to the first line, which meant an interval of about one-fourth of the width of the first line. The fill-up *tagma* or *tagmata/banda* were posted between the two divisions. We find this general version of the battle formation already in use for example in the *Ektaxis kata Alanôn* of Arrian,[17] but with the difference that during Arrian's day the cavalry front line was organized according to the 'Alan Drill' system into *koursores* and *defensores*, while at some point in time during the third century the system may have followed the Illyrikian Drill' formation.

Diagram 3.9 of the *Strategikon* is clearly defective. It has several details which are at odds with the text, while it also lacks other important details. According to the text portion of the *Strategikon*, if the first line was forced to retreat, it was to race towards the second line and its fill-up *tagma*, and when this happened the standard-bearer or the commander of the *tagma* shouted the command 'Admit' ('*Suscipe*'). This implies that the fill-up *tagma* was initially posted level with the first rank of the support line and only then retreated to open the formation for the retreating soldiers of the first line. However, in the accompanying diagram 3.9 the *tagma* is posted along the rear of the second line, which does not fit the description. It is possible that both variants were actually used, because in the *Sylloge Tacticorum* (46) the fill-up *banda* were posted along the rear of the second line. In that case it would probably have been roughly the same in its purpose as the third line rear guards for the large cavalry army. In the accompanying diagrams I have included both alternatives, although the placing of the fill-up *bandon/banda* level with the first rank of the second line is likelier on the basis of the use of the order 'Admit!'.

The *Strategikon* also states that the diagram depicts a single *meros* (max. 6,000–7,000 men) which can also be interpreted as the first line being deployed as a single

The cavalry formation for armies of 5,000-15,000 men

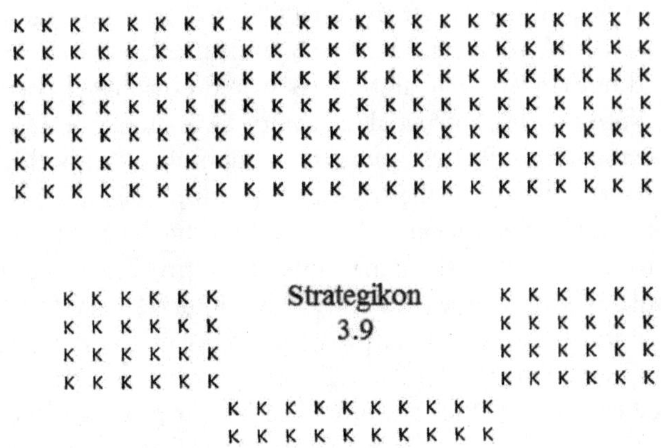

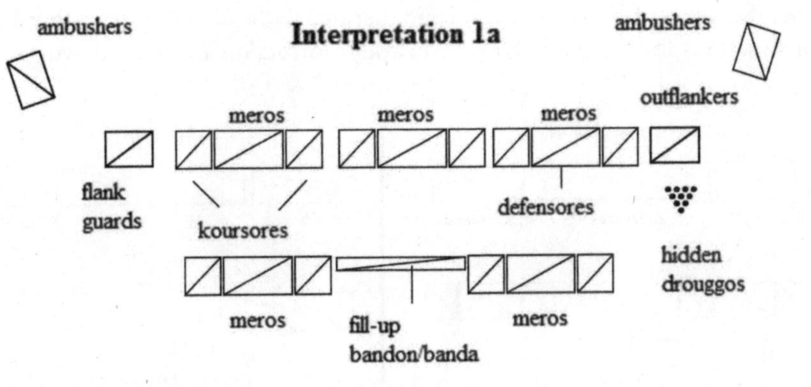

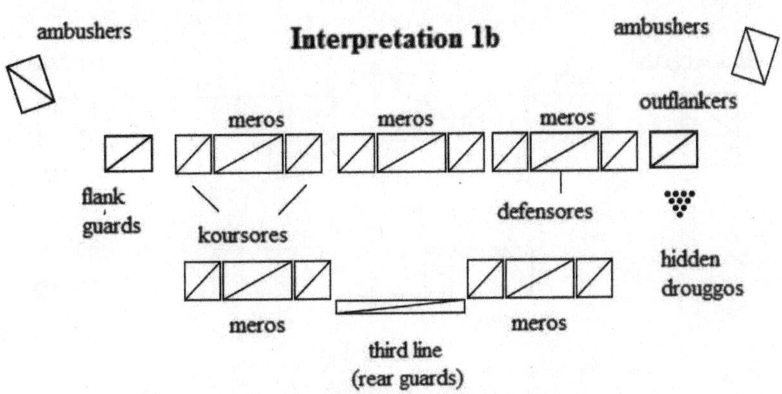

meros with three *moirai*, with the *defensores* in the centre and *koursores* on the flanks. This does not fit the rest of the instructions for various parts of the 'Italian Drill Formation' which include also the outflankers and flank guards. It is therefore probable that the front line was actually divided into three *mere*, each of which had its own three *moirai* of *koursores* and *defensores* while the flanks were protected by the flank guards and outflankers. The ambushers would obviously have also been used when feasible. Note that in these smaller cavalry formations the *meros* is a division without any specific size while the *moirai* are its subunits of unspecified size.

The interval of one-fourth of the front line is also problematic, because it is not found in diagram 3.9 of *Strategikon* and it would not fit the description of the front line retreating to the interval – the flank units of the first line would have extended beyond the flanks of the reserve divisions. If the interval would have been only a fourth of the width of the first line, then the flanks of the first line would have actually retreated either to the left or right of the reserve divisions so that there would not have been any need for the order 'Admit!'. In that case, the order 'Admit!' would have been appropriate only for the centre of the first line. I would suggest that the likeliest place for the *mere* of the second line were behind the flanks of the first line so that the instruction would fit the situation.

On the basis of the text in combination with diagram 3.9, the likeliest reconstructions for the moderate-sized cavalry formation are the following: 1a) the

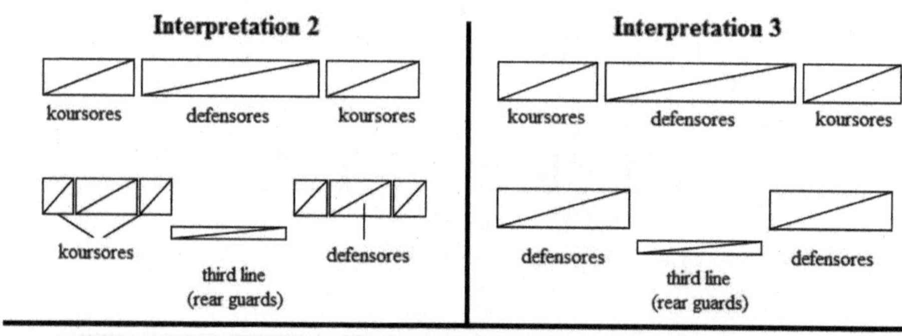

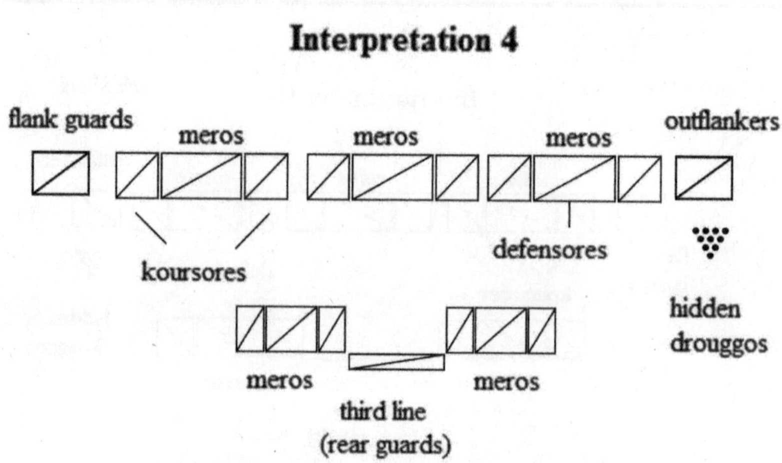

first line always divided so that it had three *mere* (each divided into *koursores* and *defensores*) and wing units (flank guards and outflankers) and the second line had two *mere* (each divided into *koursores* and *defensores*) deployed behind the flanks of the first line so that the fill-up *bandon/banda* were posted in line with the second line, or (1b) alternatively as depicted in the diagram 3.9; 2) the first line deployed as a single *meros* with wing units, and the second line as two *mere* (each with *koursores* and *defensores* and the fill-up *bandon/banda* as above) posted behind the flanks of the first line; 3) the use of the *meros* structure for the first line and with only *defensores* for the second along with the fill-up *bandon/banda* (could be in line with the reserve or behind); and 4) the first line deployed regularly but the second line (fill-up *bandon/banda* either in line with them or behind) so that it provided a place of refuge only for the centre of the first line, while its wings would have retreated left or right. The likeliest of these alternatives is the first one (1a-b), but 1a is probably the one that Maurice preferred.

5.4. The Cavalry Formation for Armies Less Than 5,000/6,000 Men[18]

The smallest cavalry armies envisaged by Maurice consisted of 2,000–5,000/6,000 men or even less. Such armies were required to possess only one *meros* in the second line, which means that if the first line was forced to retreat, it retreated to the flanks of the support line and regrouped there. The overlap of the 5,000/6,000 men once again demonstrates the flexible approach to the number of reserve divisions. Diagram 3.10 of the *Strategikon* depicting this array is once again problematic, because it leaves out many details. For example, it does not include the rear guards, flank guards, outflankers and ambushers. In my view, in this case it is actually probable that the small cavalry army did not contain any rear guards because they would have been useless when the support line consisted of only one division. Conversely, the presence of the other detachments is probable. Just as it was with the medium sized force, this array did not give the first line the same measured protection as the large cavalry formation. Therefore, it is not surprising to learn that by the tenth century all of the variants of the smaller cavalry armies had more reserve divisions in their support line or lines.[19] This means that during the Late Roman period the Romans still stressed the importance of the effectiveness of the initial frontal charge and the width of the

The cavalry formation for armies less than 5,000/6,000 men

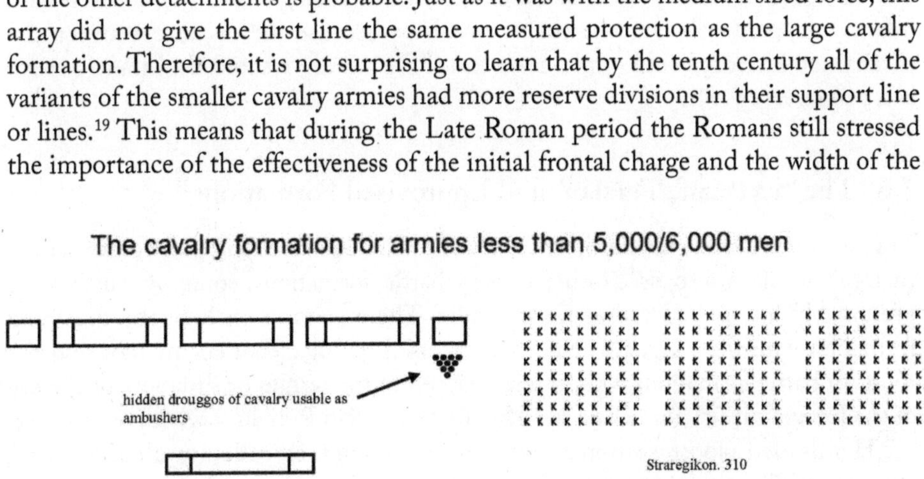

Single Line Formation

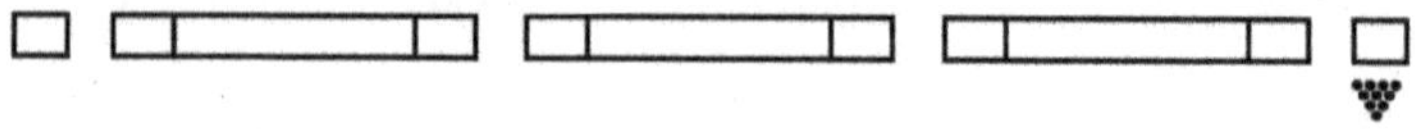

first line so that the enemy would not be able to surround it easily. In short, when the Romans only had small numbers available they wanted to get the maximum effectiveness out of the lancer charge.

5.5. The Single Line Formation[20]

The *Strategikon* and the narrative sources prove that Romans also used the single line formation without reserves, and a version in which the initial formation had reserves, but these were then transferred to the flanks to outflank the enemy formation so that the formation was in practice a single line formation. The pure single line formation appears to have been used mainly by the foreign *foederati* in Roman service (e.g. the Ostrogoths of Theodoric I the Great), while the version in which the reserves could be posted on the flanks to outflank the enemy appears to have been primarily used by the regular Roman cavalry. As already noted, Maurice in his *Strategikon* opposed both as unsafe and required the commanders to keep their second line intact as a support for the first line. It is in fact quite probable that the use of the single line (except when combined with infantry, see below) was abandoned as a result of the efforts of Maurice. It had several weaknesses which Maurice explained well in the *Strategikon*. See the discussion below. Note, however that the single line cavalry formation continued in use in three different forms: 1) It was used in drilling and training; 2) it was sometimes used when the rear was protected by infantry; and 3) it was used in hunting exercises (*Strategikon* 12.D). The battle of Yarmuk in 634 demonstrates nicely that the employment of the single cavalry line, even with infantry support, was not without its problems because the distance between the cavalry and infantry could not be too great if the cavalry wanted the infantry to act as a real reserve for it.

5.6. The 'Scythian', 'Persian', and Improvised Formations[21]

The narrative sources and the alternative drills included in the *Strategikon* demonstrate that the Romans also used other cavalry battle formations, some of which were presumably improvised rather than regular. These variants include the marching formation with the vanguard that could be used to engage an enemy blocking the route, or battle formations which were adapted to the terrain or situation, or the use of the foreign cavalry formation used by the *foederati* in Roman service. For example, the Heruls used a loose, scattered cavalry formation in their galloping attacks against the enemy. The typical example of the improvised formation is obviously the use of the ambush, which will be dealt in greater detail in its own chapter, but there were instances in which the terrain or the situation dictated the formation which could be used. The best examples of this are the use of the scattered formation by Sittas against the Armenians at the Battle of Oinochalakon/Avnik in 539. This was not the recommended tactics for cavalry, and Sittas used it only because the Armenians managed to place him in such a position. The Battle of Satala in 530 and Belisarius's cavalry skirmishes and battles in front of the city of Rome in 537 demonstrate nicely how the walls of a city or fortified camp together with terrain could result in an improvised cavalry formation. This demonstrates how able Roman commanders were ready to improvise, especially when the situation forced them to do so.

The *Strategikon* and the narrative sources (e.g. Procopius) show that the Romans also used the so-called 'Scythian Drill Formation' during the Late Roman period. In this array, all troopers were equipped alike without any separation into *koursores* and *defensores*. This was clearly copied from the nomadic mounted archers, with the Huns being the likely source of inspiration but not conclusively so because the Romans had obviously encountered nomadic mounted archers before their arrival. As already noted in the context of unit orders, in this formation all of the troopers were equipped in like manner as multipurpose mounted archers and lancers so that they were arrayed in a single long line with the idea of encircling the enemy with either crescent or *drouggoi* unit orders. The best example of the use of this system is the cavalry skirmishes of Belisarius's bodyguards in front of Rome against the Goths: all of the men were equipped in like manner, employing bows and lances as needed with the aim of decimating the Goths by arrows. Belisarius's *bucellarii* were clearly employing the *drouggos*-order in irregular fashion as swarms of horsemen. They exploited high ground, hills, feigned flight, encircling movements, and reserves posted behind the walls of the city of Rome. In addition to this, the Roman cavalry also employed the Persian cavalry tactic against the Germanic peoples, which was to place their mounted archers in close order on some hill or rough ground from which they then bombarded the enemy with arrows, or alternatively the Romans advanced slowly by bombarding the enemy with arrows.

5.7. Regular Cavalry Formation with Infantry Support[22]

Maurice's *Strategikon* has two versions of how the cavalry army was to be arrayed for combat when it had infantry as a reserve force. This combined formation of cavalry and infantry was to be used in cavalry battles when the Romans had more horsemen than footmen and the number of their cavalry was still at least equal in number to the enemy cavalry. The first variant was the posting of the regular two-line formation, but without the rear-guards in front while the infantry was posted behind as a rear-guard and last reserve. The second variant had only a single line of cavalry and it is this variant that we find the Romans using in the decisive battle of al-Yarmuk in 634. In both variants, the infantry was formed right in front of the camp as a rear guard and place of refuge. See the attached illustration overleaf.

In contrast to the regular cavalry formations, when the infantry accompanied the cavalry the battle lines were to be posted only an arrowshot (ca. 330m) apart from each other. It is possible that this results from a scribal error or lacuna, because this

Cavalry Formations with Infantry Support

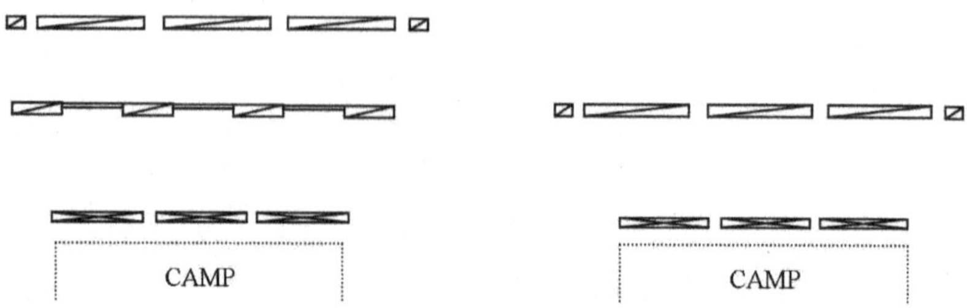

is in direct contradiction with all of the other allowed distances in the *Strategikon*, but I would still suggest that the distances are not mistakes. It is clear that when the infantry followed the cavalry to the battlefield the principal aim was to avoid the risk of the cavalry and infantry being separated from each other by too great a distance. The shorter distance between the cavalry lines ensured that the entire cavalry line was close enough to the infantry so that the enemy could not safely come between the cavalry and infantry. The infantry was so close to the rear of the cavalry formation that any enemy that would have dared to come between would have been immediately attacked and destroyed. The cavalry could therefore concentrate on defeating the enemy cavalry to their front. This in its turn required a change to the expectations regarding the first cavalry line. The distance between the first and second line was now so short that the first would have to seek support from the second immediately if it was forced to flee. The sad truth about the distances between the different battle lines is that this is only mentioned by the *Strategikon*, so we do not know if this was true for the entire Late Roman period. On the other hand, since it is clear that Maurice's instructions were based on Roman military traditions, one may expect that these reflected earlier Roman practices too.[23]

The Romans recognized the effectiveness of infantry vis-à-vis cavalry in close-quarters combat, the best proof of this being the instructions of Maurice not to use large numbers of infantry and the referrals to the effectiveness of well-ordered infantry formations against cavalry nations such as the Persians and Avars. In emergencies the cavalry was actually expected to dismount to fight more effectively against enemy cavalry. Therefore, this formation was actually a strengthened version of the regular cavalry formation, because it now had the infantry behind it. Regardless, when the array was adopted with too small an infantry force, it was actually dangerous for the Romans. The best example of this is the major battle which took place in front of Rome in 537 (Syvänne, *MHLR 6*, 145–50). The reserve infantry was annihilated by the Ostrogoths.

In addition to this, the *Strategikon* (12.A.7) also contains a variant of the above, but in which the intention was to engage the enemy cavalry with infantry rather than with cavalry in a situation in which the enemy was reluctant to engage the Roman infantry. In this case the Romans still had large numbers of cavalry present, but their infantry force was more numerous. The instruction in the *Strategikon* was that the *strategos* was to post three cavalry divisions in front (presumably with their wing divisions for safety), and have the infantry battle line follow about one to two miles (ca. 1.5 to 3km) behind. If the cavalry was then forced to retreat, it was expected to retreat by way of the flanks and rear of the infantry battle line and not by the front so that they would not disrupt the infantry phalanx. The same kind of instruction for the use of cavalry to hide the presence of infantry can also be found in the *Peri strategikes* (36) and it is clear that Arrian's *Ektaxis kata Alanôn* had a similar idea, namely the idea of inducing the enemy to attack the Roman infantry formation because it initially faced only Roman cavalry in front. The Romans exploited this type of tactic for example at the Battle of Hippis River in 550 and also against the Avars in 599.[24] This variant was clearly more dangerous than the use of the actual cavalry formation with infantry because the distance between the single cavalry line and infantry was greater than in the variant in which the cavalry did the fighting. The distance between the lines, one or two miles, gave the enemies far greater chances of getting between the cavalry and infantry than a gap of only a bowshot. This is what happened when the Muslim cavalry got between the infantry and the cavalry in the Battle of Yarmuk in 634. Although it has to be remembered that in the Battle of Yarmuk the Romans were actually using their cavalry purposefully for combat in an infantry battle, contrary to the specific instructions in the *Strategikon*, while it is also clear that the distance between the lines was too great. Therefore, this is a poor example of the use of this tactic. In the Battle of Yarmuk the Romans did most of the things contrary to their military doctrine. When employed correctly, as for example the *strategos* Priscus did against the Avars in 599, the stratagem worked perfectly.

5.8. The Moral Implications of the Cavalry Formations[25]

The various types of cavalry battle formations and unit orders had a direct bearing on the way battles progressed, which were based on the moral implications of each vis-à-vis the other types of formations. The *Strategikon* (2.1.1–2.10) summarized the advantages of the use of reserves and the disadvantages of the use of the single line as follows (not a quote but my summary). When reading the list, it should be kept in mind that Maurice also appears to have equated the sending of the second line to the flanks for the purpose of outflanking the enemy to mean that the formation consisted of a single line, as it actually was. Maurice advocated the keeping of the second line intact as a reserve for the first line in all circumstances. Even if it is clear that there had always been Roman commanders who acted in the manner advocated by Maurice, there had also been commanders who either used the single line cavalry array or who prematurely sent their second line reserves to the flanks to outflank the enemy. Therefore, the military doctrine that Maurice instilled into the Roman army improved its efficiency by removing these weaknesses.

1) The use of cavalry reserves was a mark of an experienced commander, while the use of a single line was the mark of an inexperienced and absolutely reckless man.
2) The battles were not decided by numbers or boldness alone, but by strategy and skill.
3) The use of the reserves was a sign of strategy and skill, and it was because of this that the earlier military writers had organized their armies into *drouggoi* (irregular sized units and units in irregular order), *mere* (divisions), and *moirai*, which Maurice likened to the tactics used by the Avars and Turks of his own day.
4) The Avars and Turks used two or even three battle lines, unlike the Romans and Persians who could stake the fate of tens of thousands of horsemen on a single throw.
5) The use of a single battle line by a large cavalry army was particularly dangerous when one used lancers, as the Romans did. The huge length of the battle line would result in disorder, not to mention the fact that it would be difficult to control because there could not be any coordination between the units. The end result of this could be that the line would break up even before contact with the enemy.
6) If the single line was outflanked or surprised by the enemy, it lacked the reserves needed for the protection of its rear or its flanks, and without those, the entire force would have to flee in disorder.
7) It was impossible for the general to supervise the entire battle if it consisted of a single line, because it was so spread out, with the result that the troopers making up the line could desert from their units unnoticed, with the result that others would be infected by the flight.
8) If the single line formation was forced to flee, it was impossible for it to regroup because there were no reserves behind. This view was obviously correct in most cases, but it should be noted that there are examples of the Romans, nomads and Germanic peoples regrouping their cavalry even when deployed as a single line.
9) The use of the single line could also result in defeat even when it has reached the enemy formation intact, as it would seem as if it is winning because it forced the enemy into flight, but then ends up losing the battle because the formation has been disordered as a result of the melee and pursuit – which the enemy can then exploit. This danger was particularly acute when the Romans faced the so-called Scythian nations who were in the habit of using feigned flight, or when the enemy had posted an ambush. When this happened and the Romans lacked reserves, it was impossible for them to turn the tables against the enemy.
10) The use of the single line had only two advantages, which were that the Roman cavalry force looked very large and threatening from a distance, while it also eased the encirclement of the enemy, but these two advantages paled in comparison with the disadvantages, because it was also possible to outflank the enemy more safely using reserves.
11) It was because of the above that the *Strategikon* stated that there should always be two lines, divided into *drouggoi*, *mere* and *moirai* and the other components of the 'Italian Drill Array'. The first line fought more eagerly because the troopers

Frontal view of Roman cavalry in close order

The illustration depicts the cavalry in the *pyknosis*-order when it carried spare lances behind their back in a slanted position. The example depicts four files of a fill-up *bandon* deployed two ranks deep. The shields in this example belong to the smaller *clipeus/aspis*-variant.

Below left: Legionary horseman (1st cent.) (Museum of Saint-Germain). Source: Duruy.

- It is probable that the positioning of the oblong shield in this relief is purposefully placed so that the rider's head is visible whereas in practise during the charge the oblong shield would have been used to protect the head of the rider and the head of the horse.

Below right: Roman auxiliary horsemen with Thracian sword (1st-2nd century AD, Zemalisky Museum, Sarajevo). Drawn after D'Amato / Norbury.

- Note the placing of the oblong shield so that it protects the horse's head. During approach it would also have been used so that it protected the trooper's head.
- The image depicts the trooper in such a position which would have taken place immediately after the throwing of the javelin or spear. The trooper attempts to deliver a cut against an enemy on his right side.

knew that their rear was protected by the second line and their flanks by the flank guards. The troopers of the first line were also less likely to flee when they knew that the men of the second line would be able to see who deserted their posts. The presence of the second line also enabled the rallying and regrouping of the fleeing units of the first line, so that these could be used again against the enemy. The presence of the reserves made attacks and pursuits more secure, because these could protect the first line. Furthermore, if the first line was actually routed so that it could not be regrouped and be brought back into action against the enemy, it was still possible to engage the enemy with the second line which had retained its good order. In contrast, the enemy they would now face would be disordered and their formation broken up thanks to the fact that it had just engaged and pursued the first Roman line. The well-ordered second Roman line would have moral ascendancy in such a situation, which would in all probability result in the flight of the enemy.

12) According to Maurice, there appears to have been some military men who thought that if the first line was thrown back by the enemy, then the second reserve line would be useless and would only join the first in the rout. Maurice's answer to that was that if victory was uncertain with two lines, it was even more uncertain with only one line.
13) The general was instructed to use the prescribed cavalry array with reserves while using his intelligence to avoid having to fight against an army greatly outnumbering the Romans, in particular in cases in which the enemy fought in an organized fashion.

The above summary of Maurice's thoughts shows nicely that the Romans debated the advantages and disadvantages of the use of the single line versus two lines cavalry formations, and that the ancient authorities that Maurice cited had already found the use of reserves to be the correct answer. Unsurprisingly, we find the same view also for example in Vegetius's *Epitoma rei militaris* (3.26), which states that it was always better to have more reserves behind the battle line than to extend the frontage. The fact that Maurice had to reiterate the arguments proves that the debate continued even until his reign. It required the imperial authority of Maurice to correct the situation. Maurice reiterated the old wisdom that the reserve line was never to be used to lengthen the first line. The ironic part of this is that the battle tactics of hiding the presence of two lines from the enemy demanded that the enemy could be fooled into believing that they faced only a single line and would deploy their forces accordingly, which in its turn required the use of the single line at least occasionally. This was obviously no longer possible when enemies no longer believed that the Romans used only a single line, and indeed on the basis of the descriptions of cavalry battles after Maurice wrote the *Strategikon* it suggests that enemies could no longer be fooled into the belief that the Romans used only a single line. When the enemy knew that the Romans used two lines, they could be surprised only with ambushes and surprise attacks.[26]

Chapter Six

The Cavalry Battle

6.1. Forming the Battle Line[1]

When the generals and soldiers had followed combat doctrine to the letter and there were no unhappy coincidences or human mistakes so that the decision to fight a battle was made by the Romans themselves and not by their enemies, then the preparations for battle followed certain routines and precedents. Families and personal possessions had been left behind in a separate base camp so that the soldiers would not fear for their families and property. The *Strategikon* advised the general to postpone any punishments, to leave the unfit and cowardly, if still present, in the camp and to restrict the use of wine in warm weather. The flags of the units had already been 'blessed' when the campaign began.

Life inside the marching camp and preparation for fighting the battle followed the following regulated routines, which also helped the soldiers to alleviate psychological stress: 1) the general and his officers held a council in the previous evening and made their plans; 2) the soldiers prepared their weapons and equipment for combat; 3) on the morning of the battle there were religious ceremonies (in pagan times sacrifices and in Christian times prayers led by the general and accompanying priest) to alleviate the prospect of being killed in combat because, as the saying goes, there are no atheists in foxholes; 4) the commander was expected to give a speech to encourage the soldiers; and 5) officers also encouraged the men independently. The routines and all pre-battle formalities were designed to instil confidence in the soldiers, but it is clear that soldiers and officers still felt pre-battle stress. In fact, the sources do occasionally mention the poor sleep of the soldiers and general the night before the battle.

According to the *Strategikon*, when the battle lines were aligning themselves for combat, the Romans were expected to have one to two (or even more) *banda* a mile or two (c.1.5–3km) in front of the main body. The same procedure was undoubtedly followed before Maurice, as the Romans possessed scouting units for this precise purpose, but there are instances in which the commanders failed to follow this precaution. The intention was twofold: the presence of the scouting units/vanguard prevented the enemy from observing the Roman formation, and also prevented the enemy from attacking the Roman army before it was ready for combat. If the enemy did not attack the vanguard, the *Strategikon* instructed them to raid the enemy's baggage train, or the rear or wings of the enemy line. It is quite clear that this followed military doctrine as it had always been. The *Strategikon* expected the vanguard to advance as close as a bowshot from the enemy to see whether the enemy was digging any ditches or planning any other trick. Even if there is a temptation to think that this instruction was added to military doctrine after the Persians had managed to surprise the Romans with the ditches and trenches at the Battle of Thannuris in 528, one must resist this temptation because the trick of using hidden traps was age-old[2] and was also included in the collections of stratagems (e.g. *Excerpts of Polyaenus* 37; Frontinus, *Stratagems* 2.17). In other words, there is every reason to expect that Roman combat doctrine always called for a thorough investigation of the battlefield prior to the battle, even if this was not always followed by negligent commanders.

The *Strategikon* also expected the heralds who were posted at the front before the battle to scout the area where the battle was expected while the lines were being formed up. This added another layer of security, but is unlikely to be a new invention either, because heralds had been used throughout antiquity to distribute the words of the commander to the troops. The use of specially-trained heralds was beneficial, because their voices carried further than the commands of the officers. When the cavalry had then been arrayed in battle order, the heralds were expected to encourage the soldiers further by recalling their previous victories. After this, the herald or heralds shouted (translation based on Dennis): 'Silence. Nobody hesitating. Nobody ahead of the *bandum* (flag/standard). Advance level with the front rank. Eyes on the *bandum*. Follow it with your *bandon, miles* (soldier). This is how a *miles* should act. If you abandon the *bandum*, you will not be victorious. *Miles* keep your position. *Bandifer* keep your position. Whether fighting, or pursuing the enemy, or in the front ranks, do not charge out impetuously and cause your ranks to be broken up.'[3] The most obvious purpose of this list was to stress the importance of maintaining order, but it is probable that the secondary purpose for this tedious litany of words was to direct thoughts to the training field and instil a sense of routine to the men.

When the Roman army left their marching camp, they kept absolute silence, so that the commands of the officers and the herald could be heard. The secondary purpose for the use of silence was to instil a sense of impending doom on the enemy. It is obvious that the Roman troopers found the keeping of silence unnerving and difficult, but the enemy would have found it still more unnerving. Prior to the writing of the *Strategikon*, Roman cavalry had been in the habit of using their battle cry and shouting and cheers, as can be witnessed from the battles that took place during Belisarius's day, but the *Strategikon* instructed the soldiers to use these only sparingly, and just before making contact with the enemy to unnerve them and stir up one's own troops. Maurice's opinion was that the use of the battle cry, '*Nobiscum*', was detrimental because it sometimes caused the ranks to break up when the bolder men and horses galloped forward while the more timid men and horses remained behind. It is clear that this concern was the result of actual events that had taken place in the past, so it is undoubtedly an accurate observation of reality. In the opinion of Maurice, the better the soldiers observed the silence, the less disturbed and excited were the young men and horses accompanying the veterans, and the better the commands were to be heard.[4]

Combat doctrine expected the cavalry to be arrayed in its battle formation at a safe distance from the enemy so that it initially advanced towards the enemy in open order (*araiôsis*) as the commander gave the order: 'Open order. March!' ('*Largiter ambula*').[5] If there was a need to stop, the formation halted with the order 'Halt!' ('*Sta*'). The order to continue the advance was 'March!' ('*Move/Kineson*').

The narrative sources prove that before pitched battles there were also sometimes a number of single combats/duels fought by commanders themselves, generals, officers or individual champions/duellists. The idea behind this was to demonstrate the personal bravery of the champion. The outcome of these duels sometimes had a direct influence on the outcome of the battle because these affected either positively or negatively the morale of the army. It was because of this that Roman combat doctrine opposed duels. In fact, this was forbidden by capital punishment. Regardless, as stated, the practice continued, presumably because the death penalty was not used in practice when the champion had been victorious not to mention the fact that the commanders themselves sometimes advanced in front of their army to fight a duel.[6]

6.2. The Components of the Cavalry Battle Formation

The strategos and officers[7]

Unsurprisingly the most important person in the cavalry force was its commander, the general (could obviously also be the emperor or anyone else the emperor put in command). It was the general who decided when and where to fight and how. See also Chapters 3.4, 10 and 12. Combat doctrine expected the commander to act in such a way as to be able to retain the trust of the soldiers under any circumstances.[8] It was because of this that Graeco-Roman military theorists produced lists of the qualities required from a general. The most famous of these was obviously Onasander, and it is because of this that we find his list copied by generation after generation of military writers. Combat doctrine expected that the commander was able to rally his men in case of a reverse, and this was possible only when the soldiers trusted him. Obviously, the reputation of the commander was the most important thing that decided the ability of the commander to retain the trust of his men. Only those generals who were repeatedly successful and so were able to provide the soldiers with loot retained the full trust of their soldiers even when they suffered a temporary setback.

The soldiers estimated their chances in combat on the basis of the appearance of the general, which meant that he had to instil courage and trust in the outcome through his own demeanour, actions, and speeches. The commander was also required to know the personal qualities of his officers so that he could assign them to the right places, understand the quality of forces under his command, know the terrain and the weather, and the quality of the enemy forces and their commanders, and how to act in these circumstances. The commander was expected to be able to place the right men, both officers and troopers, in the right places in the battle formation, each according to their qualities (note for example how the depth of the unit formation was varied according to the quality of the unit). The general was anticipated to be able to adapt his battle plans according to these known variables and he was expected to make his plans in a calculating manner, avoiding any rash decisions. However, when the general had formulated his plan, he was expected to act quickly and decisively so that the enemy would not be able to change their plans.

The choosing of the battlefield required a considerable amount of thought because, for example, the large cavalry army depicted in diagram 3.8 of the *Strategikon* required a battlefield which had a minimum of about 3km of open, unobstructed and level plain in width and several miles in depth (the pursuit of the enemy required this) to be suited to the demands of the lancer charge. The narrative sources prove that in most cases generals were indeed able to find terrain well-suited to cavalry battles, with the implication that most of the Roman commanders lived up to these expectations. Regardless, the narrative sources and common sense demonstrate nicely how the way in which the commanders led their troops depended on their personal and physical properties. Some of the commanders were eager to demonstrate their personal combat abilities, and some of these can actually be considered hot-headed and incautious in their approach. However, the readiness of the commander to engage the enemy at close quarters was not simply a sign of bravado or incaution. Roman combat doctrine actually expected that, in times of crisis, or when the soldiers hesitated, the commanders lead from the front to instil courage in the soldiers, or that they showed their personal ability with the weapons in some other way. The ability to show off obviously depended upon their personal skills in the use of weapons and on the age and physical condition of the commander. The *Strategikon* did not recommend this, but even the extremely cautious commander Belisarius was forced to act in that manner by circumstances.[9]

Roman generals and officers used vocal commands, trumpets, whips (and obviously also hand, sword and spear, even if the military treatises fail to mention these) and standards or flags to lead their men. Each commander and unit was expected to have a distinctive flag, which differed in size, form and colour, with the result that higher level commanders and their units had larger and more distinctive flags. The flag of the *strategos* was expected to be the most eye-catching of the lot because it was expected to serve as the rallying point of all units. Flags and standards had always been important as rallying points, and this did not change during the Late Roman period. The disappearance of the commander's flag signalled to all that the battle was lost and it was because of this that flags and standards had special guards, and it was also because of this that the commander's flag/standard was sometimes targeted. The best example of this comes from the battle of Satala in 530.[10] The Persians lost this battle because a group of Romans charged against the Persian centre and managed to throw the commander's standard to the ground, which panicked the Persians.

Roman combat doctrine required the higher-ranking officers (*merarchai* and *hypostrategos*) of the first line to lead the cavalry forces from the front until the charge was just about to begin. When the cavalry formation was about one- to two-bowshots (ca. 330–660m) from the enemy, the *hypostrategos* and *merarchai* stationed themselves in the same rank as the standards/flags. Just before the Romans launched the charge, the files on both sides of the *merarchai* and *hypostrategos* advanced to create a protective screen for them. It was these elite soldiers – the best of their respective *mere* – that then engaged the enemy in melee on behalf of their commanders. The *Strategikon* fails to state how the files were arrayed at the front, but one may make the educated guess that these could have been arrayed, for example, with a depth of four ranks to provide extra safety for their commander, or even as a small wedge.

Battle doctrine anticipated that the *strategos* would not overburden himself with trivial matters on the day of the battle. The *strategos* was to concentrate only on really essential matters. He was expected to lead, direct and supervise the entire battle formation up to the moment of the charge, after which he was to take his place among the soldiers of his own *tagma* in the middle of the second line. This enabled the *strategos* to make last minute changes to tactics if this was necessary.[11]

After the order to begin the charge had been given, and the *strategos* had taken his place in the second battle line, the battle progressed according to its own internal logic and the commander could affect the outcome only by a well-judged use of his reserves. Those rash generals who led from the front obviously could not do even this. In other words, after the order to charge, the commander was largely at the mercy of the events and actions taken by the enemy and his officers and troopers. The subordinate officers of the first line (in particular the commanders of the flank guards, outflankers and flank *mere*) were obviously required to use their own initiative if this was required. The same was true of the commanders of the ambushers, who had to time their actions according to the situation. The *moirarchai* and *merarchai* of the second line were also expected to be able to react to the changing situation without waiting for the order to do so if the situation demanded this. This could happen when speed was of the essence in a situation where the first line or a single front line *meros* galloped in retreat towards them, or when the enemy had managed to ambush the Romans. In short, it is clear that Roman cavalry tactics demanded highly skilled troopers and officers who could coordinate and cooperate as the situation required.

> … Cavalry service requires more judgment and presence of mind on the part of the officer, and more speed, bearing and skill in manoeuvre and tactics than in the work of infantry, which fights in a slower and more mechanical way, and whose success depends merely on its endurance and cohesion. Duffy's quote (p.230) of Warnery, 1785–91, III, 119–20

The narrative sources prove that the subordinate officers did indeed act on their own initiative, but this is usually visible only when they acted imprudently. In other words, the sources show a clear tendency to blame the subordinate commanders for the defeats suffered – this results undoubtedly from the tendency of the generals to blame their subordinates rather than accepting personal responsibility for the defeat. In contrast, when the sources specifically mention the subordinate commander in a favourable light, this usually results from the author's purpose of either pleasing the subordinate commander or his relative, or from the purposeful denigrating of the overall commander. Since subordinates became visible usually in these cases, it is quite obvious that they usually performed their tasks as expected. The above shows the ideal.

In practice there were also situations in which the army lacked a single clear overall commander, or the commander lacked authority over his subordinates, or the subordinate commanders held a personal grudge against the overall commander or the emperor. Unsurprisingly, in these cases the effectiveness of the Roman war effort was seriously weakened. There were also other problems which affected the ability of the Roman generals to lead their armies well. Many of the Roman generals appear to have been drunkards, in particular during the fourth- and fifth-centuries. It is because of this that Procopius praised Belisarius' prudence in the use of wine. This is undoubtedly one of the reasons why he did not take unnecessary risks while his colleagues did. Unsurprisingly, Maurice warned against the use of wine before and during the battle because it dulled the mind. Regardless of these instructions, wine caused problems even during Maurice's reign. For example, when the soldiers had captured one Slavic encampment in 593, they celebrated this with wine and so neglected guard duty during the following night and were lucky to survive the Slavic counter attack in that situation.[12]

The first line[13]

In the *Strategikon*, Maurice's description of the use of the first line in combat concentrated mainly on its role in the lancer charge in which all units were initially in close order, but we know on the basis of the drills he described and on the basis of the narrative sources that the Romans also used a skirmishing tactic in which the *koursores* skirmished with the enemy before the actual lancer charge and then took their places on the flanks of their respective *mere* in irregular order.

The officers and units of the first line were always expected to adjust their movements to the movements of the centre *meros*, which was usually commanded by the *hypostrategos*. The troopers initially had pennons on their lances to make their battle line look more impressive and menacing to the enemy, but when they were about a mile away from the enemy, the lance pennons were removed so as not to hinder archery fire. The weapons were also expected to be shiny and polished to make a visual impact on the enemy.

Immediately after this, the battle line began its preparations for actual combat with the order 'By the Flank. Close! To the Dekarchs! To the Pentarchs!' ('*Ad latus stringe, ad dekarchas, ad pentarchas!*'),[14] which in this case meant that the files of cavalrymen closed towards the centre *meros* in such a manner that the small intervals between the *mere* and *moirai* were retained.[15] This manoeuvre was performed while continuing the advance. The tightening of the width of the formation was clearly performed first to ensure that the widths of the intervals between the *koursores, defensores* and *mere* were appropriate and not too wide. If there had been too-wide intervals between the units, the enemy could have exploited this. At a distance of three or four-bowshots (ca. 990–1,320m) from the enemy the formation closed up from the rear towards the front for the lancer charge while still continuing to advance. The order for this was the 'Close ranks!' ('*Iunge*'). When the entire formation had assumed close order, the archers opened fire and the commander ordered the troopers to charge with the command 'Charge!' ('*Percute*'). The two front ranks leaned forward and covered their heads and part of their horses' necks, and held their lances high at shoulder height. From this position, the troopers could use their spear either for thrusting or throwing. The shields held in this manner provided protection against enemy missiles for the troopers themselves, their horses and the men behind. Leaning forward also helped the troopers to withstand the impact of their own lance thrusts against the enemy and by the enemy against them. The archery was meant to cause wounds and casualties among the enemy which would then disrupt the cohesion of the enemy formation before the Roman lancers reached them. The better men posted in the front ranks led the men forward while the file closers prevented the men from hesitating or fleeing.

The use of close order and canter were meant to ensure the entire first line charged the enemy line at the same time all along the front in order to achieve maximum psychological impact. It is because of this that, for example, the *De militari scientia* (7) reminded the *dekarchai* not to charge faster than the canter. The use of the gallop could have resulted in the breaking up of the ranks before coming into contact with the enemy. However, as already discussed, there were periods and occasions in which the Romans used the impetus of the gallop to achieve the psychological advantage that the speed produced in those facing this sight. This was obviously done at the risk of breaking the cohesion of the formation. As we have seen, this was the way the *Peri strategikes* envisaged the cavalry charge. Therefore, it is not surprising to find Syrianus's contemporaries Belisarius and John Troglita using the gallop in cavalry charges. Similarly, it is clear that the Germanic and nomadic federates (*foederati/foideratoi*) used it. In fact, it is clear that the Romans had never entirely abandoned the use of the gallop in cavalry attack, even if it is clear that the Romans had adopted the tactic of employing both the gallop and canter in their attack very early in their history. The best evidence for this is the use of the *koursores* and *defensores* system in the Alan, African and Illyrikian Drills. The *koursores* used the gallop when ordered to do so, while the *defensores* always used the canter.

When the cavalry line was divided into *koursores* and *defensores* and then performed the charge, the units assigned as *defensores* obviously had to regroup and reform their order after the charge was over, while the *koursores* still continued their attack at a gallop in the *drouggos*-order as they were pursuing the fleeing foe. This obviously caused a slight delay before the *defensores* could follow their *koursores*, but probably not too great a delay because there were military thinkers and commanders who preferred the use of the gallop in attack. This question clearly divided the military community in Rome just like it divided the military thinkers in modern day Europe. Some of the military theorists preferred the cautious, safer approach with the canter while others preferred the gallop because of its greater psychological impact on the enemy, even with the risks involved.[16]

The above describes the basic method of cavalry charge, whether this was done by using the so-called 'Italian Drill Formation' (dated probably from the reign of Gallienus) or by using one of the alternative versions such as the Alan, African or Illyrikian Drills. However, there also existed other variants, because tactics were adapted to the type of enemy the Romans faced and because the Romans possessed many different types of units. For example, it was possible to order all horsemen to employ the bow initially, including the two front ranks of the first line, and change those into spears only when this was needed. We find the Romans using this system in particular against the Germanic peoples who lacked adequate numbers of mounted archers. In such cases, the Roman first line operated either: 1) like the Persians, by placing their battle line on some rough ground so that it was either completely stationary until it was time to charge, or the battle formation advanced methodologically either at a trot or canter while peppering the enemy with volleys of arrows; 2) or it fought like the nomads by avoiding contact with the enemy by employing units in irregular order and gallop so that archery was used to decimate the enemy before engaging the enemy at close quarters only when it had suffered terribly from the archery. The existence of different types of equipment both among the Roman regular cavalry and among the federate and allied cavalry also meant other differences to the descriptions given above. Since these included units that had different types of armament, it is clear that the instruction for the two front ranks to hold their shields and spears in the manner described was not true of all battles, even for the reign of Maurice. In fact, this was taken into account by Maurice too, because the foreign troops were posted either separately from the Roman formation or as *koursores*.

Regardless, there is still one major difference between the era of Maurice and the period before him, which is that before he unified the equipment and training to follow a single scheme there is evidence for the existence of different types of regular Roman units which included units of javelineers and troopers without shields employing the Sarmatian style *contus* with two hands, while it is also clear that some of the men who had shields still favoured the use of the two-handed fighting technique with the *contus*. The way these units were deployed was apparently up to the commander, because both types of forces could be employed either as *koursores* or *defensores*. Similarly, the fact that Maurice demanded the troopers to use their spears in the Germanic manner, shoulder high, obviously also means that there were other ways that the spears were held.[17]

In addition to the above, there are also instances in which the Roman cavalry charged against the enemy using swords, without using spears or javelins or bows. The best example of this phenomenon is the wild cavalry charges of the unarmoured Herul contingents. In Greek military theory these lightly-equipped swordsmen who charged into contact with the enemy were called *elafroi* (light-equipped). However, even if this happened rarely, there were also occasions in which the Roman regular cavalry did the same and used only swords from the start of the battle. In normal circumstance the swords were used only when the trooper had lost his spear(s) or was so close to the enemy that the spear was useless. The idea behind the use of swords in the initial stages of the battle was that it forced the men to seek hand-to-hand combat with the enemy immediately while demonstrating to the enemy that the Romans were desperate and ready to fight to the death. It was indeed a sign of desperation when the regulars did this, but for the Heruls this was a sign of personal bravado and bravery. The use of swords from the start could actually be also beneficial for the following two reasons: 1) the enemy could be so frightened by the sight that they fled; and 2) if the enemy did not flee and the swordsmen were able to begin the melee at close quarters, swords were better suited to this than spears.[18]

According to standard Roman military doctrine regardless of period (Alan, African, Illyrikian and Italian Drills), if the lancer-charge by the first line was successful, the *koursores* were ordered to pursue the fleeing foe right up to their marching camp (or similar) with the command 'Gallop, Threaten!' ('*Cursu mina*'). The *defensores* retained their close order and followed the *koursores* at the canter, given the order 'Orderly Pursuit!' ('*Cum ordine seque*'). The distance between the *koursores* and *defensores* was not to grow greater than a mile or two during such pursuit. Maurice instructed the pursuit to be relentless so that the enemy would be unable to regroup. However, we know from other sources that some of the commanders, such as Belisarius, sometimes preferred to allow the enemy to flee without pursuit. This was particularly true in situations in which the enemy outnumbered the victorious Romans. On the basis of the battle narratives and instructions concerning fighting against Light-Haired Peoples in Maurice's *Strategikon* (11.3), it is clear that the pursuit was sometimes performed by the entire first line with all units retaining their close order and canter (e.g. when the units were employing the Persian style tactics), or alternatively that all units galloped after the foe in irregular order (e.g. when the units were using nomadic tactics).[19]

If the enemy exercised too much pressure against any single *meros* in the first line or against all of them, the unit(s) in question was commanded to 'Give ground!' ('*Cede*'), after which these retreated at a gallop about a bowshot or two (ca. 330–660m) towards the second line as if the entire unit consisted of *koursores*.[20] This presumably meant that each individual in the formation turned towards the rear or retreated at a gallop as it was impossible to wheel the unit when the enemy was already near. At the distance of one- or two-bowshots, the commander of the retreating unit was instructed to attempt to regroup the men for a counter attack with the order 'Turn back, Threaten!' ('*Torna mina*'). By this time the retreating unit had become irregular in formation (*drouggos, drouggisti*), which meant that each individual, wherever he was in the formation, turned as an individual towards the enemy. There was every chance for the counter attack to succeed if the pursuing enemy had by then lost cohesion of their formation. If the regrouping and counter attack was still unsuccessful, they were to try it once more if possible. When the units of the first line were attempting to regroup, the second line could be ordered to shout two or three rousing cheers. The way the *Strategikon* envisaged the progress of the battle is also proven by the narrative sources, which describe cavalry units performing charges, retreats, about turns, wheeling-abouts, and counter charges during prolonged battles. Cavalry battles were fluid by nature and could involve several advances and retreats by different units, even when the battles themselves lasted less time than infantry battles.

If the retreating units had not been able to regroup and counter attack the enemy, the required units of the second line were to charge forward while the retreating *meros* or two *mere* or the entire first line retreated into the space between the second and third line, where they were to regroup and in irregular order support the charge of the second line. It was very likely for the counter-attack of the second line to result in the flight of the enemy as by the time the enemy reached the Romans they were completely disordered, and when such forces faced a perfectly-aligned enemy in front of them it was probable that they would flee immediately before the Romans could make contact. This is the reason why cavalry combats were so fluid. The units advanced, retreated, and counter-attacked without actually encountering each other in melee. The disordered side simply fled when it faced a new threatening reserve formation in front.

As already implied, the success of the charge could vary from one *bandon* and *meros* to another, so that while one or several were successful some others were not. The *De militari scientia* (9–11) deals with this issue in the context of the units of the first line. Firstly, the victorious *archôi* of the *banda* were not to pursue the enemy but were to assist the *bandon* of the *merarchês* if there was a need. Secondly, the *merarchai* of the victorious flank *mere* were not to pursue the defeated enemy, but were to assist the centre *meros* if there was a need. Thirdly, the *stratêgos* of the victorious centre *meros* was not to pursue the defeated enemy, but was to assist the two other *mere* if these needed it. The pursuit was allowed only after all *mere* of the first line were victorious. This is not specifically mentioned by Maurice in the *Strategikon*, which can be considered an oversight even if one can expect that the assumption was that the victorious *banda* would assist any neighbouring *banda* in their own *meros*, while the neighbouring *meros* would also assist the one posted next to it. Maurice (*STR* 3.5.72ff.) states this in a very roundabout way by stating that all *banda* were required to become used to cooperating with each other. It is probable that the first line *mere* were arrayed so close to each other for this precise reason, namely that they could assist each other in the attack. Furthermore, we know that in practice the able *archoi* and *merarchai* acted in the manner described by the *De militari scientia* and we also know that the expectation of mutual assistance existed in the minds of the commanding officers and readers of books of history. Good examples of this are the Battles of Tricamarum in 533, Melitene in 576 and Solanchon in 586, and in fact it is possible or even probable that the *mere*-divisions provided mutual assistance also in the battles in which all divisions of the first line were successful – the mutual assistance would not be visible from the sources which only note the success of the attack. In the Battle of Tricamarum all divisions, *moirai* and *banda* of the first line cooperated. In the Battle of Melitene the Romans expected that the units of the second reserve line would have supported the first line in a situation in which the second line wing had outflanked the Persians. In other words, the expectation was that, regardless of the place of the unit in the battle formation, it would support the other units which were not as successful. In the Battle of Solanchon, the Romans once again expected that the wings would have supported their centre, but this did not happen in practice before the *strategos* Philippicus sent a specific order to the right flank, which was pillaging the enemy camp, to return to the combat. In fact, on the basis of the extant battle narratives (note, however, the above caveat) it was far more typical for the divisions of the first line to act independently of each other without providing any mutual assistance, so that it was left to the reserve units to support the *meros* that had been unsuccessful. The same tendency can be found already in at least one cavalry battle that took place in the third century. This implies either that the commanders of the first line divisions were often carried away by their rash enthusiasm, so they forgot to support the neighbouring divisions, or that the standard combat doctrine actually did not expect that the *mere* of the first line to support each other, even if this expectation can be found from some of the narrative sources and in the *De militari scientia*. It is therefore possible that in the standard combat doctrine the support of the *mere* of the first line was solely the duty of the *mere* of the second line. Regardless, it is still clear that in practice the expectation of mutual help existed even before the *De militari scientia* put it in writing. Therefore, it is not surprising to find the same expectation in, for example, the *Taktika* of Leo VI the Wise (12.67).[21]

Flank guards (plagiofylakes)[22]

As already noted, the left flank of the first line was protected by the flank guards (*plagiofylakes*) which consisted of two or three *banda* (or more when fighting against nomads). The flank guards remained close to the left *meros* until near the enemy, the idea being to hide the place of the Roman left flank from the enemy, after which the flank guards assumed a position about a bowshot to the left. The set of commands for this would have been '*Ad scuto clina*' ('Shieldward face'), '*Move*' ('Move'), and '*Redi*' ('Return'). In other words, the troopers turned individually towards the left, moved in column formation to the desired location and then assumed their original facing.[23] The *Strategikon* does not mention any overall commander for the flank guards, but it is clear that there had to be a separate commander who issued the commands given here. Their primary mission was to prevent the outflanking of the Roman left flank, if the enemy's right flank extended beyond the Romans' line. This worked, usually because the presence of the flank guards about a bowshot to the left of the Roman first line threatened the flank of any enemy unit that would attempt to attack the Roman left flank.

If the enemy attempted to outflank the flank guards by lengthening their line even further, the flank guards were to change their front to the left one *tagma* at a time with the command 'Wheel Left!' ('*Depone senestra*'). The other *tagmata/banda* of the flank guards coordinated their position according to the situation, so that together they faced the outflanking enemy with a united front. This turning movement could also be performed by the *moira*, *meros*, or even by the entire cavalry formation if needed. The secondary purpose of the flank guards was to outflank the enemy formation in those situations in which the enemy line was shorter. In this case the flank guards were expected to outflank the enemy line immediately using the crescent formation before the left *meros* launched its attack. If the lines were of equal length, they were to stay in their own position and join the rest of the *meros* in the charge. It should also be noted that the flank guards or left-wing *meros* could include a hidden *drouggos* or *drouggoi* if the *strategos* intended to ambush the enemy on their right flank.

The key to the success of the above manoeuvres was timing. If the enemy wing was forced to engage the flank guards first, it was possible that the Romans would manage to defeat the enemy with the cavalry charge. On the other hand, if the enemy crushed the flank guards quickly, then the Roman fortunes rested on the timely arrival of support from the second line and so forth. Cavalry combat required quick thinking officers.

Outflankers (hyperkerastai)[24]

The Romans posted a *bandon* or two of mounted archers (or more when fighting against the nomads) on their right wing known as the outflankers (*hyperkerastai*). Their primary function was to outflank the enemy formation, as their name indicates. The secondary function of the outflankers was to prevent outflanking on their side in a similar manner to the flank guards on the left flank. According to the *Strategikon* (3.8, 3.10, 3.14), the outflankers followed the right wing until the distance to the enemy was about two or three bowshots (ca. 660–990m).

If the enemy line was longer than the Roman line, and the Romans had at least two *banda* designated as outflankers, then the outflankers first moved about a half bowshot (165m) to the right by using column formation with orders: 'Spearward face!' ('*Ad conto clina*'), 'Move' ('*Move*'), and 'Return!' ('*Redi*')[25], after which they outflanked the enemy by using the crescent formation while the right wing *meros* slowed its speed to enable the outflankers to perform their attack. The other *mere* in the first line would probably have followed their example. When the outflankers were in the process of encircling the enemy wing, the order 'Head out' ('*Exi!*') was given to the hidden *drouggos*, which then charged out suddenly with great speed.[26] The sudden unexpected attack was bound to surprise the enemy. If the *hyperkerastai* did not include hidden *drouggos* behind them, then they simply outflanked the enemy by using the crescent formation. The *merarchês* of the right *meros* was to time his attack against the enemy according to the outflankers' actions. The cavalry charge was to be launched just at the moment when the outflankers were beginning to cause confusion among the enemy ranks

If the enemy line was shorter, the outflankers were expected to move out of the right wing *meros* and envelop the enemy's left wing by using the crescent formation. If the lines had equal width, the outflankers were to extend their lines a bit further to the right by moving there in a column formation, followed by outflanking in the crescent formation. The *Strategikon* (8.B.83) for good reason warned against overextending the battle line,[27] hence the instruction to extend the line only a short distance to the right. If the enemy attempted to counter this by doing the same, the outflankers were to launch their charge immediately, because when the enemy was still moving towards the left two things happened: 1) they had to expose their shieldless right side while also being unable to shoot arrows (unless the enemy consisted of the Persians); and 2) the movement towards the left broke up the cohesion of their formation so that it became loose.

In situations in which the enemy forces were able to outflank the Roman right, the outflankers turned their front towards the enemy with the command 'Wheel right!' ('*Depone dextra*'). As before, this turning could also be performed by the *moira*, *meros* or the formation, if needed.

The principal reason for the preference to outflank the enemy's left wing by the right wing was that the right-handed archers could not shoot effectively to their right while they could shoot effectively towards left. It was because of this that the Persians posted left-handed archers to their left flank. The other side of the coin was obviously that it was of utmost importance to avoid being outflanked on one's right wing, firstly because the right-handed archers could not shoot effectively towards one's right if the enemy approached from this direction, and secondly because the right side was not protected by shields. Most of the men were right-handed and the military formations reflected this reality.

The ambushers (enedroi)[28]

In the *Strategikon*, Maurice stressed the importance of the ambushers. He quite rightly considered perfectly timed attacks against an enemy's flanks and rear far more effective and decisive than direct frontal charges and attacks. This was particularly true if the enemy force was not large. When the situation and terrain allowed the use of the ambushers, the ambushers were either sent in advance to the rear of the enemy formation, or were positioned on the flanks of the battle array, or were behind the first line as hidden *drouggoi*, as described above. The ambushers had a double function. They were to prevent ambushes, flank attacks and rear attacks against the Romans. The ambushers themselves were to send patrols ahead for this purpose and were to be wary of the enemy ambushes. If the enemy had two lines or ambushers of their own behind the front line – and according to the *Strategikon* they often had – then the ambushers were not to attack the rear of the first line, but were to attack the flank of the enemy's first line. If the Roman flank guards or outflankers were attacking the flank of the enemy's first line, the ambushers obviously targeted the flank or rear of the enemy second line, or their baggage train. The ambushers were expected to coordinate their movements at all times with the main formation. When the Romans posted the ambushers on the flanks of the main battle formation, the main body was to advance a bit in front of the ambushers to divert the attention of the enemy from the presence of the ambushers, but the ambush was still to be launched a bit before the two armies made contact so that their attack would produce the maximum amount of confusion.

The second or support line[29]

The support line adjusted its movements according to the movements of the first line and the centre of its own line, where the *strategos* directed the reserves. Initially, the support line was a mile (c.1.5km) or more behind the first line or directly behind it to hide its presence from the enemy. During the battle the second line was about three- to four-bowshots away from the first line, so that it appears to have assumed close order at the same time as the front line. If the enemy fled, the support line followed the first line while retaining its close order so that it could protect it in case of need.

If a single *meros*, or two *mere*, or the entire first line was forced to retreat, the second line was to shout encouragement in an effort to help them to regroup and turn to face the enemy. If the units of the first line were unable to regroup, depending on the situation the entire second line or selected units from it were to move forward in close order to protect them, while the units of the first line were expected to regroup in the intervals and then follow the second line in irregular formation. If the fleeing units required instant support, the units of the second line could also use the gallop and irregular order so that help would not arrive too late. However, the support line was still instructed not to engage the enemy prematurely and it was not to get too close to the first line, because this could result in confusion and defeat. If the second line engaged the pursuing enemy too soon, it was still likely to be in close order, whereas if the second line waited long enough before engaging the pursuers it was likely that these were already disordered and thereby more easily scared when they suddenly faced a well-ordered Roman reserve in front of them.

In cases where the enemy had managed to outflank the first line by their detachments or by ambushers, the *Strategikon* instructed the *strategos* to immediately send some of the *banda* from the flanks of the second line against them to lend support, except when the Roman ambushers on the flanks of the formation already doing that. One may assume that this usually meant that the outermost *moira* consisting of the *koursores* charged at a gallop against the outflanking enemy unit. This could happen for example when the outflanking had not yet caused the flight of the Roman flank *meros* of the first line, or when the flank *meros* or its outer *moira* was trying to regroup in the empty space between the two Roman combat lines. It is probable that the entire flank *meros* of the second line charged in its entirety while maintaining close order in such situations in which it was apparent that the fleeing front rank *meros* would not be able to regroup and would continue its flight to the interval in the second line, where it would then try to regroup. The centre of the second line appears to have performed similarly when the middle *meros* of the frontline was forced to flee. In other words, the *strategos* could send chosen *banda* at the gallop to assist the fleeing units if he considered it advantageous, and both mere from the middle in such situations in which he expected that the fleeing *meros* would be forced to flee to the interval between.

If the second line was forced to flee, it was to retreat as far as the rear guard and then the remnants were try to regroup. It is probable that this last attempt at regrouping was largely theoretical, because if the second line fled together with the first it is likely that it fled for good.

The Romans made their second line double-fronted so that it would be easy for it to face threats from behind. If the enemy attacked from the rear with a small force, the rear guard was considered to be sufficient for the task, but if this was not enough the *strategos* had several different alternatives: He could order the *banda* from the intervals of the second line to help the rear guards; and/or he could send some *banda* from the flanks of the second line (or the flank *mere*) to the rear. If the *strategos* used these alternatives, he still retained half or most of the second line as a reserve force for the first line. However, if the enemy suddenly attacked with a large force from behind and the third line was insufficient to deal with them, the *strategos* ordered the entire second line to face the rear. There were two different types of unit manoeuvres that could be used to achieve this (about turn, counter-march). If a large enemy force approached from the rear and there was enough time, then the second line performed the counter-march with the order '*transmuta*' ('across'/'through' 'change' = 'counter-march')[30] so that the facing of the formation was changed to the opposite. If there was no time to perform the counter-march, then the second line was turned to face the rear with the command 'About face!' ('*transforma*') so that each trooper remained in place and turned to the rear while only the *bandoforoi* (flag-bearers) and *archoi* (commanders of the *banda*, *moirai* and *mere*) moved around to the new front at the rear. If the enemy that had attacked the rear was forced to retreat, the *strategos* was not to send more than one or two *banda* from the rear guard to pursue. It is probable that when the second line turned to face the rear that the rear guards became its flank guards and outflankers.

***The rear guard* (opisthofylakes)**[31]

The Romans posted a *bandon* or two, or more, a bowshot behind the flanks of the second line as a third line for the purpose of acting as a rear guard. It was formed from the end divisions of the support line. The principal intent was to guard the rear of the second line by preventing the enemy from outflanking or ambushing it and by facing any threats from the rear. If needed, they got help from the fill-up *banda* of the second line, but if a large force appeared behind the Romans, then the entire second line turned to face the rear and assisted them to perform their duty. The rear guards were expected to coordinate their movements with the second line. One can also make an educated guess that in practice the rear guards could also be used as a reserve for the second line, outflanking any enemy force that advanced as far as the second line. The narrative sources suggest that in practice the rear guards were not usually required to face attacks coming from the rear, largely thanks to the fact that the Roman spies and scouts had performed their duties so well that the enemies were unable to surprise the Romans from the rear.[32]

The baggage* (touldon/touldos) *and the spare horses[33]

When the spare horses and the baggage train accompanied the cavalry army to the battlefield, these obviously provided added security for the rear of the formation and also provided the chance of renewing the fight. In normal circumstances, however, these were left behind in a marching camp. The *Strategikon* advised the *strategos* to limit the number of horses taken to the battlefield because this could result in confusion if there were large numbers of spare horses present. In other words, even when the spare horses accompanied the army to the battle the expectation was that most of the spare horses would still be left in the marching camp. The Romans used their medics not only to transport men to be treated but also to transport unhorsed men back to the rear or camp to collect spare horses when needed. The use of unit pennons to separate the spare horses of the different units eased the finding of one's own horse. It is also likely that servants brought spare arrows to the troops when the battles were prolonged.

Special Cases: Foreigners, Ethnic Units, Mounted Infantry, Dismounted Cavalry[34]

As already discussed, the Romans employed a great variety of different types of ethnic units (e.g. Persians, Armenians, various Germanic tribes, Moors, Arabs, Huns etc.) among their regulars and among their *foederati/foideratoi*, not to mention the actual foreign allies. These units usually employed their own fighting methods. This means that the foreign cavalry units could be deployed, for example, as wedges or rhomboids instead of the rectangular formations. It was because of this that even Maurice found it necessary to include special provisions for their use in the *Strategikon*. Maurice recommended that these were to be used either as *koursores* or as ambushers, or alternatively deployed separately outside the main formation. It is just common sense that when the ethnic units consisted of mounted archers, such as the Huns or other nomadic groupings, and were used as *koursores* or as ambushers, they did not endanger the cohesion of the main Roman battle line. In fact, these would have improved the efficiency in some cases by providing large amounts of expert mounted archers for the Roman cavalry force which did not always include adequate numbers of these. The different ethnic groupings had different advantages: the nomads were particularly valuable as mounted archers; The Moors were valuable as fearless javelineers; the Arabs provided superbly mobile light cavalry lancers that could operate very effectively in desert conditions (they were particularly good as vanguard forces, *koursores*-pursuers, or as ambushers); the Persians and Armenians provided superb cataphract cavalry;[35] the Germanic tribes provided very high quality lancers thanks to their warrior culture.

However, it should still be noted that the ethnic units were not always reliable and they were also sometimes poorly disciplined. For example, the Battle of Tricamarum in 533 (Syvänne, *MHLR 518–565*) demonstrates the unreliability of the Huns, while the battles between the Romans and Muslims demonstrate that the Arabs and Armenians in Roman service could be very unreliable indeed (Syvänne, *MHLR 8*). In fact, one can say with very good reason that the principal reason for the Roman defeats against the Muslims was that they were repeatedly betrayed or deserted by the Arab and Armenian contingents. The ethnic units could also be unreliable in other ways. For example, the Moorish bodyguards betrayed Gratian to the usurper in 383 (Syvänne, *MHLR 2*, 229) not to mention the behaviour of the Germanic Federates in the fifth century (Syvänne, *MHLR* vols. 3–5). It was largely thanks to the actions of Ricimer and his barbarian supporters that West Rome entered its terminal decline (Syvänne, *MHLR 5*). The ethnic units could also cause a defeat without actually betraying their masters, as happened for example when the Heruls charged at the Battle of Anglon in 543 (Syvänne, *MHLR 6*, 237–9).

Regardless, the recruitment of ethnic units both into the regulars and into the allied forces was necessary because it provided the Romans with access to trained military manpower which they sometimes lacked. This was particularly true for the cavalry. Unsurprisingly, there are therefore also examples of the ethnic units being the key element in Roman successes. These include, for example, the use of the *Unnigardae* in Libya (Syvänne, *MHLR 3*, Appendix 3), the use of Ostrogothic cavalry against Isaurians and Odoacer (Syvänne, *MHLR 5*), and the use of Huns at the Battles of Dara in 530, Ad Decimum in 533, and Rome in 537 (Syvänne, *MHLR 6*, Index). Similarly, even if the Ostrogothic federates in East Roman service proved troublesome, they were often still the key to military successes during the fifth- and early-sixth century (e.g. Syvänne, *MHLR 3*, 257, 265–72, 285–8; *MHLR 5*, 73–5, 140–3, 148–61, 168–71, 232ff., Appendix). Similarly, it is clear that even if the newly-recruited Germanic federates caused multiple problems in the latter half of the sixth century, they also provided the Romans with the necessary manpower to succeed against the Avars, Slavs and Persians.

The ethnic units, in particular the Germanic peoples and Huns, brought with them another peculiar weakness, which was their fondness for wine and alcohol in general. It is clear that the Germanic peoples, Huns and others used alcohol before the battle to relieve pre-battle stress and to instil courage (*spiritus fortius*). This caused occasional problems, but was not a major factor as far as we know in wars. Furthermore, the regular Roman forces could also be guilty of this at very inappropriate times. As already noted, in 593 the Roman soldiers celebrated their victory by drinking wine and were nearly defeated as the Slavs returned. Fortunately for the Romans, their enemies were also prone to misuse wine. The Romans appear to have exploited this in particular when facing the Germanic and Slavic peoples. Unfortunately for the Romans, the Muslims did not use wine. Instead of this, the Muslims used *khat*, which was actually very beneficial as a combat drug when used in the right circumstances. It acted as a stimulant and alleviated hunger and thirst so that the soldiers could continue fighting through nights.[36]

The sixth century saw a peculiar new development that resulted from the attitude of Belisarius and other cavalry officers towards infantry. They considered the footmen of that period to be poorly trained and of very little use in pitched battles. It was thanks to this that during the re-conquest of Italy the Roman infantry (Isaurians and regulars) obtained horses, so becoming cavalry (Procopius, *Wars* 5.28.23–27, 5.29.42). It is clear that these forces cannot have been as efficient as the regular cavalry or the foreign units, but this is not immediately visible from Procopius's text, with the implication that the transition from infantry to cavalry was mostly successful even if it took place in the middle of a campaign. This means that the footmen knew how to ride and were able to fight mounted in some manner.

The Romans also expected their cavalry to be able to dismount and fight as infantry against an enemy force consisting of either cavalry or infantry. This was the recommended course of action if the enemy consisted of Persian cavalry deployed on rough ground, or against nomads in the right circumstances, or if the enemy force included a significant number of infantry. In addition, Roman combat doctrine expected the cavalry to be able to dismount when forced into retreat so that they could fight back more effectively against pursuing enemy cavalry. It is clear that the dismounted cavalry was in general not as good as the regular infantry (when it had been drilled thoroughly and did not consist of unreliable soldiers, as it did for most of the period between 468–530) in their style of fighting, but the evidence still suggests that dismounted Roman cavalry fought very well and in all known instances defeated their enemies with this style of fighting.[37]

6.3. Cavalry Grand Tactics[38]

The Roman *strategos* could employ his cavalry in many different ways. The two main variants of the Roman cavalry battle tactics, with or without the second line, consisted of the charge with the first line as a whole, or of the charge with the first line after the *koursores* had already skirmished. Both of these could involve outflanking on one or both flanks, or the use of the centre *meros* before the wing *mere*. In all of the variants the role of the second line was restricted to being a reserve line which protected the flanks of the first line and provided a place of refuge for the *mere* of the first line. The flank guards, outflankers and ambushers also protected the entire formation against outflanking, while the third line together with the second line and its fill-up *banda* protected the rear if there existed a threat from that direction. If possible, the flank guards, outflankers and ambushers were also used against the flanks and rear of the enemy formation. The outflanking actions, whether by the outflankers, flank-guards or by the ambushers, were likely to cause the enemy force to panic, since horsemen were unable to form a circle or square immediately like the infantry could. The horsemen could only retreat and then try to reform and face the pursuing enemy. The chances of this happening improved if there were reserves behind. There existed one method for the outflanked cavalry unit to continue the combat without being assisted by a cavalry reserve, which was to dismount the men on the spot or some other place to the rear before the arrival of the enemy and face the enemy with a barrage of arrows and rows of lances. However, on the basis of the extant evidence, the retreating cavalry dismounted usually only when it reached some reserve force behind them rather than entirely on their own. The usual formation that the dismounted Roman cavalry used in such instances was the rearward-angled formation (*epikampios opisthia*). If the battle became prolonged, with each side having conducted their initial attacks without result, it was possible that the rest of the battle evolved in a very fluid manner, with some of the units advancing and some of the units retreating and wheeling back to face the enemy, or at times involving the reserve units in support, until something decisive happened, for example the enemy commander was killed or his standard thrown to the ground. It is because of this that the *Strategikon* and Roman military doctrine stressed the importance of having fed the men and horses prior to the battle and the importance of carrying a sufficient amount of fodder, food and water for the men and horses. The soldiers and horses would need these if the battle became prolonged.[39]

In addition to this, Roman cavalry also employed feigned flights with ambushes, surprise attacks, stratagems, improvised formations, nomadic formation, and the Persian style formation when the situation required this. Furthermore, the Romans could also employ ditches, ravines, caltrops, and horse-breakers as part of their battle tactics. However, the principal form of cavalry tactics was still the cavalry charge with lance, conducted in close order either at the gallop or at the canter, depending on the time period and type of cavalry employed by the Romans.

The attack with* koursores *deployed as skirmishers[40]

On the basis of the training systems in the *Strategikon* and battle narratives, it is clear that one of the standard tactics employed by the Romans included the use of the *koursores* as skirmishers to disorder the enemy force before the actual main charge. The vanguard[41] and ambushers could also be used in like manner. The idea behind this skirmishing was to weaken the enemy force, attempt to induce them to charge prematurely, or induce them to leave their advantageous ground. The Battle of Tricamarum in 533 demonstrates that the *koursores* of the centre *meros* could also be used in this fashion on their own. See the short summary below. When the *koursores* were used in this manner they separated from their *defensores* and galloped ahead in irregular formation. Then, after riding forward not more than a mile or two (c.1.5–3km), they either forced the enemy into flight or they retreated about half that distance, made three or four quick charges to the right and to the left against the enemy, and then circled back again. After this skirmishing was over they assumed

Koursores skirmishing

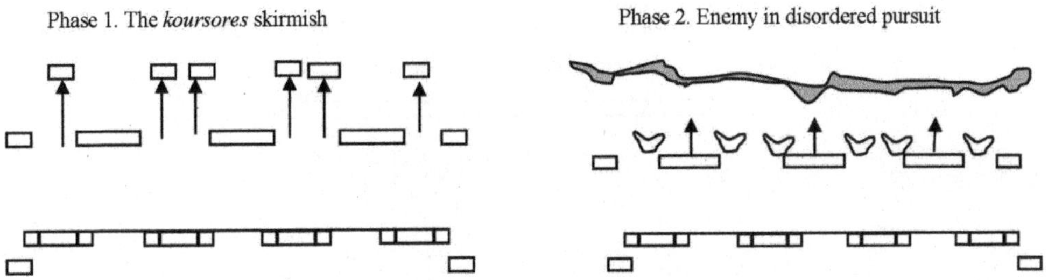

their original position on the flanks of the *defensores*, and then charged together with them against the pursuing enemy. This was essentially a form of feigned flight, and it is also very likely that when the enemy pursued the *koursores* that they also disordered their own line in the process, so that it was easy for the Roman *defensores* to force them into flight.[42]

The Enemy Line with Equal Width[43]

If the Roman first line was equal in length with the enemy line, the standard tactic was to outflank the enemy on the right by using the outflankers. However, it was also possible to outflank the enemy on the left by using the flank guards if the situation suited this. This tactic had two standard variants: 1) the outflankers were all posted along the right flank *meros* so that there were no hidden ambushers in *drouggos*-formation behind them; and 2) one *bandon* (or more *banda* if the enemy were nomads) of outflankers was posted in front alongside the right wing *meros*, so that one *bandon* of outflankers was concealed as a *drouggos* behind them as ambushers so that when the outflankers posted in front started encircling the enemy wing by using the crescent formation the concealed *drouggos* suddenly charged out of the formation to surprise the enemy.

In attack the different officers in charge of the outflankers, right wing *meros*, centre *meros*, left *meros*, flank guards and ambushers (when present) all had to judge the proper time for their respective actions. The first attached diagrams are taken from the *Strategikon* (3.8–10), after which follow my reconstructions of the battle arrays in such a manner that the missing information taken from the text is included. The *Strategikon* includes in the diagrams only the combat tactics employed by the first line, which is the reason why in the reconstructions I add the missing elements for a battle formation that consisted of two lines. It should be noted that the grand tactical manoeuvres depicting only one line in the *Strategikon* were obviously equally usable by cavalry formations that had only one line.

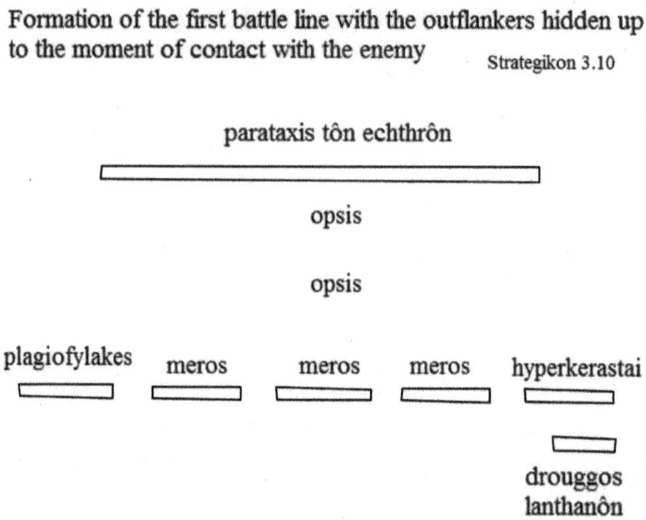
Formation of the first battle line with the outflankers hidden up to the moment of contact with the enemy
Strategikon 3.10
parataxis tôn echthrôn
opsis
opsis
plagiofylakes
meros
meros
meros
hyperkerastai
drouggos
lanthanôn

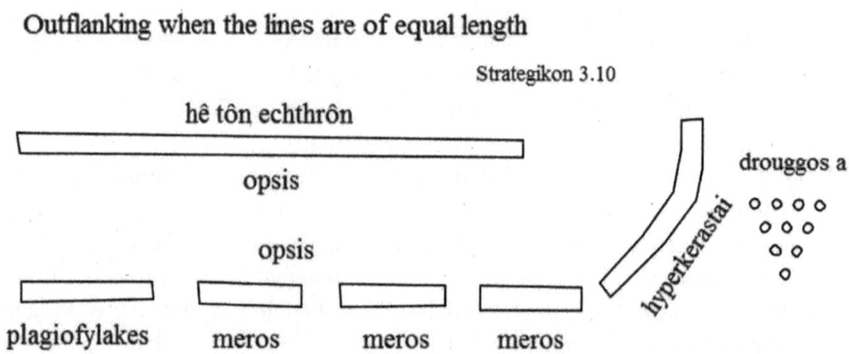
Outflanking when the lines are of equal length
Strategikon 3.10
hê tôn echthrôn
opsis
opsis
drouggos a
hyperkerastai
plagiofylakes
meros
meros
meros

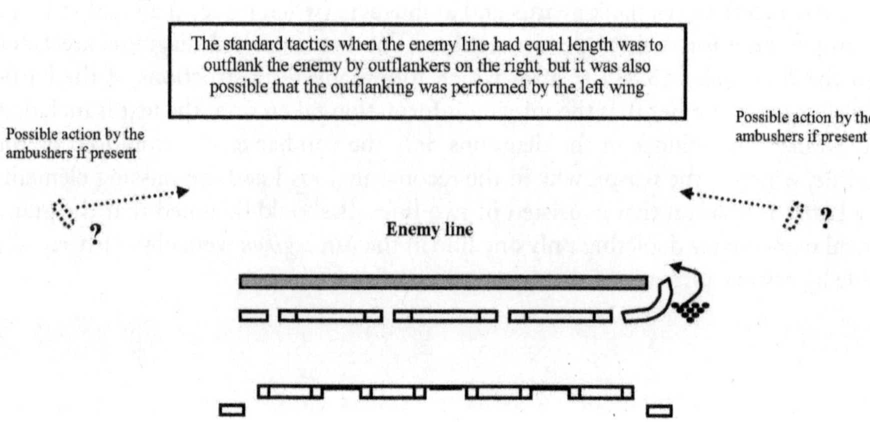
The standard tactics when the enemy line had equal length was to outflank the enemy by outflankers on the right, but it was also possible that the outflanking was performed by the left wing
Possible action by the ambushers if present
Possible action by the ambushers if present
?
?
Enemy line

The enemy with wider frontage[44]

If the enemy formation was wider than the Roman line and the Romans did not have time to extend the line to equal it, the centre *meros* advanced against the enemy before the flanks. The idea was to defeat the enemy before the flanks were in contact with the enemy. If the wings of the enemy line attempted to outflank the Roman centre as by advancing it had exposed its flanks, the Roman flank *mere* could attack those in flank, as happened in the Battle of Tricamarum in 533. The same battle demonstrates that the same method could also be combined with the use of the *koursores* as skirmishers. It is probable that the elite units were posted in the centre *meros* of the first line for situations like this. In the sixth century, the Romans employed this tactic successfully at least twice. The associated diagrams depict the one in the *Strategikon*, followed by my reconstruction of it with the missing units, after which follows a set of diagrams depicting the Battle of Tricamarum between the Romans and Vandals in 533.

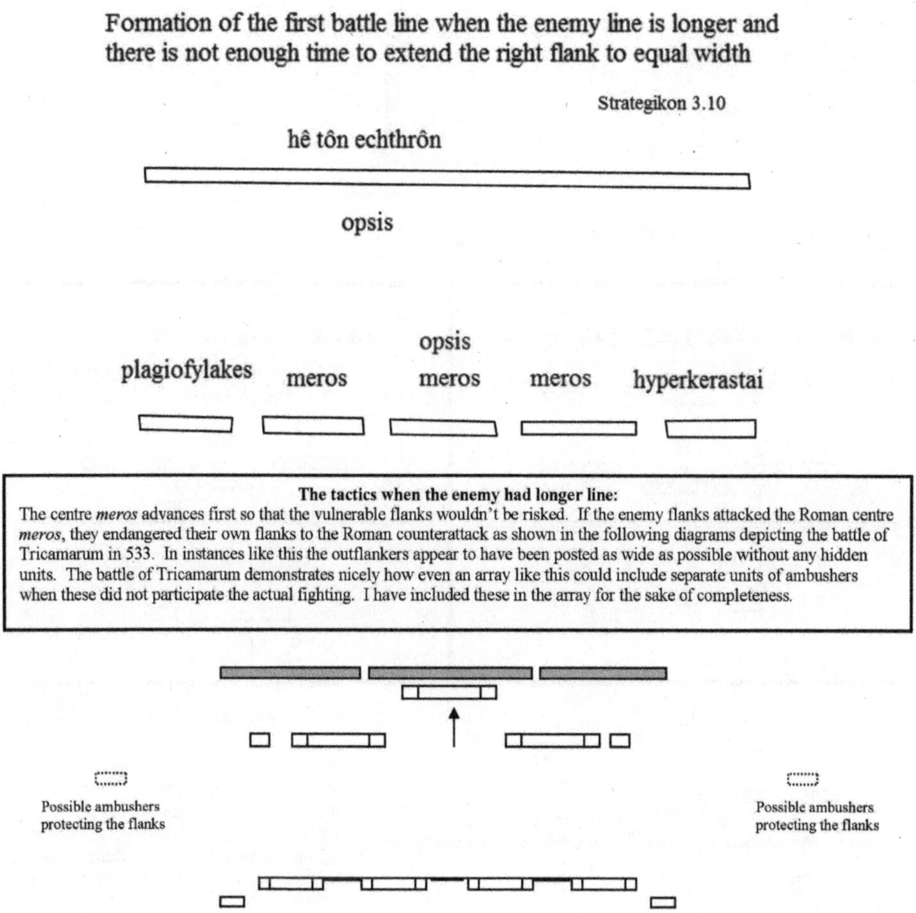

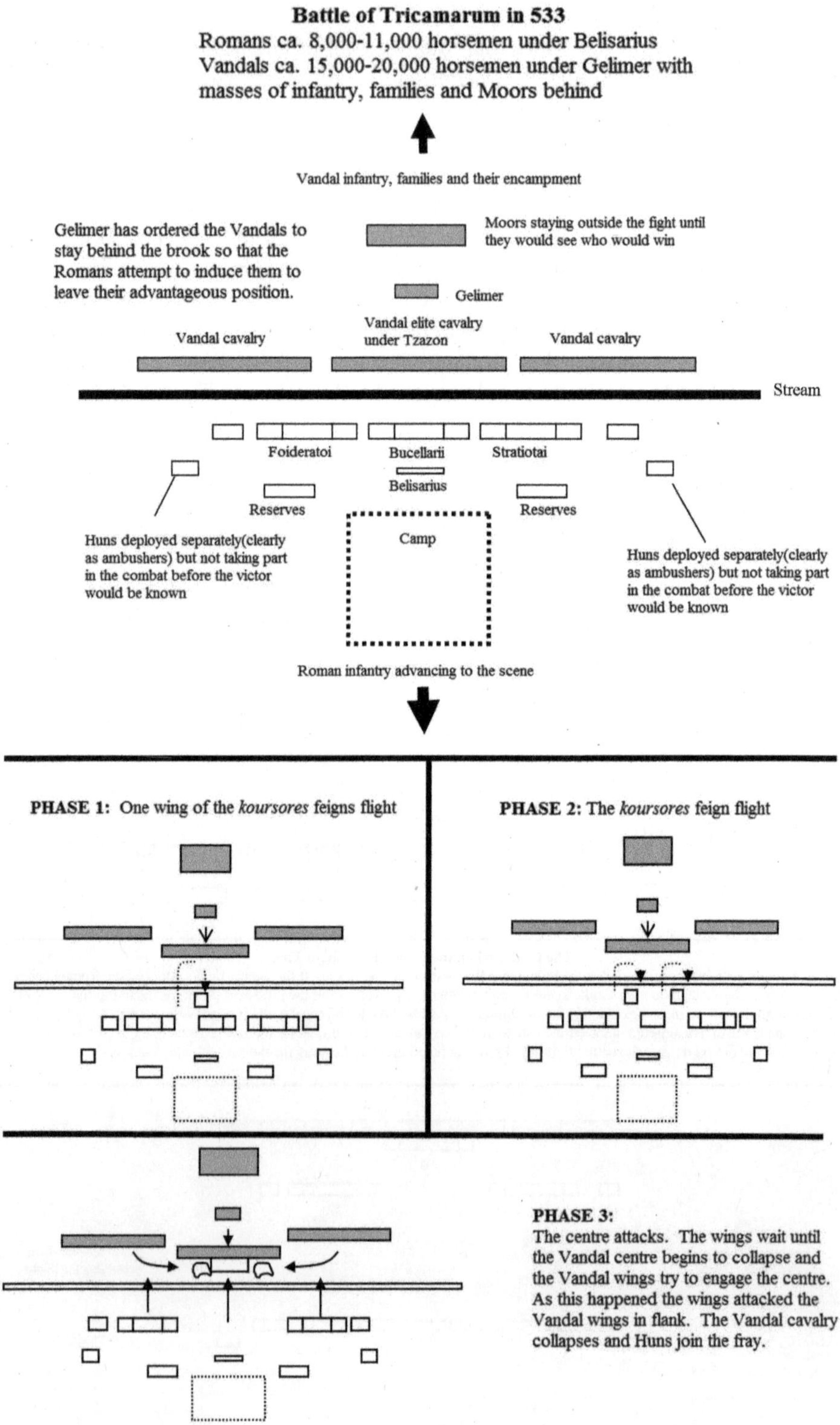
Battle of Tricamarum in 533
Romans ca. 8,000-11,000 horsemen under Belisarius
Vandals ca. 15,000-20,000 horsemen under Gelimer with masses of infantry, families and Moors behind
Vandal infantry, families and their encampment
Gelimer has ordered the Vandals to stay behind the brook so that the Romans attempt to induce them to leave their advantageous position.
Moors staying outside the fight until they would see who would win
Gelimer
Vandal cavalry
Vandal elite cavalry under Tzazon
Vandal cavalry
Stream
Foideratoi
Bucellarii
Stratiotai
Belisarius
Reserves
Reserves
Camp
Huns deployed separately(clearly as ambushers) but not taking part in the combat before the victor would be known
Huns deployed separately(clearly as ambushers) but not taking part in the combat before the victor would be known
Roman infantry advancing to the scene
PHASE 1: One wing of the koursores feigns flight
PHASE 2: The koursores feign flight
PHASE 3:
The centre attacks. The wings wait until the Vandal centre begins to collapse and the Vandal wings try to engage the centre. As this happened the wings attacked the Vandal wings in flank. The Vandal cavalry collapses and Huns join the fray.

The enemy line shorter at both ends[45]

When the enemy line was shorter at both ends, the three *mere* slowed down their advanced a bit to allow the flank guards and the outflankers to get around the enemy line in crescent formation before coming into contact with the enemy. This had the aim of disordering the enemy line before the main forces engaged them. This could actually result in the rout of the enemy line even before the main formation reached it, so the *koursores* could begin their pursuit immediately. Even when this did not happen, it was very likely that the enemy line would collapse soon after the main force reached it, so the enemy flight would have started from their flanks, from which it would then have spread like wildfire towards the centre. However, the sources give us some instances in which this happened only on the enemy's flanks and their centre stayed in the fight. The reason for this appears to have been the fact that the Roman wing *mere* failed to support their centre *meros* and instead chose to pursue the fleeing enemy wings. Good examples of this phenomenon are, for example, the Battles of Constantia/Monocarton in 581 and Solanchon in 586 (Syvänne, *MHLR 7*, 154–8, 212–6).

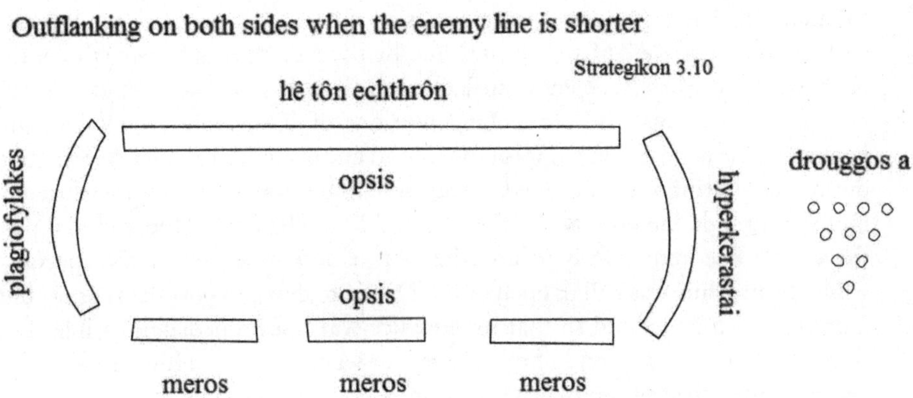

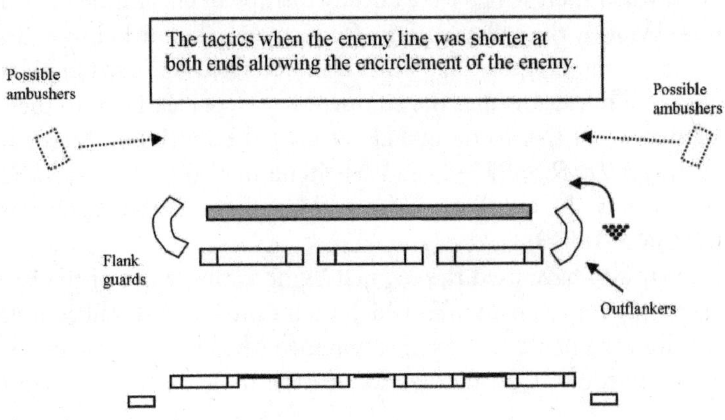

It is of note that when the Romans used this system the concealed *drouggos* was still free to attack the rear of the first line, or the enemy's second line, or any units coming to the rescue of the first. Therefore, the system of hiding one *drouggos* behind was not only useful in outflanking the enemy but also in providing additional protection for those doing it. In situations in which the Romans used separate units of ambushers the situation was even better, as these not only protected the Roman flanks but could also target any enemy force that attempted to assist their first line. This scenario clearly demanded independent initiative from the commanders of the flanks and from the *hypostrategos* or *strategos* who had to slow down the advance of the main line to give the flanks enough time to perform the outflanking. The first of the accompanying diagrams depict the version contained in the *Strategikon* and the one after that is my reconstruction with all of the missing units included.

The Second Line in cavalry Combat[46] *(see also Appendix 2)*
The Romans used the second line in four different manners:

1) When the presence of the second line was hidden from the enemy the idea was to induce the enemy to place their reserves on the flanks to outflank the Romans so that the wing *mere* of the second line or their outermost *banda* (the entire *meros* was used when the enemy outflanked the Romans in great strength) served as ambushers against the outflanking portions of the enemy line. When this happened, the Roman wing units of the second line encircled the enemy formation during the pursuit of the defeated enemy units. The Romans used three different methods to hide the presence of their second line. The first of these was to post the second line immediately behind the first in close order during the approach while the first line was still in open order. The second was to post the support line more than a mile behind so that its presence was not immediately visible. The third was to post the second line in a forest or woods, behind a hill, in a valley or behind some other terrain or artificial feature (e.g. behind walls, as Belisarius did during the Siege of Rome in 537). When the use of the wing reserves was well-timed, the success of this action was certain because the enemy would be totally taken off guard when their wings were already disrupted, but if timed poorly the end result was obviously the collapse of the Roman wings before help could arrive. The evidence, however, suggests that when this stratagem was used in practice it always succeeded. This means that the commanders were clearly up to their task. Examples of the use of this tactic include at least the Battles of Mesopotamia in 422 (Syvänne, *MHLR 3*, 275–8) and Melitene in 576 (Syvänne, *MHLR 7*, 103–12), and possibly also the battle between Heraclius and the Persians in 625 (Syvänne, *MHLR 8*, 168–9).
2) The Romans may also have used the feigned flight of the centre of the first line to induce the enemy into a position in which their centre and overall commander were targeted from the flanks and front simultaneously. The probable example of this tactic is the Battle of Constantia/Monocarton in 581 (Syvänne, *MHLR 7*, 154–8).

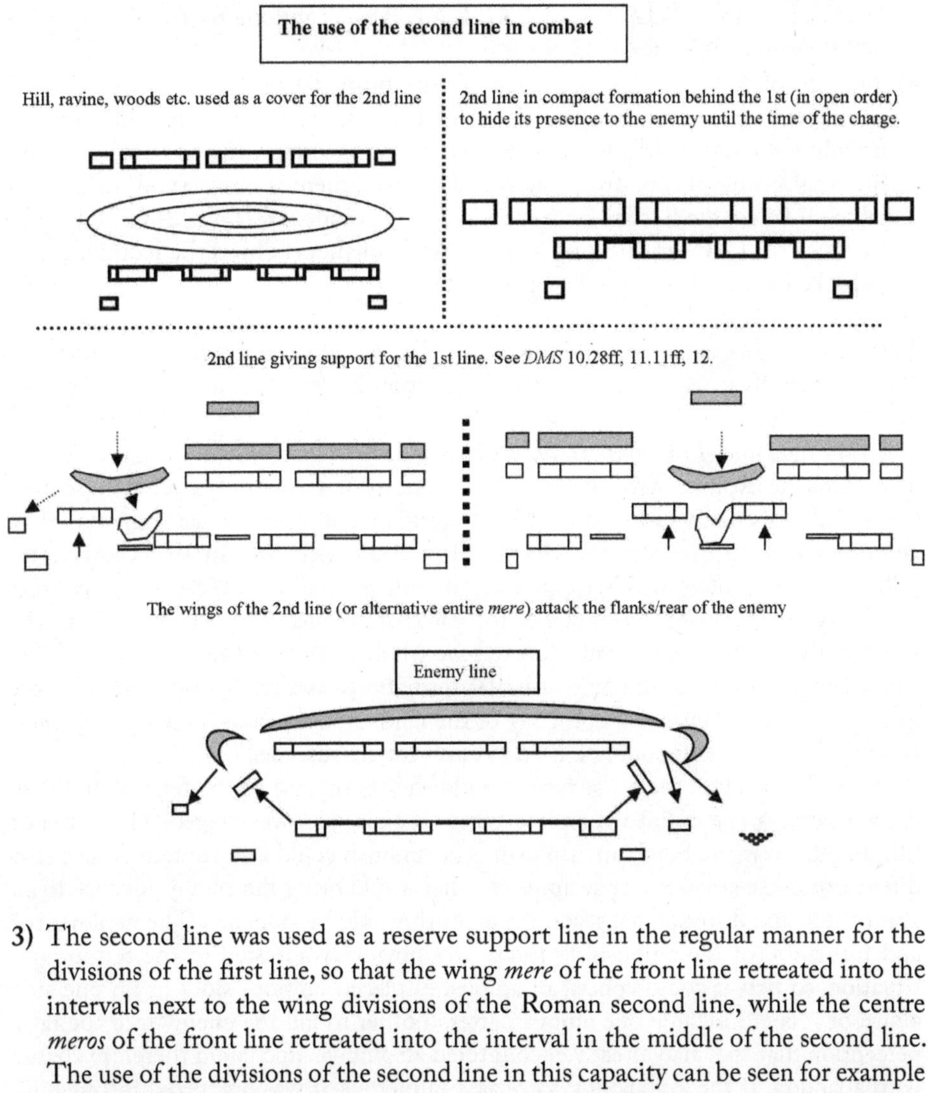

3) The second line was used as a reserve support line in the regular manner for the divisions of the first line, so that the wing *mere* of the front line retreated into the intervals next to the wing divisions of the Roman second line, while the centre *meros* of the front line retreated into the interval in the middle of the second line. The use of the divisions of the second line in this capacity can be seen for example in the following battles: Philip the Arab against the Carpi in about 247 (Syvänne, *Gordian III and Philip the Arab*, 171–3), Constantia/Monocarton in 581 (Syvänne, *MHLR* 7, 154–8), Solanchon in 586 (Syvänne, *MHLR* 7, 212–6), Lake Urmiah in 593 (Syvänne, *MHLR* 7, 276–7) and Plain of Canzak in 593 (Syvänne, *MHLR* 7, 277–80). A variation of the use of the second line in support of the first was to post the reserves behind a wall, as Belisarius did during the Siege of Rome in 537 (Syvänne, *MHLR* 6, 143ff.). Belisarius used this combination in such a manner that if the cavalry that had advanced from the city to fight against the Ostrogoths was forced to retreat, he could dispatch reserves from inside the walls to its support. The use of reserves usually worked as expected, but there were also instances in which the reserves did not ensure success for the Romans. These include for example the Battles of Placentia in 271 (Syvänne, *Aurelian and Probus*,

79–81), Faventia in 542 (Syvänne, *MHLR 6*, 246–7) and the Battle of Mugello in 542 (Syvänne, *MHLR 6*, 247–8).

4) The use of the second line against threats from the rear. Cases in which the enemy managed to surprise the Romans from the rear were rare and there are usually not enough details for us to know how successful the use of the second line was against threats from the rear. However, there is one certain instance in which it failed, the Battle of Faventia in 542 (Syvänne, *MHLR 6*, 246–7). In that instance the Ostrogothic king Totila sent ambushers behind the Romans, who panicked when these attacked the rear.

The attached diagrams, borrowed from my *The Age of Hippotoxotai,* show how the Romans used their second line in combat to ambush the enemy and also as reserves.

Feigned Flight and Ambushes[47] *(see also Chapter 12)*

The favourite Roman cavalry tactic was actually not to fight a regular pitched battle with the enemy in the open. The preferred tactics were to use feigned flights, ambushes and surprise attacks, which will be dealt with later in a special chapter (Chapter 12) devoted to this topic. Here it suffices to discuss these in the context of cavalry tactics. The importance of the topic of ambush is also obvious from the fact that the *Strategikon* not only devoted the whole of Book 4 this but also included ambushers even among the regular battle formations, as already discussed. The key to using cavalry ambushes was the lay of the land. Dense woods, valleys, steep hills, ravines, and mountains could be used as cover for ambuscades.

In cavalry combat, the most typical ambush was the use of the feigned flight to draw a careless enemy into the place where the ambush was posted. The name of this ambush was the Scythian Ambush. The ambush could also contain caltrops, or ditches, or horse-breakers, or swamps etc. that would bring the enemy advance to an abrupt stop and disorder just when the ambush would be launched. The deployment and formation of the ambushing forces was obviously adjusted to the terrain and situation, so that it could consist of ambushes placed on both sides or on one side alone, or it could include one ambush after another to lull the enemy into the false perception that they had already encountered an ambush and could therefore charge with abandon. If the ambushing force was numerous, it was organized into regular units (*tagmata, moirai, mere),* but if the ambushing force was small it was expected to be deployed in irregular formation as *drouggoi.* It was also possible to perform the feigned flight in Scythian formation, so that the units closest to the enemy continued to flee while the rest turned left or right and attacked the flanks and rear of those following. After this, the centre (with possible reserves) also turned around to attack the pursuers.

As already noted, the regular cavalry array had a variant formation in which horse-breakers (or caltrops) were placed between the battle lines so that the first line was placed a mile (c.1,480m) in front of the obstacles and the second line two to three bowshots (ca. 660–990m) behind the obstacles. The Romans left four to five gaps in the obstacles, each clearly marked (tree branches, or spears, or rocks, or earth, or other marks) and each 300- to 400-ft. wide, so that the first line could flee

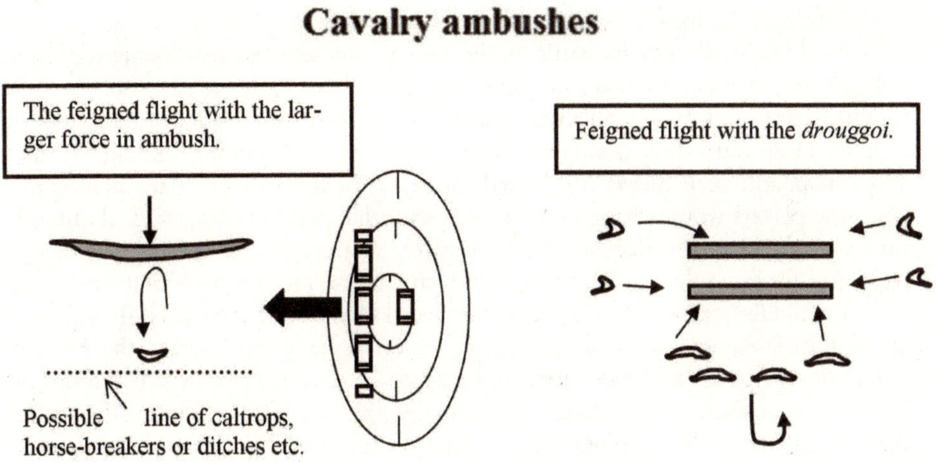

through these safely. The retreat was to be performed by *banda* in irregular *drouggos*-formation. The markers were to be removed by the last units. If necessary, the second line could also advance through the gaps to help the first line by attacking the enemy. When the enemy ran into the obstacles, the Roman ambushers were to attack them from both flanks. The other alternative was to place the entire Roman array behind the obstacles so that they were about three bowshots (ca. 990m) behind these. In this case, the intervals between the obstacles were to be smaller to maximize the damage to the enemy and the Romans were to time their attack to the moment when the enemy horses were falling into the pits or being disabled by caltrops or horse-breakers.[48] The Battle of Adrianople in 587 is a good example of a Roman feigned flight (Syvänne, *MHLR* 7, 251).

Dismounting the Cavalry[49]

The horsemen were not only required to be adept at cavalry combat proper, but also as infantry. If the enemy was posted in difficult terrain, or if the enemy force included significant numbers of footmen, or if the Romans needed to storm a fortified enemy camp, the troopers could be ordered to dismount either to fight as infantry on their own or to stiffen the numbers of Roman infantry. As already noted, the standard infantry formation for the dismounted cavalry was the *epikampios opisthia* (rearward angled formation). It is quite clear that the Roman cavalry was the most versatile and able force of its day. They were true multipurpose soldiers who could fight as mounted archers and lancers just as well as acting as light and heavy infantrymen. In Roman service even the Huns could be ordered to dismount to fight as infantry.

6.4. Length of Cavalry Battle[50]

The typical cavalry battle did not last long. Most of the cavalry battles were very soon over unless one had the support of infantry contingents, or some of the men were dismounted, or the men were able to retreat into a fortified marching camp, fortified city, or fort. The reason for this was the physical limitations of the horse. We can calculate the typical length of a cavalry battle on the basis of the rules and requirements of horse polo, because it was designed to imitate the demands of high-paced cavalry combat. On the basis of this we can calculate that the horse could endure a maximum of 12–13 minutes of all-out effort, during which it could cover a distance of 4.8km (3 miles) at a gallop, and even this is considered excessive for horses in modern polo games. The polo players would change their horses after that period of time.[51] Excluding the time periods when the Romans used the gallop as their principal form of attack, as happened for example during the reign of Justinian, we may estimate that the Roman cavalry could fight a bit longer because they paced their attack and used only the canter in the initial stages, and the use of the gallop was usually restricted to the *koursores*, retreat or emergencies. We also need to take into account that if one side did not flee before contact then the two battle lines usually halted in front of each other, which meant that the horses were not used to their maximum. This means that the horses could endure the fight a bit longer than in a polo game.

Perhaps the best way to estimate the length of the cavalry battle when the Romans used the canter would be to compare it with a modern cavalry parade. Such cavalry parades which include the gallop, resting, simulated attacks and flights, parade ground manoeuvres and standing still, can go on for an hour or so.[52] However, it is clear that the distance the polo horses cover is roughly the same as we should expect from the *koursores* when they pursued at a gallop. In other words, whenever the Romans used the gallop their horses could not endure more than 12–13 minutes of action before becoming exhausted. This set the limits for the length of the cavalry combat. If neither side had achieved victory within these parameters, the troopers needed to change their horses, which then enabled them to launch a new round of attacks and retreats. If this did not bring victory to one side, the situation continued as long as there were spare horses. It was because of this that all armies possessed spare horses. On the basis of the referrals to spare horses, we know that each Roman trooper had at least one spare horse, but it is likelier that the typical number of horses was three because we find the Ostrogoths in Roman service using this number. This means that the maximum length for the intense cavalry battle would have been one hour if all of the spare horses were used, but as already noted at least during the reign of Maurice the Romans did not bring all of their spare horses to the site of the battle, which definitely shortened the length.

The nineteenth-century French savant Ardant du Picq (Greely and Cotton tr. 219–20) aptly characterized cavalry combat when he stated that the cavalry always advanced further in attack than infantry and it exposed its flanks by doing so, with the result that cavalry needed more reserves than infantry to protect its flanks and rear. In short, the pursuers always needed reserves. The victory usually belonged to the side which had the last intact reserves, because with cavalry one could take the offensive only when one had reserves left. As a result of this, cavalry engagements lasted only a moment and regrouping had to be performed immediately. This required quick-thinking cavalry officers. The cavalry battles were usually characterized by sharp action which did not last long.

When the battle became prolonged it is clear that parts of the battle line became stationary due to the horses needing to be rested. As already discussed, the military treatises of Arrian (second century) and *Sylloge tacticorum* (tenth century) include a cavalry unit order called 'tortoise'. There is no definite evidence for its use during the Late Roman period and it is not mentioned by the *Strategikon*. However, one may make the educated guess that this extra-tight version of the cavalry formation may have been used in cases where the cavalry was stationary for one or another reason (note for example the Battles of Satala in 530 and Callinicum in 531) and needed protection against enemy missiles. What is certain is that the front rankers would have at least adopted a stationary variant of their charging unit close-order in that they would have placed their shields in front to protect themselves and their horses' heads when peppered by missiles.

However, there are very good reasons for the omission of this unit order and stationary tactics in the *Strategikon*. Stationary cavalry lost its main function, which was its psychological impact on the enemy. In order to be effective, cavalry had to charge, retreat and repeat the charge. It is therefore clear that in the *Strategikon* the cavalry was not expected to be stationary at all. If the cavalry was unable to achieve a victory in the manner described by the *Strategikon* it was either to dismount or it was to retreat to some safe location, such as a marching camp or behind an accompanying infantry formation. See the discussion in the next chapter.

The battle, however, could also result in a stalemate, meaning that the availability of spare horses did not have any influence on the length of the battle, when one side dismounted, or sought safety in difficult ground, or a fortified marching camp, or a city or fortress. In such cases, cavalry battles could last even for days. The best example of this is the battle fought between the Romans and Chosroes II at the Battle of Arzamon in late-604 or early-605 (Syvänne, *MHLR 8*, 51–3). The Roman cavalry won the first day but could not inflict a decisive defeat, with the result that Chosroes's elephants gave him the victory next day. Regardless, it is clear that the typical cavalry battle did not last for long because it was far more typical for one side to win either outright or after the reserves intervened in combat. This means that the typical cavalry battle was over in minutes so that most of the time was actually spent in pursuit, which did not last more than a few minutes before the horses needed to be rested or changed.

Obviously the physical condition of the horses or their physical abilities (horses could not be used like footmen in a stationary position) were not the only matters that influenced the length of the battle. The most important were actually the morale, the numbers, the amount of food and water intake, the individual initiative of the officers, the type of enemy, the weather, and the terrain. Roman fighting methods and their cavalry battle formations were designed to get the maximum benefit out of the psychological impact that these had upon their own and their enemies' soldiers. If all of these were in order and the Romans followed their combat doctrine, they were likely to win within minutes, because as the most versatile cavalry force on earth they were always able to pit their strengths against the enemy's weaknesses, but as always there were individuals who acted contrary to the combat doctrine. Furthermore, there was always the unfathomable thing present in wars: the influence of chance on the result, which after the adoption of Christianity the Romans called the Will of God. This was the reason for Maurice's recommendation for the use of stratagems, surprise attacks and ambushes rather than pitched battles if possible. When employed well, all of these shortened the length of combat even further.

6.5. After a Victory[53]

If the Roman cavalry attack resulted in the flight of the enemy, when conducted by the book the pursuit required that the pursuit of the enemy was to be relentless if this was possible while still taking precautions against possible reverses and the presence of enemy reserves and ambushers. It was because of this that there were the *defensores* and the second line. Maurice recommended unrelenting pursuit of the enemy with the aim of destroying them completely. The pursuers were to seize the opportunity and not give the enemy any chance of regrouping or claiming victory. The *koursores* were to follow the enemy at the gallop to their marching camp, while the *defensores* and the rest of the Roman formation followed.

The *koursores* were to try to regroup on their own first before seeking help from the *defensores*. Under no circumstances were the troopers to loot the dead enemies, or raid the baggage train or camp, or plunder the belongings of the enemy until the enemy were completely destroyed. The reason for this was simple. If the soldiers scattered to pillage, this disordered the army, prevented effective pursuit of the enemy and gave the enemy chance of regrouping for a counterattack. The looting of deceased enemies and enemy camp was to be conducted by the medics (eight to ten of the less skilled fighters in each *tagma*) posted behind the battle line. They had a triple duty: they evacuated the wounded, took unhorsed soldiers back to their spare horses, and looted the enemy. This system was meant to ensure that the soldiers posted for combat were able to do so without distractions. In short, Roman military doctrine took into account all of the necessary precautions needed. However, in practice these precautions were sometimes thrown to the wind when the soldiers started to loot indiscriminately. Good examples of this bad behaviour are the Battles of Tricamarum, Melitene and Solanchon.[54] This resulted mainly from two things: 1) the soldiers were not only motivated by their salaries and military glory, but also by the availability of war booty because the army consisted of mercenary professional soldiers, which at times when the salaries were in arrears could cause a collapse of discipline; and 2) when the army included large numbers of foreigners who fought for financial gain, the likelihood for the pillage was greater than with native soldiers. However, since the untimely looting of the enemy marching camps is not mentioned in most cases, it is clear that the Roman soldiers followed their combat doctrine most of the time.

The use of the *koursores*, *defensores* and support line were perfectly designed for the maintenance of safety in pursuit, but obviously with the caveat that it was always possible that the destruction of the pursuing *koursores* could also infect the *defensores* and second line with panic so that these failed to act as combat doctrine expected. The sources provide us with three examples of this happening. Firstly, the emperor Aurelian lost at the Battle of Placentia in 271 when the invaders used feigned flight so that his first line charged into an ambush posted in the woods. Secondly, when the entire first line charged into trenches prepared by the Persians in front of the city of Thannuris in 528. And thirdly when the *koursores* charged into an ambush posted inside the town of Anglon in 543.[55] In each of these instances the reconnaissance was done poorly and the pursuit of the enemy was conducted in a reckless manner contrary to military doctrine. However, in most cases the pre-battle reconnaissance and the pursuit of the defeated foe was conducted by the book. When the Romans followed their combat doctrine, they were the most efficient force in existence.

If the defeated enemy had retreated inside fortifications, Maurice proposed three schemes: attack, cut off supply lines, or agree to an advantageous treaty. It was not an easy task for the cavalry force to attack a fortified city or fortress or a fortified camp. The narrative sources demonstrate this well. Sometimes the Romans were successful while at other times they were not. When the enemy was unable to close the gate behind them, or was too demoralized thanks to the effectiveness of the pursuit, the storming of the place (this often involved the dismounting of cavalry to fight as infantry or the bringing of infantry force to the scene) was always successful: but when the enemy had not lost its morale completely, or possessed infantry reserves, the Roman attack following the pursuit usually failed. The main problem for the cavalry was that if the enemy was not completely demoralized, the cavalry was not well-suited to the storming of fortified positions or places of refuge like hills and mountains. Fighting in these conditions required the cavalry to dismount or it required the presence of infantry reserves. It was in such situations (when there was not enough infantry present or the general did not want to dismount his troopers) that the Romans were to cut off the lines of supply or were to negotiate with the enemy. One of the main advantages of Roman combat doctrine was indeed its adaptability to different situations.

One of the ancient stratagems that the Romans used in such cases was to leave one route open to induce the enemy to flee rather than fight, but this was obviously not shared by all Roman thinkers. As we have seen, Maurice preferred relentless pursuit of the foe. It was easier to fight a fleeing foe rather than to fight a foe who fought back from an advantageous position. This was used in particular when there was a danger that the enemy could collect a relief force, which was usually the case when the Romans faced Persians (e.g. after the Battle of Solanchon in 586). Sometimes it was also preferable to allow the enemy to flee. For example, if the defeated enemy outnumbered the Romans, they could be forced to fight back if pressed too hard during pursuit. However, when there was no danger of a relief army, like happened in the case of the besieged Vandal king Gelimer in 533, the Romans settled on besieging the place until the enemy ran out of supplies. If the Romans considered it impossible to capture the place where the enemy had sought refuge, they abandoned the attempt altogether (e.g. after the Battle of Nineveh in 627), or the battle was renewed next day (e.g. after the Battles of Melitene in 576 and Arzamon in late-604 or early-605).[56] The principal difficulty for the cavalry arose from the fact that enemy camps and places of refuge such as hills and mountains were unsuited for cavalry action.

The relentless pursuit of the enemy was important because the enemy was at its most vulnerable precisely during the pursuit. When the pursuit was pressed home, the fleeing horsemen were unable to cooperate together against those chasing them. The pursuit was helped by the tendency of the defeated crowding together in the initial stages of the routs, but it was typical for even the later stages of the flight. The fugitives who crowded upon each other always made themselves easy targets. Those who scattered away from the formation had better chances of survival but as lonely targets they were still easy targets for larger groups of pursuers. If the battle had included encirclement or ambushes, the situation for the fugitives was even worse. An effective pursuit could turn a minor victory into a major victory. Most of the casualties among ancient and medieval armies resulted from disorderly flights. It was typical for panicked soldiers to throw away their weapons and shields and sometimes even their helmets and armour in an effort to speed up their flight. They were often so panicked that they rode their horses to death. The panicked Romans did the same for example after the defeats at Anglon in 543 and in Armenia in 577. Excluding situations in which the route of retreat was blocked by the encircling Romans, or by ambushers, or by the presence of a terrain obstacle such as a river, cavalry battles were still unlikely to cause the same amount of casualties as infantry battles.

It should be noted that in practice the Romans did not always conduct their pursuit relentlessly, even when the soldiers obeyed and followed all orders. There were occasions in which the Roman commander chose not to pursue the defeated enemy. This was particularly true in situations in which the enemy outnumbered the Romans significantly, so that the Roman commander feared the prospect of counter attack if the enemy were pressed too much. Furthermore, there existed another school of thought in Roman military circles concerning the advisability of pursuit, which was that a defeated foe should not be pursued too effectively to ensure that it would not be forced to fight back. See for example Frontinus (*Stratagems* 2.6). Maurice did not share this view. Consequently, we find the Romans following two entirely different approaches to the pursuit of a defeated foe. Which of these two approaches was adopted depended on the personality of the commander (cautious vs. bold), on the personal views of the commander regarding the right policy, and on the situation.

6.6. After a Defeat[57]

Roman combat doctrine also took into account the possibility of defeat, so it is not surprising to find instructions on how to act in such situations in Frontinus (*Stratagems* 2.7–8, esp. 2.10 and 2.13), Vegetius (*Epitoma rei militaris* 3.25), *Peri Strategikes* (38), *Excerpts of Polyaenus* (46–8) and Maurice's *Strategikon* (7.B.11).

The extant instructions of Frontinus obviously concern the use of stratagems to restore the Roman position, but it is clear that his work on tactics included instructions that were similar to the ones that we find in Late Roman works. If the enemy had inflicted a defeat, the first order of things was to attempt to save the remnants of the army. As noted by Vegetius, depending on the situation, this could be done by retreating into a camp, rough ground, or a hill. Vegetius also notes that it was sometimes possible to reverse the situation if brave men were still left in the Roman army who could exploit the disorder of the pursuers. This was indeed a distinct possibility. Vegetius also recommended the regrouping of the remnants, after which they were to be exhorted and new recruits and reinforcements sought, from when the enemy was to be engaged with surprise attacks and ambushes which revived Roman morale. All of this is just common sense which we also find in the narrative sources.

The *Peri strategikes* advised the commander to post a detachment of experienced horsemen two- to three-miles behind the phalanx in a hidden location (behind a hill or trees, or on a river bed etc.) so that these could then show themselves to enemy if the army was forced to retreat. Their sudden appearance was bound to frighten the enemy with the prospect of ambush or with the arrival of enemy reinforcements. If terrain was unobstructed, the detachment was to be posted three or four miles away. The *Peri strategikes* also instructed the rear guards to carry caltrops which they could then throw behind to stop the pursuers.

It is very likely that these methods were also used in other circumstances. For example, one could easily use the vanguard mentioned by the *Strategikon* that had retreated behind the battle formation prior to the actual battle in this manner. The *Strategikon* does not state where the vanguard *banda* were posted during the battle. A place two- to three-miles (terrain allowing the hiding of the men) or three- to four-miles behind (open terrain) could easily have been this place. *The Excerpts of Polyaenus* naturally concentrated on the various stratagems that could be employed to enable the Romans to retreat which are too numerous to include here.

The instructions in the *Strategikon* are also known to reflect older Roman military practices, not only the methods of Maurice's day, which means that it is probable that similar methods were used for most of the Late Roman period. As above, the first order of things was to secure the remnants of the cavalry force. If the victorious enemy consisted mainly of infantry, then the Romans simply withdrew on the double on horseback while maintaining good order to such a place where they could build a marching camp safely. If the enemy consisted of cavalry, for example the Persians or Scythians, the *Strategikon* recommended the abandonment of superfluous property and slower horses. While a small detachment of mounted men was retained, the rest of the troopers dismounted and were deployed on foot as a double phalanx or as a hollow square/oblong. The horses and the baggage were placed in the middle. The soldiers were deployed so that the outermost layer of the infantry formation consisted of archers on foot. The use of the dismounted men then enabled the Romans to retreat because cavalry formations were usually unable to threaten well-ordered infantry formations. The instructions concerning the marching camp and infantry make it clear that the Roman cavalry could also retreat to their camp rather than dismount in the manner described. On the basis of the narrative sources, we also know that Roman cavalry did not usually dismount in the manner described by Maurice, not even during his reign, but usually retreated either to their marching camp (as also instructed by Maurice) or to some higher ground where they could attempt to regroup and attack their pursuers. Obviously there were also instances in which the Romans were so completely defeated that regrouping could only take place several days later.

According to the *Strategikon*, if the first day of battle had ended in defeat, the same defeated soldiers were not to be used for combat within the next few days. It was a cultural trait for the Romans not to be able to retrieve a defeat immediately after a defeat, whereas the so-called Scythians had made a habit of it. It is well-known that the role of morale was decisive in deciding who would win. Therefore, Maurice instructed that no immediate offensive against the enemy was to be tried unless it was absolutely necessary. The recommended course of action was to build up new confidence in the troops before engaging the same enemy. This was done by the use of stratagems, deception, carefully timed surprises, and fighting while fleeing, until the troops forgot their defeat. Maurice and Vegetius clearly agreed. If morale had recovered and the commander considered it advisable to fight another battle with the same enemy, then Maurice advised to post the former second line in front and the first as its reserves, while retaining selected *banda* from the former first line, because the second had in most cases been smaller than the first line.[58] This safety precaution probably had a double intent: it removed those soldiers from the front line that may have still felt insecure, while it also served as a humiliation and punishment for the previous defeat. The *Strategikon* also recommended that the commander was not to hesitate but was always to act in a decisive manner immediately after a defeat, because delay only demoralized the defeated soldiers further – they would simply have had more time to think about the threats still facing them.

Just as Vegetius advised, Maurice also expected that the officers would immediately exhort the troops to restore the morale. On the other hand, the victors only gained in confidence when they witnessed the inaction of the enemy. The only excuse for delay in the opinion of Maurice was that the commander expected the arrival of help, or if the enemy had made overtures of peace. If the situation was opportune for revenge, then the *Strategikon* recommended the use of the regular cavalry battle formation. However, if the situation did not allow this, it was still important to appear bold both to the enemy and one's own soldiers because this influenced both.

The fact that Roman combat doctrine also took into account the possibility of defeat enabled the Roman commanders to draw on their memory for instructions on how to act, even in such situations. This could be very beneficial indeed when reverses occurred.

6.7. Cavalry vs. Infantry[59]

We also have to ask ourselves the question: 'What was the combat potential of Roman cavalry against infantry?' There is no easy answer to this, because the effectiveness of the Roman cavalry charge against enemy infantry varied from one place and time period to another. As a general rule of thumb, one can say that the psychological and physical limitations of man and horse played the decisive role in all encounters. The horses and men will usually not charge against a solid obstacle such as a close order infantry formation bristling with spear points, while the man at the receiving end is usually unwilling to stay in place when 500+ kilograms of horse with the rider pointing their spear towards them canters or gallops directly towards him.[60] The galloping horse and man was mechanically equal to ten footmen in depth. It was and is possible to change these limitations through training. Men and horses can be taught to charge into solid looking objects, while footmen can be taught not to be frightened by the sight of approaching cavalry. The Romans also made flight more difficult for the infantry by employing the kneeling type of *testudo* or *foulkon* against cavalry. It is therefore not surprising to find that it was still more usual for the horsemen and horses to shy away from the infantry formation because, despite all training to the otherwise, the fear of death is usually the prevailing feeling in such conditions. It is because of this that soldiers have used alcohol and other substances to alleviate this fear, and it was because of this that the *Strategikon* recommended that the generals were not to use great numbers of cavalry in infantry battles and recommended the use of close order infantry formations against horsemen.

The typical cavalry charge against infantry had the following stages: 1) the use of the gallop by the *koursores* when pursuing the defeated enemy cavalry up to their infantry support, or the use of the gallop by the main force (not used during the reign of Maurice) against an intact enemy formation with the idea of scaring the enemy with the impetuosity of the attack so much that they break their formation before contact; and 2) the steady charge at a canter which preserved order and looked threatening to the enemy because of this. The use of the gallop was good against inexperienced or disordered enemy forces, while the use of the canter was better against steady enemy force in situations in which the charge was carried home in the cataphract style, because it was likelier for the cavalry formation to smash into the enemy formation if it retained its cohesion and order. If neither of these tactics worked, then the cavalry usually resorted to the use of missiles to break the cohesion of the enemy array in the hopes that an opening would appear in the enemy formation, or, alternatively, the cavalry attempted to break the enemy formations by feigning flight, or the cavalry dismounted to fight as infantry.[61]

The effectiveness of the cavalry against infantry also depended upon the terrain and the type of infantry and cavalry, and on the type of unit order and tactics used.[62] It is clear that in open terrain, the expectation was that the heavy cavalry would easily defeat the light-armed infantry because these were usually deployed in open formation. This meant that lightly-armed footmen needed heavy-armed infantry in close order to support them in open terrain. Heavy infantry in close order equipped with spears or javelins could defend itself successfully against cavalry, especially when it also included light-infantry so that enemy cavalry could not use missiles with relative impunity. The cavalry could counter this by attacking the flanks and rear of the enemy infantry formation, but if the enemy managed to form up in a battle formation which faced all directions (hollow square or oblong, circle) then the situation was the same as previously. If the infantry did not possess enough missile-armed footmen, and the enemy had enough mounted archers or javelineers, then the cavalry was in a position to pepper the infantry formation with almost impunity, with the result that it would break up eventually opening the way for the cavalry to charge into the formation and break it into pieces.

In order to succeed, the infantry also needed adequate numbers of cavalry, because if it did not possess a sizable cavalry force to help, then the enemy cavalry was in a position to cut off its lines of supply with the result that the infantry force would be defeated if it did not find a place of refuge soon enough. There are such instances in which Late Roman infantry was indeed in this position, and had to retreat without adequate cavalry support. The narrative sources prove that in most cases (there were obviously some defeats too) the Roman infantry was able to retreat to some safe place even when harassed by enemy cavalry, but usually with the result that the Romans then had to negotiate a truce or peace settlement with the enemy. Most of the instances in which this happened took place when the Romans fought against the nomadic mounted archers – the multipurpose cavalry employed by the Persians and Muslims gave the Romans fewer chances, because they could disorder the Roman formation with missiles while the resulting openings were exploited by their cataphracts or lancers. The same was obviously true for Roman cavalry. As multipurpose cavalry the Romans could harass the enemy infantry with their mounted archers and javelineers and then exploit any signs of disorder with their heavy cavalry. In contrast to the Persians, the Romans were also prepared to dismount, which made them actually combat-superior against all types of infantry forces. This, however, was no longer the case when the Romans faced the Muslims. Then the Roman cavalry did not dismount and fight as footmen when this would have been necessary and paid a heavy price for this neglect.[63]

The terrain also mattered. The infantry could use wooded, hilly, broken, swampy and mountainous terrain to its advantage when facing cavalry, because all of these types of terrain broke the cohesion of any cavalry formation attempting to approach, with the result it was unlikely that the cavalry would reach the enemy infantry formation if the footmen used missiles at the same time. The same could be achieved with field fortifications, caltrops or fortified positions. The infantry had a combat advantage in such circumstances. As already noted, Maurice instructed the commanders to dismount their men in such circumstances so that the battle could be fought on almost equal terms. In sum, when the motivation, leadership, physical condition and numbers were adequate, the Roman cavalry could expect to face any type of enemy in any circumstances. No type of enemy had defence superiority when it faced Roman cavalry operating in the manner described by Maurice, but when an enemy infantry force faced Roman cavalry forces that were not prepared to dismount, as happened when the Romans fought against the Muslims, they had defence superiority over the Roman cavalry.

The Romans also possessed a type of cavalry, the super-heavy cavalry, the *cataphractarii*, *equites catafracti*, and *clibanarii*, which was designed not only for cavalry combat, but also as a specialist cavalry force against enemy infantry. It is because of this that we find Vegetius (*Epitoma rei militaris* 3.23) stating that the cataphracts were safe from being wounded thanks to their armour, but with the caveat that the heavy armour limited their ability to endure long combat. The emperor Aurelian had exploited this very fact at the Battle of Immae in 271 by having his more lightly-equipped cavalry conduct a long feigned flight, at the end of which the Palmyran horses were completely exhausted.[64] According to Vegetius, in battles the *equites catafracti* were used against dispersed footmen, but at the same time he noted that the cataphracts could also be used for breaking enemy battle formations (*acies*) when they were posted in front of the legions or between the legionary soldiers (see Chapter 6.2.). In other words, the Romans employed the cataphracts in the same manner as the Persians in frontal charges against enemy infantry. This was disapproved of by some military thinkers, like Maurice, but the fact remains that there were always military thinkers and practitioners in Rome who employed the cataphracts in this manner. In fact, the use of the cataphract wedge against enemy infantry was considered particularly useful during the tenth century AD and we find the great military emperors Nikephoros II Phokas and John I Tzimiskes both using the cataphract wedges with great efficiency against infantry. In situations in which the cataphracts charged against infantry, it should be kept in mind that the use of the canter or trot lessened the probability of the infantry spears actually piercing the armour, which in its turn increased the willingness of riders and horses to trample the infantry under foot. We should remember that horses and riders are able to engage even bulls in modern bullfights. According to Nazarius, the fourth-century *cataphractarii/clibanarii* cavalry were taught to preserve the course of their attack even after they had smashed into the enemy line, which means that they were taught to charge through the infantry formations.[65] The best weapon for use against the infantry at close quarters was the lance because of its greater reach, but the troopers obviously used whatever weapons they had. In fact, the specialist infantry weapon against the cataphracts was actually very short. It was the wooden club or iron mace, which was employed by specialist units trained for anti-cataphract warfare. The decisive element in all such fights was not the weapons system but the will to win.

In short, there is no simple answer to the question of the best use of cavalry when it faced an infantry formation ahead of it. The effectiveness of the frontal cavalry charge depended on the time period and there was great variety from one time period and place to another. Regardless, it is clear that the instances in which the cavalry could defeat a sizable well-motivated infantry arrayed in close order were rare indeed.

The use of cavalry against infantry in frontal combat was not restricted solely to the super-heavy cavalry *cataphractarii* and *clibanarii*. The Romans had always used their cavalry to trample down enemy infantry if the situation appeared favourable and the enemy showed the signs of disorder that allowed the Romans the chance of pressing home their attack, or if the infantry force was in loose order. Good examples of this can be found from the first century onwards, and in fact there are several examples of Roman cavalry just overrunning even legions with a frontal charge in the third century. It is uncertain what proportion of the Roman force in these instances consisted of the super-heavy cataphracts and what proportion consisted of regular cavalry, but what is clear is that these cavalry charges succeeded even against legions. Regardless, it is still clear that even Gallienus, the foremost user of cavalry in the third century, recognized the fact that it was preferable to employ infantry against an enemy force that included infantry and was posted in an advantageous position with a wagon laager. There was a limit to what even the best cavalry of the era could achieve. This is what separates the emperors Heraclius and Gallienus from each other. Heraclius recklessly attempted to overrun enemy infantry formations. At the Battle of Nineveh in 627 he attempted to overrun the Persian infantry when he pursued the defeated Persian cavalry up to them – and failed. This was still excusable, because it was expected that cavalry would attempt to overrun the enemy's infantry reserve when it pursued the defeated enemy cavalry. However, Heraclius's instructions for his brother Theodorus for the Battle of Yarmuk in 634 were inexcusable, and on the basis of al-Tabari's narrative we know that Theodorus acted as Heraclius instructed him to act. In this battle we find Theodorus using the cataphract cavalry against the Macedonian-style pike phalanx employed by the Muslims. Despite the daunting disadvantages that cavalry had when it faced such a formation, the Romans came very close to defeating the Muslims which speaks to the very high quality of the Roman cavalry at the time. This, however, does not change the fact that Maurice's instructions not to use cavalry against infantry were far more sensible than the attempt to break steady infantry formations with a frontal charge. As if this was not enough, the Romans repeated the same mistake in the following battles, using their cavalry in frontal charges against the Muslims.[66]

In Heraclius's defence one can say that he was not alone in his attitude, because the Romans had used their cavalry for frontal charges against infantry throughout the Late Roman era, even if there were commanders who disapproved of this, such as Maurice. The reason why some commanders persisted in their use of cavalry for frontal charges against infantry was that this was sometimes found to be very successful. The best examples of this are the Battle of Siscia in 388 – but with the caveat that in the next battle Theodosius I the Great did not attempt to use his cavalry against the enemy's infantry squares – and the numerous successful cavalry charges of the Ostrogothic cavalry, both in Roman service and on their own, during the period 484–537. It should also be noted that Narses employed cavalry armed with *sarisae* against the Frankish infantry wedge at the Battle of Casulinus River in 554, which strongly implies the use of *clibanarii* cavalry with two-handed *sarisae*. Regardless, throughout the period in question the infantry forces were usually employed against enemy cavalry as the last line of defence with the expectation that they could protect their cavalry against enemy cavalry. This, if anything, stands as proof of the recognition of the general superiority of tightly ordered infantry formation against cavalry, excepting the Roman cavalry in those periods in which it was prepared to dismount and fight on foot.[67]

6.8. The Weaponry, the Individual and the Unit in Combat[68]

The way the battle progressed from the point of view of the individual and his unit depended mainly on the type of unit and the chosen tactic. As we have seen, these could also vary greatly in the course of the battle depending on the success of the tactics. The first stage of combat typically involved the use of bows by selected men during the approach, after which followed the use of spears and swords if contact with the enemy was made. However, if the unit had initially been designated to a skirmishing role, then most of the fighting was limited to the use of the bow in combat, which included the use of retreats, wheelings, about turns, feigned flight and counter attacks. The following discussion analyzes the different stages of combat from the point of view of the individual and his unit.

The Main Charge

When the Romans opted to use the cavalry charge straightaway, the archery was started at about the range of a bowshot (ca. 330m), which was the maximum range for mounted archery.[69] On the basis of the training scheme, the demand to remove lance pennons before combat and the specific instructions for the cavalry charge in the diagrams suggest that the Romans had two different modus operandi for their cavalry charge. Firstly, it was possible that all troopers carried the entire panoply of arms, so that each trooper had two *kontaria*, a *skouton*-shield, a *spathion*-sword, a bow, and a quiver of arrows, so that during combat the troopers designated as archers simply placed the shields behind their backs, which already had the two *kontaria* hanging in slanted upright position. This system may have protected the rear ranks from enemy arrows in the same manner as the pikes in the Macedonian infantry phalanx while interfering only slightly with the Roman archery because it was easier to aim between the spears when these were near. The men designated as archers simply changed their equipment to the spear and shield when close to the enemy and then grasped their bows again later for pursuit. The second of the alternatives was that only the two front ranks and the rear rank carried spears so that each trooper in those ranks had only a single spear, the intention being to facilitate the effectiveness of the archery by the rear ranks prior to the melee. This is the version we find also in Procopius's texts.

On the basis of the training system it is probable that the troopers who were designated as archers shot at least one or two arrows per man in a high trajectory before their unit made contact with the enemy, after which the archers probably grasped their melee weapons. Considering the distance, it is likely that this was the Roman archery technique which stressed the power of the bowshot, which the Romans had used as their only archery technique at least until the reign of Justinian, because we find Procopius contrasting the powerful Roman technique with the weaker Persian one.[70] By the reign of Maurice, the Roman troopers were also required to be able to shoot in the Persian manner with weaker bows, which in their turn enabled a faster shooting rate. We do not know when the Romans adopted the Persian technique because it is possible that it took place before the writing of the *Strategikon*, but we certainly know that there had always been units in the Roman army that used Persian archery techniques because there had always been Parthian or Persian units in the imperial Roman army. By using the Persian technique the archers were therefore able to shoot at least five to ten arrows before the unit reached the enemy line.[71]

It is clear that the archers used the command to charge and the lowering of the lances by the two front lines as indicators of the approximate range to the enemy. The archery was also facilitated by the fact that the front ranks leaned forward and put their shields in front of them and held their spears shoulder high. This gave the men of the third and the fourth ranks a clearer view of the enemy by removing the obstacles in front. When the troopers started using stirrups in greater numbers (in other words in the sixth century), the troopers could stand on these to obtain a better view over the front ranks. In those cases where all men were commanded to use bows, as would have been the case for example in some of the battles fought during the reign of Justinian, the front-rankers obviously aimed their shots at the chests, legs and heads of the horses, while the rear rankers aimed at higher trajectory. The men behind the fourth ranks simply had to shoot by using instinct acquired through training. The bowshots in higher trajectory were obviously aimed at an area so that these dropped from above and hit whatever was below. This was not a particularly effective way of using mounted archery, but we have to remember that the main aim

MOUNTED ARCHERY TECHNIQUES

- The three basic forms of mounted shooting (Saracen Archery 71-72):
a) downward; b) upward; c) shooting horizontally.

Ten Forms of Shooting by right-handed archer from Horseback according to Saracen Archery (pp.80-81):

1) To left flank, forward and downward parallel with thigh.
2) To left flank, forward and upward.
3) In horizontal plane (target roughly level with left shoulder).
4) Ahead with bow, upper limb to right, canted above horse's neck.
5) To right flank, forward and downward.
6) To left flank rear, upward or downward, with bow canted over horse's croup.
7) To left flank rear, upward or downward, with bow vertical.
8) Jarmaki of two kinds, each with four shots.
9) Beneath horse's neck from right to left.
10) Beneath horse's neck from left to right.

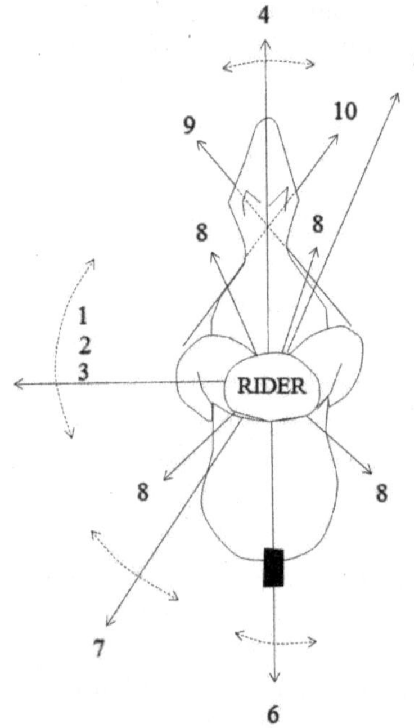

of this archery was not to kill but to cause wounds and confusion among the enemy ranks before the contact was made between the main forces.[72]

The archers which outflanked the enemy formation obviously used their bows according to their relative position vis-à-vis the enemy and according to the type of unit order and formation they were using, so that when the unit had ranks and files, the archers were posted behind the two front ranks while in the case of the units using the various irregular orders this depended on their place in the different variations. The human physique and the horse set up the limits the ability of the mounted archer to deliver his bowshots. The accompanying diagram shows the various forms of shooting used by the right-handed Mamluk horsemen. This demonstrates what was also possible for the trained Roman horsemen. The ability of the right-handed archers to shoot towards their right was very restricted. The cavalry tactics were based on this fact. It was better to outflank the enemy on the right wing. The diagram also shows the other types of bowshots that a right-handed mounted archer could use during skirmishing or retreating. The shot towards the rear was the famous 'Parthian Shot'.

The use of the canter ensured that the unit was able to retain its cohesion and so that the entire front rank made contact with the enemy at the same time. In contrast, the use of the gallop could result in the formation breaking up. As we have seen, both were used during this era for different purposes at different periods of time. The idea behind the use of close order and spears by the two front ranks was to ensure that these dual-purpose soldiers would not concentrate on long distance combat with arrows in the same manner as the Persians, while still increasing the effectiveness of the lancer charge due to the indirect fire provided by the rear ranks. The use of

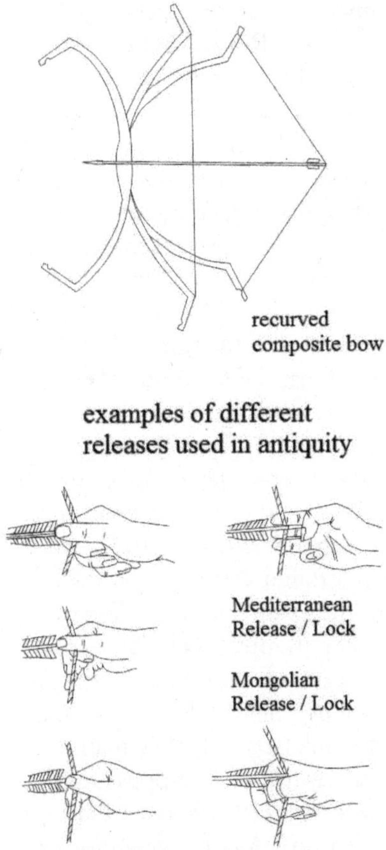

The Romans employed a great variety of different types of bows ranging from the wooden self-bows to the re-curved composite bows (short Scythian/Alan, long Parthian / "Sasanian", asymmetric long "Hunnic"), but so that the standard type of bow was the re-curved composite bow shown here. The names of the bows are misleading as all of the above-mentioned models were in use from the 1st century AD onwards in the Mediterranean. The Romans also used a great variety of different locks / releases. The various different thumb releases had been copied from the Parthians, Sarmatians and Alans while the Mediterranean Release was probably common among the peoples of the Mediterranean. The usual mistake is to assume that the thumb-lock would have allowed more powerful shots with the bow or that it would have been impossible to use shower archery with the Mediterranean Release. The Mediterranean Release was equally suited to both even if each of the different locks had its own benefits and drawbacks. The archers had to use different releases, arrows and types of bows according to their physical characteristics (height, length of arm and fingers) and type of armour worn and the great variety of bows and releases made this possible. The stiffer the bow, the stronger the man, the longer the arms and fingers, and the heavier the arrow, the more powerful would be the arrowshot. This illustration and text follows the interpretation published in the *Historia i Swiat 2015*. The original concept, however, was developed for the forthcoming *Military History of Sasanian Iran*.

the canter allowed the mounted archers the opportunity to launch more consecutive volleys of arrows towards the enemy than the use of the gallop, but the downside was that it also subjected the Romans to increased numbers of enemy volleys. In contrast, the use of the gallop diminished the time when the Romans were under enemy archery fire. This, however, did not necessarily mean that it was advantageous to use the gallop against mounted archers such as the Huns, Avars and Persians, because the use of the gallop had the tendency to break up the cohesion of the unit, with the result that the Roman cavalry formation would not charge into contact with the enemy. It was more likely that a formation carried its charge home when the formation retained its order and cohesion, hence the use of the canter rather than the gallop in the *Strategikon*.

The effectiveness of the bowshots depended obviously on the composition and experience of the Roman and enemy forces and on the direction of the wind and sun. Adverse wind hindered archery while sunshine in front hindered the aim. For example, if the enemy force consisted of lightly-armoured troops which lacked mounted archers, then the chances were that the Roman archery would disrupt the enemy sufficiently for the lancer charge to be successful. The chances increased even more if the enemy did not use close order. Conversely, if the enemy employed close order, were well-armoured and had large numbers of mounted archers (e.g. the Persians or other Romans), then the chances were that the Roman archery was less effective in disordering the enemy formation in advance of the melee phase.

The protective value of the Roman cavalry shield varied according to the type of shield. The *Strategikon* favoured the use of larger shields because, according to Maurice, the Roman cavalry shield (*skouton/skoutarion*, in other words presumably rounded/oval version of the *scutum/thureos* rather than rectangular one because the latter was not usually used when mounted) was large enough to protect the head and part of the horse's neck during the charge. In other words, the *skouton* protected the most vital areas (i.e. the face, upper body and perhaps a part of the horse), the rest being protected by the frontal armour carried by the horses of the front rankers.

On the basis of the narrative sources and works of art, the size and type of the cavalry shield appears to have varied greatly from one unit and time period to another. Some of the cavalry shields were large, like the one depicted in the *Strategikon*, but other sources show greater variance. Some depict large (*scutum*, large oval or oblong shield), or unspecified oblong or round shields (*clipeus* and *aspis*), or small round shields (*pelte, parma* and *aspis* of Procopius). The meaning of *clipeus* for this period is particularly problematic, because it originally meant a round shield but on the basis of John Lydus it meant the oblong *thureos* during the sixth century.[73] This causes problems for the analysis of the shape of the shield. For example, Nazarius mentions the golden *clipeus* of Constantine the Great (oblong or round); Julian names the *aspis* (round shield); Ammianus the *scutum* (large oval or oblong shield); Claudian mentions the *clipeus* (oblong or round), Procopius describes the small *aspis*-shield (small round shield) attached to the shoulders of some of Belisarius's *bucellarii*; Corippus mentions the light *clipeus*, the regular *clipeus*, the *scutum* (in this case the *clipeus* was presumably round in contrast to the *scutum*, but not conclusively so); Agathias mentions the *pelte* (small round shield, presumably the same as the ancient *parma equestris*). And this is not even the entire extent of the problem, because we know that the shield could even be octagonal or sexagonal in shape during the fourth- and fifth-centuries and possibly also during the sixth. However, the most typical shapes for the cavalry shields were the oblong, oval and round, and these appear to have prevailed over the other variants by the sixth century, so that at the time of Maurice the preferred shield type was the *skouton/skoutarion*. In sum, there were several different sized and shaped cavalry shields during the Late Roman period.[74]

One may make the assumption on the basis of the descriptions of the cavalry shields in the sources that these had the same size differences as in the tenth century. According to the *Sylloge Tacticorum* 39, the *katafraktoi* carried 'oblong' shields measuring about four and a half *spithamai* in length (105.3cm long, the width probably being 93.6cm, the width of the cavalry file in close order),[75] the *doryforoi* cavalry carried similar shields (their armour was lighter and their horses unarmoured), the *akontistai* (javelin-armed cavalry) were equipped with an oblong shield measuring four *spithamai* in length (93.6cm) or a round shield measuring three *spithamai* (70.2cm), and the mounted archers did not carry any shields. In sum, it is likely that the large Late-Roman cavalry *scutum* had a length of 105.3cm and width of ca. 94cm,[76] which corresponds with the width of the infantry *scutum* and width of the cavalry file in the *pyknosis* order, and the round shield *aspis* and *clipeus* width of ca. 70.2cm. However, it is probable that the *parma equestris* and *pelte* meant round shields smaller than this. My best educated guess is that those could be about 64cm (this is the size in Hellenistic works of art for the *pelte*) in diameter.

In cavalry combat the shield was at its most useful against arrows during the charge phase and then in the initial contact with the enemy when spears and javelins were used. The value of the shield during the charge phase is clearly in evidence in the instructions of the *Peri strategikes*, because Syrianus instructed the foot archers to aim their arrows both at the feet of the horses and high at the men so that the enemy troopers would not be able to protect both with their shields. See Chapter 1.8. In the melee phase of the combat the shield was no longer as effective, because effective riding required the use of reins and legs, and at close-quarters fighting the spear and sword were at their most useful when directed against the opponents on the right hand side, the side which lacked the shield. Regardless, the shield was still used for blocking/parrying and bashing, even on horseback. It was still possible to unseat the enemy with a combination of moving horse and well-placed shield bash.

The horseman who did not possess a shield at all because he used the Sarmatian style heavy *contus* that required the use of both hands was at a distinct disadvantage when facing a shower of arrows in comparison with the trooper using the shield, but this disadvantage disappeared once the horseman was so close to the enemy that the rear rankers could no longer target him. If he faced an enemy who employed bows initially, the situation was the same, because the bow-armed person needed to change his equipment before the *contus*-bearer reached him. The rider who used the two-handed lance could protect his face only by attempting to hide behind the horse's head. This made him and his horse extremely vulnerable to missiles.[77]

The shieldless heavy cavalry clearly had to minimize the time they spent under archery fire or they needed to be heavily protected with armour. The Sarmatians, Alans, Saka and Parthians had already found a solution to the problem. They had used separate units of mounted archers, cataphract lancers, and less-well armoured lancers. It is not a coincidence that the lance that was used with two hands was called *Contus Sauromatus*. It had been invented by the Sarmatians and it was the perfect weapon against mounted archers once contact had been made. Their tactical method was to harass the enemy first with the mounted archers (the *koursores*), after which the cataphracts (the *defensores*) with their two-handed *conti* charged against the disordered enemy. By then the archery fire was less effective because the enemy unit had lost its cohesion, while the cataphract armour protected the lancer and horse from arrows. The Roman cataphracts appear to have charged using the canter, but if the lancer unit (i.e. the *kontoforoi* = *contus*-bearers) was not as well-armoured as the cataphracts, the charge appears to have been conducted using the gallop, which diminished the time the attackers was under archery fire.

The Goths and Romans copied the mounted archers, lancers and cataphracts from the Sarmatians and Alans, meaning that the Romans added to their cavalry a unit equipped like the Persian *clibanarii* in the third century. The unifying principle for the Gothic and Roman *kontoforoi* (*contus*-bearers), *katafraktoi/cataphractarii* and *clibanarii* cavalry was that all of these were equipped with the *contus*-pike/lance.

However, we do not know how many of these carried the Sarmatian *contus* that required a two-handed grip, and how many of these carried the Gallic *contus* with shield because the use of this did not require the two-handed grip. We also do not know if the Romans themselves made a distinction between these two types in practice, because even if there are sources that separate, for example, the *cataphractarii* from the *clibanarii*, there are also sources which equate the two.[78] Regardless, it is still clear that the use of cataphract armour did not fully compensate for the lack of a shield. The shield gave the troopers more confidence during the charge and helped in bringing the soldiers into contact with the enemy.

When deployed by the book, the armour of the Roman troopers and horses provided them with adequate protection against arrows. The protective value of the armour was improved by the use of the protective jacket (*subarmalis*) beneath the armour, which also had the added benefit of helping in the removal of those arrows that had penetrated the armour. However, in the right circumstances the regular armour could still be penetrated by archery that used stiff bows. The best proof of this is the text of Procopius, who states that the Romans were able to shoot through the Persian shields and armour thanks to their stiffer bows and stronger men. In contrast, the Persian shower-archery technique with weaker bows was unlikely to be able to penetrate the armour, even if it inflicted wounds when it hit unarmoured parts of the trooper or horse. Regardless, the evidence still suggests that in most cases even Roman archery lacked the power to penetrate the combined protection of shield and good quality armour at those distances at which it was usually used. In order to be able to penetrate the combination of the shield and double armour, the arrow had to be a bodkin-type of arrow designed to penetrate armour, and it had to be shot at a relatively close range from a very stiff bow and at the exactly right angle to the shield and armour.

In fact, the principal contribution of archery at distance was its moral effect rather than its penetrative power, and its ability to disorder the enemy formation when some of the horses became uncontrollable thanks to the wounds they received. The best evidence of this is the great many instances in which the cavalry lancers were either able to force the enemy mounted archers into uncontrollable flight or were able to charge into contact with them. However, in order to achieve this, the cavalry lancers had to maintain an even and unbroken front up to melee range. In fact, in the context of fighting against the Persians, the *Strategikon* specifically noted that if the Roman cavalry formation lost its cohesion, they would suffer serious damage and would not be able to come to close quarters.[79] This means the following things: 1) the close-order formation and the use of the canter during the charge together with the use of the shield and armour were usually all that was needed to counter enemy archery so that the Roman cavalry line would carry their attack to the enemy; and 2) if the unit order and formation broke, the enemy archers could kill both the men and horses with greater ease because the vulnerable spots (in particular in horses) became exposed, which in its turn made it unlikely for the Roman attack to reach the enemy lines. In sum, when the lancers conducted their attack by the book it was very likely that they reached the enemy lines in a relatively unscathed condition which was certain to lead to the defeat of the enemy archers who were psychologically less well-prepared for the melee stage.

During the charge the lancers vibrated their lances to counterbalance the movements of the horse, which also had the benefit of giving them better control over the lance.[80] The 3.74m *kontos/kontarion* (Gallic *contus*) was used both for thrusts and throws, but probably throwing was preferred in such situations in which the troopers carried two *kontaria* while 'lancing' was probably preferred when they carried only one *kontos*.

On the basis of the ancient narratives and modern observations, cavalry combat typically had several stages depending on the type of unit and unit formation. When two close-order formations faced each other, the most typical phenomenon was that one of the two turned about and fled before contact. However, if both had the same will to fight then the formations typically halted just in front of each other so that only the front ranks could claim immediate contact with the enemy. In this situation, some of the Roman front rankers would have already thrown their *kontarion*-lance at the enemy (the leather thong in the middle of the spear assisted both in the handling of the spear and in throwing it) before they reached a close distance so that they could then grasp either the other *kontarion* or their *spathion*-sword (*spatha*), or they retained the *kontarion* for thrusting and stabbing. After this, the individual fighters fought in whatever manner they preferred, while the rear ranks may have supported them by advancing into spaces the front rank had vacated if they managed to penetrate the enemy formation, or they threw their spears or javelins over the heads of the front rank men.

According to the *Strategikon* (6.9–16), the lances of the ranks behind the fourth could not reach beyond the front, which means that the men behind them would not have thrown spears or javelins in support of the front rank. The ranks beyond the fourth influenced the combat only through the fact that their presence hindered any attempts to flee. If the thrown *kontaria* and javelins or spear thrusts and stabs then created an opening in the enemy formation, the bolder Roman troopers were likely to exploit this by advancing into the opening to try to enlarge the gap and eventually break the enemy formation and ensure its rout.[81] If the Romans or their enemies (e.g. the Persians, Avars or Muslims) used one of the unit formations designed for penetration of the enemy formation, namely either the wedge or rhombus, the chances of this obviously increased. However, simultaneous attack with the entire line at the same time was at least as effective as phased attacks with the individual units arranged as wedges or rhomboids.

The requirement of Maurice to use the lancing technique of the Light-Haired peoples and the extant illustrations demonstrate well that there existed also other methods of using the lance/spear in combat. There were four basic techniques to hold the lance/spear while mounted (see p.232) and innumerable variations of these four which I will not detail – I only include an illustrative example depicting the use of spears by mounted gladiators in the accompanying drawings. The four basic variants were: the holding of the spear like a javelin roughly at the height of shoulder, which is the technique recommended by Maurice and which was used for both throwing and thrusts; the low position, which was used for thrusts; the underarm technique, which could include the tucking of the spear in the armpit; and the two-handed technique (this version enabled most of the different variations depending on the side of horse where the lance was used and by what grip – with this technique it was also possible to use the lance like a quarterstaff with a blade).[82] All of these variants were used before Maurice's reign, and on the basis of the extant works of art the Romans continued to use at least the low position and underarm techniques alongside the javelin-like technique.

If the two opposing cavalry lines stopped just in front of each other so as not to slam into each other in a suicidal manner, as usually happened when two close order formations attacked each other frontally, then what happened next depended on how close these had come to each other when the spears/lances were thrust at the enemy. Obviously, there were also instances in which not all of the horses and men were always able to stop their advance in time, with the result that men and horses collided with each other and thereby created openings and opportunities for expert horsemen to exploit, which could lead to penetration of the formation.

If very close to each other, then the possibility of the spear breaking or becoming stuck in the enemy became greater, if further away then the spear was less likely to break or stick in the enemy or his horse. The underarm- and two-handed methods were the likeliest to result in the breaking of the spear or it becoming stuck in the enemy so that the spear became useless. If the trooper had used the low position, it was more likely that the trooper retained the ability to draw back his spear and use it again for stabbing. As long as the enemies were at the distance of the spears, the best way to use the spear was with a two-handed grip which enabled the trooper to use it like a bladed staff with feints and parries. For the basic combat variants of the two-handed *contus*, see the accompanying illustration. When employed in this manner, the fighting with spears resembled a fencing match or stick fight with long wooden poles. It was because of this that the Mamluks favoured the two-handed grip over the Damascus-style armpit attack. The principal target areas for the spear thrusts and throws were the head, chest and legs of the horses (it was because of this that the horses of the front rank carried armour) and the face, chest and the legs of the riders.[83] Regardless, if the two formations were at that distance, it was likely that most of the front rankers lost their spears sooner or later, so they had to use their swords. In fact, at close-quarters fighting the use of the sword was advantageous if the intention was to break the enemy formation because once past the enemy's spear-point the swordsman had the advantage. In that situation, the mounted swordsman had combat advantage even when he did not himself possess a shield (for example if he used the Sarmatian *contus*), because once past the enemy's spear-point the enemy could not harm him except by changing weapons. It was because of this that we find the Heruls using swords and we find the Vandal king Gelimer ordering his men to employ only swords. The idea was to charge past the spear points into close contact to break the enemy formation. During the fighting, regardless of which weapon the trooper used, he had to maintain a calm pose and keep his eyes on the target to be able to hit it. It was because of this that the better fighters were posted in the front. During modern times, the European military treatises instructed the troopers to direct their lance attacks towards the left hand side and sword blows towards

- The four principal variants for the use of the shafted weapons on horseback before the introduction of stirrups. The shield could be used in various different ways and the illustrations show only some of these.
- The illustration below (drawn after a painting in Pompeii) shows two mounted gladiators as an illustrative example of how the shafted weapons could be used in many different manners besides those depicted above.

The over-the-shoulder technique with the spear was used for both downward thrusts and for other forms of stabbing and for javelin throws.

When the spear was placed in the low position it was used for downward and/or straight thrusts

Underarm technique with the spear tucked between the arm and side, or in the armpit was used for shock and thrusts.

Two-handed grip used when employing the Sarmatian *contus*. It could be employed with different grip variations on both sides of the horse. This technique enabled the rider to use his spear like a bladed quarterstaff when needed.

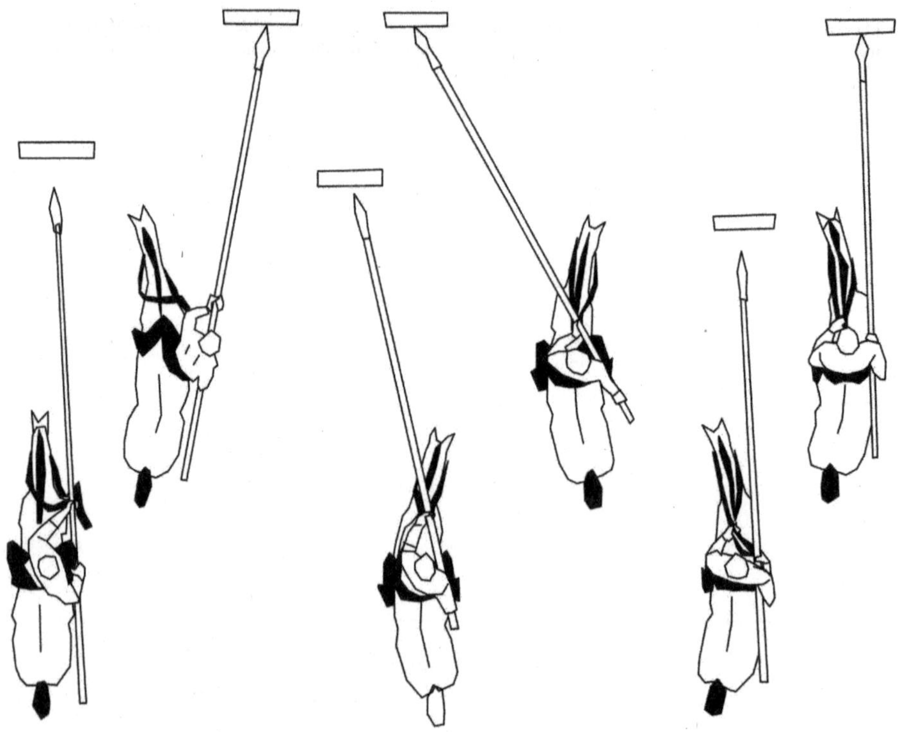

The basic combat techniques with the two-handed Sarmatian *contus*. This is the way how the *clibanarii* would have used their *conti*. Note that the lance could also be used with a single hand grip by using the underarm or couched lance technique. Drawn after Gawronski's illustration based on Simonenko (2010, 84 fig.56).

the right hand side even when the trooper carried a shield. This was undoubtedly true also in Roman times, but with the addition that the Romans would also have lanced between the ears of the horse because they held the lance at the height of the shoulder.[84]

It was also at this close range that warhorses were used as weapons and the larger heavier mounts could be used to push over the smaller enemy horses, while the men could hit their enemies from the higher advantageous position. The well-trained warhorses, such as, for example, used by Belisarius, could also support their rider by moving in the right manner to enable the rider to use his weapons better.[85] If the horse was trained to use kicks and jumps at close quarters, and many of the horses undoubtedly were, the rider could use his mount in this manner against both the enemy horse and rider.

If the Romans managed to fight their way into the enemy formation, the likelihood for the enemy force fleeing increased as in both opposing forces the best men were always posted in front. In other words, the deeper the troopers advanced, the more likely it was that the enemy would flee. In fact, if the enemy managed to penetrate the two front ranks of the Roman formation, they came face-to-face with those troopers that had been deployed as archers and these men did not expect to engage the enemy in melee. When the enemy then fled the flight began from the rear of the formation and spread from there forward. It is very likely that the situation was the same also in enemy formations, namely that the two front ranks (and the rear rank) were the men who were expected to decide the battle. It is therefore clear that in prolonged battles the amount of elite veteran troopers diminished and the role of the less experienced men rose in proportion, and if the role of the less experienced men increased the situation became more volatile and less predictable. It was in such situations that the importance of heroism and personal fighting skills of individual troopers increased. However, in most cases the decisive elements were not the fighting skills of individuals, but the morale of the force in general, its quality and type of equipment, unit order, the battle formation, and the tactics at army level. For the psychological importance of unit order, see the above discussion of unit orders on cavalry combat on pp.151–65.

As far as the armament of the soldiers is concerned, the general rule of thumb is that the more heavily-equipped horsemen would defeat the lightly equipped in melee unless the latter had better morale. The earlier and later observances of cavalry combat also suggest that the deeper column formations usually had an advantage over the shallower formations (e.g. one to three rank formations) but not necessarily over the standard Roman formations that had five to ten ranks.[86] The reason for this was that it was easier in the shallower formation to flee, and secondly because the deeper formation had more inbuilt reserves with which to penetrate the shallower formation once these halted in front of each other. It is also clear that success varied from one *bandon* to another and from one *mere* to another. It was because of this that units were required to support each other.

Similarly, it is clear that the individuals in each *bandon* fought with varying success, some being able to kill or wound their opponent(s) while others suffered wounds at the hands of the enemy, hence even as each trooper fought as an individual he was also required to support his comrades in arms both during the battle and pursuit. Regardless, the battles and pursuits, in particular prolonged battles, could also result in great confusion of ranks and units so that individuals lost contact with each other. It was because of this that the Romans had battle standards and passwords for each day. The standards enabled the scattered troopers to regroup and find their own unit, and the passwords allowed them to recognize friends from enemies. After the battle lines had clashed, confusion reigned.[87]

In the past it was typical to claim that the introduction of stirrups was the reason why cavalry dominated the battlefields during the Middle Ages. This has been demonstrated as being wrong, for it is clear that the success of Macedonian and Parthian cavalry are entirely comparable with the success rate of medieval cavalry. Furthermore, as I have shown, the Roman cavalry had seen its first period of battlefield dominance in the third century AD and it demonstrated all the same capabilities as the medieval cavalry of Europe by being able to mow down even legions with a frontal charge. Regardless, it is still clear that the adoption of metal stirrups helped the troopers both to retain their seat and in the use of archery while mounted. The Romans had known the use of stirrups at least since the first century and some of their units had even used them, but the whole-scale adoption of the use of these appears to have taken place only in the sixth century, so that we find the demand to use these in Maurice's *Strategikon*. The probable reasons for the slowness in adoption are the conservatism of the military and the need to be able mount and dismount the horse quickly. The best evidence against the claim that stirrups were the decisive technological breakthrough that enabled cavalry to become the dominant arm is the fact that the Muslims did not use stirrups during the seventh century and still defeated the Roman and Persian cavalry that used them.[88]

As already implied, the result of the cavalry combat depended also on the horses and the riding skills of the rider. The saddle and other equipment had to be in order and the warhorse had to be well-natured and trained for war. We know that the Romans put a great deal of effort into the breeding (note e.g. the imperial stud farms) and into training in all kinds of weathers and terrains. This gave them a distinct advantage over most of their enemies. I have already given the instance of the role of Belisarius's horse in combat, but the sources also mention other well-trained horses by name. A well-trained warhorse enabled its rider to use the weapons in the most advantageous way as the horse turned left or right as the situation required, while he could also push over the opposing mount or the rider could order the horse to perform one of the jumps or kicks that we find in use for example in the modern era Spanish school.[89]

At close quarters, once the troopers had advanced past the spear-points to sword range, they acted instinctively just like they always did when they reached that stage. There was no time to wait for orders, even when these could be heard. As stated above, it was preferable to use the sword towards the opponent on the right so that the shield protected the left. This meant that the sword was sometimes used for both offence and defence on the right hand side. However, this is not the whole picture because the actual battle descriptions demonstrate that the shield was used for parries and blocking on both sides of the horse just like the sword was used against enemies on the left and right and on foot – although it is clear that the spear or javelin was preferable if the target was really low.

The swords used by the Romans, Persians, Germanic tribes and nomads were usually double-edged and pointed during this era, so the troopers could use both the point and the edge.[90] The troopers could use four basic ways of attack – the cut, the reverse, the downward blow and the thrust, plus their variants. It was easier to use the cut/slash when mounted because this did not demand the same amount of manoeuvring with the horse to be effective as the thrust. The cut/slash was also safer to use because it did not transfer the same amount of impact to the attacking rider as the thrust at speed. This could be important when the saddle did not provide adequate hold on the rider or when stirrups were not used. This was not really a problem for most of the Romans, because prior to stirrups they usually used the horned saddle which gave the riders a safe seat during such encounters, but it could influence the outcome in such cases where the horseman did not use a saddle at all or used some other less safe version. Regardless, in mounted combat the expectation was that the sword technique utilized the impetus of a horse's movement in order to achieve the maximum impact on the enemy, the favourite technique even then being the cutting technique. See the image opposite of mounted fencing. The detailed cavalry fighting

Top Left: A legionary equipped according to one of the third to early fourth century fashions with the Pseudo-Attic helmet, Newstead *lorica segmentata*, socketed *pilum/spiculum*, *spatha*-sword, old-style *pugio*-dagger, and rectangular cylindrical *scutum*. However, other types of helmet, armour, shield and spears were also used. (*Author's drawing*)

Top Right: A legionary equipped according to one of the third and early fourth century fashions. The helmet is my interpretation of the helmet shown (Arch of Constantine) above the legionary so that it combines the information included in the image with the Imperial Italic H-helmet. The man also wears the Newstead *lorica segmentata*. However, other types of helmet, armour, shield and spears were also used. (*Author's drawing*)

Above left: The right foot forward fencing stance for thrusts favoured by Vegetius and some other Romans. The soldier has thrown his javelin (*spiculum, lancea*) or spear (*hasta, lancea*) prior to this and now uses the sword. He wears Berkasovo Type-1 helmet (ridge-helmet) and uses the hoplite-style round *aspis/clipeus/parma*. (*Author's drawing*)

Above right: A soldier belonging to the elite *Cornuti*, which are prominently depicted in the Arch of Constantine the Great. The Arch of Constantine depicts the *Cornuti* without armour, but with helmets and shields. The *Cornuti* belonged to the *auxilia palatina* category of auxiliaries, which essentially fought as heavy infantry despite their lack of armour. (*Author's drawing*)

Left: Photograph depicting Constantine the Great's and Maxentius's cavalry at the battle of the Milvian Bridge in 312.

Constantine's cavalry is depicted unarmoured and without shields while wearing Pseudo-Attic helmets. They are clearly equipped with spears for combat, which are likely to have been similar to the ones depicted below in the Sarcophagus of Helena. In other words, it is likely that Constantine's cavalry charged straight into contact and then used their spears for thrusting and as quarterstaffs.

Maxentius's cavalry in this battle consisted mainly of the Praetorians so that it is likely that those are depicted here. His cavalry is depicted wearing scale armour, shields, swords and Pseudo-Attic helmets, but it is possible that the relief depicts them in a situation in which they have already used their spears. Note in particular the use of the Pseudo-Attic helmets, which suggests variation in the preferred type of helmets in the different portions of the Roman Empire. In the east (Arch of Galerius) we see the Roman cavalry accompanying Galerius equipped with scale armour, shields, spears, swords and segmented helmets. (*Public domain*)

Below: Two riders from the Sarcophagus of Helena. Note the lack of armour and the type of spear used. The spear clearly has a sort of apple at the rear end of the shaft suggesting the use of this type of spear as a sort of quarter staff in combat. In fact, I would suggest that it was because of this that we see the riders sometimes without shields because the effective use of the spear by both ends required the use of two hands. The fact that we find these images depicting unarmoured cavalry in the works of art connected with the wars of Constantine the Great implies that Constantine preferred lightly-equipped cavalry that was used like heavy cavalry because this made them more mobile and faster on the battlefield, both qualities that made his cavalry forces more mobile than the enemy cavalry. The lack of armour and shield can be connected with the manly show of bravado also adopted by the Herulian cavalry in combat. It is of note that the men are also shown without swords in both instances, in the Arch of Constantine and in the Sarcophagus of Helena, implying that Constantine's elite cavalry forces did indeed fight by using the spear alone for the reasons stated. The fact that most of the infantry forces of Constantine are also depicted without armour suggests that Constantine preferred to increase the mobility of his forces at the expense of protective gear and secondly probably also his policy of economizing in that respect. (*Author's drawings*)

Above left: The right foot forward fencing stance for cuts favoured by Vegetius and some other Romans. The soldier has thrown his javelin (*spiculum, lancea*) or spear (*hasta, lancea*) prior to this and now uses the sword. He is unarmoured and wears an *Intercisa*-helmet (ridge-helmet) and uses the oblong *scutum*. The equipment worn does not define the unit in this case because the legionaries could also be equipped without armour just like the soldiers of the *auxilia palatina* units. Note the use of the rectangular cylindrical *scutum* which appears to have been used as one shield variant at least until the tenth century AD. (*Author's drawing*)

Above right: *A clibanarius* charging against enemy infantry. The *clibanarii* could also be equipped without bows. The fully armoured *clibanarii* were the favourite cavalry type of Constantius II, which had been introduced into the Roman armed forces during the reign of Alexander Severus. The *clibanarii* were typically used to break up enemy infantry formations and were therefore usually placed next to the infantry wings or in the middle of the infantry for this purpose. (*Author's drawing*)

Below left: Roman soldiers equipped with Pseudo-Attic ridge helmets in the San Maggiore Mosaic (ca. 432-440).

Top left: A member of the *Ioviani Seniores* equipped with a helmet (Der el-Medineh helmet consisting of six plates, classed as *Spangenhelme* by Bishop and Coulston), scale armour, a sword, a spear and a shield. The shield emblem has been taken from the *Notitia Dignitatum*. It is not known what type of helmet was worn by the *Ioviani* which means that the reconstruction should be viewed tentatively. Similar equipment could be worn throughout the late Roman period.

Top right: A slinger equipped with a staff sling (could be a legionary or auxiliary). The trousers, shirt and haircut date the soldier to the fourth to sixth centuries. (*Author's drawing partially after Phil Parker*)

Below left: Late Roman *skoutatos* equipped for fighting in difficult terrain. (*Photo © Jyrki Halme*)

The equipment consists of: a round shield, ridge helmet, knife, *spatha*-sword, and shorter spear (*kontarion, hasta*). The shoes worn were well-suited to fighting on foot.

Below right: Late Roman footman in gear that he could have worn in a city. (*Photo © Jyrki Halme*)

Skoutatos, *aspidoforos*, hoplite using the right foot forward guard with the short *lancea*-spear and *aspis/clipeus*-shield. He wears muscle armour and Deurne-helmet. The *spatha*-sword hangs from leather straps placed below the muscle armour.

Above right: Roman cavalry officer (early fourth century) wearing muscle armour, *spatha*-sword, Berkasovo-helmet, greaves and a mace. The officer is a pagan and he advertises this fact with the *Sol Invictus* decorations. The golden armour means that he belongs to the *Scholae*.

Bottom left: A fourth century footman based on the painting of a warrior in the Via Latina Catacombs, Rome. He is equipped with a round shield, chainmail armour, *spatha*-sword, two *lanceae* (lances) and a ridge helmet. Similar equipment remained in use throughout the late Roman period. By faith he is Christian. This fact is advertised by the Chi-Rho symbol placed on the helmet.

Hottenroth's (1891) reconstruction of Aetius based on the Ivory Diptych of Stilicho or Aetius.

Hottenroth's (1884) reconstruction of an emperor in military gear (400–500) based mainly on the no longer extant Column of Theodosius I. The late Roman generals could wear similar equipment.

Hottenroth's (1884) reconstruction of a fifth century Roman officer based on the Ivory Diptych of Stilicho or Aetius.

Below left: Soldiers in the San Maggiore mosaic (ca. 432–440) depicted in late Roman equipment.

Note the continued usage of eagle and Phrygian helmets, and the use of the spears like javelins for stabbing and thrusting. Note also that at least two men depicted in front of the rest have their right foot in front as Vegetius expected when the javelin/spear was used for stabbing. (*Public domain*)

Above: Two scenes depicting soldiers in the San Maggiore mosaic (ca. 432–440). The men are portrayed in late Roman equipment. (*Public domain*)

Note that the soldiers could vary the number of spears carried. It is probable that when they carried two spears the intention was to throw at least one of those before the melee phase.

Below: Soldiers in the San Maggiore mosaic (ca. 432–440) portrayed in late Roman equipment. (*Public domain*)

Note the continued usage of the eagle and Phrygian helmets, and the spiked shields.

Above left and right: Standing and walking Roman heavy infantry *skoutatos* (shield-bearer) equipped for fighting in difficult terrain. (*Photo © Jyrki Halme*)

The equipment consists of:
a round shield, ridge helmet, knife, *spatha*-sword, and shorter spear (*kontarion, hasta*). The shoes worn were well-suited to fighting on foot.

Bottom left: Cavalry in the San Maggiore mosaic (ca. 432-440) depicted in late Roman equipment.

Note the use of various kinds of spectacular helmets. (*Public domain*)

This image depicts one of the typical cavalry variants used by the Romans from the third century until the 580s. The Romans knew the use of the stirrups from the first century onwards but their use became common only during the latter half of the sixth century. This rider is therefore depicted without stirrups.

A fully equipped multipurpose trooper from Thrace who could be a regular *stratiotes/kaballarios* from the *katalogoi*, or *bucellarius/hypaspistes/doryforos* in the service of some officer.

I have depicted the man in such gear as can be seen in the period works of arts which I have combined with the information provided by the textual sources.

An Avar cataphract:
Horse: iron stirrups; saddle with a high pommel; lamellar armour for the horse with Chinese style metal chamfron.
Man: a segmented *Kuppelhelm*-helmet; lamellar cuirass with shoulder pieces, cuisses and brassards of iron strips; straight single-edged sword; a spear with a pennon; a shield behind the back which was rarely used; a quiver; and an unstrung bow in a bowcase.

Adapted from Gorelik as a composite of a Turk and Avar (14-7) on the basis of the fact that the Avars were originally a subject tribe of the Turks. See also the plate sections of *MHLR* Volumes 7 and 8.

Note that the late sixth century Roman cavalry could also be equipped in a similar manner because Maurice considered the Avar-style equipment as a suitable alternative for the Romans too.

A Lombard knight:
Horse: unarmoured, no stirrups.
Man: a segmented *Kuppelhelm*-helmet; lamellar cuirass with chainmail, shoulder pieces, cuisses and brassards of iron strips; straight double-edged sword, and a *contus*-spear.

The Lombard *foederati* in Roman service could be equipped in a similar manner.

Above left: According to Hottenroth (1884, 97) a marble figure from the burial chapel of the exarch Isaac the Armenian (ca. 620–37 or 625–44).

Below left: Hottenroth's (1884) reconstruction of an East Roman officer based partially on the above figure.

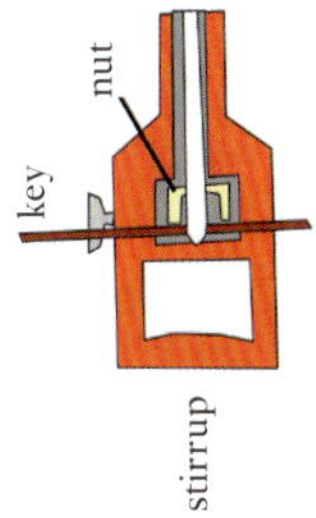

A stirrup-guide according to Latham and Paterson, p.147. Details of the mechanism unknown.

There were several different types of arrow-guides in existence. These are listed by Latham and Patterson (145–51). The accompanying illustrations show some of these variants. The use of the arrow-guide was very advantageous because it doubled the length of the shots and enabled the archers to carry more ammunition than would have been possible with regular arrows. It was also advantageous because the enemy could not reuse the darts. The darts could be square, round, hexagonal or octagonal in shape, each of which required a corresponding type of arrow-guide.

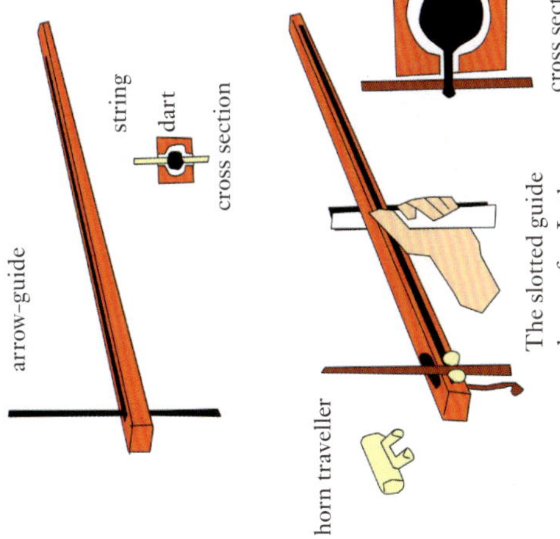

The slotted guide drawn after Latham and Patterson, p.150

plumbata tribolata

Roman footman using darts. The *Strategikon* (12.2.16.39–46) instructed those of the *skoutatoi* (shield-bearer) footmen who had lead-pointed darts (*mattiobarbuli*, *martiobarbuli*, *martzobarboula*, *plumbatae*) to throw these at the right distance (ca. 40–60m) from the enemy line. The darts were usually placed inside the shield five apiece. In this example I have assumed that the footman was not among the front ranks, but was one of the rear-rankers who used a javelin instead of *kontarion*-spear. He has thrust the javelin into the ground so that he can grasp it immediately after he has thrown the darts, but he could equally well have held it in his left hand or placed it on the ground. The footman does not wear regulation footwear because the *Strategikon* instructed the foot soldiers to use Gothic shoes and not boots like this man. He uses the above head javelin/dart technique, which was safer to use in a phalanx than the use of the softball/cricket throwing style. The *skoutatoi* equipped with the *kontarion* probably placed their spears on the ground. The *plumbata tribolata* was designed to act as a sort of caltrop so that when large numbers of these were thrown the darts formed up a defensive field of caltrops. (*Author's drawing*)

Below: Three differently equipped Roman heavy infantry *skoutatoi* in readiness to fight in regular terrain (sixth to seventh centuries).

A lightly-equipped Hunnic mounted archer. The late Romans did not only face the Huns as enemies but also used them as regular *foederati*, *bucellarii* and as foreign allies (*symmachoi*).

A *skoutatos* equipped lightly with the 'larger' shield for difficult terrain.

A slinger. Note that all footmen were trained as slingers.

A *skoutatos* equipped regularly for a pitched battle in open and level terrain. Note the use of the wooden shin-guards and the helmet taken from the David Plates.

A *skoutatos* equipped lightly for difficult terrain. Note the use of the oblong cylindrical *scutum/skouton* and the padded armour and boots.

A fully equipped multipurpose trooper from Thrace who could be:
1) regular *eques / stratiotes / kaballarios* from the *katalogoi*;
2) or *optimus / optimos / aristos* from the *optimatoi / optimates*;
3) or *boukellarios / bucellarius / hypaspistes / doryforos* in the service of some officer.
equipment:
man:
- a chain mail armour reaching down to the ankles and a plumed helmet with an aventail
- a padded armour reaching knees.
- a cavalry *kontarion / contus* (ca. 3.74 m) with an Avar-style spearhead and a leather loop to ease the handling.
- a javelin-quiver
- a round shield
- a *spatha*-sword
- a composite bow and a quiver for arrows
- a hooded cloak used to cover the armour from sunshine when necessary to hide the trooper.
horse:
- *katafrakta neurika* = padded / quilted leather armour (in this case chamfron, crinet, peytral, but could also include a flancard and crupper.
- a saddle, stirrups, blanket etc.
- a saddle-back to carry provisions, water and fodder for emergencies.
6th century *bucellarius* with small pennons attached to the shoulder
The horseman wears:
- a variant version of the Phrygian helmet
- a large *scutum* shield
- a chain mail armour reaching ankles
- mail mittens
- a padded armour under chain mail
- a *kontarion / contus* with Avar-style spearhead.
- a bow and quiver
- a *spathion / spatha* (not shown)
The horse is protected with lamellar armour of the Persian type.
© Dr. Ilkka Syvänne 2023

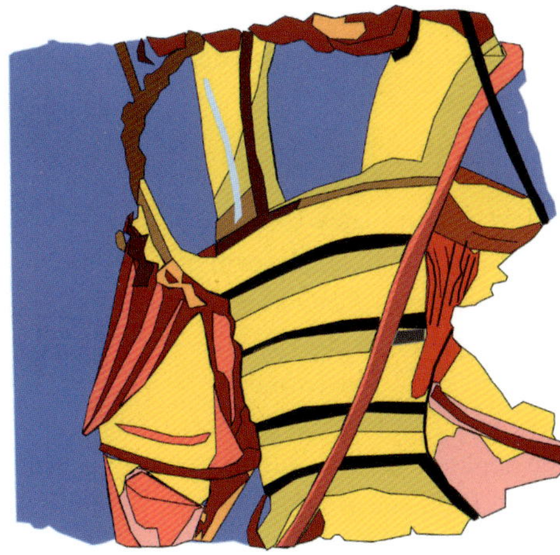

Above right: Umayyad era *lorica segmentata*? If the Umayyads used the *lorica segmentata* it is possible that the late Romans also continued to use it at least until the eighth century. The first person to suggest the use of the *lorica segmentata* by the Umayyads is Tawfiq Da'adli and the image is drawn after his study 'Reconstructing the Frescoes of Khirbat al-Mafjar', *Jerusalem Studies in Arabic and Islam 46* (2019). Please note that my reconstruction changes the colouring in places.

Above left: Walking late Roman soldier as a hunter or scout. The late Romans could employ infantry as lightly-equipped scouts in advance of the army when the terrain was difficult as depicted here. (*Photo © Jyrki Halme*)

Below right: Standing late Roman soldier as a hunter or scout. (*Photo © Jyrki Halme*)

Note the equipment worn by the soldier: a knife, *spatha*-sword, throwing axe also usable for close quarters fighting, throwing / thrusting spear and a segmented *Spangehelme*. The footwear is particularly well-suited for fighting on foot.

Top right: Medallion of Justinian I the Great depicted in parade armour. British Museum. (*Photo by Ilkka Syvänne*)

Below: David Plates. Combat scene. Metropolitan Museum of Art, New York. (*Public domain*)

Note the use of the covers for the Pseudo-Corinthian helmets, the hoplite style *aspis*-shield and spear, chain mail armour with shoulder pieces and boots.

Mounted fencing techniques against cavalry and infantry

The illustrations are adapted from the 19th cavalry treatises for the conditions of the late Roman period.

sequences provided by Corippus and by other sources support this conclusion. It is therefore not at all surprising that the Romans later adopted the sabre as one of the cavalry swords. The principal reasons for also using the shield when mounted was that the *spatha* was not particularly good for parrying, and as noted, most of the wounds at really close quarters were caused by sword strikes. It is probable that it was because of the demands of close quarters combat that the favourite cavalry shield was the round shield during this era. It was lighter and smaller and therefore handier at close quarters than the large *scutum*, which was very difficult to move from left to right over the horses' heads when the battlefield was crowded.[91]

The European Renaissance-era martial artists stated that if the rider had penetrated the enemy cavalry line, he was to attempt to gain the rear of the enemy or to gain the enemy's left hand side and to prevent the enemy from doing the same,[92] while during the pursuit it was preferable to approach the fugitive from the left so that he could not use his sword against the pursuer which was towards his left. The same renaissance authors recommended the use of the sword against the horse, and the neck and face of the rider. Close-quarters mounted combat could also include the use of mounted wrestling, in which the movement of the horse was used to one's own advantage. We possess plenty of evidence for this in sources describing mounted combat from different eras.[93] There seems to have been a remarkable continuity in fighting techniques.

The element of surprise was also important in the field of equipment used. If the other side was unfamiliar with some particular type of equipment and the tactics it required, it was probable that the side which faced a novel way of fighting lost the initial encounters. It is because of this that when the early-sixth century Vandals were unfamiliar with the fighting methods of the Huns, the latter routed them easily. The same holds true for any period or place. However, when the weapons were relatively similar (i.e. bows, spears, lances etc.) and as the human physique is what it is, the initial shock of an unfamiliar system wore off quickly and methods of dealing with it were soon found if the men were ready to learn from their enemies.

In true Roman tradition, the Late Romans were prepared to borrow from their enemies. When facing an unfamiliar foe, the *Strategikon* recommended the use of minor skirmishes before the battle to familiarize the troopers to the fighting methods of the enemy and to improve their morale.[94]

Descriptions of Fighting Techniques

The period narrative sources do not provide any detailed analytical descriptions of how the men fought in the tight cavalry formations as this was not their purpose, but we possess plenty of poetical descriptions of combat or descriptions of how soldiers fought in single combat. Unfortunately, even these are incomplete, because it is unlikely that all of the movements, feints etc. are described and it is also clear that some of the narratives have been invented when the event took place long in the past or the participants themselves no longer remembered all of the details. As I noted in my doctoral dissertation, what we have left is the equivalent of a radio broadcast of a boxing match in comparison with the live-televised version that can be reviewed in slow motion. However, since most of these tales were recounted during the lifetime of the participants, it is clear that when the hero demonstrated exceptional ability with his weapon it is likely that this was really the case. These demonstrations of personal prowess were the things that the participants remembered most clearly. It is therefore not surprising to find that the narratives give us descriptions in which the hero is able to cut off the head, or cleave the torso etc. All of these were and are demonstrations of perfect cutting technique with the sword, while also demonstrating the quality of the sword in question.

The same was true for demonstrations of personal prowess with the bow or lance or javelin. The accuracy of the javelin-throw or spear-thrust demonstrated the skill of the wielder, and the ability to puncture the enemy armour and shield demonstrate the strength of the wielder. The skill in the use of these weapons was a demonstration of personal fighting prowess, which was highly esteemed both among the military classes and among the populace in general. However, that said, it is still likely that in most cases the first strike with a sword or thrust with a spear was not usually deadly thanks to the use of armour and helmets, so that in practice the first strike that hit the mark would have only left the enemy stunned or unable to parry, requiring the troopers to continue to hack, thrust and hit repeatedly until they were sure that the enemy was dead and would not rise to continue the fight. And we should not forget the rage commonly associated with fighting, and the need to assure oneself that the enemy was dead often resulted in 'overkill' of the enemy. This means that those soldiers who had received a concussion hit or wound would be hacked to pieces or stabbed to death, unless saved by a friend or comrade-in-arms. Obviously there would also have been those cases in which the well-trained or lucky trooper managed to kill his opponent immediately with a thrown javelin, or with a spear or a sword. If a trooper fell from his horse, he was usually killed on the spot unless saved by his comrades.[95]

When troopers charged into the intervals between enemy files or into the space vacated by the death or flight of an enemy horse or trooper, the fighting was truly intense and fast paced. The men faced the prospect of being hit simultaneously by several enemies. They had to put their trust in their armour, shield, helmet, comrades and luck. The following quote from the pen of Procopius demonstrates nicely how the hits and wounds would simultaneously come from different directions, even when champions fought in front of their armies, and how important the protective equipment was for the survival in this brutal action. The amount and quality of armour often decided who would prevail and survive.

> ...But there was a certain Artabanes in that Roman army, a Persarmenian by birth, ...This Artabanes in the present battle placed himself in the space between the armies, taking with him two of the Roman soldiers, and thither came some of the enemy also. Artabanes charged these men, and engaging with one of the Persians who was a man of high valour and great bodily prowess, he straightaway slew him with his spear and throwing him from his horse brought him down to the ground. But one of the barbarians standing beside the fallen man smote Artabanes on the head with a sword, but not with a mortal stroke. Then one of the followers of Artabanes, a Goth by birth, attacked this man, and while he still held his hand at Artabanes' head, smote him with a well-directed blow in the left flank and laid him low. Thereupon the thousand being terrified at what had taken place, began to withdraw to the rear....
>
> Procopius, *Wars* 8.8.21–8, trans. by Dewing, 125–7.

The following quote from Corippus's text provides us with a description of the different stages of the cavalry attack. In Corippus's case, the Roman and Moor battle lines both contained infantry, but the text still gives us all of the stages of the cavalry battle: first the initial charge towards the enemy; then the pursuit; then the regrouping of the pursued and the counter attack; then the retreat of the Roman *koursores* up to the *defensores*; and then the counter attack by the *defensores*. The spears were clearly used for both thrusts and throws, as instructed by the *Strategikon*.

> They urged their horses on and hurled weapon after weapon from their hands. Indeed, the sunlight itself grew dim beneath the thick flight of spears, and black night rushed down beneath their javelins… The clashing of the swords and shields was hampered by the arrows they sent flying first from one side then from the other… [*The order of the events is slightly off the mark. The arrows were the first that hit both formations followed up by the throwing of spears at a closer distance, after which the men grasped their swords and advanced to close quarters with their shields and swords.*] and each time a weapon was thrown, a man on one side or the other fell wounded, for all the incoming missiles threatened death… Often, when a weapon was thrown, it would strike an enemy javelin in its course and, linked with weight of its own would fall to the plain… The air whistled with the spears that were thrown. War raged on and, catching the horses up in the slaughter of men, sent them wallowing along… [*In other words the horses were also targeted.*] and the men, blinded by the confusion and the frenzy, exposed their breasts to the enemy weapons. And sometimes soldiers preparing to attack others would feel the steel in their own bodies and let their sweet life flow from their wound. Fierce Ricinarius rushed forward, breaking through the enemy as they pressed on. He put the hostile lines and standards to flight by killing Eilimar… The great hero faced him and pierced his breast with a steel as he came on. Breaking his ribs, he cut open his body and ran the spear through his back with all his might. [*This demonstrates the use of the spears for thrusts and the breaking of the enemy formation.*] Then the Mazax turned their horses in flight as cold terror took hold of them. A band of Romans followed them and by keeping pressure on threw them into confusion. Galloping in hot pursuit over the plains… [*This is the pursuit at gallop by the Roman koursores.*] But when the broken cavalry saw their own infantry standing steady and facing them, valour returned to their hearts and they turned their horses around… [*The Moorish cavalry regrouped and counter-attacked, forcing the Roman* koursores *to seek the protection of their* defensores.] But our general blocked their way… breaking through the centre of the ranks in his excitement, gave the whip to his horse and rode through the enemy troops, … His brave army followed with serried weapons. [*This describes the counter attack by the Roman* defensores *which was supported by the regrouped* koursores.] Everything was hidden by the dust, and a storm of iron rained down on the battlefield as flying shafts were shot from bowstrings. First, our general sent Mantisyan to the world of the dead, slicing off his head with his sword. ... Nor did the sword linger in the bones after it was stained with his blood. The man's head lay in the

> grass, its eyes still open, and his swift horse galloped over the plain carrying his trunk. [*This demonstrates the perfect cutting technique of John Troglita.*] ... Next the commander stabbed Laumasan through the temple with his sturdy blade. The bone was shattered and the sword passed through the soft brain as well. [*This shows how the sword could also be used effectively for thrusts.*] Then, aiming at his helmet and mantle, John cleft his forehead in two ... Throwing a javelin at close range, he laid low the swift horse of Guarsutia. [*John used one of the javelins from his holster.*] ... Then, striking the neck of Iartus, John dislodged the hand which held the man's weapon. ... Mazana ... As he came on, he hurled the missile at the general, then pulled up his horse and turned away. But our brave commander, ..., took the hurtling spear on his shield and deflected it. [*Shows how the shield was used in defence.*] Fearlessly he made for his enemy and, undaunted, struck him down. .... He hesitated no longer but straightened himself in his saddle and made for Cullan with poised spear. [*John had presumably grasped one of his spare-spears from his back.*] He pursued for some time, and the Moorish horseman, fleeing his mighty foe in terror, retreated in every direction. But the commander followed him in his rage, hurled his spears at him and finally pierced his back ... And now the bodies fell on top of one another as the enemy were swept backward over the battlefield. ... all of his ranks began to tremble.
>
> Corippus, *Iohannis,* 5.51–155, translation excerpted from the excellent translation of George W. Shea, 131–134.

We find similar descriptions from the Georgian Chronicles that depict combat techniques used during the fourth- and fifth-centuries in that area.[96] Once again we find all of the principal elements of a cavalry fight present: the approach stage with one or both parties being shot at by bows, the use of shield and lance, the two-handed use of the *contus*, the initial encounter with lances, the results of the initial encounter, the follow up by manoeuvring the horse to a halt or around the enemy, and the use of the sword or the continued use of the lance. All of the accounts capture the fast pace of the cavalry action well. The troopers could only make one or two feints and then attack and hope for the best. The quotes also demonstrate the effectiveness of pairing the shield and the sword at close range, while also demonstrating how the two-handed *contus* was used. The quotes also depict one important detail, which is that the Georgian king changed his weaponry to meet the different kinds of enemies. We find the Romans using the same approach when they changed their combat tactics according to the type of enemy. The use of a shield was particularly important when the enemy employed bows, but the other methods could be employed when the enemy did not use bows.

> Now Ossetes had Xazar auxiliaries, among whom was a giant by the name of T'arhan. This Xazar T'arhan came forth, ... King Vaxt'ang had Persian auxiliary troops, among whom was a man called P'arsman-P'arux. ... He came forth to do battle with T'arhan. They both shouted out and rushed on each other. At the first encounter (T'arhan) brought down his sword on the helmet of P'arman-

> P'arux, and split his head down to his shoulders. Juanšer, 151–2, Georgian version, Thomson tr. 166–7.
> Then Vaxt'ang descended the hill and stood on the bank of the river. In his hands he held a lance. … At the first encounter Vaxt'ang struck his lance at (T'arhan's) belt. His strong armour was unable to protect him; it came out through his back and he died.
>
> Juanšer, 153, Georgian version Thomson tr. 168–9.
> Underlining added to show the use of two hands.

> The next day another champion came forth from among the Ossetes: he was called Baqat'ar … he had killed all his adversaries. For the length of his bow was twelve palms, and his arrow was six palms. … Vaxt'ang disposed his troops … He mounted his horse, which was covered with chain-armour; he took up his shield of tiger skin, which a sword could not cut; then he descended the hill and stood close to the river. … Baqat'ar crossed the river and began to shoot arrows. By the sharpness of his eyes, the keenness of his mind, and the agility of his horse Vaxt'ang avoided the arrows. For he saw from afar the arrow coming; he would jump away and nimbly run forward. … Baqat'ar was not able to shoot more than two arrows at Vaxt'ang's shield, and he did not hit it. Then he shot another arrow at Vaxt'ang's horse, which penetrated it. While the horse was still falling, Vaxt'ang rushed on Baqat'ar, brought his sword down on his shoulders and penetrated to his heart. At that moment Vaxt'ang's horse fell. He quickly put out his hand and grasped Baqat'ar's horse. First he fell to the ground … Then he mounted Baqat'ar's horse.
>
> Juanšer, 154–5, Georgian version Thomson tr. 169–170.

> Here is the combat between Vaxt'ang Gorgasali and Polykarpos the Greek general, and the killing of Polykarpos by Vaxt'ang. … Polykarpos rushed on him, but Vaxt'ang cautiously and deliberately approached. They both shouted out with a fearsome cry of warriors, and there was a noise like the sound of thunder, from which the ground shook. With his lance Polykarpos smote Vaxt'ang's shield, which was of tiger-skin; it penetrated the shield about a cubit, for the thickness of the lance was that of a man's arm. The king left his shield to the lance and rushed forward face-to-face. He struck his sword on (Polykarpos') helmet and sliced his head in two down to his shoulder-blades.
>
> Juanšer, 174–5, Georgian text Thomson, 192–3.

The following quote from the same source gives us some examples of how the horses were manoeuvred in cavalry combat to achieve advantage in a single combat. However, similar attempts to gain an advantageous position vis-à-vis the enemy would also have taken place when the battle formations had become mixed or when the formations were loose to begin with.

> Then the king of the Sinds came forth, and … Vaxt'ang went out, and both of them held lances. They began to run with a circular motion, and each was

> seeking a way to stick the point of his lance (in the other). Then the king of the Sinds seized an opportunity to avoid the tip of Vaxt'ang's lance, and rushed up in order to strike him with his lance. But with a brave and adroit manoeuvre Vaxt'ang turned aside the lance and attacked. Like a whirlwind he went around him and struck his lance into the left shoulder-blade of the king of the Sinds. His strong armour was unable to protect him, and he was severely wounded; for it came out in front a cubit's length. The king of the Sinds was brought down; Vaxt'ang went up to him, stretched out his hand and seized his foot, and dragged him in front of the Persian king.
>
> Juanšer, 194, Georgian version Thomson tr. 210–1.

The military manuals fail to mention the use of the lasso in the context of cavalry combat proper, but we find descriptions of this in the narrative sources. The Huns in particular were famous for their use of the lasso against the Romans, which could be an effective way to break up the cohesion of the static formation. The following description shows how it was used in a single combat by a Gothic *comes foederatorum* in Roman service. Regardless, it is still clear that the lasso was not used by the Romans in cavalry combat proper, because the cavalry was not well-suited to this. The Romans did not fight in the Hunnish manner by using lassoes. In Roman use the lasso was restricted for the capture of prisoners for interrogation or for the capture of fugitives in general – although one cannot entirely preclude the possibility that the lasso was used by some Huns or Goths in Roman service in their native manner too.[97]

> When these terms had been agreed, the emperor of the Persians chose a Persian named Arzanes from the division known as the Immortals, while the Romans selected Areobindus, a Gothic *comes foederatorum*. The two came out on horseback fully armed. Areobindus also carried a lasso, as is the Gothic custom. The Persian charged at him first with his lance, but Areobindus, bending down to his right, lassoed him, brought him down off his horse and killed him.
>
> Malalas, tr. By Elisabeth Jeffreys, Michael Jeffreys and Roger Scott, 14.23, p.199.

Attack with Javelins

The narrative sources also suggest that the Romans continued to use the technique of attacking the enemy by only using javelins. This tactic appears to have been usually used against static enemy infantry forces, but there is at least one example of the Romans using this against Hunnic cavalry. When the Romans opted to use javelins, they obviously carried shields too. On the basis of the narrative sources and Arrian's description of cavalry drills, the standard ways of using this method consisted of the following: 1) charge forward as a unit while the men threw as many javelins as possible at the enemy in the course of the charge, after which they either charged into the enemy formation or retreated and repeated the attack; 2) a charge by units so that these charged close to the enemy and then threw their spears/javelins at a single spot to create an opening in the enemy formation for the troopers to penetrate; and 3) the sending of individual files from the units to skirmish so that these advanced in various manners and then returned to their own units, after which the next set of files advanced forward to skirmish.[98]

The Nomadic and Persian Ways of Using Cavalry

As already noted, most of the Roman cavalry could also be deployed without separation into *koursores* and *defensores* so that they fought in the Scythian manner, all equipped alike, or in the Persian manner, as close-order forces all employing bows for shower-archery. When used in the Scythian manner, the troopers used the irregular *drouggoi* and the gallop, and when using the Persian tactic they either employed the close order with the trot or canter, or they assumed a stationary stance in some advantageous position. The *koursores* were obviously deployed like the Scythians when they pursued the defeated enemy. It is clear that even their front ranks employed the bow when the distance required this. Both of these tactical variants were principally used against the Germans during the sixth and seventh centuries, probably as a result of the great successes of Belisarius. However, it is clear that both of these systems were used well before this, because the Romans had employed Persian auxiliaries, Huns (from the late-fourth century onwards) and other nomadic Federates (e.g. Sarmatians) throughout the Late Roman period.

One can reconstruct the capabilities of the mounted archers on the basis of the later Mamluk archery manual of Taybugha. According to Taybugha, the Mamluks' training scheme consisted of shooting at the gallop on courses that varied in length between 120–260m. The expectation was that the Mamluks would need about 40–87m to get their horses into a full gallop going in a straight line. The horse had to go in a straight line, because the rider dropped his reins to shoot with the bow. The horse was also trained not to change direction in response to leg pressures because some of the bowshots would have generated leg pressures which the horse could have misunderstood. It was during the second third of the course that the rider prepared and shot his bow and arrow, which according to J. M. Smith Jr.'s calculation allowed 40m between shots for the skilled mounted archer. This means that the proficient horse archer who had started his attack from a range of 260m to the enemy could have shot three times before being forced to return. This in its turn means that the best horse archers were able to shoot three shots by the time they had galloped 161m, after which they had to turn back if they did not want to make contact with the enemy. Taybugha gave the troopers 40m of distance for turning in front of the enemy.[99]

If we then place this in a Roman context, the Roman horse-archer would clearly have started his archery at maximum range (the bowshot ca. 330m), meaning that he was able to shoot at least 2–3 times before turning when employed in the Scythian manner. The Roman system gave each trooper more time and space for making the retreat than the Mamluk training scheme, and it was actually more practical than the Mamluk training scheme which did not take into account a possible counter attack made by the enemy, which made the Romans' system far more cautious. During the retreat both the Mamluks and the Romans obviously shot additional shots by using the so-called Parthian shot. The number of arrows shot during the retreat obviously depended on the length of the retreat. If it was a strategic retreat it could take days to complete, but if it was a tactical retreat then it was continued up to the reserves, or to the place of ambush, or until the enemy showed signs of being tired or disordered.

The number of archery shots obviously increased if the Romans used the Persian-style attack with the trot or canter. If we assume that the use of the trot increased the number of Roman style shots only by one, so that they could have shot three or four arrows, this means that the Romans could shoot fifteen to twenty shots with the Persian archery technique. The latter option was obviously available only from the end of the sixth century onwards, or before this for the Persian and Armenian units in Roman service. If the archery duels became prolonged, then the troopers either retreated to the rear to replenish their supplies of arrows (the usual method according to the sources) or the servants brought these to the front. When employing either the Scythian or Persian tactic, the Romans obviously charged to close quarters to finish of the enemy when it was weakened enough.

In his study of the Mongol tactics Smith Jr. considered the above practice ineffective, because the first bowshots were shot at a great distance.[100] This is taking the evidence a bit too far, because it is clear that the period warriors and soldiers clearly considered shooting at long distance to be useful. In fact, when the Romans attacked in units the typical manner of using the bows was to shoot arrows at a high trajectory. The *Strategikon* acknowledged that the higher trajectory shots were not as effective as direct shots, but they were clearly useful. The successive volleys of arrows falling upon the enemy forced them momentarily to concentrate upon deflecting the arrows rather than on shooting at the Romans, while also causing wounds in those places that were poorly protected. It is clear that the bowshots at a higher trajectory had only a limited effect upon the enemy, but the use of the Persian shower-archery technique demonstrates that sheer volume also mattered. Furthermore, we known that the Mongols, Romans and Persians all had arrows designed for different ranges and for different uses. This demonstrates beyond doubt that it was worthwhile to employ the bows also at longer ranges. Not all of the arrow shots were intended to be penetrating power shots because this was possible only at close range for the troopers of the front ranks. When the Romans used their cavalry in Scythian or Persian manner this was possible.

When the Roman archers were close to the enemy they would have used their armour-piercing bodkin-type arrowhead or other types of armour piercing arrowheads if the enemy wore armour, and other types of arrows meant to maximize bleeding if the enemy was unarmoured. The principal target areas of the power shots at close range would have been the chest, legs and the head of the horse and the visible parts of the enemy horseman. In sum, the Romans had long- and close-range combat methods for every imaginable purpose, so there did not exist any type of enemy that they could not have faced at least on equal terms. The Roman approach to fighting was highly effective.

Chapter Seven

Late Roman Cavalry Tactics[1]

The Late Roman cavalry was the most flexible and efficient force of its era. It was also the most efficient cavalry force in existence when it followed its combat doctrines to the letter. The cavalry was required to be able to fight at long distance with bows, and at short distance with javelins, spears and swords. The cavalry was required to be able to fight even as infantry.

Most of the combat methods in use had been developed in the course of the first three centuries of our era so that the Roman cavalry formation had achieved its final formation by the end of the third century. The Romans used several different unit orders for different purposes, which included the open (*araiôsis*), close order (*pyknosis*), tortoise (*synaspismos*), interlocking of two front ranks, irregular *drouggos*-order, crescent (*menoeides*), and different variants of the rhombus and wedge (*embolos*) formation. Each of the different unit orders had its particular place in the standard cavalry battle array or in cavalry tactics in general. The most flexible of the unit orders was the *drouggos*-order, because it could be used in almost any circumstances even if the other unit orders were the principal ones for the actual regular pitched battle.

The Romans had already adopted the use of cavalry reserves during the Republican period. The addition of the third line probably took place during the reign of Gallienus as a response to the defeat of Decius at the Battle of Abrittus in 249, which resulted from the use of the double ambush by the Goths. It demonstrated the need for a third line. In short, it is very likely that the Italian Drill system, which had three divisions, flank guards and outflankers in the first line; one, two or four divisions in the second line, with fill-up *banda* when needed; and the use of the two units of rear guards, were all adopted by Gallienus' reign. These components were adapted to different states of affairs, so the Romans had a solution to any situation they could face on the battlefield.

The Romans had also adopted the use of the *koursores* and *defensores* system well before the third century, because we find the Romans using the Alan Drill during Arrian's time, but we do not know for certain when the other variants of this system (African, Illyrikian, Italian) were introduced, but it is very likely that all of these variants were in use by the late-third century, because the African Drill already includes the Italian version of the *koursores/defensores* system. The Scythian Drill formation was probably adopted at the latest in the fifth century, while it is likely that the Persian formation was adopted during the sixth century. However, it is also possible that both were adopted well before this, because the Romans had employed nomadic groupings and Parthian/Persian units well before the Late Roman era. These two unit orders and formations added further flexibility to the Roman combat system. In addition to this, the Romans specialized in various forms of feigned flights, stratagems and ambushes.

The use of a great variety of unit orders, formations, and tactics; the requirement to be able to fight at long distance with bows and at melee with spears and swords; and the ability to fight as infantry required a well-trained professional force, which the Romans actually possessed for most of the Late Roman period. It was not the cavalry that suffered from poor quality during the late-fifth and early-sixth centuries, but the infantry.

In combination with the above, the ability to dismount and fight as infantry made the Late Roman cavalry force the most flexible and efficient fighting force of its era. There were of course periods in which the ideal was not achieved or when the Romans employed their cavalry in different manner, for example by using the cataphracts in greater numbers. A good example of the former is the supremacy of the Ostrogothic cavalry during the late-fifth and early-sixth centuries. This was alleviated by the fact that the Romans also employed Ostrogothic cavalry of their own, and by the fact that the Ostrogothic cavalry abandoned the use of mounted archery in the early decades of the sixth century, probably as a result of their successes with the lancer charge. This made the Roman cavalry superior to them by the time Belisarius began the reconquest of Italy.

The end of the supremacy of Roman cavalry came when it abandoned the tactic of dismounting to fight on foot when the Muslims began their conquests. In the Battle of Yarmuk in 634, the Roman cavalry still fought well while mounted, but thereafter its effectiveness also diminished as a mounted force because of three things: the successive defeats demoralized the Romans; the commanders placed in charge of the cavalry were incompetent; and the cavalry forces included disloyal elements (Arabs and Armenians) which caused a succession of defeats. It was largely thanks to this that the Muslim cavalry, especially the cavalry under Khalid ibn al-Walid, can be considered to have been the foremost cavalry force of its era.

Chapter Eight

The Infantry Unit Orders[1]

Roman infantry units were trained to use four basic unit orders: the open order (*araios/araiôsis*), close order (*pyknôsis/puknôsis*), several variants of shield-interlocking (*chelônê, testudo, foulkon, synaspismos, syskouton*), and the irregular (*drouggos*) order. Open order was used for marching and the initial approach to the battle site. The close-order and shield-interlocking orders were used in regular combat and used primarily used by the heavy (*hoplitai, skoutatoi*) and medium infantry (*peltastai*). However, the light-infantry *psiloi* could also assume these when the situation so required. The irregular *drouggos* order was used mainly by the light infantry skirmishers (*psiloi*) outside the battle formation, but it could also be employed by heavy infantry. The unit orders were basically the same as had been used since the dawn of regular warfare, so we can already find these same unit order variants in use during Early-Republican times and in the later treatises such as the *Sylloge Tacticorum*.[2]

As already noted in the discussion of cavalry unit orders, the tenth-century *Sylloge Tacticorum* (43.6–7) gives us the best description of the space occupied by each footman and horseman in a rank-and-file formation, which corresponds with the figures given by the *Strategikon* and earlier sources. In short, each footman (and horseman) occupied in width an *orguia* in open order (*araiôsis*) during march (4 *pêcheis* = 187.38cm), half an *orguia* in close order when battle was expected (*puknosis* = 3 *podes* = 3 x 31.23cm = 93.69cm), and a third of an *orguia* during battle in shield-interlocking order (*sunaspismos/chelônê/suskouton* = 2 *podes* = 2 x 31.23cm = 62.46cm). The last detail of course refers only to the situation in which the infantrymen used the pike phalanx or tortoise array. The close order *puknosis* was used when there was no need for the extra protection that the interlocking shield wall (*sunaspismos*) offered. In practice, the width and depth obviously varied according to the size of the shield, and similarly if the men used javelins the depth of the array was adjusted to enable their use.[3]

Vegetius (*Epitome* 3.14–5) provides us with a different scheme, better suited for javelin-armed infantry forces. According to him, each file of footmen (all arrayed in six ranks) was deployed so that each file occupied three feet (i.e. were deployed in rim-to-rim *pyknôsis* order) while each rank (each rank with a depth of one foot) had six feet between them. The reason for this was that his footmen, both the heavy-armed and light-armed, were equipped with missile weapons. In fact, ranks three and four consisted of lightly-armed men, which typically required more space to throw their missiles. However, it is clear that in practice ranks one and two were more tightly arrayed in depth during the actual melee phase, because they were required to stand like a wall and fight together when the enemy came close – ranks three and four would obviously have retained their depth to enable them to continue to throw their javelins and darts or use their bows. It is therefore probable that in actual combat the two front ranks either threw their javelins at the approaching enemy, after which the second rank advanced closer to the first to support it with swords when needed, or that the second rank retained its spears/javelins so that it could support the first with javelin/spear thrusts by advancing close enough to the first rank.

As the above has already implied, the close order and shield-interlocking were primarily intended for the heavy infantry, but it is still clear that when the light infantry was deployed alongside the phalanx that it too had to assume the same file width even when it retained its greater depth between ranks. However, since the light infantry demanded greater amount of space to be effective it was more typical for it to be deployed separately from the heavy infantry phalanx. This enabled them to use looser formations and irregular order when necessary. The slingers and staff-slingers in particular required a fair amount of space around them to be of any use in combat.

The small wedge, *caput porci(num)* and *globus* (an independently- and separately-operating massed group), can also be considered to have been a specialized unit orders.

8.1. The Open Order (*araios, araiôsis*)

Open order (*araios, araiôsis*) meant an order in which the ranks and files had wide intervals separating them from each other. According to Asclepiodotus (4) and Aelian (Matthew ed. 11), in open order the files and ranks occupied four *pêcheis* of space, and according to the *Sylloge Tacticorum* (43.6–7) the files occupied an *orguia* of space which equalled four *pêcheis*. The *pêchus* varied greatly from one place and time to another, so that the ancient *pêchus* had three basic variants (Attic 44.4cm; Alexandrian 45.12cm; Doric 48.72cm), while the *pêchus* in the *Sylloge Tacticorum* appears to have been ca. 46.8cm. The ancient variants would therefore mean that each file and rank occupied about 177.6cm, or 180.48cm, or 194.88cm, while the tenth-century *Sylloge Tacticorum* stated that each file occupied about 187.4cm in width. One may therefore assume that the ranks in open order varied in practice roughly from 178 to 195cm, because we need to understand the figure of four *pêcheis* only as an approximation.

On the basis of the *Strategikon*, the files were usually only four men deep in open order before a tightening of the formation for the combat. Open order was used during marching and approach to the battle because it did not tire out the soldiers like close order. Open order enabled the soldiers to use their natural gait, which was impossible in the closer orders in which the men had to adjust their steps according to those in front of them. The use of *pyknôsis* and *synaspismos* demanded that each soldier would march in cadence, which is determined by the men composing the front ranks, and which forces each soldier – including the front ranks – to conform his marching step into such length and rhythm that each one with differing physical dimensions can follow. The use of open order for the approach march was therefore very important if one wanted to bring the army to the battlefield in full fighting condition.[4] In contrast, the use of the *pyknôsis* or *synaspismos* were essential for actual combat, because these unit orders gave each soldier the physical and emotional support that they needed for the eventual battle. Open order was used by both the heavy and light infantry. It is also clear that the slingers used this unit order or the *drouggos* order in combat because the use of the sling or staff-sling required room to use.

8.2. Pyknôsis: The Infantry Close Order[5]

When the Roman combat formation got to the distance of about two or three bowshots (ca. 660–990m) from the enemy's line and was preparing for combat, the commander was expected to give the order 'Unite!' ('*Iougge/Iunge*'). This was the last safe opportunity for the assumption of close order because the enemy cavalry could cover this distance in about two- to three-minutes at the gallop. The soldiers then closed towards the centre until the shields of the men in the front rank touched each other and those lined up behind them would be almost glued to one another. The file closers were expected to force anyone who hesitated forwards while ensuring that the line would be straight. This unit order could be assumed both by stationary troops and by advancing troops. The ancient military theory called the resulting rim-to-rim formation with the name *pyknôsis*. In military theory (e.g. *Sylloge Tacticorum* 43.6) the width of each file in this formation was a half *orguia*/three feet (ca. 94cm), but it is clear that in practice the width occupied by each file in the formation varied according to the width of the shield used and the widths of the shields varied from ca. 0.80 to 1m.[6] I would also suggest that in practice the shields were also held above

The close order (shields rim-to-rim formation) *pyknosis* (infantry)

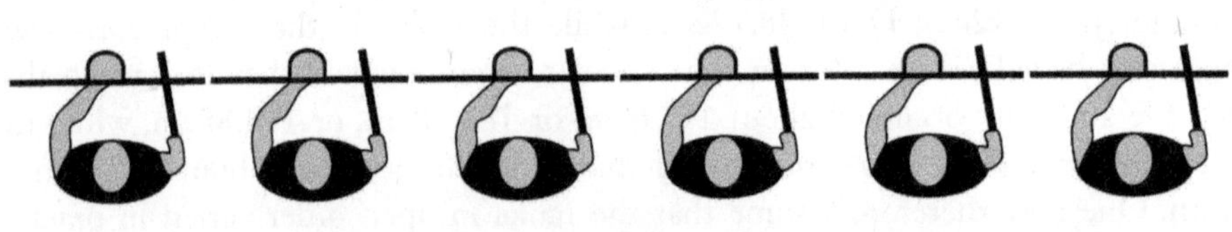

The rim-to-rim (*pyknosis*) unit order in the Siege of the Citadel of the Faith.

Note that the shields as depicted are clearly too small to be of any use in the *pyknosis* order. It is therefore clear that the artist has depicted the shields purposefully too small. In truth the shields would have been at least twice as large to be of any use in the *pyknosis* order.

Drawn after a photo of the Siege of the Citadel of Faith, a Coptic wood carving (ca. fifth to seventh centuries). The carving was destroyed during the WW2. My drawing is slightly simplified and in places slightly emended.

Pyknosis in works of art

The auxiliaries in the Column of Trajan employing the rim-to-rim unit order (*pyknosis*) in combat against the Dacians. Source of the drawing Bernoulli. Note that the shields are depicted slightly too small also in this relief.

Pyknosis in the Joshua Roll, usually dated to the 10th century.

Note, however, that the equipment is exactly the same as we find in the seventh century David Plates celebrating the victories of Heraclius over the Persians.

Public Domain.

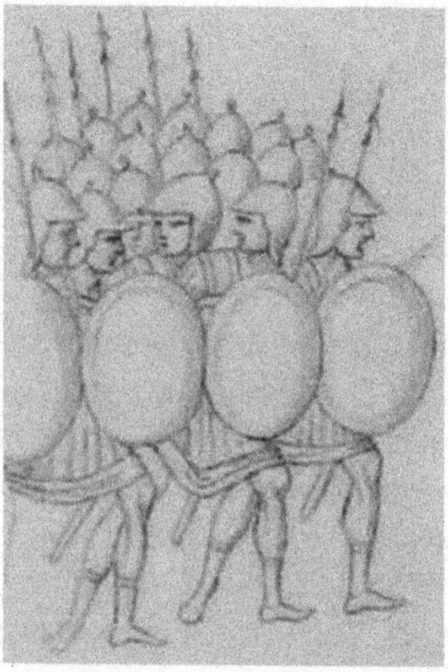

the heads by the rear ranks when needed. It is just common sense that the soldiers used their shields defensively when the enemy used missiles against them. This unit order gave each soldier the freedom of individual action, as well as the support of his comrades next to him. It enabled the soldiers to use their sword fighting techniques as individuals, which the *foulkon/testudo* order denied. It also facilitated the ability of the soldiers and units to manoeuvre with greater speed and agility than was the case with the *foulkon/testudo.*

8.3. The shield-interlocking orders: *synaspismos/syskouton*, *testudo/chelônê*, *foulkon/fulcum*[7]

The ancient and medieval military treatises and narrative sources equated several different kinds of unit orders under the same category of interlocking shields (but with various different terms), because in each of these shields were interlocked either in width and/or in depth, the only unifying factor being the fact that the shields were interlocked in some manner. *Synaspismos* (*syn* + *aspis*) and *syskouton* (*syn* + *skouton/scutum*) simply meant the interlocking of shields and were used to mean both the shield-wall (shields rim-to-boss in width) and the tortoise formations (shield roof). The tortoise formations (*testudo/chelônê, foulkon/fulcum*) usually meant those unit orders that had a shield roof. The name *foulkon/fulcum* was a nickname for the Roman unit order, which was apparently borrowed from the Germanic word *volk/folk* at some point in time because the Germanic soldiers in Roman service used this term to denote those unit orders in which the shields were interlocked.[8]

The basic shield-interlocking order variants were: 1) a shield wall in which the shields were interlocked in width usually rim-to-boss (width of a file in formation ca. 2 *podes* = 2 x 31.23cm = 62.46cm), but which did not necessarily include a shield roof, in such a manner that the long spears of the cavalry-type (ca. 3.74m *contus, kontos, kontarion*) of the first three- to four-ranks protruded in front of the phalanx; 2) shields interlocked rim-to-boss, both in width (width of a file ca. 62.46cm) and depth (i.e. it had a shield roof), so that the entire formation knelt to receive the enemy's missiles (the enemy could consist of either cavalry or infantry), and then rose up to engage the enemy when it approached close enough; 3) the advancing *foulkon/testudo/chelonê* order with a shield roof (shields interlocked rim-to-boss in width and depth), which was used to attack enemy infantry; 4) the stationary *foulkon/testudo* used to face the charging enemy cavalry with sloping shields (front rank kneeling, second crouched, third stood almost upright, fourth was upright) so that the shields were interlocked in depth roughly rim-to-boss but not in width (the resting of the spears against the ground by the two front ranks means that these were between the shields of two files); 5) siege *testudo* with a sloping shield roof against the wall; 6) the 'Macedonian' spear phalanx as described in the *Peri strategikes*; and 7) the infantry phalanx against cavalry in the *Peri strategikes.*

8.4. *Peri strategikes*: *Synaspismos/syskouton* 'Macedonian' spear phalanx and infantry phalanx vs. cavalry

Syrianus describes in the *Peri strategikes* (16) a spear phalanx which he called by the name 'Macedonian', in which the files were arrayed with a width of a *peêhus/pêchus* (62.46cm) and in which the spears of the first four ranks protruded in front of the formation. The actual Macedonian phalanx was obviously different. In the variant versions, the spears of the second rank were longer so that the spears of the first two ranks reached the same distance in front of the array, or the ranks behind the fourth had javelins or light spears or other missile weapons. For an illustration of this array, see the analysis of the military treatises (pp.18–34). On the basis of the description of infantry equipment in the marching camp (*PS* 27), it is clear that Syrianus expected all footmen to possess also bows, which means that the other missile weapons referred to were arrows.

According to Syrianus (*PS* 36.1–8), the infantry phalanx was deployed against cavalry so that the first and second ranks kept up continuous archery fire against the feet of the horses (should be interpreted to mean the entire frontal area of the horse) while the rest of the men aimed at a higher angle so that their arrows dropped down from above (this implies that the first two ranks aimed their shots directly at the horse and not only against the 'feet'). This forced the enemy horsemen to use their shields either for their own protection or for the protection of their horses. The fact that both the first and second rank were able to shoot directly may imply that the formation was originally deployed in the *pyknosis* formation for the archery stage and then closed in width and depth for the final phase with spears. The second version (*PS* 36.9–20) adds details and a variation to the above. According to this version, the first three ranks placed theirs spears on the ground and kept up continuous archery fire against the approaching enemy cavalry. Since the soldiers placed only their spears on the ground the shields must have been hung from leather straps against the shoulder and back. When this archery had then slowed down the enemy, the infantry picked up their spears and advanced against the enemy. The referral to only three ranks in the latter version could be interpreted either as a mistake, or to mean that only the first three ranks had bows beside their spears, or to mean that actually all three frontal ranks could aim their shots directly at the horses, or that the three frontal ranks were used like the three to four ranks in the *foulkon* vs. cavalry in the texts of Arrian and Maurice, but with the difference that in this case the men also had bows. On the basis of Procopius's text, it is likely that the second version is either a mistake, or that only the first three ranks were also equipped with bows, or that either two or three ranks aimed their shots directly and the others with a higher trajectory. The likeliest alternatives are that Syrianus was just careless or that either two or three ranks aimed their shots directly. The reason for this is that Syrianus (*PS* 27) expected all footmen to possess both spears and bows. We can also preclude the use of the *foulkon* vs. cavalry à la Maurice on the basis of the fact that in Procopius's text the multipurpose footmen were using the *sunaspismos/suskouton* interlocking of shields just like the phalangites of Syrianus (*PS* 16).

Philip Rance (2004, 304) notes that we can find the shield-interlocking array (*syskouta*) without the shield roof also in the *Strategikon* (12.D.82–3) in the context of using large-scale hunting as a form of military training. It is therefore quite clear that the *synaspismos/syskouta* shield-interlocking array was also used without the shield roof. It should also be noted that the shieldwall without the shield roof was also used when facing enemy cavalry, which is not in the least surprising because the traditional Macedonian pike phalanx did not have a shield roof – the pikes were considered adequate defences against enemy missiles.

We can use the text of Procopius to shed further light on the issue of how the Romans employed the shield-interlocking *synaspismos/syskouton* array against cavalry:

> ... it was at this point that the fifty took up their position, standing shoulder-to-shoulder [*shoulder-to-shoulder means that the men interlocked their shields in width, which means that this is not the* foulkon/testudo *used against the cavalry as described by Arrian and Maurice*] and arrayed in the form of a phalanx as well as the limited space permitted [*en chrô men xuniontes allêlois, es falagga de ôs en stenochôria*]... The [*Gothic*] horsemen accordingly charged upon them with great hubbub and shouting, intending to capture them at the first cry, but the Romans drew up together into a small space and making a barrier with their shields and thrusting forward their spears [*oi de eis oligon xyntetagmenoi kai tais aspisi fraxamenoi, ta de doratia epanateinamenoiestêsan*], held their ground... the fifty, pushing with their shields and thrusting very rapidly with their spears [*tôn te aspidôn tô ôthismô kai tôn doratîon të epibolê pyknotatê*], which were nowhere allowed to interfere one with the other, defended themselves... and they purposefully made a din with their shields, terrifying the horses,... and the men... with the points of their spears... two of them distinguished themselves... in this action, Paulus and Ansilas, who had leaped out from the phalanx and made a display of valour surpassing all other for they drew their swords [*of the generic* 'xifos'*-type*] and laid them on the ground [*it is probable that these two men had lost their spears as a result of the previous Gothic cavalry charges and were now ready to demonstrate their personal bravery by other means in the hopes of receiving promotion*] and then stretched their bows and kept shooting with a most telling aim at the enemy. And they destroyed many men and many horses as well, as long as their quivers still held arrows... when their missiles had now entirely failed them, seizing their swords and holding their shields [*of the generic* 'aspis'*-type*] before them, all by themselves they warded off the assailants. And whenever any of their opponents on horseback came at them with their spears [*of the generic* 'dory' *type*], they immediately broke off the heads of the spears with a blow of their swords... it came about that the sword of one of them (this was Paulus) was bent... and so was utterly useless. This he then immediately threw on the ground, and seizing the spears with both hands he would wrench them from the enemy... he made himself the chief cause of their abandoning their attempts.
>
> Procopius, *Wars* 8.29.15ff., tr. by Dewing with my additions in Italics.

The above quote clearly shows Roman infantry using the Macedonian-style phalanx as described by the *Peri strategikes.* It is also unlikely to be a coincidence that the footmen were equipped with bows and spears, as expected by Syrianus. Procopius clearly shows the footmen using the tight formation with interlocking shields by different files (men standing shoulder-to-shoulder), the din of the shields and the thrusting of spears by several ranks simultaneously, as the principal reason for the inability of the Gothic cavalry to break through the Romans. Totila actually employed two groups of horsemen who repeated their charges several times before giving up. The structure of the 50-man array was either ten files with five ranks, or twelve files with four ranks so that there were two extra men left. The latter is likelier as the two extra men would have been Paulus and Ansilus.[9] It is a pity that we do not know if the footmen placed their shields above their heads (in a formation with only five or four ranks the spears of the rear ranks did not provide cover for those in front), but in light of the fact that the Gothic horsemen did not use missiles this is not likely in this instance. In short, it is likely that during the reign of Justinian the Roman infantry fought as spear phalanx with four ranks while being multipurpose troopers equipped with bows too.

> Belisarius with some few men remained there, ... then at length he too fled with his men and came to the phalanx of infantry, who with Peter were still fighting, although not many in number now, since most of them too had fled. There he himself gave up his horse and commanded all his men to do the same thing [*i.e. cavalry dismounted to fight as infantry, as expected in emergencies*] and on foot with others to fight off the oncoming enemy. And those of the Persians who were following the fugitives, after pursuing for only a short distance, straightaway returned and rushed upon the infantry and Belisarius with all the others. Then the Romans turned their backs against the river so that no movement to surround them might be expected by the enemy...And again the battle became fierce, ... for foot-soldiers, and very few of them, were fighting against the whole Persian cavalry. Nevertheless, the enemy were not able either to rout them or in any other way to overpower them. For standing shoulder-to-shoulder [*shoulder-to-shoulder means that the men interlocked their shields in width, which means that this is not the* foulkon/testudo *used against the cavalry as described by Arrian and Maurice*] they kept themselves constantly massed in a small space, and they formed with their shields a rigid unyielding barricade, so that they shot at the Persians more conveniently than they were shot at by them. Many a time after giving up, the Persians would advance... but they always retired... For their horses, annoyed by the clashing of the shields, reared up and made confusion for themselves and their riders.
>
> Procopius, *Wars* 1.18.41ff., tr. by Dewing, 173
> with some changes and author's comments inside square brackets

Procopius here once again describes a similar shoulder-to-shoulder formation with shields interlocked in width in use in the Persian theatre of war. In this case, the same defensive array was adopted by both the regular Roman infantry and dismounted cavalry against the Persian multipurpose cavalry which also employed bows. The unfortunate thing in Procopius's description is that he once again omits details concerning the rear ranks and their shields. Considering the fact that these men faced the prospect of being hit by Persian arrows, it is quite possible that the men placed their shields above their heads, at least when not using bows. In this case, however, the use of the shield roof would not have been necessary because in this case the infantry phalanx appears to have had enough depth so that the spears of the rear rankers could have been projected forward in such a manner that this provided adequate protection for the entire formation. Notably, Procopius is once again describing a situation in which the Roman infantry and dismounted cavalry were using bows against enemy cavalry. This suggests the prospect that the multipurpose

Above: Roman soldiers with in the lost Column of Arcadius (16th century drawing). Note the two different ways of interlocking the shields in width.

Below: Dura Europos exodus fresco (mid- 3rd century) drawn after Bishop & Coulston with some minor emendations. Note the different ways of interlocking the shields within the same formation.

Synaspismos in the Joshua Roll, usually dated to the 10th century.

Note, however, that the equipment is exactly the same as we find in the 7th century David Plates celebrating the victories of Heraclius over the Persians.

Public Domain.

infantry and dismounted cavalry acted as instructed by Syrianus, namely that the front ranks shot arrows until the enemy came so close that it was advisable to form up the interlocking shield wall with spears protruding from it. What is certain is that Procopius is not describing the *foulkon/testudo* as used against cavalry in the works of Arrian and Maurice. The referral to the shoulder-to-shoulder array precludes this.

8.5. The advancing *Foulkon* against Infantry and the kneeling *testudo*[10]

If the front-rank men were not wearing coats of mail or shin/knee guards, Maurice recommended the forming of the advancing *foulkon* when the enemy's line was at about the distance when archers would start to shoot. The command for this was: 'Form *foulkon*!' ('*Ad fulcum, Ad foulkon*'). In this unit order, the front ranks closed in width until their shields touched the shield-bosses (rim-to-boss/shoulder-to-shoulder order) so that they completely covered their midsections almost down to their ankles. The men standing behind the front rank held their shields above their heads, and rested their *skoutaria* (shields) on the *boukola* (shield-bosses) of those in front so that these formed a sort of roof. The resting of the shields (*skoutaria*, sing. *skoutarion*) on the shield-bosses (*boukoula*, sing. *boukoulon*) of those in front means that the shields were placed rim-to-boss in depth[11] with the rim rested/pushed against the top of the shield-boss.

In ancient military theory this was the equivalent of the *synaspismos/testudo/chelone* unit order, which according to the *Sylloge Tacticorum* (43.6) had the width of a third of an *orguia*/two feet (62.46cm). This implies the use of the extra-large infantry shield, which according to the tenth-century *Sylloge Tacticorum* (38.1) was rectangular/oblong in shape with a length of 140.2cm – in the tenth century it could also be triangular, but this was apparently not yet the case during the Late Roman period. The formation is the equivalent of the Roman *testudo* or tortoise formation. The fact that the shields covered the men to their ankles confirms the above suggestion that the shield had to be at least 140.2cm long. It is in fact possible that the shield could be even ca. 160 long, as we find in the *Peri strategikes*. The use of the larger shields by those who were not equipped with armour and shin-guards corresponds with the requirements of the *Strategikon* for the shield-bearers (*skoutatoi*) in difficult terrain. When deployed for combat in difficult terrain the *skoutatoi* were to be equipped without armour and helmets and were to use short spears, not the kind used by cavalry, and 'larger' shields. The last mentioned detail suggests that the footmen varied even the size and type of their shields according to the situation envisaged so that it was expected that the shield-bearers (*skoutatoi, hoplitai*) without armour had larger shields. It is therefore not surprising to find that the *Strategikon* required that the light infantry (*psiloi*) also varied the type of shield, and were required to use smaller shields in difficult terrain. This obviously implies that the width and depth of the heavy infantry (*skoutatoi, hoplitai*) close-order and *foulkon* formations varied in practice according to the type of shield used, because it is clear that the *foulkon/testudo* order was also adopted when the men did not possess the 'larger' shields in situations in which there was a need for extra protection against volleys of arrows and missiles. The soldiers were not always equipped as the situation required.[12]

It is also of note that in this variant the rear rankers rested their shields on the shields of the men in front of them rather than placed them below their shields. See the attached illustration overleaf. This method restricted the ability of the men behind to operate independently, as they could regain full control of their shields only when the man directly in front of him took down the shield and placed it in front. We find both methods, the placing of the shield below the shield in front and the placing of the shield on top of the shield in front, in the columns of Trajan and Marcus Aurelius (see the attached images of the siege *foulkon/testudo*), which proves that the Romans varied their methods according to the needs. In practice, when employing the large shields each rank would therefore have occupied about 80–90cm

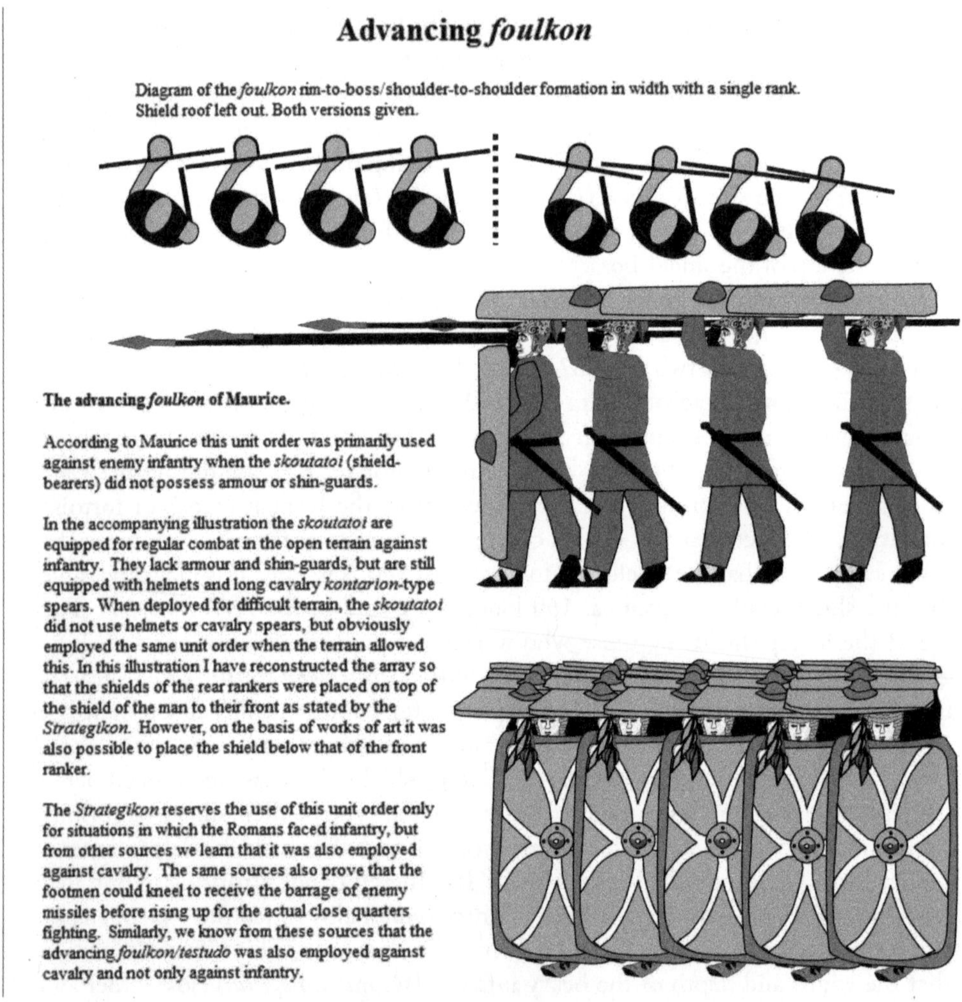

in depth. The depth of each rank in the formation was obviously diminished when the soldiers had smaller shields.

This unit order restricted the ability of individuals to fight as individual swordsmen and also the ability of the unit to manoeuvre. The use of the advancing *foulkon* meant that the individual soldiers lost their ability to fight as individuals, and this was replaced by the use of mass combat as a shield wall in which the soldiers fought from behind their interlocked shields or used the 'mass-shove' *othismos* with shield bashing and pushing. If the unit became even more crowded than this, it lost its ability to fight. It is therefore not surprising to find in the *Strategikon* (12.B.16.91–6) the order '*Largia ad ambas partes*', which was used to open up the formation (either by *meros* or by battle line) either when marching or at the halt if the officers detected that the men had crowded each other.

It is in fact possible that the Roman commanders gave this order just before engaging the enemy in melee if it was considered desirable. We can find them using somewhat similar tactics with their *testudo* even before this. In the past when there existed a need for extra protection against missiles for example because the infantry had smaller, less tall shields or the men needed extra rest while the enemy cavalry bombarded them with arrows, the Romans knelt down in this formation to receive the showers of missiles and when the enemy horsemen then came close enough, the Roman infantrymen rose up and charged at the enemy – which implies a spreading out of the formation.[13] One may imagine that this happened also during the Late Roman period when the Romans could not immediately attack enemy infantry thanks to the presence of enemy cavalry archers, or when the footmen possessed smaller shields so that it was preferable to kneel to protect the legs. See the quote of Dio with analysis of its contents and the attached drawing of the unit order.

Notably, the *Strategikon* also instructed the front rank to protect themselves with their shields during the approach to the swordfight while the rear ranks kept their shields high to protect their heads. In other words, the intervals between the files could also increase during the advance, or perhaps the tightness and cohesion of this formation could be partially lost during the advance. The fourth-century author Ammianus gives us one instance of such an occurrence which happened during a siege.[14] In that instance, the shields became separated from each other and left openings for the enemy arrows to penetrate.

It is also worthwhile to note that the *foulkon* could also be ordered to face all directions (see later) while the soldiers used the shield roof. This demonstrates nicely that the soldiers were able to turn while in the *foulkon* formation, even if the array still restricted the ability of individual- and unit-manoeuvre in comparison with the regular rim-to-rim close order.[15]

The advancing *foulkon/testudo* when the men kneeled

In this example the soldiers vary the shields in the same manner as Dio states so that the rectangular/cylindrical *scuta* are placed on the edges and the flat round shields are used for the building of the roof. The shields were place rim-to-boss both in width and depth. The soldiers are depicted with the long cavalry spears, but the soldiers could equally well be equipped with javelins if the intention was to engage infantry forces. The use of the rectangular *scutum* prevented the pointing of the spears outside the array when the men assumed the kneeling version (left knee against the ground) so that the spears would have been kept inside the formation. However, when the men stood upright the spears were held like javelins to present a wall of spears against the enemy.

The *foulkon* advanced always in rim-to-boss formation up to the javelin- and dart-range. If the men had javelins or darts, then the *foulkon* order was opened up in depth to allow the men to throw their missiles, after which the ranks were once again brought together to protect the soldiers during the final approach. It is clear that the standard expectation when assuming the advancing *foulkon* was that it was also to be used in close quarters combat as a *syskouton* shieldwall, because the unarmoured men needed this extra protection also in melee. The *foulkon* sacrificed the freedom of individuals to fight as swordsmen for the safety of the unarmoured men.

Unfortunately, the *Strategikon* fails to state how the shields were interlocked in width and the artistic evidence is also unhelpful because it provides evidence for both methods of interlocking shields, namely for the placing of the left rim/side of the shield on top of the right side of the shield of the man standing on the left and so forth throughout the line, and vice versa the placing of the right side of the shield on top of the shield of the man standing on the right. The former method of interlocking gave the soldier the advantage of having the protection of the shield against the opponent to his right if he decided to advance out of the formation against the opponent in front of him. This was beneficial because the sword-hand side was more vulnerable to attacks while the opponent on the left had more difficulty attacking his left without exposing his whole right sword-hand side. In this position the shields would also have directed any frontal enemy attacks towards the right against the shield of the comrade on the right. When the soldiers placed the right sides of their shield on the shields of the men standing to their right, the men could advance forward from this formation by moving their shields toward left (this protected the left side of the soldier) to strike the enemy. This enabled the soldier to use his weapon (usually sword or spear) more effectively but it also opened the soldier to counter attacks on the right side, which was no longer covered by the shield. It is probable that this was the typical way for the interlocking of shields, because it gave the individual fighters more freedom for attacks than the other alternative, but is not conclusive because there exists artistic evidence for both uses. There are great numbers of Roman works of art that depict the shields interlocked in both manners and therefore one cannot make a certain conclusion on this question. See the attached sample of illustrations. It is in fact probable that both versions were used, just like there were two different ways of forming up the shield roof of the *testudo*. Both versions had their advantages and disadvantages. Furthermore, at very close range these things lost their importance.

In my doctoral dissertation I asked the question of how long the soldiers would be able to hold the shields above their heads. I came to the conclusion that, as the shields were relatively light in weight and the physical demands of the battle restricted the time that hand-to-hand combat could continue, it was entirely plausible that the soldiers would have been able to hold their shields up until they were so close to the enemy that one could attack them and one or the other retreated.[16] It was also possible to rest their hands and shields against the crests of their helmets to take some of the weight away. It should be noted that the shield roof of the tortoise formation was not unique to the Romans. Anyone possessing a shield would have instinctively held his shield above the head when facing a shower of missiles. In fact the 'shield castle' was the standard defensive formation amongst the Germanic peoples.[17]

The shield roof gave the soldiers the illusion of invulnerability, which in its turn made it easier for the officers to order the men forward, but obviously this was achieved at the cost of depriving them of the sense of sight. It is possible that in certain circumstances the latter may have caused panic among the ranks when commotion arose in some quarter, but it is clear that this happened very rarely because its continued use proves that in general it gave the men the illusion of safety and thereby made them more willing to move forward.

Philip Rance (2004b, 298–300) suggests that the following quote from Plutarch's pen is actually a description of the type of *foulkon/testudo/chelônê* meant for use against cavalry as depicted in the texts of Arrian (*Ektaxis kata Alanôn*) and Maurice. Rance also suggests that in his description of the same tactic of Antony's legionaries, Cassius Dio has misunderstood the structure of the *chelônê* (tortoise) when he claims that all troops knelt down (quoted below after Plutarch). This is possible, but still fairly unlikely on the grounds that Cassius Dio knew very well how the Roman army of his own era operated and was therefore in a better position to interpret the material than Plutarch. We should remember that Dio had more experience of Roman military methods than Plutarch. It is rather the other way around, namely that it is likelier that Plutarch would have made a mistake, but on a closer look it is actually very uncertain if Plutarch is even describing the sloping version of the *foulkon* vs. cavalry as depicted by Maurice and Arrian. However, I have here given the benefit of doubt that Rance is correct in his interpretation of Plutarch because his description does resemble what we find in Arrian and Maurice, and so I am suggesting that the Romans used two different versions of the *testudo/foulkon* against enemy cavalry, one in which both the files and ranks interlocked their shields in width and depth already described above, and one in which the shields were interlocked only in depth (Maurice's *foulkon* vs. cavalry). This is actually also confirmed by Arrian, who has two different versions of the *chelônê/testudo* in his own military treatises. In the *Technê taktike* (11), Arrian equates the *testudo* with the Greek *synaspismos* and states that it could be square, oblong or circle in shape and that it could withstand even wagons and rocks without breaking its cohesion, while in the *Exktaxis kata Alanôn* 16–7, 26) Arrian describes the same type of *testudo* used against cavalry (with shields roughly rim-to-rim in width and rim-to-boss in depth) as we find in the *Strategikon*. In short, the Romans used two different variants of the *testudo* against cavalry.

Plutarch:

> ... as the Romans were descending some steep hills, the Parthians attacked them and shot at them as they slowly moved along. Then the shield-bearers [*thureoforoi = legionaries equipped with the long oval* thureos-*shield*] wheeled about, enclosing the lighter-armed troops [*psiloi*] within their ranks [*i.e. formed a hollow square/oblong*], while they themselves dropped to one knee and held their shields out before them. The second rank held their shields out over the heads of the first, and the next rank likewise. The resulting appearance is very like that

Lipsius' (16th cent.) reconstruction of the Roman *testudo/chelônê*, which Arrian (*Techê taktika* 11) equated with the Greek *sunaspismos* (shields interlocked) unit order. According to Arrian, the *testudo* was usually square in shape, but circle and oblong formations were also used.

of a roof, affords a striking spectacle, and is the most effective of protections against arrows, which glide off from it. The Parthians, however, thinking that the Romans dropping on one knee was a sign of fatigue and exhaustion, laid aside their bows, grasped their spears by the middle and came to close quarters [*note the double armament of the Parthian cavalry*]. But the Romans with a full battle cry, suddenly sprang up, and thrusting with their javelins [*(h)ussoi = pila*] slew the foremost of the Parthians and put all the rest to rout. This happened also on the following days as the Romans little by little proceeded on their way.

Plutarch, *Antony* 45, tr. by Perrin 241 with my comments inside square brackets.

Dio:

… One day when they [*Mark Antony's men when they were retreating from Parthia in 36* BC] fell into an ambush and were struck with fast-flying arrows, they suddenly made by joining their shields ['*synaspisantes*'] the *testudo* [*chelônê = tortoise*], and rested their left knees on the ground. [*Note that this corresponds with Vegetius's instructions, 1.20, regarding the fighting stance with the gladii and pila when those were used for thrusting. The placing of the left knee against the ground means that when the soldier stood up to fight the right foot was forward so that the* pilum *would have been thrust at the enemy. When the pila were thrown the left foot was placed forward. The right foot/leg forward stance was the more aggressive of the two and enabled the soldier to deliver the thrust faster.*] The barbarians had never seen anything of the kind before and thought that they had fallen from their wounds and needed only one finishing blow; so they threw aside their bows, leaped from their horses, and drawing their Akinakes-daggers came close to put an end to them. At this the Romans rose to their feet, spread out the phalanx at a word [*this implies that the shields were interlocked in width before this*], and each one attacked the man nearest and facing him; thus they cut down great numbers since they were contending armed against an unprotected foe, men prepared against men off their guard, heavy infantry against archers, Romans against barbarians. All the survivors immediately retired and no one followed them for the future. This *testudo* and the way in which it is formed deserve a word of explanation. The baggage animals [*skeuoforoi*], the light-armed troops [*psiloi*], and the cavalry [*hippoi*] are marshalled in the centre of the army [*strateuma*]. Those infantrymen who use the long oblong, hollow, cylindrical shields are drawn up around the edges, making a rectangular figure [*plinthion*]; and, facing outward and holding their arms ready for combat [*I have here followed Cary's more accurate translation for the* 'ta hopla' *instead of Forster's 'with spear-points projecting'. However, it is still clear that Forster's translation expresses what Dio's text means, namely that the spears were pointed towards the approaching cavalry*], they enclose the rest. The other infantrymen, who have flat shields, form a compact body in the centre and raise their shields above themselves and above all the rest, so that nothing but shields can be seen in every part of the phalanx alike and all the men by the density of formation are under shelter from missiles. It is so marvellously strong that men can walk upon it, and whenever they get into a hollow, narrow passage, even horses and vehicles can be driven over it. Such is the method of this arrangement, and this shows why it has received the title of *testudo* [*chelonê = tortoise*], – with reference to its strength and to the excellent shelter it affords. They use it in two ways: either they approach some fort to assault it, often even enabling men to scale the very walls, or where sometimes they are surrounded by archers they all bend together – even the horses being taught to kneel and recline [*the Roman cavalry horses were clearly superbly trained for various tasks*] – and thereby cause the foe to think that they are exhausted; then, when the others draw near, they suddenly rise, to the latter's great alarm. [*It is probable that the kneeling of the footmen on the outer edges of the hollow square was also meant to make it easier for the archers, slingers, staff-slingers and operators of carroballistae to shoot over their heads straight at the approaching enemy. The kneeling of the horses would have enabled their riders and others to form a sloping tortoise for the protection of the horses, such as described by Arrian in his* Ektaxis kata Alanôn, *for which see Syvänne (2023a), and the* STR *12.A.7, for which see Syvänne, 2004.*]

Dio 49.29–30, tr. by Forster 249ff. with comments and changes.

8.6. The sloping *foulkon/testudo* vs. cavalry in Arrian and Maurice[18]

> The first, second, and third in each *akia* [*battle line, which means that the first, second and third rank men in each acies rather than a file, as translated by Dennis and Rance, but obviously the meaning is still the same*] form a *foulkon*, interlocking their shields, fix their spears firmly in the ground, holding them slanted forward and directly outside their shields, so that anyone who dares come too close will quickly experience them. They also lean their shoulders and put their weight against the shields to resist any pressure from the enemy. The third, who is standing nearly upright, and the fourth hold their *kontaria* like javelins, so when the enemy gets near they can use them either for thrusting or for throwing and then draw their swords. The *psiloi* and the *kaballarioi* use the bow.
>
> *Strategikon* 12.A.7.52–60, translation based on the tr. of G.T. Dennis with some changes including e.g. the meaning of *akia* changed from file to *acies*. Underlining and brackets added.

This stationary sloping *foulkon/testudo* order was used against a cavalry charge.[19] The idea behind the kneeling and interlocking of the shields in width was to present a solid wall of shields and spears to the approaching enemy cavalry while also preventing one's own men from fleeing. The archery and missile fire by the rear ranks reduced the enthusiasm of the enemy cavalry to come into close quarters. On most occasions the effectiveness of the array was improved further by having the soldiers banging their shields while they also shouted. This loud cacophony was used to scare off the enemy horses.[20] However, as noted above, the Romans also used the standard *synaspismos* and *foulkon/testudo* (shoulder-to-shoulder) orders against horsemen in which the spears were used to thrust.[21]

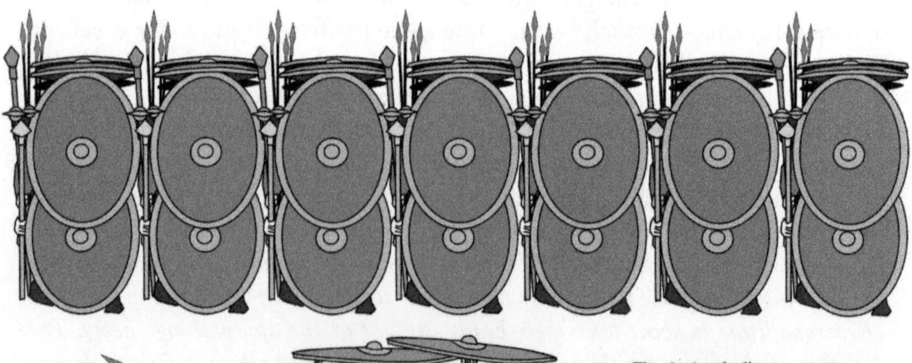

The sloping *foulkon* vs. cavalry à la Arrian and Maurice

It was very difficult for the third ranker to place his shield rim-to-boss with the second ranker so that I have here made the assumption that he just brought the shield above him. On the basis of the fact that the two front rankers placed the butts of their spears against the ground means that the shields of the files did not overlap but were brought only almost rim-to-rim. The files were therefore ca. 94 cm wide.

8.7. The Tortoise for the Storming of the Fortifications[22]

The Late Romans also used a variation of the tortoise formation for the storming of enemy fortifications, just as their ancestors had. The array was obviously not only used in sieges but also in battles when the soldiers had to storm a fortified marching camp or city in the course of combat. A good example of the use of this array is Theophylact's description of the storming of Beïuades, the man climbing the wall had directly under him his fellows in the tortoise formation, which indicates that he had climbed on top of them. Notably, the men who assumed the sloping siege *testudo/foulkon* were dismounted cavalry which demonstrates nicely the very high quality of the sixth-century Roman cavalry forces. However, the best descriptions of the array come from the pens of Livy and Ammianus.[23]

> Among other diversions, the director of the games used to introduce about sixty young men [*iuvenes*] in arms, sometimes more, whose performances were partly a representation of troops going through the military exercise, and partly a display of more accurate skill than appeared in the practice of soldiers, and which approached nearer to the mode of fighting used by gladiators. [*The example of Livy shows the Roman youths demonstrating their skills in the forming up of a sloping testudo, first in the circus and then on the battlefield in 169 BC.*] After performing various evolutions, they formed in a square body with their shields raised over their heads, and closed together, the foremost standing upright, the next stooping a little, the third and fourth lines more and more, and so on, until the hindmost rested on their knees, thus composing a covering in the shape of a tortoise-shell, and sloping like a roof of a house [*i.e. a sloping testudo*]. Then two armed men, who stood at the distance of about fifty feet, ran forward, and after some menacing flourishes with their arms, mounted over the closed shields, from the bottom to the top of this roof; and treading as steadily as if on solid ground, sometimes paraded along the extreme edges of it, as if repelling an enemy, and sometimes engaged each other on the middle of it. On the present occasion they raised the same type of [*sloping*] *testudo* against a part of the wall, and the soldiers standing thereon [*propugnatores*] mounted until they were as high as the defendants on the battlements; these they soon beat off, and the soldiers of two *signa* [*units of unspecified size*] climbed over to the town. The only difference between this and the playful contrivance was that here the outside men in the front and in the two flanks did not raise their shields over their heads, lest they should expose their bodies, but held them before them as in battle; so that the weapons thrown at them from the walls as they advanced, did them no injury, while those that were poured in showers on the roof slid down the smooth slope to the bottom, without doing any mischief.
>
> Livy 44.9, tr. by Baker 309–10 with changes and comments.

> He fastened together three boats, and built upon them a protective covering after this fashion: in front stood armed men on the thwarts with their shields held close together above their heads, those behind them stooped down

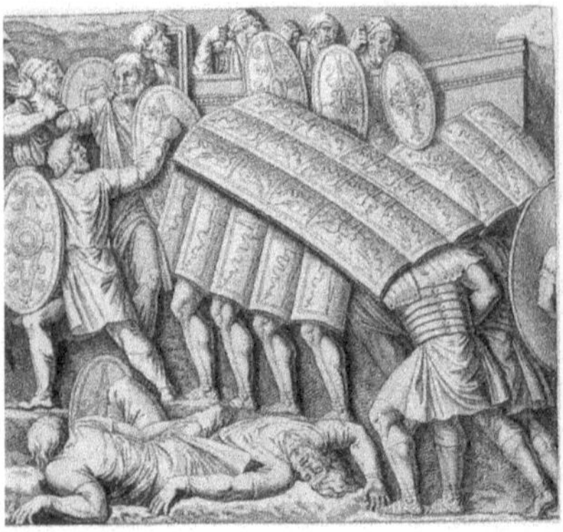

Above: Siege *testudo* as depicted in Roman art

Left: Siege *testudo* in the Column of Trajan. **Right:** Siege *testudo* in the Column of Marcus Aurelius. Source: Montfaucon. The illustrations do not depict the sloping of the unit order accurately, but should still give an idea of how the order was used in the storming of forts.

Note the difference in the way how the shield roof is formed in these two images. On the left the rear shields are placed on top of the front one (not shown clearly in this drawing) as suggested by the *Strategikon* for the moving *foulkon* against infantry while on the right the rear shields are placed underneath the shields in front. When used in actual field battle the version in which the shields were placed on top of the front ones removed the freedom of using shields individually towards the front so that when one wanted to get out of the *testudo* array it had to start from the front. In other words, if the front ranker died or fell, the man directly behind him had to bring his shield down really fast. When the shields were placed underneath the shields in front of those, the men retained better control over their shields when the enemy appeared in front because it was not "tied" in place by the shield of the man in front of it.

Below: Lipsius's reconstruction of the siege t*estudo*.

Both show the principle of using the *testudo* in the storming of the city/fortification, but leave out the sloping of the rear portion of the siege *testudo*.

somewhat lower, and those in the third rank gradually lower still, so that, since the hindermost rested on their hams, the whole gave the appearance of an arched building. This kind of device, used in battles against walls, has this form in order that volleys of missiles and rocks, gliding down the sloping side, may flow off like showers of rain.

Ammianus 26.8.9, tr. by Rolfe Vol.3, 629.

8.8. The irregular order (*drouggos*)[24]

The light infantry *psiloi* used the irregular *drouggos*-order in difficult terrain and also in regular battles when the situation required this. The same appears to have been true also for the heavy infantry (including the so-called *skoutatoi* of Maurice) when they were required to fight like light infantry – the best proof of this is their training scheme, which included the use of javelins, bows and slings. The only extant description of the use of the *drouggos*-order in difficult terrain is in the *Strategikon*, but it is clear that similar tactics had always been used when the terrain required this. For example, it is quite obvious that the light-infantry *psiloi* could not have been deployed in ranks and files in broken mountainous or wooded terrain which clearly required irregular groups of men. According to Maurice, the light-infantry was in such situation drawn up in *drouggoi* (irregular) four- to five-man groups, with three or four of the men armed with javelins and shields, and one with the bow. The archer was expected to provide covering fire for the rest. These little *drouggoi* were not to advance far from the main body of men, which were in such cases drawn up in two- or four-phalanxes each consisting of lightly-equipped *skoutatoi*. The *drouggoi* were deployed one after the other so that they could protect each other's back.[25]

The *drouggos* was the most manoeuvrable of the infantry unit orders, just like their cavalry counterparts in the *drouggos* formation. It is quite obvious that, when the situation required, the light infantry *psiloi* also used the same irregular order when deployed alongside the regular phalanx. The *psiloi* were expected to be versatile and fast and the use of this unit order gave them this ability. It should be noted that this unit order was the most demanding of the unit orders for soldiers, because they did not have the psychological benefit of having their comrades right next to them. The soldiers had to trust in each other, and they were also required to be able to improvise at a moment's notice. This was possible because the use of the tent group (*contubernium*) system ensured that the men knew each other. The use of the four- to five-man groups means that at least two neighbouring groups consisted of the men of the same tent, which helped to build trust. This ensured that the groups of men could also feel trust in those following them. However, just like it was with the cavalry, it is still probable that the typical unit size for the infantry in irregular order was the *tagma*, so that the men followed it in combat and obeyed commands as required. The following quote from Julius Caesar's *Civil Wars* demonstrates how the irregularly-operating forces could cause problems even in regular combat, especially if it took place in difficult terrain. In the example given, the Pompeian legionaries had adopted from the Lusitanians an irregular fluid combat tactic in which there was no need to maintain ranks and files.

Caesar:

> The method of fighting adopted by the enemy's troops was to charge at first at full speed, boldly seize a position, take no particular trouble to preserve their ranks, but fight singly and in loose order; if they were hard pressed they did not consider it a disgrace to retire and quit their position, for, waging a continuous

warfare against the Lusitanians and other barbarous tribes, they become used to a barbarous kind of fighting, as it usually happens that when troops have spent a long time in any district they are greatly influenced by the methods of the country. It was this system that now threw our men into confusion, unaccustomed as they were to this kind of fighting; for as the enemy kept charging singly they thought that they were being surrounded on their exposed flank. As for themselves, they had judged it right to keep their ranks and not desert their standards, nor give up without grave cause the position they had taken. And when the vanguard was thrown into confusion the legion posted on that wing could not stand its ground and withdrew to the nearest hill.

Julius Caesar, *Civil Wars* 1.44 (tr. by A.G Peskett pp.65, 67).

Cuneus-wedge, rhombus, caput porci(num), globus, rhombus

The Romans definitely used two variants of the infantry unit wedge (*cuneus, kounion/kouna, embolos*): a small ca. 200–300 strong wedge (could also be e.g. a *tagma* or a cohort) without any hollow space inside (which looked like 'Δ') and a larger wedge formed out of two larger units, such as for example 4,000-strong phalanxes (which looked like 'Λ'). The question of the different versions of the wedge will be

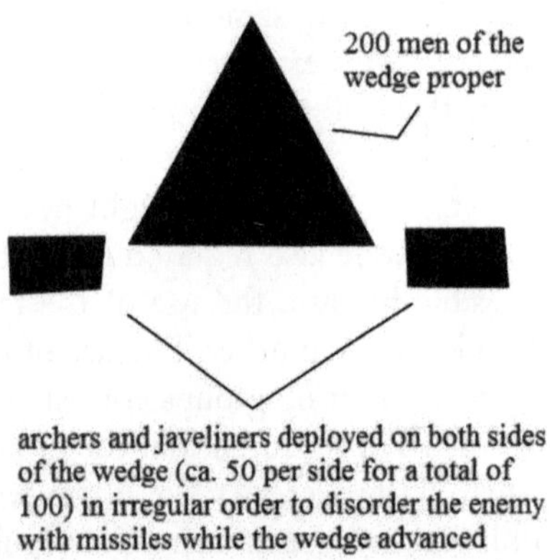

Roman Wedges

Above: small unit wedge consisting of 300 men: 200 in the wedge and 100 as supporting archers and javeliners.

Below: large wedge consisting of sizable units formed up as a lambda.

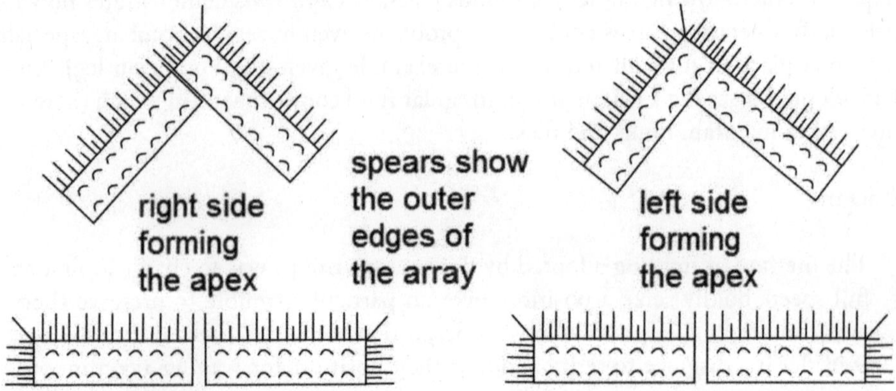

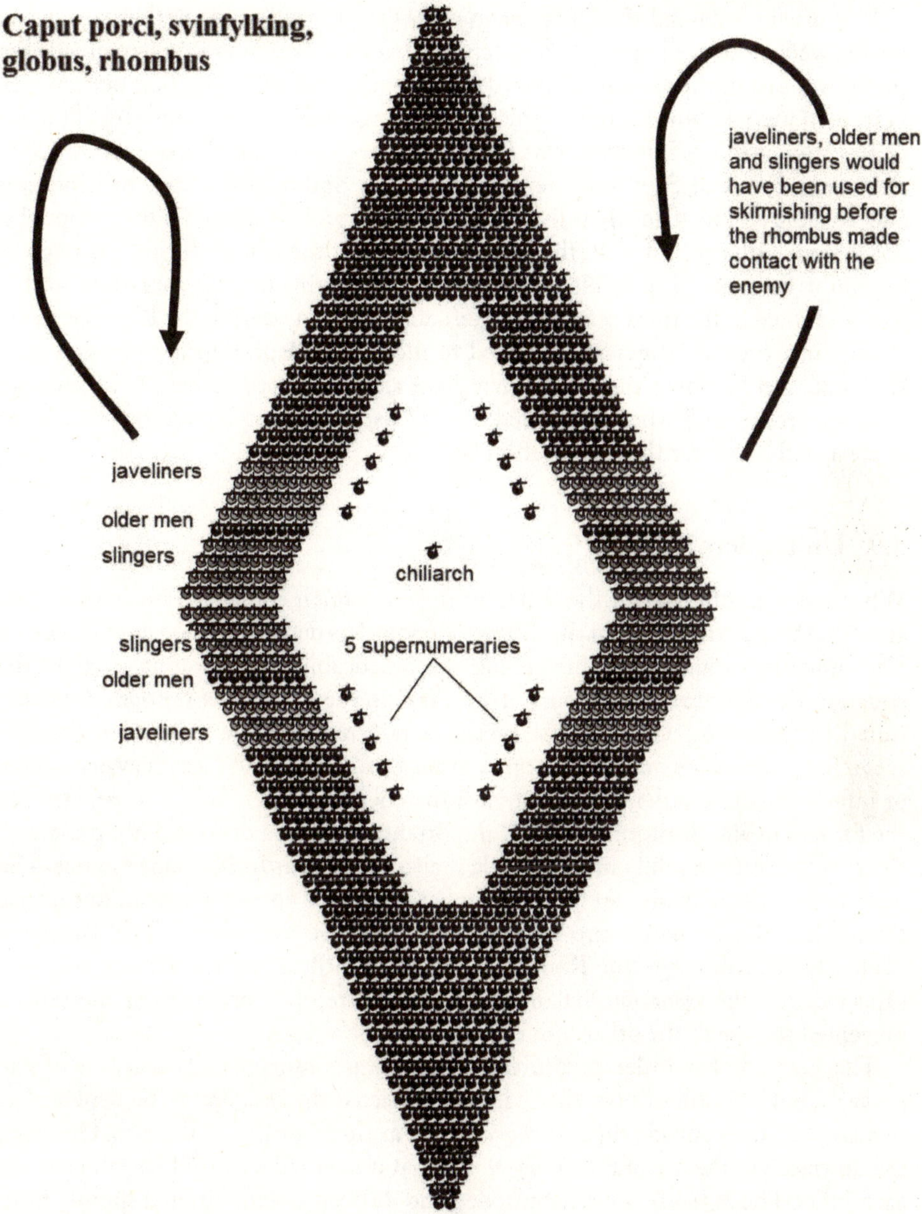

elaborated further in the analysis of the wedge as a battle formation. On the basis of the 300-strong wedge of Praetorians, of which 200 performed the actual attack in the city of Rome in 193, it is possible to reconstruct the likely shape of the smaller-unit wedge. The front consisted of 24 ranks deployed as a wedge (1, 2, 3, 4, 5 etc. per rank) for a total of 200 men, behind which were the rest, the remaining 100 men, who presumably acted as light-infantry archers for the heavy infantry wedge in front of them.[26] In combat, the wedges proper obviously used either close order (*pyknôsis*), or *synaspismos*, or the tortoise order.

It is usually assumed that *caput porci(num)* ('pig's head') was another new name for the wedge, adopted by the fourth century, but in my opinion it is actually likelier that it was one of the variants of *globus* (globe, independently operating unit), which in its turn was a hollow rhombus which the Vikings were to call *svinfylking*. The size of the Viking *svinfylking* conforms (1,110) to the size of the Roman cohort (555) after the reforms of Septimius Severus and as we find it in the texts of Modestus and Vegetius.[27] The date when the hollow *rhombus/globus/caput porci* was adopted is not known with certainty. On the basis of the size of the cohort during the reign of Septimius Severus, it is possible that it was adopted during his reign or before, or that it was adopted in the third century, or alternatively already during the Republican era at the same time as '*globus*' was adopted to mean an independently-operating unit. It should also be noted that *globus* may have also been used to denote a defensive infantry circle (*orbis*) when it adopted the defensive kneeling *testudo*, but this is not as clear as the case for the hollow rhombus.

8.9. Unit orders in battle[28]

What was the influence of the different unit orders on combat? This is not easily answered because the effect of the different unit orders depended upon circumstances. The most important factor influencing the practicability of the unit order to the situation was the terrain. In open and level terrain the open order (*araiôsis*) was only suited for marching. Open, unobstructed terrain required the use of the *pyknôsis*, *synaspismos* or *foulkon* orders. The open order was ill-suited to face a cavalry charge or infantry in close order, because it gave the enemy a better chance of penetrating the formation and disrupting it. Most importantly, the open order did not present to the enemy cavalry a solid obstacle, while it also gave cowards the chance to flee. The open order was therefore not used in combat unless the enemy managed to surprise them while the Romans were marching or otherwise unprepared. The same was obviously not true when the Romans marched in difficult terrain. In fact, Maurice expected that the *skoutatoi* phalanxes used the more open order when the terrain prevented the use of the other unit orders.

The best combat order in difficult terrain was unsurprisingly the use of the *drouggoi*, but it is still of note that Maurice expected the *skoutatoi* to be deployed as regular phalanxes outside these *drouggoi* to act as their last line of defence. However, the downside to the use of the *drouggoi* was that it required elite soldiers who trusted each other. The *pyknôsis*-order for wooded and difficult terrain differed slightly from the standard compact order because the *skoutatoi/hoplitai/legionaries* were then more lightly equipped and their phalanxes had only four ranks of men. When employed in regular battle, the light-infantry *drouggoi* were also useful because these were particularly well-suited to skirmishing and fast movement, the last of which could be particularly useful when there was a need to bring a reserve force forward.

As noted, unobstructed terrain required the use of the *pyknôsis*, *synaspismos/suskouton* and *foulkon/testudo* unit order. The use of these orders ensured that all of the men would arrive simultaneously against the enemy, which increased the effectiveness of the attack. Furthermore, the use of these unit orders meant that the enemy had far fewer chances of penetrating the Roman formation and creating disorder. The men were also better-protected and psychologically more ready to face the enemy. The closeness of these formations also presented a visible physical obstacle, both to horses and their riders, which lessened their chances of defeating the Roman infantry. The stationary kneeling *foulkon* against cavalry was particularly good for its purpose, because it presented a visible obstacle to the enemy while making it difficult for the Roman soldiers to flee. The *testudo*, used in the storming of fortifications, in its turn enabled the soldiers to storm enemy fortifications on a moment's notice, which could be very useful during the final phases of a battle when one stormed the enemy camp. The two wedge formations and the *caput porci* were used mainly for breaking through enemy formations, or alternatively for the breaking up of an enemy attack in front of the main battle line, or as marching formations that protected the flanks. On the basis of Ammianus's text, the *caput porci* appears to have been the favourite breakthrough tactic of Roman infantry during the fourth century. Whether the case remained the same in the fifth- and sixth-centuries is not known, but it is probable that it was no longer in fashion during those centuries because the principal infantry combat formation was the hollow rectangle, and Maurice pays very little attention to the wedge and none to the *caput porci*.

In sum, the Romans had unit orders for all kinds of eventualities that might arise on the battlefield.

Chapter Nine

Infantry Battle Formations and Tactics

9.1. The battle formations of the third century

The principal infantry combat formation of the ancient and medieval periods was the lateral phalanx (*plagia falagx*). By the late-third century it had many different forms and variants, some of which were based on Greek military theory and some of which were based on traditional Roman cohort tactics. See Chapter 1. In the late-third century the Romans used both versions, the Greek style phalanx tactic, which is nicely described in the texts of Arrian, and the cohort version. The Roman legions followed two different organizational schemes so that the internal structures of the new legions after Septimius Severus at least until the reign of Diocletian were slightly different than those before him.[1]

In the older system, the legionary cohorts from two to ten were organized as centuries of 80 men, two of which arrayed for combat as a maniple of 160 men, three of maniples forming a regular cohort; or as a milliary 1st cohort, which consisted of five centuries each with 160 men. The cavalry component of these older legions consisted of units of about 500 or 600 horsemen. The auxiliary units were organized similarly. In addition to this there were mixed units of infantry and cavalry, but these followed the same basic unit structure as the above. In combat these infantry units were organized either as cohorts or as phalanxes.

The new legions after Septimius Severus until the reign of Diocletian were organized as follows. Each of the cohorts from two to ten consisted of 555 footmen (360 heavy infantry deployed four deep and 180 light infantry deployed two deep, in addition to which were five optiones, five standard-bearers, and five centurions). The first cohort consisted of 1,105 footmen (720 heavy infantry deployed 4 deep and 360 light-infantry deployed two deep, in addition to which came ten optiones, ten standard-bearers, and five centurions). The cavalry component had been increased to 22 *turmae* of cavalry for a total of 704 horsemen plus 22 decurions, 10 musicians, five standard-bearers, and five centurions for a total of 746 horsemen. In combat these legions were usually organized as six-deep phalanxes.

In the older version, the legionaries (organized as centuries, maniples and cohorts) were typically organized with unit depths of four or eight heavy-armed men (each equipped with a *pilum*-javelin or two *pila* or a *hasta*-spear; a *gladius* or *spatha*-sword, *scutum*-shield, a helmet; and chainmail, or scale, or *lorica segmentata*, or muscular armour), but this varied in practice according to the availability of soldiers. The auxiliaries were organized in a similar manner. This was basically the same file structure as in the Greek style phalanx, which the Romans never stopped using. If there were extra men, then these could be added to the ranks so that there could be ten

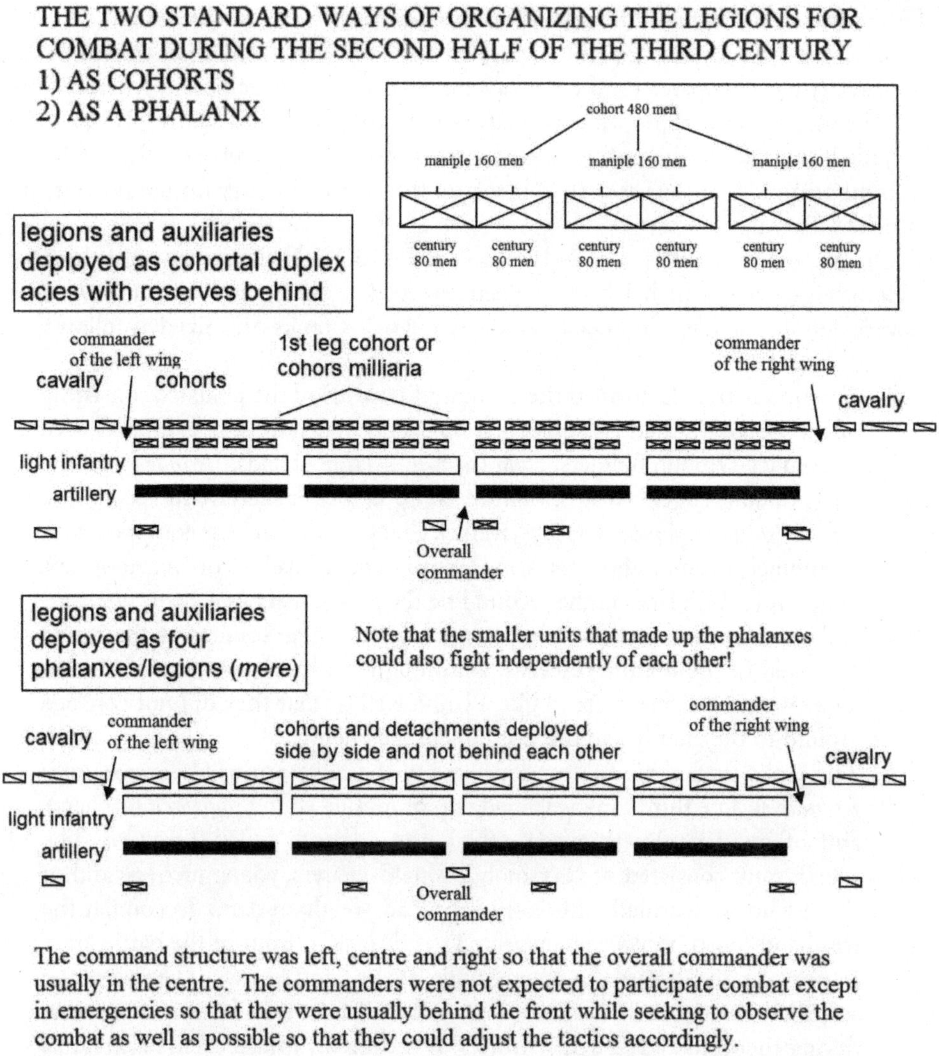

ranks, as Pompey had at Pharsalus, and if there were too few men then these could be organized with a shallower depth. In normal circumstances the light-armed slingers, javelin-throwers and archers were auxiliary units, meaning that these were deployed separately from the legionaries in their own units – for example behind the cohorts (these could remain there and shoot over the legionaries, or they could advance in front to skirmish before returning to their place), or on the flanks. However, since a quarter to a third of the legionaries were trained as archers and all were trained as slingers and javelin-throwers, it was also possible to form the four- and eight-deep formations so that one of the ranks, usually the last, consisted of the lightly-equipped legionary, or alternatively so that the heavy-armed legionaries were deployed as either three-deep or six-deep formations while the lightly-equipped legionaries were deployed elsewhere. When the old-style legions and auxiliary forces were deployed for combat in the latter half of the third century these were usually deployed either as phalanxes (usually a single or double phalanx) or as a double line of cohorts, with the cavalry usually posted on the flanks while the reserves were posted wherever the commander thought appropriate (usually behind the flanks and centre). If the army included cataphracts, these were posted either next to the infantry on the flanks, or in front or middle of the infantry to break up the enemy infantry formation. See the attached diagrams.

On the basis of Modestus (12–4) and Vegetius (3.14–7) the structure of the new legionary phalanx (which I have conjectured to date from the reign of Septimius Severus) up to the reign of Diocletian consisted of six ranks organized as follows:

1) The first two ranks formed the 'stonewall' (*murus*) of the phalanx, consisting of the heavily-armed (*gravis armatura*): The first rank was equipped with cataphract-armour, helmets, *scuta* (*scutum* = large shield), *spathae* (*spatha* = long, double-edged sword), *lanceae* (sing. *lancea*, lance/javelin/spear) and *spicula* (*spiculum* = *pilum*/heavy-javelin). The second rank was equipped with cataphract-armour, and was armed with *spicula*-javelins or *lancea*-spears. Vegetius (3.14) places in the second line the cataphracted *sagittarii* (archers) and the best men armed with *spicula* or *lanceae*. The soldiers of the *murus* consisted of the mature veterans, who fought with *pila/spicula* and swords. They were expected to stand like a stone wall so that they did not concede ground to the enemy and did not pursue the enemy.
2) The ranks from three to four consisted of the light-armed (*levis armatura/ferentarii*). The third rank was made up of mobile young *sagittarii* (archers), and of good javelin-throwers, which the ancients called *ferentarii*. The fourth rank consisted of very mobile shield-bearers, young archers, and of those who were armed with javelins or lead-weighted darts. In combat the third and fourth ranks were expected to advance in front of the battle array to provoke and harass the enemy with javelins and arrows. They were the so-called *antesignani*. The lance-armed skirmishers were known as *lanciarii* during the third and fourth centuries. If the enemy forced them back, ranks three and four retreated behind the first and second ranks, and if the enemy fled, they and the cavalry pursued them. The slingers of the fifth rank could join ranks three and four in the above action.
3) Ranks five and six consisted of the so-called heavy-armed, even if rank five was equipped lightly. In practice this meant that the fifth rank was equipped as *ballistarii/balistarii* (probably users of *manuballista* or *arcuballista*),[2] *funditores* (slingers), *tragularii* (in this case probably users of *tragula*-javelin), and *fustibalatores* (staff-sling users). The sixth rank consisted of strong shield-bearing soldiers equipped with the full panoply of equipment. The last rank made the formation double-fronted, which enabled the array to face both the front and rear simultaneously. The fact that the fifth rank consisted of the 'heavy-armed' in light equipment enabled the phalanx to also target enemies behind with long-range missiles.

New type of century from the reign of Septimius Severus until the reign of Diocletian

1	1	1	1	1	1	1	1	1	1	1	1	1	1	1	1	1	1 C
2	2	2	2	2	2	2	2	2	2	2	2	2	2	2	2	2	2 V
3	3	3	3	3	3	3	3	3	3	3	3	3	3	3	3	3	3
4	4	4	4	4	4	4	4	4	4	4	4	4	4	4	4	4	4
5	5	5	5	5	5	5	5	5	5	5	5	5	5	5	5	5	5
6	6	6	6	6	6	6	6	6	6	6	6	6	6	6	6	6	6
								O									

See the attached diagram for one possible way in which soldiers in a century of the new type (numbers 1–6 show ranks), *vexillarius* (V), *optio* (O) and *centurio* (C) could have been deployed for combat.

When deployed for combat the new type of phalanx was deployed so that the cavalry formed the wings of the infantry phalanx, with the *loricati* and *contati equites* posted next to the infantry and the *sagittarii equites* further out. The *catafracti equites* were often posted in front of the infantry or between legions to break up enemy infantry formations. One may make the educated guess that the *loricati* and *contati equites* could also include these. The reserves of infantry and cavalry were placed behind the flanks and centre. The overall commander was on the right flank between the infantry and cavalry, the second in command was in the centre of the infantry phalanx, and the third was posted on the left flank. The other sources obviously prove that the overall commander could choose his place as he thought necessary. The new phalanx was basically only a variant of the old with a different armament.

The basic tactics in this system were the single- and double-line formations, and the use of the hollow square/oblong and wedge formations. See the attached diagrams. The single- and double-line formations were basically variants of the lateral phalanx/oblong rectangle formation used in frontal battle with the enemy, while the hollow square/oblong was primarily a defensive formation used in marching and combat, but which could also be used offensively. The hollow square/oblong formation will be dealt in greater detail in a separate chapter devoted to that array. See Chapter 9.7. As already discussed, the wedge had two different variants, the unit order and grand tactical formation, both of which could be used for breaking the enemy formation. The grand tactical formation could actually consist of the entire army or of the different divisions of the army. The large wedge used by the Franks at the Battle of Capua/Casilinum in 554 is a good example of the use of a single wedge by the entire army, and the invasion of Germania by Germanicus in 14 AD and Roman operations against the Quadi in 358 are good examples of how the army could be deployed as several divisional wedges to cover a long stretch of enemy territory for the purpose of maximizing destruction.[3] The unit and divisional wedges receive further attention later in this chapter.

The more advanced set of combat tactics consisted of six unit tactics (*cuneus, forfex/forceps, caput porci, orbis/rotundus, globus/drungus, serra*) and seven combat formations.[4] Vegetius includes also the *epikampios opisthia* formation (see Chapter 9.6), but he restricted its use to marching.

The Principal Roman Combat Formations according to Modestus in ca. 275

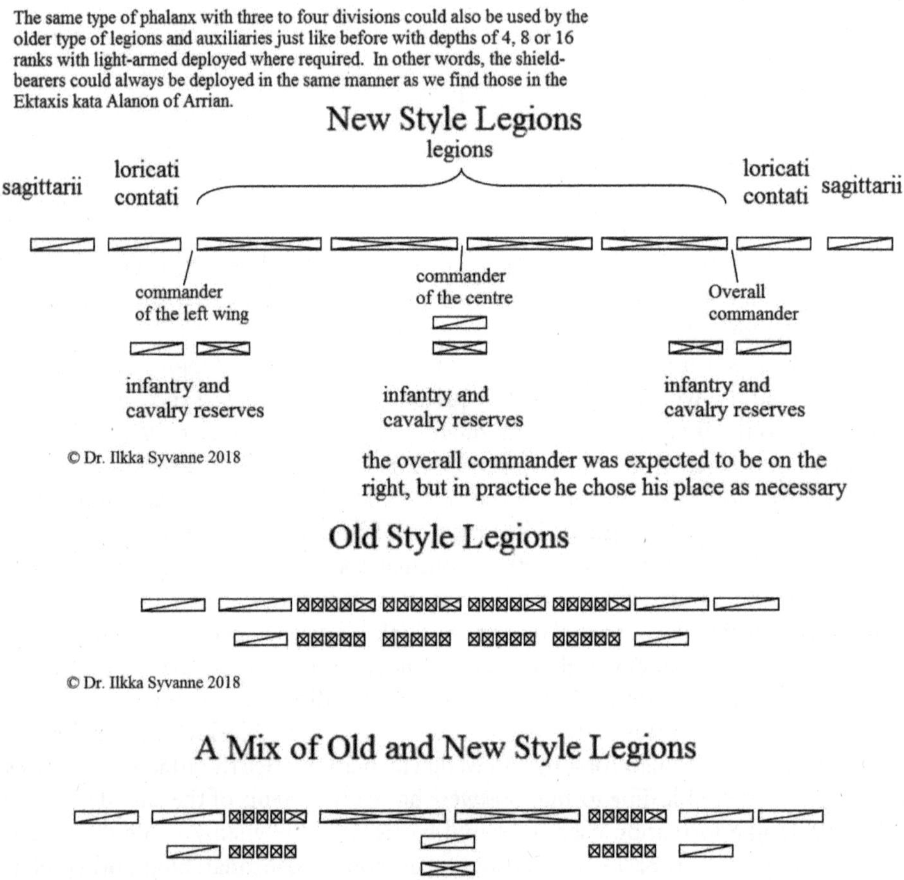

The six unit tactics were:

A) The *cuneus* was a wedge which was formed of infantry units to break through the enemy formation. According to Vegetius, the commander required reserves behind the phalanx so that he could form the wedge or the pincer (hollow wedge) against the enemy wedge.

B) The *forfex* (shears, claw)/*forceps* (pincers) was used as counter-formation against the wedge. It was basically a hollow wedge and/or *antistomos difalaggia*. It had at least two different versions, a unit formation meant to outflank the enemy (a counter tactic against wedge and column) and a battle formation used in ambush. The latter will receive further attention later.

C) The *caput porci* (swine head) appears to have been a hollow rhombus that separated from the infantry formation for the same purpose as the *cuneus*, while having the advantage of reserves behind.

D) *Orbis/rotundus* (circle, round-shaped) was a defensive formation that faced the rear like the two-faced (*amfistomos falagx*), but which could also be an actual circle to offer a front against all directions.

The Standard Legionary Combat Formations in 275 according to Modestus

Vegetius' Seven Tactics 3.20, 3.26:

1. Oblong rectangle array was the standard array, but it was considered unsafe because the length of the line could cause a breach of the line during advance. It was also necessary to use reserves to protect the flanks. Its use was recommended only in such instances where there were enough brave troops to surround the enemy formation on both sides (*hyperfalaggesis*).

2. Oblique array vs. enemy's left wing. This version of the oblique array was considered better since the attack was directed against the shieldless side. The purpose was to place the best inf. and cav. on the right and use them to surround the enemy.

3. Oblique vs. enemy's right wing. Note the wedge shape of the tip of the oblique array. It seems probable that the reason for calling Epaimenondas's array both a wedge and oblique has resulted from this. This was considered weaker than the 2nd tactics and its use was recommended only when the left wing was much stronger than the enemy's right. In that case, the commander posted his reserves on the left. The general was also to take special care that enemy wedges wouldn't penetrate the array.

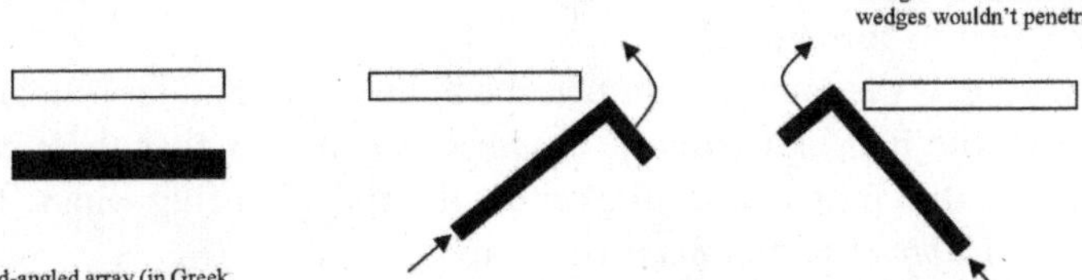

4. The forward-angled array (in Greek *epikampios emprosthia*) was used to outflank the enemy suddenly on both sides when he did not expect it. When the army was 400-500 paces away from the enemy, the wings suddenly charged forward to surprise the enemy. This was dangerous tactic, if the enemy managed to hold its own, because the wings were separated from the centre.

5. The fifth formation was an improvement of the fourth. In this array the light-armed and archers were placed in the front of the 1st line (*ante primam acie*: this proves that Vegetius's source envisaged the presence of at least two lines). The presence of these protected the centre from the failure of the wings. This tactic was used by Scipio Africanus at Ilipa in 206 BC.

6. Outflanking on one side (hyperkerasis). According to Vegetius, this array was the best formation for those who were outnumbered by the enemy. When the army advanced towards the enemy, the general suddenly sent the right wing consisting of his best cavalry and very swift infantry against the enemy, while the remaining part of the army stayed behind and lengthened the line to a straight javelin-like line. This method of attack was often used while on the march (i.e. vanguard was sent against one enemy wing).

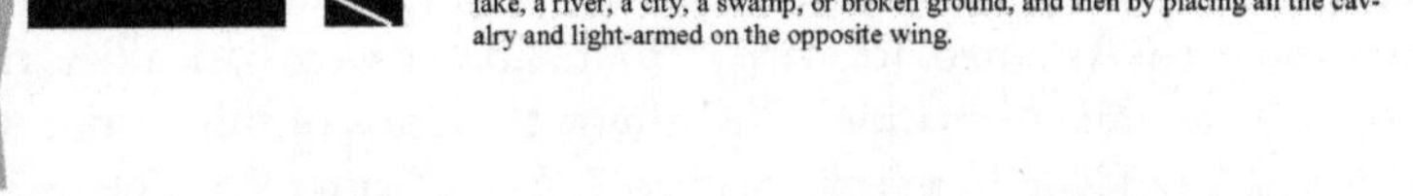

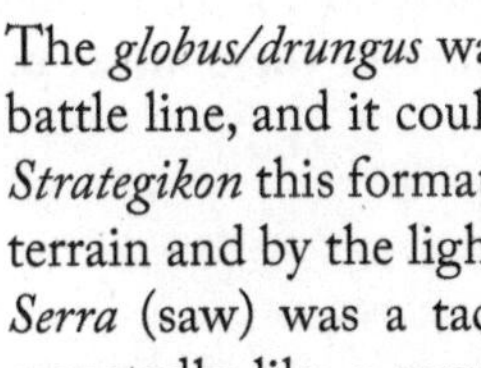

7. When the army had fewer men (including cases with fewer brave men), it was possible to even out the odds by resting one wing against a mountain, a sea, a lake, a river, a city, a swamp, or broken ground, and then by placing all the cavalry and light-armed on the opposite wing.

E) The *globus/drungus* was an irregularly arrayed unit that advanced in front of the battle line, and it could consist of either cavalry or infantry. On the basis of the *Strategikon* this formation appears to have been employed also in difficult/broken terrain and by the light infantry.

F) *Serra* (saw) was a tactic in which the separate units advanced and retreated repeatedly like a saw to disorder and disorganize the enemy line. It had two versions: units advancing and retreating, and a phalanx deployed as wedges. Discussion follows later in this chapter.

The seven advanced combat formations were:

1) rectangle with the long side facing the enemy (lateral phalanx), used primarily to outflank the enemy and in which the reserves were primarily used to counter the enemy's breakthroughs and outflanking moves.
2) oblique formation, in which the right flank cavalry and infantry were used to outflank the enemy's left wing. The best cavalry and infantry were in this case posted on the right. This formation could be used to outflank the enemy when the Romans had fewer men by keeping the left wing out of the combat. If the enemy forestalled the Romans, then the commander was to use his reserves. It was vulnerable to being attacked by enemy wedges, but the reserves could be used to counter these.
3) the third formation was similar, except in this case the oblique formation was used to outflank the enemy's right wing. This was a weaker formation than number 2, because the shield was carried on the left side.
4) when the formation was four- or five-hundred paces from the enemy, the flanks were sent forward to outflank the enemy on both sides. This is a variant of the *epikampios emprosthia* formation.
5) the fifth formation is the same, except that the light-armed and foot-archers were sent in front of the first battle line ('*primam aciem*' implies that there could be several battle lines) to protect both the centre and the advancing wings. This is a variant of the *epikampios emprosthia* formation.
6) if there were not enough men for the double-outflanking and the commander lacked confidence in his army, he dispatched his right wing (consisting of his best cavalry and swift infantry) against the enemy's left wing while keeping the rest of the army behind (the formation was also simultaneously extended to protect its flanks).
7) the seventh formation was the use of the terrain to one's advantage in situations in which the Romans had fewer men or were able to exploit the terrain against the enemy. In the example given by Vegetius, when there was a mountain, a sea, a river, a lake, a city, a swamp, or broken ground on one side of the army, then one was to post all the cavalry and light-armed on the open side while the rest of the army was posted as a straight line.

The attached diagrams of the seven battle formations of Vegetius explain how these were deployed and used. As noted, formations four and five were basically variations of the *epikampios emprosthia* formation. The different variants of this version will be analyzed in greater detail later in a separate chapter. See Chapter 9.5. The rest of the variants depicted are rather straightforward and do not require additional comments, even if these will receive further light in this same chapter in the context of the analysis of the commands preserved by the *Strategikon*.

9.2. The phalangial battle formations from the fourth century to the seventh century[5]

The organizational structure based on auxiliary and legionary cohorts, maniples and centuries appears to have been abandoned soon after the reign of Diocletian (or perhaps even during his lifetime?), because when towards the end of the fourth century Vegetius wrote his treatise on the topic he no longer understood the meaning of *ballistarii* in the context of the infantry phalanx and interpreted it to mean the users of *carroballistae*, while in 275 Modestus had still understood the term correctly to mean footmen using either *manuballista* or *arcuballista*. The likeliest date for the abandonment of this system is the reign of Constantine the Great, who created new types of units that appear to have followed the Hellenistic phalanx unit system. This would have been the legionary system which had exactly 256 *skoutatoi* (shield-bearers) in each *tagma* (presumably the cohort in Constantine's system), deployed sixteen deep, referred to by Maurice in his *Strategikon* (12.B.8.1–9), so that in the ideal theoretical army there were in total 64 *tagmata* for a total of 16,384 men in, addition to which there were 8,000 (in truth 8,192) light-infantry *psiloi* and 10,000 *kaballarioi*-horsemen. This structure is clearly based on the theoretical Hellenistic numbers, but the referral to the legions (*STR* 12.B.8.1) and the size of the cavalry component (in Hellenistic theory the ideal cavalry strength was 4,096 horsemen) prove this to be the ideal Late Roman version of the phalanx structure. It should be noted, however, that the figures represent the ideally proportioned army, which was used as a model for the actual army which each commander had. If he had more men, then he adapted it to follow the same system and if he had fewer he did the same. The key point here is that the Romans abandoned the earlier cohort tactics in the course of the fourth century, so by the time the *De rebus bellicis* was written its structures were based on the Hellenistic phalanx (the hollow squares and oblongs were phalanxes) so that existing older types of legions and auxiliary units were simply permanently organized for combat as phalanxes as they had always done when required (note e.g. Arrian's *Ektaxis kata Alanon*). This was easily done because in the Roman system the legions had ca. 5,000 to 6,000 footmen who could easily be deployed as phalanxes (divisions, *mere*), so that there was ca. 4,096 heavy-armed in front and ca. 2,048 light-armed behind – the strength of the division remained basically the same. A good example of this phenomenon is the battle in front of Ctesiphon in 363. Julian organized his forces as three successive lines of phalanxes, not as a *triplex acies* of cohorts. If the commander still decided to employ three lines of cohorts as in the old system *triplex acies*, it would have consisted of four cohorts in the first line, four cohorts in the second line and three cohorts in the third line, but when deployed according to the phalanx system each of the three lines would have required four cohorts. However, in the phalanx system the typical *triplex acies* would not have consisted of cohorts, but of entire legions, so that each of the three battle lines had the same numbers of legions.[6]

With the exception of the antiquarian *Epitoma Rei Militaris* of Vegetius we find the 'new' Constantian phalanx structure in all of the subsequent military treatises (*Definitiones*, *Hermeneia*; reproductions of Asclepiodotus, Aelian and Arrian, *De rebus bellicis*; Urbicius' *Taktikon* and *Epitedeuma*; *De scientia politica dialogus*; Syrianus Magister's *Peri strategikes*; Maurice's *Strategikon*) and narrative sources (see the *MHLR* series). The Constantian phalanx structure was based on the Macedonian model, which had been modified to take into account the different armament used by the Romans and the different-sized units which were then either united or divided according to the unit sizes in the Hellenistic phalanx (see below). The key difference between the Roman and Macedonian phalanx was that the Macedonian phalanx was based on the use of the five to seven metre *sarisa/sarissa*-pike, while the Roman phalangites were equipped either with the shorter 2.5–3.74 metre spear (*lancea, hasta, contus*) or with the javelin (*pilum, spiculum*). What is notable about this is that the Late Romans trained or used all of the different tactical variants that were included in the treatises depicting the uses of the Macedonian phalanx.

However, we still find two new developments for the period of ca. 366 until the late-fifth century in the texts of the *De rebus bellicis* and *Vegetius*. Both of these suggest that the Romans introduced new combat methods and equipment to counter the war elephants and cataphract cavalry fielded by the Sasanians, while the *De rebus bellicis* also suggests that the Romans needed to adapt their battle tactics for difficult wooded terrain. The likely reasons for this was the ability of the Persian war elephants and cataphracts to penetrate the Roman hollow square formation during Julian's ill-fated Persian campaign in 363 and the wars between the Romans and Germanic peoples in the early-fourth century. The issue of fighting in difficult terrain has already received attention in the context of the *De rebus bellicis* and will also be dealt in greater detail later in the chapter devoted to tactics in difficult terrain.

The *De rebus bellicis* altered the structure of the lateral phalanx so that each of the 2,000 strong units (*moirai* of Maurice) belonging to the 6,000 strong 'legion' (*meros* of Maurice) were deployed as a hollow square/oblong so that there these were able to operate independently of each other while also providing support for one another. This enabled each of the 2,000-strong *moirai* to face enemies simultaneously on all sides, which meant that if enemies managed to penetrate the Roman phalanx, they could not outflank them in practice. We find the Romans using this battle formation at the Battle of Poetovio (Syvänne, *MHLR 2*, 245–51), but with the difference that in this case the legions were deployed as hollow squares/oblongs. The subsequent descriptions of Roman battle formations (the legions operating independently of each other without reserves) support its use up to the latter half of the fifth century (see Syvänne, *MHLR*, vols. 3–5). However, we do not find this system in any of the subsequent treatises (Urbicius, Syrianus, Maurice) and the narrative sources do not offer any definite evidence for its continued use in the sixth century. This implies that this deployment pattern (*moirai* or *mere* deployed as hollow/oblong squares) was abandoned by the early-sixth century, probably as a result of the ever-worsening quality of infantry and the officer cadre.[7]

In addition to this, the lateral battle formation in the *De rebus bellicis* was protected by various types of machines (e.g. *tichodifrus* and *ballista quadrotis*) while scythed chariots were used in pursuit of the enemy (see Chapter 1). As already noted, the same text also promoted the use of a version of the wagon laager (*carrago*) as a method of marching and combat. We find this same method of combat also in

Urbicius's *Epitedeuma*. The use of the various kinds of machines and wagon laager for the protection of the infantry suggests that the general quality of the Roman footmen worsened thanks to the increasing corruption of the top brass. For this development, see in particular Syvänne, *MHLR*, volumes 3 and 4.

The quality of the Roman infantry arm reached its low point in the early-sixth century when Urbicius no longer trusted the Roman infantry to be able to use the battle formations included in his *Taktikon*, nor did he trust in their ability to withstand the impetus of the 'Scythian' (i.e. Gothic) cavalry charge without the addition of extra anti-cavalry devices, even when deployed as a wagon laager. The reason for this lowering of quality was the destruction of several Roman field armies, meaning that there were no longer adequate amounts of experienced officers and men in the ranks. The lowering of the quality of the infantry started when the East Romans lost their expeditionary army at the naval battle of Cap Bon in 468 and reached its low point as a result of the massive losses of infantry in the battles fought between the forces of Vitalianus and Anastasius. See Syvänne, *MHLR 5*, 80ff.

We find in the texts of *De rebus bellicis* and Vegetius information that is relevant in the context of how the Romans opposed the Persian war-elephants. This is not surprising because Julius Africanus and Ammianus were impressed by the pachyderms. In fact, the third-century military theorist Julius Africanus described the Sasanian war elephants as virtual super weapons of antiquity (*Kestoi* 1.18).[8] The reasons for their effectiveness were multiple. Soldiers and horses feared elephants if they were not familiar with them. The towers carried by the elephants added to the terror felt, because these added height. In battle the elephants formed a kind of rampart that advanced in front of the phalanx, while their trumpeting frightened the enemy. The elephants were protected by their thick skin, trunk and by their spear-like tusks. Julius Africanus went so far as to claim that the elephants dominated the battlefields with their presence. He likened the pachyderms to the bronze ram of a trireme that, instead of ships, crushed the interlocked-shield (*synaspismos*) formations, and once this happened the army was as good as destroyed. According to Julius Africanus, the war-elephant looked like a mountain, while it knocked down its opponents, caught horses, men and chariots/wagons (*harma*) in its trunk and then knocked them violently down, and rolled them, trampled them, and frightened the horsemen, while from the height of its tower the archers shot at the charioteers/drivers (*harmatêlatas*) from a distance. Julius Africanus instructed the Romans to use light infantry javelineers and archers against the elephants, but opposed the use of cavalry against them because when the horses were frightened by the unfamiliar beasts this could disorder the entire army. This was no longer the case during the Late Roman period when the cavalry was perfectly trained for this task. Julius also instructed that the attack against the elephants was to be accompanied with the loud trumpeting of the trumpets to scare the beasts. The javelineers were to aim their javelins at the elephant and the archers were to aim their fire at the wooden towers. If any of the missiles managed to penetrate the skin of the elephant it would become enraged, and in rage the elephant was usually more dangerous to its own side. He also advised the use of iron caltrops in very large quantities on the expected path of the elephants. These systems continued in use.

It is of note that Julius Africanus's text also suggests that the third-century Romans continued to use chariots/wagons against elephants in battles which they had already employed against the elephants of Pyrrhus, and this same policy also appears to have been continued during the Late Roman period. It is unlikely to be a coincidence that in the aftermath of Julian's failed invasion of Persia we find in the *De rebus bellicis* (12.1, 13.1, 14.1) cataphracted scythed chariots, *tichodifrus*-obstacles and *ballista quadrirotis* (the Romans used the *carroballistae* with great success against elephants in 363) for use in combat. It is also unlikely to be a coincidence that we find both Modestus (ca. 275) and Vegetius (ca. 390s) listing counter-measures against elephants as the Persian war-elephants had clearly made an impression on the Romans. Their list includes: cutting off the trunk; the use of the chariot drawn with two cataphracted horses from which the *clibanarii* threw *sarisae* at the elephants (note the similarity with the scythed chariots of the *DRB*); the sending of cataphracted legionaries ('*cataphractos milites*') against the elephants, armoured with iron barbs attached to their arms, helmets and shoulders so that the elephants were unable to seize them with their trunks (i.e. they looked like medieval knights);[9] the use of slingers against the Indos (drivers); the opening of the *agmen*-formation to allow the elephants to pass through (this is the equivalent of the *antistomos difalaggia*), after which they would be surrounded with a *globus* (means irregular *drouggos* in this case); the use of '*velites*' mounted on horses to wound the beasts with thrown *lanceae* (*lancea* was a lance which had two variants, long and short) or '*maiora spicula*' (a long version of the *spiculum/pilum*, which means that the *pilum*-type of javelin-head was placed on a longer than usual shaft; note the implication for the *Ektaxis kata Alanôn* of Arrian); and the use of *carroballistae* (artillery mounted on wagons) behind the battle line (Syvänne *MHLR 2*, 112–3). The key improvement over the text of Julius Africanus was the improvement of the ability of the cavalry to engage the elephants. During the Late Roman period it was primarily the cavalry that engaged elephants after Julian's campaign, largely thanks to the increased importance of the cavalry in the Roman army.

Julius Africanus clearly exaggerated the effectiveness of war-elephants, but it is still clear that their presence on the battlefield was so frightening to the Romans that they developed special methods to counter the threat. Most of these systems had been known since the days of Pyrrhus and Hannibal, but the use of cavalry against the elephants proves that the Romans had improved their counter measures since those days. In fact, it is clear that from the 270s onwards the Romans expected their cavalry to able to defeat Persian war-elephants without assistance from either infantry, or wagons, or *ballistae*. Actually, excluding those instances in which the war-elephants and cataphracts managed to penetrate Roman infantry- and cavalry-formations during Julian's ill-fated Persian campaign in 363 (and even in these cases the Romans prevailed in the end), the Romans were always successful when they faced Persian elephants on the battlefield until the defeat they suffered at the Battle of Arzamon in late-604 or early-605. In that battle, Chosroes II deployed his elephants as a fortress, probably in the manner as described by Julius Africanus, and inflicted a crushing defeat on the Roman cavalry army. The probable reason for this success was that in that battle the Romans lacked men used to dealing with elephants. In contrast, in a similar situation at the Battle of the plain of Canzak in 593, the Roman cavalry had charged at the elephants causing them to flee. This proves that these horsemen were familiar with the beasts and able to isolate and wound them with shafted weapons in such a manner that they became enraged. In contrast to the cavalry, there are no known examples of the Persians being able to defeat a large Roman infantry army with their elephants. The Roman infantry had developed perfect counter-measures against them.[10]

The quality of Roman infantry was still considered suspect in the 530s by officers like Belisarius who were cavalrymen by heart. It is because of this that he posted the infantry behind trenches at the Battle of Dara in 530, did not attempt to use infantry offensively at the Battles of Callinicum in 531, Ad Decimum in 533 and Tricamarum in 533, and later used infantry against the Goths only with great hesitation during the Siege of Rome in 537. However, the actual combat performance of the infantry from that era demonstrates that there existed some high quality infantry forces, even if there are also examples of their poor combat performance. At the Battle of Callinicum, the regular heavy infantry performed well while the so-called Isaurians did not. The same was true of the battle in front of Rome. It was once again the Isaurians who failed to perform as expected. In contrast, the Roman infantry, especially its elite forces under the *Comes Excubitorum*, performed very well at the Battles of Mt. Bourgaon in 534, Mt. Aurasium in 535, and relatively well in the Battle of Scalae Veteres in 537. Thereafter the quality of the infantry improved markedly during the 540s so that infantry were not only able to withstand enemy cavalry charges without flinching, but were also able to use sophisticated tactical manoeuvres at the Battles of Antonia Castra in 547, Gallica in 547 (in fact infantry saved the cavalry that had been ambushed), Campi Catonis in 548, Hippis River in 550, Trachea in 550, Taginae in 552, Casilinum in 554, Phasis in 555, and Rhizaeum in 557. The only distinctly poor performance by the infantry in a large battle during that era took place at the Battle of Onoguris in 554, when the Roman cavalry fled and the infantry joined them in flight. This demonstrates that the quality of the Roman infantry forces improved significantly in the course of the reign of Justinian and that the officers started relying on their infantry more than was the case before the 540s. This is the period when Syrianus Magister wrote his treatise, so it is not surprising to find the Romans using the very same tactics he described with great effectiveness. The ability of the Roman infantry to fight as hoplites and as foot archers simultaneously proves the very high quality of the training at this time.[11]

Thereafter the general quality appears to have deteriorated, as during his last years Justinian I and then his successor Justin II failed to spend enough money on the army, so there were not enough men-at-arms for the Roman army to begin its invasion of Persian territory in 572. Justin II worsened the situation further by his personal interference, but the state of affairs started improving when Tiberius II, first as Caesar and then as Augustus, reinforced the army with new recruits and improved discipline. The greatest improvement both in the size and quality of the infantry (and cavalry) took place when Maurice was appointed as supreme commander of the Eastern Front in 577. He recruited natives into the army and improved both training and discipline. The situation improved even further when Maurice became

Augustus in 582 and reformed the Roman armed forces to follow the instructions in his *Strategikon*. Under Maurice, the expectation was that the infantry could use all of the battle tactics and formations found in the *Strategikon*.

According to the *Strategikon* (12), the principal infantry battle formations were: the lateral (*plagia*) phalanx; the rearward-angled-formation (*epikampios opisthia*); the double/quadruple phalanx formations for difficult terrain, but the text also includes the forward angled formation (*epikampios emprosthia*); the '*taxis allê*'; the mixed formation; and the hollow square/oblong. The most important of these were the lateral *plagia* phalanx (both the single- and double-phalanx versions) and the *epikampios opisthia* because even the dismounted cavalry was expected to be able to employ both, but in such a manner that the lateral phalanx formation was still the basic phalangial formation, which was then, when necessary, manoeuvred to form the other formations in the course of the battle. See for example, Syvänne (*MHLR 1*, 247, 249; *MHLR 6*, 281–5, 304–5, 340–4; *MHLR 7*, 307, 315). However, since Maurice (preface) considered the *Strategikon* to contain only the rudimentary basic learning and expected that the officers would improve their skills further by reading other treatises, it is not surprising to learn from the narrative sources that the period infantry was able to use also the other methods described by the military treatises in the Hellenistic tradition. The quality of the infantry was therefore extremely high during the last years of Maurice's rule.

However, quality sank once again under Phocas and Heraclius. In the former instance, the main reason for the lowering quality was the poor morale resulting from the civil war in conjunction with the Persian war – the Romans were divided into those who supported Phocas and those who supported Maurice's son. In the latter case, the main reason for the lowering quality appears to have been the fault of Heraclius and his officers, who considered infantry only as an auxiliary arm for their cavalry so it was not used offensively except in sieges. In combat, Heraclius and his officers used infantry only as a place of refuge for the cavalry. Under Heraclius, the infantry was therefore deployed for combat only as a lateral phalanx, or in the mixed formation, or as a hollow square, but the infantry was never used offensively – this task Heraclius and his commanders reserved for their cavalry.[12]

The discussion of battle formations begins with a description of the lateral phalanx formation, together with the combat manoeuvres associated with that formation. Fortunately, the *Strategikon* preserves the traditional Roman commands and manoeuvres. These commands were employed not only when using the lateral phalanx but also when employing the other formations. This information is then combined with the information preserved by Asclepiodotus, Aelian, Arrian, Urbicius, Syrianus Magister and the narrative sources to arrive at the reconstruction of the Roman phalangial tactics. By combining these sources, it is possible to analyze the development of the phalanx formation in the course of the Late Roman period.

As the following discussion of the battle formations and unit tactics will make clear, the Romans varied their infantry battle formations according to the composition of their own forces, the terrain, the situation, and to the type of enemy they faced. The general rule of thumb is that when the Roman infantry faced an infantry force in pitched battle, the favourite battleground was an unobstructed and open plain, but when they faced a cavalry force, the favoured terrain was the difficult and rough terrain which negated the advantages of cavalry in mobility, while also breaking up their formation during attack. As already discussed the favourite tactics and battle formations also varied from one time period to another. These variations will be taken into account in the following analysis of the battle formations.

9.3. The lateral phalanx (*plagia falagx*) until Maurice's reign[13]

The lateral phalanx (*plagia falagx*) was a battle formation in which the long side faced the enemy, and which consisted of ranks and files of footmen. The lateral phalanx was not only the most often-used infantry battle formation, but also the basic building block used in all of its variants. The standard depth of the formation was sixteen ranks, because this could easily be divided into two eight-rank-deep phalanxes. The depth of four ranks was considered to be the shallowest allowed, because depths shallower than this were considered too weak in combat. The maximum depth was thought to be 32 ranks, because increasing the depth beyond that did not increase the effectiveness of the formation in mass shoving. Each of the files was also divided so that there were officers to lead the men (file-leader, rear-guard) and so that every other rank consisted either of the *protostates* (first ranker; in Maurice *primus*) or *epistates* (behind-ranker, in Maurice *secundus*) to ease the division and doubling of the files. Fuller discussion follows in the context of the analysis of the *Strategikon*. We can find all of the same information in the texts depicting the phalanx (Asclepiodotus, Aelian, Arrian, Urbicius, *Definitiones*, *Hermeneia*, Syrianus Magister) until Maurice and Leo VI the Wise. However, there are still some distinct differences in the earlier treatises written before the 540s (or their later interpolations) and after (Syrianus Magister and Maurice). The first of these concerns the ideal unit sizes in the phalanx.

The treatises in the Hellenistic tradition (Asclepiodotus, Aelian, Arrian, Urbicius, *Definitiones*, *Hermeneia*) divide the phalanx into the following units:

> *Hoplitai* (hoplites, heavy infantry) total 16,384 hoplites: 16, 32, 64, 128, 256, 512, 1024, 2048, 4096, 8192 and 16384 men.
>
> *Psiloi* (light-infantry) total 8,192 light-armed: 16, 32, 64, 128, 256, 512, 1024, 2048, 4096, and 8192.
>
> *Hippikon, hippeis* (cavalry, the *kaballarioi* of Maurice) total 4,096 horsemen: 16, 32, 64, 128, 256, 512, 1024, 2048, and 4096.

These were deployed for combat as follows. The 16,384 *hoplitai* were deployed side-by-side as four phalanxes of 4,096 men each (left, left-centre, right-centre, right) with a depth of 16 ranks, alternative depths being four, eight or thirty-two ranks. The 8,192 *psiloi* were deployed either behind the phalanx consisting of four *falaggarchiai* with a depth of eight; or between the files of hoplites (means the archers and the javelineers, the slingers were deployed elsewhere); or in front of the phalanx to skirmish; or on the flanks of the phalanx (usually between the hoplites and cavalry); or behind the flanks of the phalanx; or in front of the flanks of the phalanx. The 4,096 horsemen were deployed on the flanks (the usual deployment), in front, or behind.[14] The phalanx could also be deployed into other formations, for example columns, oblique formations, crescent, convex, double phalanx, triple phalanx, and hollow square/oblong formations – in addition to which it could be made shallower, deeper, into a wedge, a hollow wedge, and two-fronted, and its men could face new directions by turning individually, and its units could be by wheeled and countermarched by rank or file.

This is the ideal deployment pattern, including the drills needed to form the various formations, but we learn from the narrative sources that from the very start (i.e., the Hellenistic period) the ideal patterns were adapted to the actual size of the army and to the types of troops encompassing it. This means that the army could be considerably larger or smaller, while the various components making up the army did not follow the ideal proportions. The phalanxes could also be deployed even three battle-lines deep so that lines two and three formed the reserves. This was particularly true for the later major battles of the Hellenistic period and also for the Late Roman period. For the former, see for example the battles of Hannibal, Antiochus the Great and Mithridates the Great against the Romans, and for the latter the battles of Constantine the Great, Julian the Apostate and so forth in the *MHLR*, volumes 1–6.

Macedonian and Roman Phalanx

The standard Macedonian phalanx was arrayed as four phalanxes each called *falaggarchia* (4,096 hoplites) each lead by a falaggarchês. The accompanying diagram shows the locations of these officers in the formation. I have drawn it after *Codex Burnley*, but I have emended eight persons to the two phalanxes on the left which in the diagram have only seven men. In the standard array the light infantry *psiloi* would have been placed behind the heavy infantry, but its place could be varied. The cavalry could be placed either on the wings or behind the infantry.

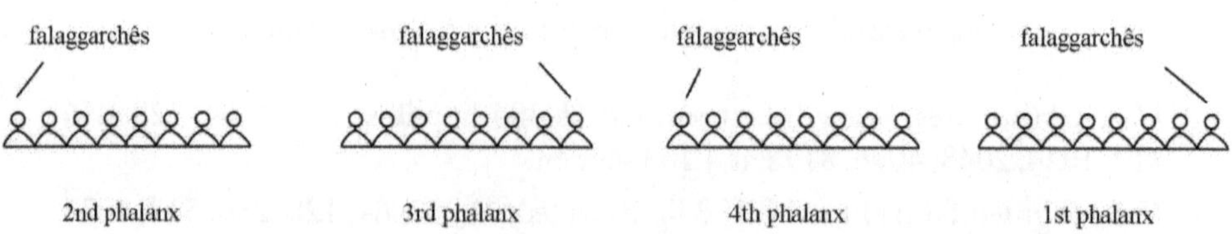

However, even if the lateral double phalanx had in practice already been used so that the second phalanx behind acted as a reserve during the Hellenistic period, the traditional Hellenistic theory still omitted this.[15] As we have seen in Chapter 1, Syrianus Magister's text (*PST* 31–2, 34) was closer to the reality because in his text the second phalanx could also be used as a real reserve force for the first phalanx.

Syrianus gives us two different ways how the double phalanx could be used in outflanking: 1) if the Romans had twice as many men as the enemy, or even more, the commander was to use the double phalanx for the double encirclement of the enemy; and 2) if the Romans had a divided double phalanx, one of these could be used frontally and the other against the enemy's flank. Syrianus does not state how the

Two versions of the double-outflanking with the double phalanx in Syrianus Magister's Peri strategikes

Double Outflanking Version 1: Battle of Verona in 312

Left: Constantine's cavalry wings charge to gain time for the infantry manoeuvre to oppose the wider enemy front. It is likely that the relief army had baggage train accompanying it, and therefore I have reconstructed their array as a single phalanx.

Right: The frontline marches to the wings to encircle the enemy and the second line marches forward. The flank reserves would have protected the inner flanks of the split up first line. Eventually these two units would have been arrayed between the split up 1st line and the advanced 2nd line to form the one continuous lengthened phalanx. The centre reserves may have been used to form an infantry wedge or deep array in the middle for the breaking up of the enemy array.

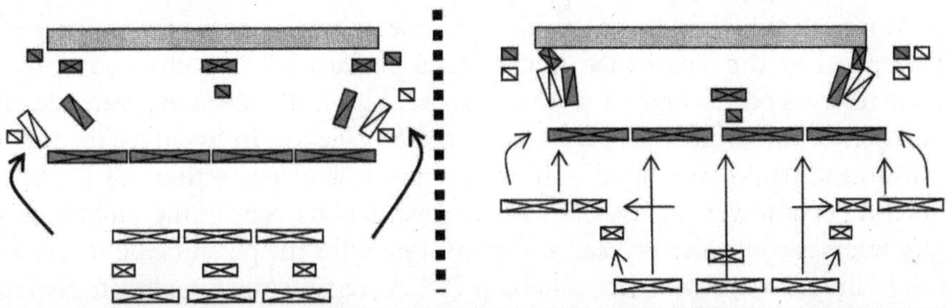

Double-outflanking with the double phalanx version 2

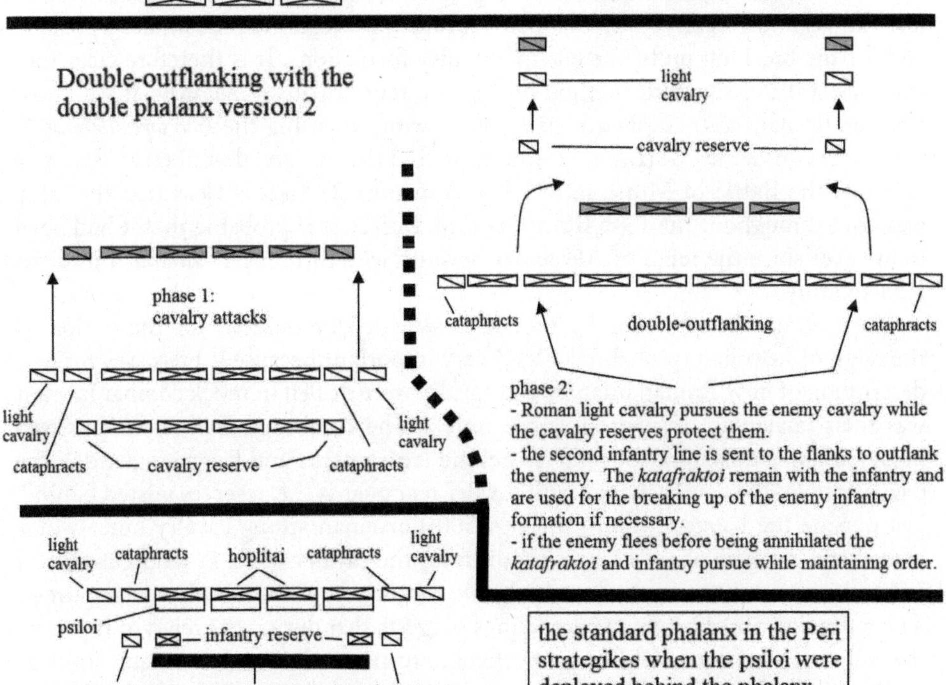

veteran cavalry posted two to three miles (visibility obstructed) or three to four miles (open terrain) behind the battle formation to protect the army if it suffered a defeat.

the baggage train is located behind the phalanx because the enemy is on the front second phalanx was used for the double outflanking, but there were two alternatives: 1) the front phalanx could be divided to outflank the enemy while the support phalanx advanced forward; and 2) the second support phalanx could be divided and sent to the flanks of the first phalanx to outflank the enemy. The former (a variant of the *epikampios emprosthia* and Vegetius's fourth and fifth formations) was used by Constantine the Great at the Battle of Verona in 312 (*MHLR 1*, 248–9), while the latter version is simply the sending of the reserves to the flanks to encircle the enemy (see pp.277 and 287). Neither of these outflanking systems is found in the *Strategikon*, because Maurice restricted the use of the double phalanx against threats from the rear just as in traditional Hellenistic military theory. See the illustrations drawn after the *Codex Burnley* on pp.292–4.

In Maurice's thinking, the outflanking of the enemy infantry formation was to be performed by the lengthened single lateral phalanx with cavalry and/or by the separate reserves posted behind for this purpose. This is discussed in greater detail in the analysis of the commands preserved in the *Strategikon*. In Syrianus, the reserves posted on the flanks were used only for the purpose of protecting the flanks (see Chapter 1) – however, on the basis of the instructions concerning ambushes, the cavalry reserves could also be used as support forces for the pursuing light cavalry as shown in the accompanying diagram on p.287. As regards the cataphracts, Syrianus, just like Vegetius, clearly restricted their use for the protection of the infantry phalanx and for the breaking up of the enemy infantry formations. It is therefore clear that this was still the standard method during the reign of Justinian, even if we know that the Romans also employed their cavalry wings by using the *koursores/defensores* system. See Syvänne, *MHLR 6*. However, we find the phalanx described by Syrianus in use at the Battle of Mursa in 351 (see Appendix 2), so it is clear that the array was used throughout the Late Roman period. In fact, it is probable that it had been in use ever since the reign of Alexander Severus, who introduced *clibanarii* into the Roman army.

The information provided by the *De scientia politica dialogus* for the period of the reign of Justinian (probably 540s) is very important because it preserves for us a description of how Roman infantry and cavalry were drilled in mock combat in what was their phalangial formation at the time.[16] The combined formation consisted of the infantry phalanx with reserves behind (rear guards and baggage guards); the baggage train; the cavalry wings with cavalry rear guards (i.e. reserves posted behind, just outside the baggage train); the two 500-horseman strong cavalry outer wings posted at a distance of a stade (ca. 184m) from the cavalry to act as flankguards and outflankers; and the commander with his *bucellarii* (who chose his place as required). The referral to 500-horseman outer wings suggests that during the reign of Justinian the military thinkers considered it preferable to post full-strength cavalry units as outflankers and flankguards and not a flexible strength *banda* (200–400 horsemen) as was the case during the reign of Maurice. It is of note that none of the other extant treatises suggest the use of outer-cavalry-wings to protect the cavalry wings of the infantry phalanx, but we still find Narses using such as ambushers in the Battle of Taginae in 552. This may suggest that the Romans were indeed experimenting with this deployment pattern during the reign of Justinian, but which was later abandoned

Phalanx in the *De scientia politica dialogus*

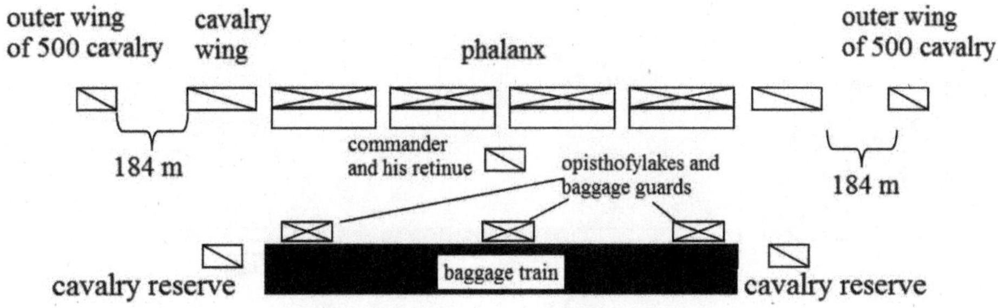

because we do not find it in the *Strategikon* – unless of course Maurice's referral to the need to post the better *banda* further out refers to this. The problem with this interpretation is that Maurice has infantry flank guards for the cavalry and not cavalry flank guards. The other important point besides the lateral phalanx formation in the *De scientia politica dialogus* is that the heavy infantry is trained to use spears for thrusting against other infantry, which means that the heavy infantry was expected to fight like hoplites with spears. In contrast, Maurice required the front rank to throw their spears when fighting against infantry and use swords in melee. This means that there was a change in preferences in the late-sixth century.

The pitched battles of the era demonstrate that the Late-Roman commanders were not bound by the military theory, but adapted the theory and training methods to actual circumstances so that the phalangial deployment pattern was adjusted to the terrain and situation. The best examples of this are Julian's use of the triple phalanx at the Battle of Ctesiphon in 363, Modares's attack with infantry in loose order downhill in 379, Belisarius's use of field fortifications and improvised battle formation at the Battle of Dara in 530 (see Chapter 9.6), and Narses's deployment of the infantry and cavalry at the Battle of Taginae in 552. The battle of Taginae in 552 is important for three reasons: 1) it shows how the military theory could be adapted to the circumstances and terrain; 2) it proves that Syrianus's text contains period material; and 3) it has extra units of cavalry posted on the left wing as ambushers, which reminds one of the extra units of cavalry in the *De scientia politica dialogus*. It is because of this that I give a summary of the main features.

In this battle Narses the Eunuch faced the Goths under Totila. Narses's plan was to exploit the terrain so that the Gothic cavalry would launch their attack against the Roman infantry phalanx in the centre and when the Goths did this he intended to exploit it by turning his wings (consisting of dismounted archers in front and cavalry behind) inwards towards the centre so that end result looked like a double outflanking with the name *hyperfalaggesis*, or a crescent (*menoeides*; this was de facto a variant of the crescent array that had originally been the lateral phalanx), depending on how one wants to describe it, but *hyperfalaggesis* would have been the correct theoretical term. This put the Gothic cavalry into a crossfire from both flanks. In addition to this, Narses had posted two ambushing units of cavalry so that 1,000 of those were

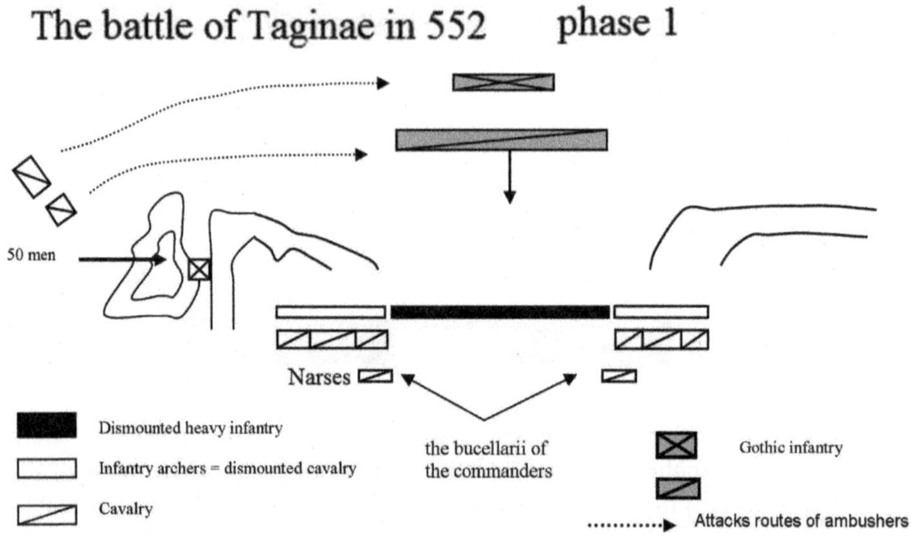

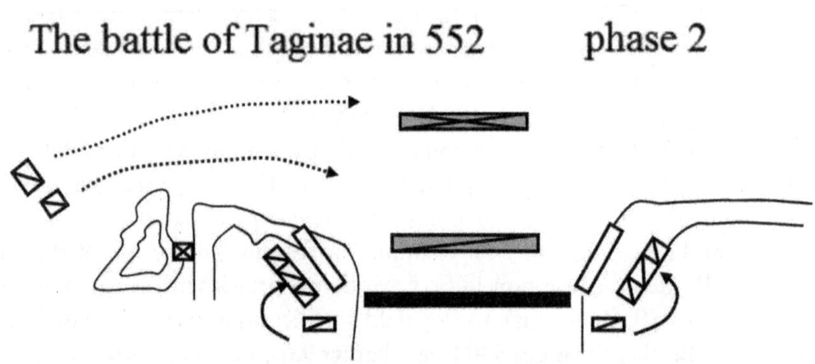

directed against the Gothic infantry in the rear to tie it in place and 500 against the flank of the Gothic cavalry. Additionally, there were fifty multipurpose heavy infantry posted to defend a narrow path between hills. The reserves were posted on the flanks. The plan worked as expected. For a fuller analysis of the different stages of the battle, see the *MHLR 6* (340–4).[17]

Asclepiodotus and Aelian (including the late-Roman and 'Byzantine' copies)[18] both include three different methods for compacting the phalanx into the close order *pyknosis* and by implication into the *synaspismos* formations. Firstly, the tightening of the formation could be performed towards the right wing so that the right file stood in place while the other files right-faced and moved towards the right. When the files had reached the correct position, these turned towards the enemy, after which the ranks moved forward to tighten the formation in depth. In the second version, the formation was tightened towards the left in like manner, and in the third version the formation was tightened towards the centre (the navel between the phalanxes). Arrian and Syrianus do not state how the formation was tightened, while Maurice simplified the system so that the closing was performed always only towards the centre of the unit or battle formation.

Aelian and its late-Roman and 'Byzantine' copies and interpolations[19] thought that the lateral *plagia* phalanx was also useful as a countermeasure against the enemy cavalry when the latter was deployed as a column (*heteromekes* formation, in which the depth was double the width). According to Aelian, the reason for this was that when the infantry was deployed in the shallow formation, the enemy cavalry charge affected only a small section of its front, even if it managed to pass through it. This smacks of theory, because in practice the Greeks and Romans always sought to prevent the penetration of their formation with counter-measures like the close order *pyknosis*, *synaspismos*, and *testudo*, or they used a hollow infantry wedge to counter the enemy's infantry wedge/column, or they posted the light-armed as a wedge or in open order in front (*protaxis, antesignani, lancearii*, club-bearers, mace-bearers) to protect the main line, or they purposefully opened up their formation by using the *antistomos difalaggia*.

In addition, we also find Aelian recommending the use of the crescent (*menoeides*) against a cavalry rhombus (in the accompanying diagram on p.293, rank-and-file rhombus) and the *epikampios emprosthia* (forward-angled formation in the diagram opposing the non-rank-and-file rhombus). In both cases it is clear that we are dealing with unit formations used against cavalry rhomboids. There is no definite evidence for the use of the crescent or *epikampios emprosthia* as unit formations outside Aelian during this period, even if these are also included in the *Peri strategikes* as military terms except as large battle formations (see Chapter 1 and below), but I would not preclude the possibility that individual units of the phalanx could manoeuvre themselves in this way in the course of battle when the enemy cavalry charged as individual rhomboids – we just do not possess enough details of the way individual units fought in battles. One may imagine that when the enemy rhomboid approached close enough and was stopped just in front of the *synaspismos* or *testudo/foulkon* formation, the individual units making the phalanx could then have easily used either the crescent or the *epikampios emprosthia* against the enemy rhombus just in front of them. If the infantry reacted in this way, Aelian advised the cavalry troopers to use javelins in the same manner as the ancient Tarentines had in an effort to disorder the crescent. This was basically all the cavalry could do, and if it did not work, they had to flee. It is therefore probable that the *epikampios emprosthia* and crescent both saw use not only as large battle formations but also as unit formations.

In fact, we find all of these countermeasures or tactics in use in one form or another during the Late Roman period. The light-armed could be send in front of the array for the following purposes: skirmish with the enemy (used against both cavalry and infantry); stop the cavalry attack with missiles (arrows, slingshots, stones, javelins) while employing either the wedge (this was used to stop the cavalry attack) or the open formation (in this formation the light-armed heroically engaged the enemy horsemen as individuals); stop the cavalry attack with an infantry wedge/*globus* (separate formations) or with the *caput porci* formation (used at least in until the fifth century) posted in front of the phalanx. The wedge formation could also be used by the heavy infantry phalanx so that it had units form up like the letter lambda

A selection of Diagrams in the *Byzantine Interpolation of Aelian* 1

Symbols used in the *Byzantine Interpolation of Aelian*

a file-leader (*lochagos*); a misleading term since in all of the diagrams the *lochagos* is clearly a higher ranking officer (probably a *falangarchês* or in some cases the *kerarchês* or even *strategos/hypostrategos*) usually posted in the front center or in the front right flank of the formation

kontos-bearing heavy infantryman (*hoplitês kontaratos pezos*), the *kontos* was a ca. 3.74 cm cavalry spear that was also used by heavy infantry as a thrusting and throwing weapon.

targeteer or light-armed slinger (*peltastês ê sfendonêtês psilos*); the 10th c. AD infantry peltast seems to have been a javelin thrower.

archer (*psilos toxotês*)

horseman with a spear (*kaballarios kontaratos*)

pezoi meta tzikouriôn êtoi peltastai (infantry armed with battle-axes, in other words peltasts.

ϲ τ ό μ α *stoma*, mouth / front of the array

Double outflanking and outflanking with overlapping wings
(drawn after Aelian, *Codex Burnley*)

hyperfalaggêsis (outflanking phalanx) means outflanking the enemy with both wings.

hyperkerasis means the outflanking of the enemy with one wing.

Antistomos difalaggia (fronts against each other facing double phalanx). This array was formed up by opening up a route for the enemy pass through so that its attack would not disrupt the formation while also putting the advancing enemy between two phalanxes facing each other. In practice this tactic was not used only against cavalry wedges which is depicted here but also against other cavalry formations meant to break the cohesion of the phalanx and also against elephants and chariots. The stoma means the front.

The square cavalry formation was opposed by infantry wedge. The Interpolation (Devine 47.3, Dain L3) states that whereas the apex of the cavalry wedge could be formed by one man, the infantry wedge needed more. Note that the Roman infantry wedge used in 193 may have had only one man at the apex unless we take the referral to the 200 men wedge only as an approximation. The Interpolation (Dain L4-5) continues that Epaminondas had used the wedge at Leuctra and that the wedge was formed up by joined the wings of *amfistomos difalaggia* together. This wedge in the text is actually different from what is depicted in the Codex Burnley shown below. The Burnley version is clearly a small unit wedge resembling the one used by the Romans in 193, but with the difference that it has three men in the front rank. This array would have been as a separate wedge in front of the phalanx or as a wedge protruding from it. See also Chapter 7.9. According to Aelian (Matthew ed. 37-8, Dain C4-D4), the square cavalry formation ('*tagma tetragonôn*') was used by the Sarmatians and Alans who fought with long spears ('*megalois dorasi*').

antistomos difalangia

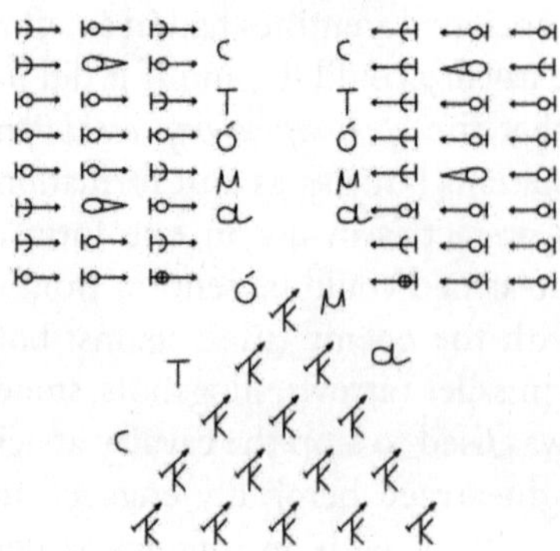

hippikê falagx tetragônos

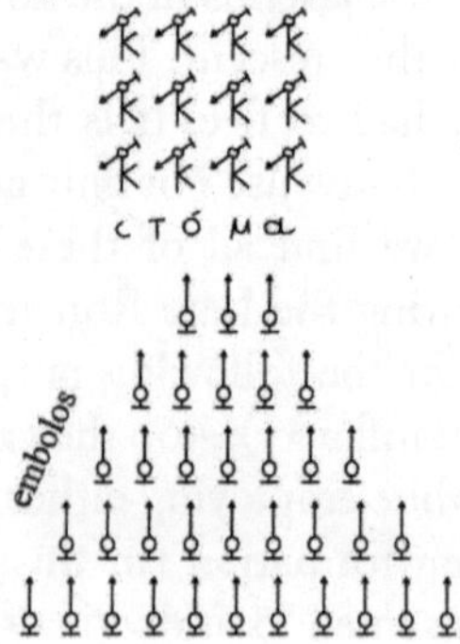

A selection of diagrams from the *Byzantine Interpolation of Aelian* : 2

The rank-and-file rhombus was opposed by *menoeides* (crescent). This was a unit formation used to subject the attacking rhombus to attacks from the flanks. The counter tactic for the rhombus was to attack the infantry wings with javelins. However, the Macedonians and Romans both also used grand tactical crescent consisting of thousands of men to outflank their enemies. Drawn after *Codex Burnley*.

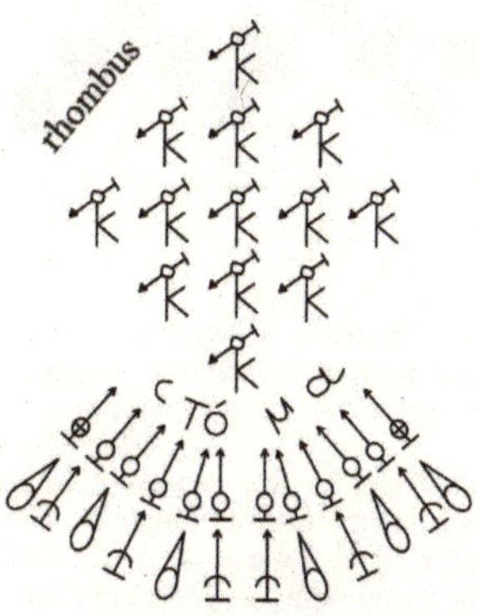

menoeides

The non-rank-and-file rhombus was opposed by the *epikampios emprosthia* (forward-angled half-square). This is once again unit formation, but the Romans also used a grand tactical version of the *epikampios emprosthia* consisting of thousands of men. Drawn after *Codex Burnley*.

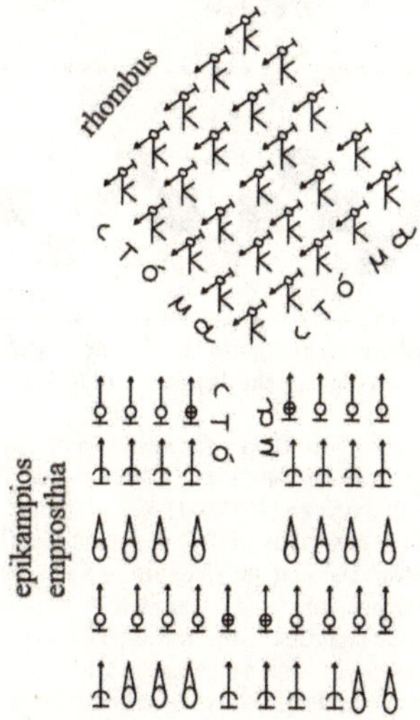

The column version of the *hippikê heteromêkês ilê* (cavalry oblong wing) had greater depth than width with the aim to break through the enemy array. This formation was opposed by the *plagia falagx* (lateral phalanx). According to Aelian, the aim was that if the cavalry managed to pass through, its charge would affect only a small section of the formation. Drawn after *Codex Burnley*.

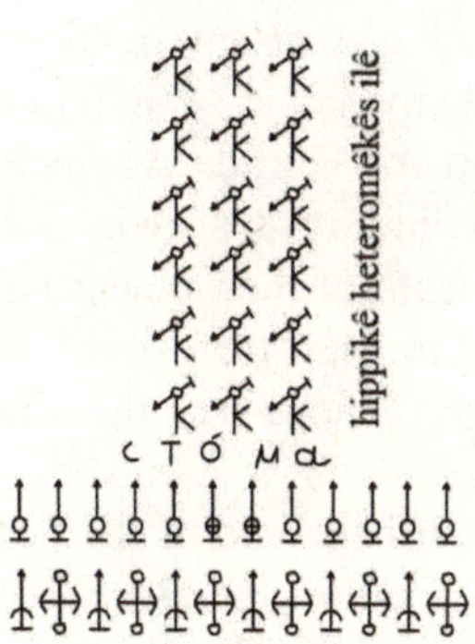

plagia falagx

Infantry column (*peri keras, peri orthês epagôgês*) was opposed by *koilembolos* (hollow wedge). On the basis of other sources like e.g. Vegetius we know that the hollow wedge was also used against infantry wedge. The counter tactic against *koilembolos* was the use of the triple phalanx in which the flanks attacked the wings of *koilembolos* while the centre remained behind. Drawn after *Codex Burnley*.

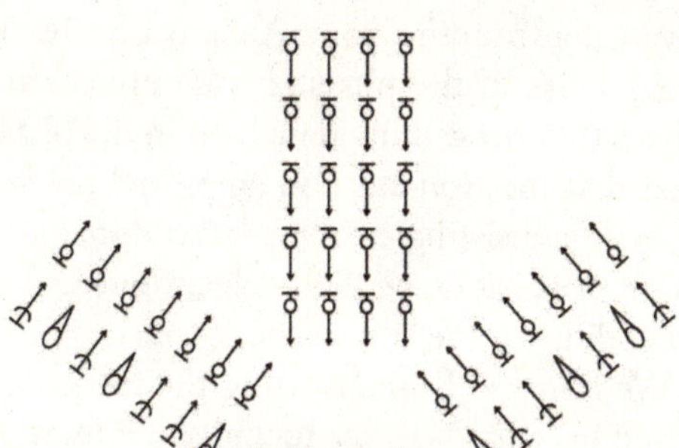
koilembolos

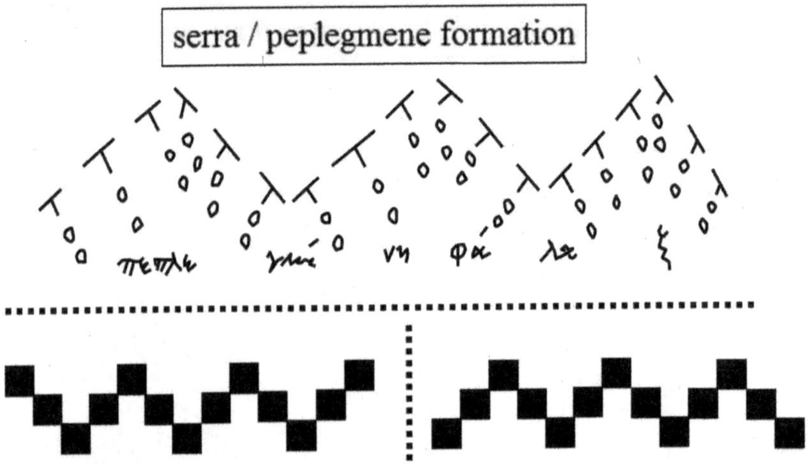

Above: ***Peplegmene falagx*** **in the** ***Codex Burnley*** **Aelian;**
Below: two versions of the ***Peplegmene falagx*** **in Asclepiodotus**

Aelian: According to Aelian, the *peplegmene* (woven) array was used against the hollow oblong formation (*plaision*) to break its cohesion. In other words, the hollow oblong formation was to be broken with wedges. Compare with Syvänne (2019b, 151ff) where the hollow oblongs are opposed by wedges. This lends further credence to the historicity of the account of Geoffrey of Monmouth.
Asclepiodotus: The text and diagram in Asclepiodotus allow two different versions for the *peplegmene* formation both of which are shown. Asclepiodotus's version (esp. the one on the right) clearly resembles Aelian's version, but there is one significant difference which is that Asclepiodotus's *peplegmene* array has separate units advance in close contact with each other like in a Roman legion consisting of cohorts and maniples while Aelian's version is clearly based on the phalanx formation with wedges. One may assume that both versions were used in practice depending on the wishes of the commander.
Vegetius: The *peplegmene falagx* is the saw (*serra*) formation of Vegetius which was similarly used to harass the enemy. It is probable that Vegetius's *serra* is close to the *peplegmene* array of Asclepiodotus while it is likelier that the late Romans usually used the phalangial version given by Aelian.

Λ towards the front. According to Aelian and its Roman and Byzantine copies and interpolations, the infantry wedge was used as a countermeasure against the cavalry square formation, but it was obviously equally good against any cavalry formation advancing from the front. Aelian's text describes only the larger version in which two larger units in the phalanx were united to form up the lambda Λ which naturally means that these units remained in the phalanx and did not advance in front, but it is clear that the Romans also employed the wedge in front of their phalanx as a *globus/drouggos* consisting of the *protaxis/antesignani*, who in their turn could assume two different versions of the wedge, the small and large versions (see the Chapter 7.9) depending on the size of the force.

We find the Romans using the infantry wedge until the end of the Late Roman period because Maurice included a referral to this formation, but as we shall see later in this chapter he restricted its use primarily for the infantry units posted on the flanks. This, however, does not preclude its use in the traditional manner, because Maurice considered his treatise to serve only as an introduction to more sophisticated tactics. We also find the Romans using the light-armed in front of their heavy infantry throughout the period, and we find them opening up their phalanx for the purpose of negating a cavalry attack, or the charge of elephants, or an enemy infantry wedge, and it is also clear that they used the hollow-wedge formation as well. The use of the club-bearers and mace-bearers as a specialist force against cavalry cataphracts (first in evidence in the Column of Trajan) is clear at least until the fifth century, but there is no longer any firm evidence for their continued use in the sixth century. However, the fact that the *Excubitores* were equipped with the iron *distria* (iron maces, sing. *distrion*) suggests that the Romans continued to use club-bearers and mace-bearers as countermeasures against cavalry cataphracts throughout the Late Roman period, with the maces employed by both Roman cavalry and infantry.[20] The attached diagrams on pp.292–4 drawn after *Codex Burnley* show how the different versions were presented in the 'Byzantine' versions of Aelian.

If the enemy approached from behind, the standard Hellenistic tactics consisted of the use of double phalanxes, two-fronted formations, about turns and counter-marches by file. For further details, see the discussion of these tactics below in the context of the commands in the *Strategikon*. As regards the counter-marches, the Romans used a far greater variety before Maurice, who streamlined and simplified these. In the Macedonian counter-march (phalanx formed in front of its former place), the file-leaders turned 180 degrees and the files behind them marched through the intervals to assume positions behind the file-leaders so that the phalanx occupied the ground previously in front of it (in other words it retreated from the direction of the enemy while doing this). In the Lacedaemonian counter-march (phalanx formed behind its former place), the file-closers turned 180 degrees so that the file-leaders and files behind them marched to the rear through the intervals so that the new file was formed in front of the file-closers (this was the aggressive form of countermarching, because the phalanx advanced towards the enemy). In the Cretan/Persian/Choric counter-march (the phalanx remained in the same position), the file-leaders marched to the position occupied by the rear-guards and the rear-guards to the position previously occupied by the file-leaders. All of these variants – even the one by ranks – were still included in the *Peri strategikes* (24), but the *Strategikon* simplified this (see below) by requiring the men to be able to perform only the Cretan/Persian version by files or the about turn individually by all men of the file if they had already assumed the close order *pyknosis*. This is yet another example of Maurice simplifying and reforming the system. It was far easier for the soldiers to learn only a single method than a great variety, which could result in confusion during the stress of combat.

Syrianus Magister includes a lateral phalanx with a stratagem that is not included in the other treatises and for which there is no definite evidence in the narrative sources. However, it is still quite possible that this stratagem was used in some of the period battles of which we know next to nothing, because it is clear that Syrianus Magister's treatise was based on the actual tactics used during the reign of Justinian – and it should be remembered that Constantine Porphyrogennetos considered this treatise to be compulsory reading for the commander on campaign. This proves its practical value. As already noted in Chapter 1, the *Peri strategikes* advised that in situations in which the Romans unexpectedly faced a sizable enemy force in open terrain and were unable to retreat to higher ground, the commander was to draw his army in

Syrianus's stratagem with a phalanx when the Romans had too few men

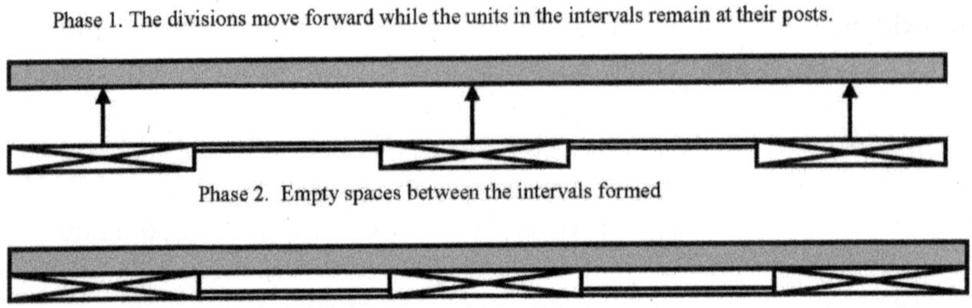

two to three phalanxes facing the enemy, posting two- to three-ranks of infantry into the intervals so that the phalanx looked as one continuous line equal in width to the enemy. The front ranks were to be on higher ground and the rear ranks on lower to cover the stratagem. When the enemy approached, the phalanx advanced a short distance so that the ranks in the intervals became level with the rear of the formation. According to Syrianus, the surprised enemies were too scared to enter the intervals because they did not want to advance into an exposed position where they could have been placed between enemies. This then meant that the enemy was able to fight only against the forward-posted phalanxes. One may assume that someone had used this in the past or that it was used during the Late Roman period in some unknown instance.

The internal structure of the heavy infantry phalanx structure was different in Syrianus Magister's treatise because he included several different versions that the commanders could choose to use according to their own preference (spear lengths varied between ranks and different compositions for the rear ranks, see Chapter 1) alongside his multipurpose heavy infantry able to use both bows and melee weapons. Syrianus Magister also does not give ideal sizes for the different arms of service (he only noted the names of the heavy infantry units from the file of sixteen up to the *falaggarchia* of 4,096 men that made up the phalanx). In sum, towards the end of the reign of Justinian the Roman infantry achieved a very high standard which allowed them to use a greater variety of different tactical compositions that had not been used before.

9.4. Lateral Phalanx in the *Strategikon*

Maurice (*STR* 12.B.8) refers to the ideal theoretical Roman army that had existed before, which consisted of 64 *tagmata* for a total of 16,384 men, in addition to which there were 8,000 (in truth 8,192) light-infantry *psiloi* (archers and javelineers) and 10,000 *kaballarioi*-horsemen, while also noting that in the past the *tagmata* had the uniform size of 256 men which was no longer the case. This ideal army was still based on the theoretical Hellenistic numbers, but the referral to the legions (*STR* 12.B.8.1) and the size of the cavalry component prove this to be the idealized Late Roman version of the phalanx structure which existed in some Late Roman period military treatise or treatises that Maurice had used as his source(s).

Maurice noted that the use of the single uniform size for the *tagma*, 256 men (the *syntagma* of the Hellenistic treatises), caused problems when in practice the *tagmata* consisted of different sized units. As stated by Maurice, if the units were divided and united to form up the *tagmata* of 256 men this would cause some of the men having to serve alongside unfamiliar men, while in other cases the superfluous men could not be fitted into the 256-man units and so would have nothing to do. This proves that the Romans had been in the habit of doing both, because the different sized new and old auxiliary units required this if the men were to be uniformly deployed as 256-man units in the phalanx. We find this instruction still in effect in the texts of Urbicius and Syrianus Magister. In contrast, Maurice recommended flexibility. The officers were advised to form either one large *tagma* or two smaller *tagmata* rather than try to force the units into the theoretical structure requiring exactly 256 men for the *tagma*. What is notable about this instruction is that it differs from the cavalry section as Maurice does not give any minimum or maximum size for the infantry *tagma*.[21] The combat effectiveness of Maurice's infantry armies bespeak the value of the reforms of Maurice which simplified and unified the Roman combat methods. See Syvänne, *MHLR*, Vol.7.

Even if Maurice's description of the lateral phalanx (*plagia falagx*) is still fully grounded in the ancient Greek military theory as practised by the Romans, its approach is still more orientated towards its practical use in combat than in any of the previous treatments.[22] Unlike in the traditional theory, Maurice's text took into account the actual size of the army. The infantry armies were classified into two basic categories: infantry forces of 24,000 infantrymen or more, and armies with less than 24,000 footmen. The armies that had 24,000 footmen or more were divided into four *mere* (divisions) and the armies that had fewer than 24,000 infantry into three *mere*

THE LATERAL PHALANX OF 24,000 FOOTMEN OR MORE

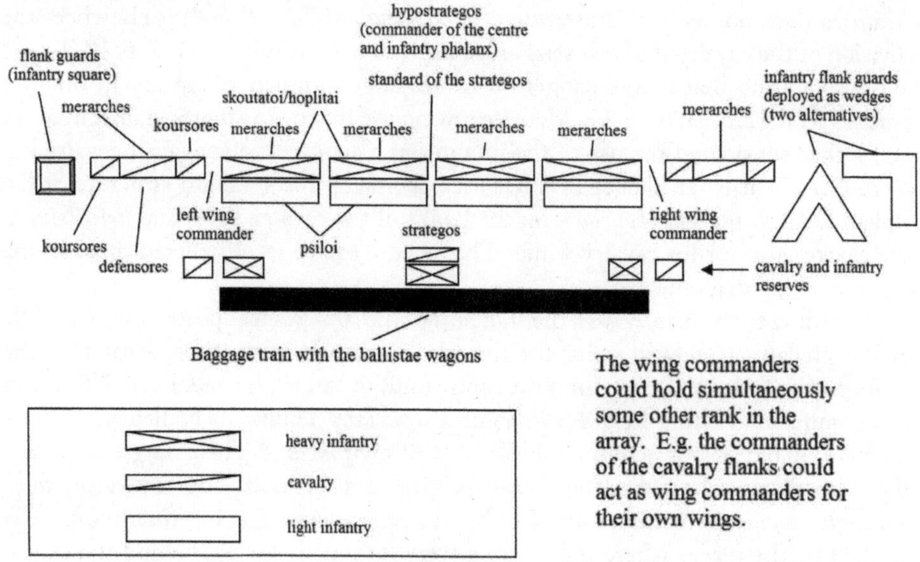

(divisions). On the basis of the marching column formations (single, double, triple and quadruple phalanxes), it is probable that the lateral phalanx was also sometimes divided into just two *mere*. In full battle readiness, the intervals between the *mere* were ca. 100 to 200 feet (ca. 30–60m) in width, so that these were clearly separated while still being able to cooperate and support each other. The number of light infantry also varied according to the size of the army. When an army had 24,000 footmen or more, half of their number were to be light infantry, but when the army consisted of less than 24,000 men, a third of the force was to consist of light infantry. The fact that all armies that had more than 24,000 were always divided proportionally into four divisions (*mere*) means that in the *plagia* phalanx formation the *mere*-divisions could be significantly larger than in the cavalry forces, where the maximum size for the *meros* was 6,000–7,000 horsemen.[23] This in its turn means that the number of *tagmata* per *meros* varied greatly from one army to another, as did the sizes of the *tagmata* in each *meros*. The only requirement in the *Strategikon* for their size was that the sizes of the *mere* were to be proportionate to the size of the army.

We do not know what the envisaged minimum and maximum sizes for the infantry armies were, but this is only to be expected. The size of the army always depended on the availability of the soldiers at each particular place and point in time. The maximum could therefore consist of tens of thousands of footmen while the minimum for some neglected theatre of war could be just a few thousands.

The baggage train with *ballistae* carts were posted behind the phalanx to protect the rear, but when it did not follow, the infantry used the double phalanx formation.[24] Separate *skoutatoi* and *psiloi* units were posted on the flanks and in the centre of the baggage train or elsewhere to serve as reserves. The Battle of the Casulinus River in 554 (Syvänne, 2004, 473–4; *MHLR 6*, 346–53) proves that the servants and camp followers could also be used to stiffen the reserves. The cavalry were posted on the flanks so that the best *tagmata* (i.e. the mounted archers) were posted farther out. This may imply the use of the heavy cataphract cavalry next to the infantry, but since Maurice does not state this anywhere in the text, while still noting elsewhere the division of the cavalry into *koursores* and *defensores* (e.g. *STR* books 1–3, 6, 12.A.1–2) and also stating that it was dangerous to use large numbers of cavalry in infantry battles (*STR* 12.B.23.14–5; i.e. Maurice saw no use for the *clibanarii* against infantry and rather considered the use of the infantry as a counter tactic against cavalry e.g. *STR* 12.A.7), it is far likelier that Maurice envisaged the standard structure of the Italian Drill formation (*koursores* on the flanks of the cavalry *meros* and *defensores* in the centre) also for his cavalry wings. This is how I have reconstructed those in the accompanying diagrams.

According to the *Strategikon*, the standard of the *strategos* was posted in the middle of the phalanx as a focal point for the whole army. It is therefore clear that the *strategos* usually led the army from the centre until the attack started. Each of the *mere* was commanded either by a *merarches* or a *stratelates*. Unlike in Hellenistic theory, their command post was in the middle of their own *meros*.[25] The *strategos* obviously chose his place as he pleased and that the situation required. The *Strategikon* does not include separate commanders for the left, centre and right, but their presence is implied by the places where the reserves were stationed, namely behind the centre and flanks. In other words, there is every reason to believe that the upper-level command structure was the same as it was when Vegetius wrote his treatise, but with the difference that now the *strategos* usually placed himself in the centre where his standard was. This conclusion is proven beyond doubt by the narrative sources, which always mention separate commanders for the left, centre and right. The wing commanders commanded the cavalry and infantry of their respective flanks together with the infantry and cavalry reserves posted behind. The *strategos* obviously did not only command the cavalry (his *bucellarii* and other bodyguards) and infantry behind the centre, but all of the reserves and other forces belonging to the army. This was possible because the narrative sources prove that there was usually a separate commander for the entire infantry line who also usually served as *hypostrategos* (second-in-command, lieutenant general) for the entire army. The commander of the infantry usually took his place in the centre/the navel/mouth of the phalanx so that he should really be considered to have been the commander of the centre, which means that the *strategos* was able to move about the battlefield freely as the situation in his opinion required.[26]

Maurice divided the heavy infantry (*skoutatoi*) file so that it consisted of eighteen men (these consisted of two *contubernia*, so that the extra two men not listed here were the servants who were posted among the baggage train) of whom only sixteen were to be used for actual combat duties. The two worst soldiers (usually the recruits) were to be posted among the wagons or elsewhere. The actual phalanx of the heavy-armed *skoutatoi* naturally consisted of ranks and files, and its ideal depth was the 16-man file of ancient times. As before, the first four and the last four ranks of the phalanx were expected to consist of 'better' men.[27] In ideal circumstances the tallest men were placed in the front, but if there were not enough tall, skilled soldiers available then it sufficed to choose the better men for the front ranks. The idea being the use of the taller men in front was to exploit the human animal instinct, which naturally considered taller men more threatening. It was for this very same reason that the men wore pointed, plumed helmets. These made them look taller to the enemy. The files in their turn were divided into first (*primi*) and second (*secundi*) rank men (every other rank in a file being one of these) to facilitate the dividing and doubling of the files. The first of the *primi* was the *lochagos* in command of the sixteen-man file, and the first of the *secundi* was the *dekarchês* (commander of ten, who commanded eight men in the file when required) who stood behind the *lochagos* in the second rank. The last of the *secundi* was the *ouragos* (rear-guard), the last man in the file. Even if these are not mentioned by Maurice, it is clear that there were also two *pentarchai* (sing. *pentarchês*, commander of five) who acted as commanders and file-leaders for those four-man files (when the depth was only four ranks) that were not commanded by the *lochagos* or *dekarchês*. The front rank man, the *lochagos*, led the whole file forward, while the last man of the file, the so-called file closer (the *ouragos*), pushed the men forward. Just as in the Hellenistic phalanx, the 'ordinary' soldiers were placed in the middle ranks, so that the 'better' men shielded and guided the weaker elements of the phalanx.

The depth of sixteen ranks gave the phalanx an inbuilt reserve to face threats from the rear and front, which in the case of the frontal threats operated so that each man in the file would take the place of his fallen comrade in front of him. However, the principal function of the depth of the phalanx was still psychological, because the phalanx would have collapsed long before the last file would have had its turn to fight. It is in fact likely that the phalanx would have collapsed when the first four 'better' men had been taken out of action (killed or wounded) since the men in the middle ranks were less steady as fighters. The depth of the array and the presence of comrades around helped to keep the men in their place in the formation. In addition, the depth of sixteen ranks was more secure than the shallower formation because it enabled the emergency division of the phalanx into double front and double phalanx, each with eight ranks. The shallowest allowed depth for the phalanx was four ranks, as the shallower formations lacked adequate depth and also placed inexperienced ordinary soldiers in the front rank. The maximum allowed depth for the phalanx was thirty-two ranks, because in Maurice's opinion the depths deeper than sixteen ranks did not really help in combat.

Maurice's *skoutatoi* (heavy infantry shield-bearers) were required to have hair cut short. They were equipped with Gothic shoes, Gothic tunics or short tunics, simple mantlets, shields (various types and sizes depending on the situation), helmets with plumes, Herulian swords (the Herul version of the *spatha* was clearly considered the best), spears (*kontarion* type for use in regular open terrain), slings, and lead pointed darts, and the chosen men or at least the first two men of the files were to have mail coats; and the first and last of the files were to have greaves of iron or wood. In certain circumstances, for example in difficult terrain, the *skoutatoi* were required to change their equipment. See later. The training of the *skoutatoi* as spearmen and javelineers enabled them to serve in both capacities. The equipment placed among the baggage train shows that when there was a need, the *skoutatoi* could also be equipped with other weapons or tools like axes and pickaxes. These, just like the knives and daggers that the soldiers carried for personal use, could be needed in combat as shown, for example, at the Battle of Antonia Castra in 547 (Syvänne, *MHLR 6*, 275). It was because of this that Maurice included those among the list of equipment carried.

In the *Strategikon*, the light infantry *psiloi* were organized into *dekarchiai* under competent *dekarchai* (sing. *dekarchês*, 'commander of ten'), so that an officer (*archôn*) called *archisagittatôr* (chief-archer) acted as their commander. It is quite likely that we should identify the *archisagittatôr* with the *primosagittarioi* and *toxotai prôtoi* (first-archers) of John Lydus (1.46, p.74.3). Neither John Lydus nor Maurice specify the exact duties of the *archisagittatôr*, but one may make the educated guess that *archisagittatôr* commanded all of the *psiloi* attached to a specific *meros* and that a separate *archisagittatôr* was named for each separate groupings of the light-armed. For example, it is probable that the *psiloi* posted on the flanks had separate commanders in situations in which there were also *psiloi* posted behind the *mere* – each of the latter would have had their own separate *archisagittatôr* as commander. In other words, we should equate the chief archer with the Hellenistic *epixenagoi* (2,048 men) or *systremmatarchai* (1,024 men), but with the belief that when the size of the army or detached light infantry component differed, so did the size of the number of men under their command. The soldiers belonging to the light infantry were clearly expected to be able to improvise on the spur of the moment, as there were no commanders between the chief archer and the *dekarchai*. However, even if Maurice fails to mention these, it is still clear that the light-infantry file also included *pentarchai* (commanders of five men), because Maurice grouped the light-armed into four-to-five-man irregular groups for combat in difficult terrain. This means that the light infantry tent group, the file, was divided into two four-to-five-man groups, and that one of these was led by the *dekarchês* and the other by this subordinate, the *pentarchês*. The ability of the light infantry to operate in this manner proves that its members were the crème-de-la-crème of the Roman army. The low ranking members of the light infantry could adapt their tactics to the situation without the presence of middle ranking officers. They were expected to be able to improvise and adapt without instructions from above.

Maurice's light-armed, the *psiloi*, were required to carry bows, arrow-guides, quivers with arrows, small shields, and those not practiced in archery were to carry small javelins, lead-pointed darts and slings. This means that the *psiloi* actually consisted of four different types of troops: archers, javelineers, dart-throwers and slingers, all of which could be deployed in different manners on the battlefield. As will be made clear below, in the case of the javelineers and dart-throwers this means that they could also be employed as if they were heavy-armed.

According to Maurice, the light infantry could be deployed in a wide variety of ways, but he mentions only the following. Firstly, the light-armed *psiloi* could be deployed at the rear of the heavy infantry phalanx, so that the sixteen deep phalanx of *skoutatoi* would have four ranks of *sagittarioi* (archers) behind: this means that when the depth of the heavy infantry phalanx was reduced to four ranks, there was still one rank of *toxotai* (archers) behind. This appears to have been the minimum depth for the foot archers, because Maurice stated that when the army consisted of 24,000 men or more, a half of the force was to consist of the *psiloi* and when the army had less than 24,000 men a third of the force was to consist of *psiloi*. This means that there could even be sixteen ranks of *psiloi* behind the *skoutatoi*. This, however, was not necessarily the ideal solution for two reasons. As we shall see, when there were large numbers of *psiloi* present some of these could be used as flank guards, in addition to which we should keep in mind that the *psiloi* also included javelineers and slingers. Maurice actually preferred to place those on the flanks of the phalanx. Regardless, since Maurice assigned all light-armed behind the *skoutatoi* when the army was drilled regularly, one may assume that the deployment of the *psiloi* behind the *skoutatoi* was the standard practice, regardless of the numbers present.

Secondly, the *psiloi* could be deployed in depth with the *acies* (battle formation or file), meaning that a *skoutatos* alternated with a *toxotês* (archer). The literal translation would mean that there was one rank of *skoutatoi* followed by a second rank of *toxotai*, and third rank of *skoutatoi* followed by a fourth rank of *toxotai*, and so forth. This does not make sense, because it would have broken the heavy infantry formation while not placing the archers in the formation in such a manner that their file leader, the *dekarchês*, could have directed their actions. Therefore, it is probable that we should interpret the deep/depth (*bathus*) in this case to mean the placing of the light-armed in the length/breadth of the *acies* (battle line),[28] so that one file of *skoutatoi* alternated with one file of *toxotai* as we find it in Hellenistic theory (the *entaxis* formation of

Aelian and Syrianus Magister; the *parentaxis* of Asclepiodotus). This system would have enabled the archers to operate with the *dekarchês* of each file directing the archery while the files of the *skoutatoi* protected them with their shields and spears. The minimum width for an array like this was to have the *skoutatoi* in the *pyknosis* order, because when the *skoutatoi* were deployed with file widths of ca. 94cm it was still possible for the foot archers to use their bows when deployed sideways towards the enemy. If deployed in this manner, the shields of the *skoutatoi* provided protection to the archers. However, it is still far likelier that when the files were alternated in this way that all of the files, the files of *skoutatoi* and *toxotai*, each had an approximate width of 94cm because this enabled both arms of service to perform their battlefield missions without hindrance. The fact that the foot archers of Maurice (*STR* 12.B.3, 5) were required to carry small shields made them combat ready also in melee if necessary.

The third alternative given by Maurice states that sometimes the *psiloi* were placed in the *acies* and on the flanks of the phalanx meaning, to the inside of the cavalry. In other words, the *psiloi* were placed in the battle line itself between cavalry and *skoutatoi*.

The fourth choice concerned situations in which there were large numbers of light-armed present, in other words, when the army was large so that a half of the infantry force was to consist of the *psiloi*. In this case some of the *psiloi*, together with a few *skoutatoi*, could be placed as flank guards for the cavalry. There is no definite evidence for the use of this option in the narrative sources, but it is clear that it must also have been used in practice for it to find its way into the *Strategikon*. Maurice does not state what battle formation the flank guards were to assume, but on the basis of the instructions for the rear guards with similar unit composition in the *epikampios opisthia* formation (*STR* 12.A.7 discussed later in this book) it is clear that the flank guards used either the hollow-square or wedge formation in such instances so that the *skoutatoi* formed the outer edges of the hollow square and the front of the wedge. Both should be seen to represent versions of the Roman *globus*-order, as these were deployed outside the phalanx proper as independently-operating units. There are two possible ways in which the wedge could have been used on the flank, both of which are shown in the accompanying diagram of the infantry phalanx. The use of the small-unit version (*tagma*) of the wedge in this context is less likely, because we are here dealing with large numbers of light infantry. The different uses of the wedge will receive further attention in the context of the wedge as battle formation.

The fifth option concerned the *psiloi* who used javelins (*bêrutta*; after Latin sing. *verrutum*, *verutum*, pl. *veruta*; shaft 1.05m and a 0.37cm iron shank, of which 12cm formed the head, for a total length of 1.17m) or darts (*martzobaroulon*; after Latin sing. *mattiobarbulus*, pl. *mattiobarbuli*, i.e. *plumbata/plumbatae*). These were to be arrayed either behind the *skoutatoi* or on the flanks, but not in the middle (i.e. not in the *entaxis* formation). One may assume that when these *psiloi* were deployed as the outermost units of the infantry phalanx, they were used as if they were heavy infantry units in close order so that these men were expected to be able continue the fight with swords once their missiles had been thrown. In fact, one may even assume that they could carry medium- to large-sized shields, because these allowed the men to place more darts in the shield.

The sixth version stated that the slingers (*sfendobolistai, sfendobolistoi*) were to be arrayed always on the flanks. The adverb 'always' appears to be an exaggeration, because all of the light-armed were still drilled in deploying behind the *skoutatoi*, not to mention the fact that the slingshots easily flew over the *skoutatoi*. One should therefore understand Maurice's text to mean that the preferred position for the slingers was on the flanks. The reason for the flank position was that the slingers were considered particularly effective when they were able to target enemy cavalry from a flanking position. The slingers also required room to operate, so they could not be posted alongside the other light-armed. The slingers needed to be posted separately in open order for them to be effective.

In sum, all of the light-armed were usually deployed behind the *skoutatoi*, but Maurice still envisaged several alternative ways for the employment of the various types of light-armed alongside the lateral phalanx, yet there is one glaring omission. Maurice did not recommend the use of the light-armed in front of the lateral phalanx which we find in Asclepiodotus (*protaxis*), Aelian (Matthew ed. 30), Vegetius, and still find in muddled form in the *Peri strategikes* (31). Syrianus Magister places some men in front of the phalanx (in the *parembole* the men in front were reinforced by bringing men from the phalanx as their reinforcements, which may also mean that the light-armed were in a skirmishing position; this has been copied from the *Byzantine Interpolations of Aelian*, Devine ed. 31.1) or in front of the flanks of the phalanx (*prosentaxis* i.e. in the outflanking position copied from the *Byzantine Interpolations of Aelian*, Devine ed. 31.2). Maurice places the light-armed in front of the *skoutatoi* only in two instances; when the army marched in difficult terrain (*STR* 12.B.20) and when the dismounted cavalry army (in this case the *skoutatoi* are the dismounted horsemen) retreated after a defeat, either in a double phalanx or four-sided formation in situations in which the enemy consisted of cavalry (*STR* 7.B.11.18ff.). One can therefore safely say that after Maurice the Roman combat doctrine, when followed, restricted the use of the light infantry in front of the *skoutatoi* to such situations. In fact, none of the narrative, albeit defective in details, suggests that the *psiloi* would have been used in any other manner than as stated by Maurice.

The place of the cavalry in the lateral phalanx was on the flanks, with the better *banda* posted further out. If the cavalry force accompanying infantry was over-large, in other words over 12,000, the depth of the cavalry file was expected to be about 10 men, but if it was less than this the depth was expected to be about five men. The diagrams depicting the infantry formations and the instructions regarding the quality and depth of the cavalry formation prove that in practice the five and ten ranks were just approximations. The cavalry wings were divided into *koursores* and *defensores*. The cavalry reserves were posted behind the cavalry wings just outside the line of wagons. These were used as a general reserve, as well as flank guards for the cavalry wings and as rear guards. The principal function of the cavalry wings was the same as always, namely the protection of the flanks, combat against enemy cavalry, pursuit of the defeated foe, and flank attacks against the enemy. If the Roman cavalry could not resist the enemy cavalry force, they were expected to flee to the area between the infantry phalanx and baggage train where they were to be regrouped for a counter attack, and if this was not possible they were to dismount and fight on foot.

As noted above, if there were large numbers of light infantry present, these together with heavy infantry could be posted outside the cavalry wings as their flank guards. In addition to the wing reserves, the *strategos* also possessed guardsmen/ *boukellarioi* of his own that he could use as an emergency reserve. The *Strategikon* did not recommended the use of large numbers of cavalry in infantry battles, only 3,000–4,000 horsemen. This probably means that only this number of cavalry could be used safely against infantry, while the rest of the cavalry were to be used against enemy cavalry or in pursuit of the defeated foe. In other words, it is possible that the inner side of the cavalry formation was to be used in the same manner as the *clibanarii* in the past against the infantry flanks if the commander so desired. This was sound advice, because the flight of the cavalry could demoralize the infantry. Notably, the earlier ideal size for the cavalry had been 10,000 for an army of 24,000 footmen. This suggests that the Late-Roman ideal proportion for cavalry in the combined army of infantry and cavalry was to be about a third of the entire force.

When the wagon train followed to protect the rear of the phalanx, it was stationed a bowshot behind the infantry force so that the wagons covered the entire width of the phalanx. The wagon train in its turn was protected by its drivers, using javelins, bows, slings, caltrops and metal darts, and by some soldiers. In desperate situations the drivers were expected to throw out a few caltrops to hold the enemy at bay. Maurice required that each wagon had its back-part covered by a heavy cloth which was used to protect the drivers and the oxen from the impact of enemy arrows. Maurice also demanded that the oxen pulling the wagons were to be hobbled or tied and positioned so far away from the infantry phalanx that these could not disrupt the infantry formation if they panicked because of some noise or arrows. The use of the heavy cloth coverings and hobbling of the oxen were the result of lessons learnt during the reign of Anastasius, when the Bulgar mounted archers in the service of the rebel leader Vitalianus destroyed the loyalist army by shooting fire-arrows at the oxen drawing the wagons. See Syvänne (*MHLR 5*, 239–41).[29]

The *carroballistae* (ballista-carrying carts) were distributed along the whole front so that the most powerful *ballistae* were stationed on the flanks. The Romans used the double phalanx in situations in which the wagons were far away or were not present at all. On the basis of Maurice's list of commands (see below) he meant primarily that the double phalanx was formed up so that the single phalanx was divided so that half of each file was marched to the rear. Maurice gives us two alternatives for the organization of such a double phalanx. Firstly, the single phalanx could be divided into a double phalanx so that half of each file was marched to the rear. Secondly, Maurice could mean a double phalanx consisting of actual separate units, because we find him using such in difficult terrain. The wagons were not used when there was a need to march quickly or when the terrain was difficult, so the latter option was particularly relevant for these situations. The Romans used the double phalanx also in support of the wagons when the enemy attacked in force from behind and the drivers and soldiers posted among the wagons were insufficient for the task.[30]

The principal function of the *carroballistae*, together with some stone-throwers/trebuchets (if present), was to terrorize the enemy with long-range shots of bolts and arrows and incendiary arrows (and small stones) that had the power to penetrate armour. Fire-bombs could also be used if stone-throwers were prepared. We can calculate the approximate size of the Late-Roman field artillery accompanying the phalanx on the basis of the information provided by Vegetius 2.25. According to him, the ancient Roman legions had 55 *carroballistae* and 10 *onagri* (torsion powered stone thrower) in ox-carts. This means that an army of 24,000 footmen had 220 *carroballistae* and 40 stone-throwers (probably trebuchets). The *carroballistae* and trebuchets (when deployed for use) were undoubtedly operated as batteries (several firing simultaneously salvoes), just as these had been used in the past.[31] The sight of the effectiveness of these shots was usually enough to keep the enemy away from the formation, and in fact Vegetius goes so far as to claim that neither armoured cavalry nor the shield-bearing infantry could stand their ground under this fire.

However, there is one major difference between the sixth-century and earlier usage of the field artillery. In the past, the field artillery was used so that it shot over the phalanx at the enemy in front. This appears not to have been the case during the sixth century, as on the basis of the *Strategikon* and descriptions of combat (e.g. Syvänne, *MHLR 6*, 183–5), the artillery was used only against enemies approaching from the rear or from the flanks – the artillery wagons were placed an arrowshot behind the phalanx, which meant that these could not be used for shots over the phalanx against the enemy in front. Regardless, I would still suggest that the Romans could employ their field artillery in the old manner when necessary. Furthermore, we know that when the phalanx was deployed just in front of a city or near the shoreline, it received supporting fire from the mural artillery or naval artillery.

In the *Strategikon* the lateral phalanx was used against both infantry and cavalry. The standard version with the baggage train and *carroballistae* following was considered suitable against both, but if the enemy's cavalry force outnumbered the Roman cavalry and the baggage train did not follow, then Maurice instructed the *strategos* to avoid open and level terrain.

In such instances the phalanx was to be deployed on rugged and difficult terrain or in swampy, rocky, uneven or wooded places. In fact, in difficult and uneven terrain Maurice actually instructed the *strategos* to leave the wagon train behind, because the terrain provided sufficient protection for the formation.[32] The narrative sources also prove that in such circumstances the Romans could post their infantry phalanx so that its rear rested against city walls or a fortified marching camp that had archers and mural artillery and soldiers to act as reserves, or that they used a river in front of the phalanx as a protective line against cavalry, or that they used field fortifications in front of the phalanx, as Belisarius did at the Battle of Dara in 530. I will deal with the improvised battle formation of Belisarius in the context of the mixed formation below, because it had cavalry deployed in a very peculiar and unique manner in the centre of the infantry phalanx. In fact, Belisarius's battle formation was so peculiar that it was neither the lateral phalanx nor the mixed formation, but an entirely unique formation in the history of mankind.[33]

deployment pattern with Roman symbols for the lateral phalanx when the infantry army was extra large and had equally large numbers of light infantry

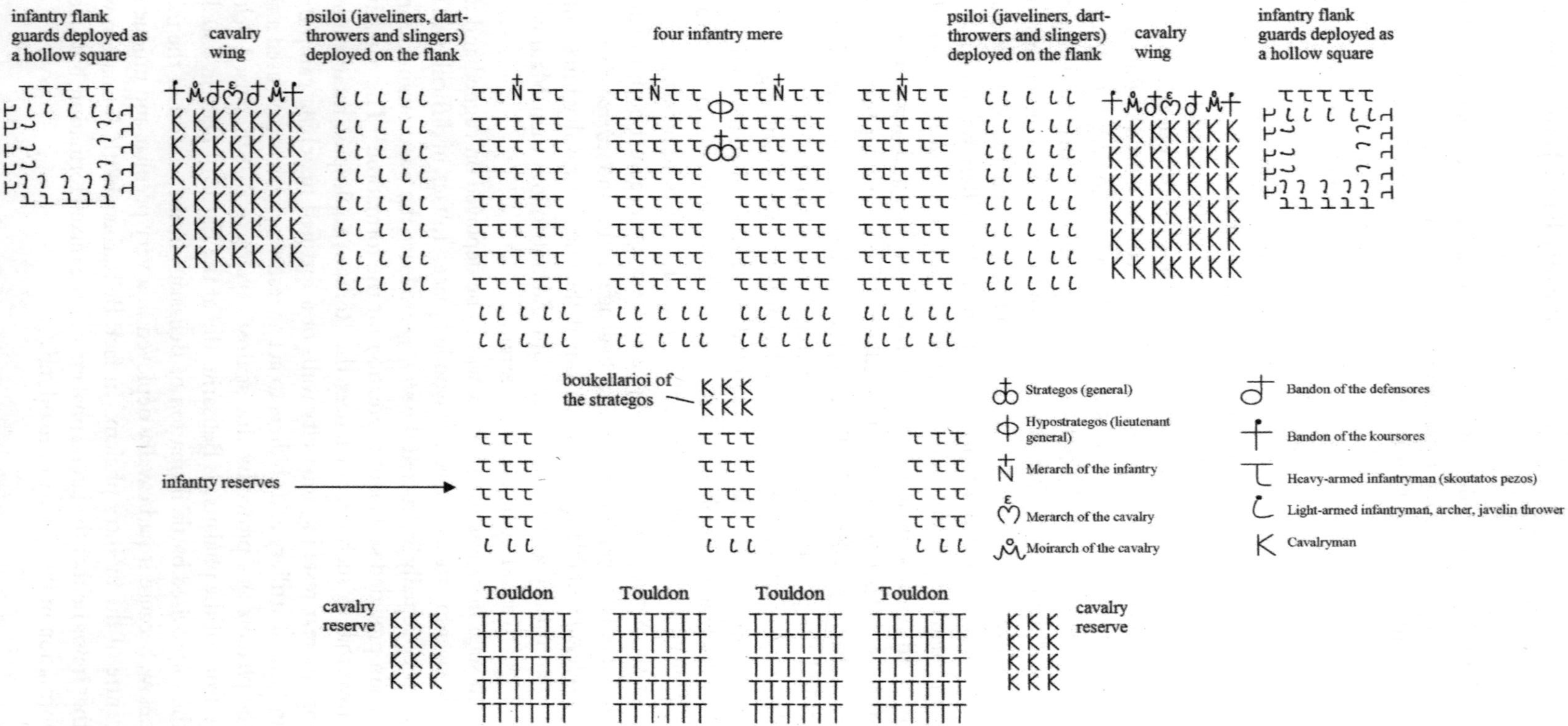

When the lateral phalanx was deployed in difficult terrain, it waited for the enemy cavalry to approach, because otherwise the advantage of the terrain would have been lost and the danger of losing the cohesion of the formation would have increased. In such situations, the lateral phalanx derived additional advantage from the use of the infantry archers with arrow-guides, because their arrow-shots outranged the regular bows (maximum range for the arrow-guide was about 680 metres[34]). It was very unlikely for the enemy cavalry formation to be able to maintain its cohesion to contact with the infantry phalanx, so if there were elements of the enemy cavalry that made their way all the way up to the infantry phalanx it was as good as lost when it came face-to-face with the Roman infantry *foulkon* formation. In fact, as I noted in my doctoral dissertation, *The Age of Hippotoxotai,* during the period of study the lateral phalanx, with its inherent manoeuvrability into the crescent and *epikampios emprosthia* formations, was the most effective infantry formation in use against cavalry. My reasoning for this was that when in the course of the battle the wings of foot archers and cavalry were used to encircle the enemy cavalry with the *menoeides* (crescent) or *hyperfalaggesis* (double-outflanking) formation, they could inflict a greater amount of casualties by crossfire than would have been the case if the enemy had encountered a straight line, or if the Romans had used the double phalanx or hollow oblong/square formations. The correct choice of terrain and the judicious use of the wagons also freed more men to fight against the cavalry in a long lateral line. In other words, the lateral formation brought more men simultaneously into the combat than any of the other formations. It was because of this that Narses had arrayed his forces initially as a lateral infantry phalanx against the Goths at the Battle of Taginae in 552 (Syvänne, *MHLR 6*, 340–4).

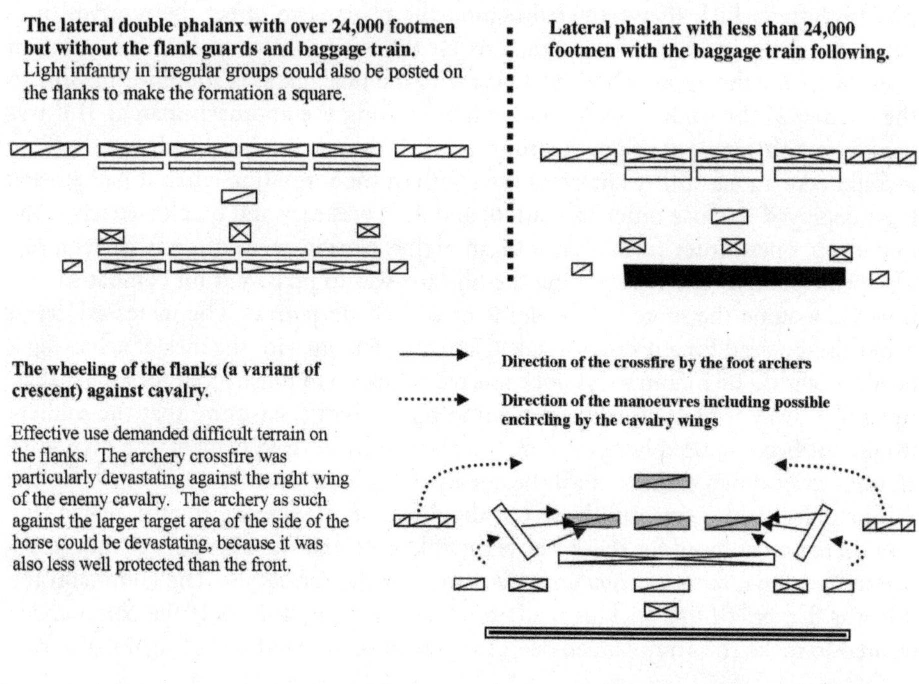

Commands and unit Manoeuvres in the Strategikon and grand tactics

Maurice's *Strategikon* lists a set of commands that the *tagmata* and *mere* were to learn during training and drilling. These commands include the thinning and doubling of the files, counter-march, about turn, straightening of the ranks, different forms of dividing the line for the assumption of the double phalanx, facing all directions (double-front), assuming close order so as to ready the men for combat, assuming *foulkon* and so forth.[35] Even though not always specifically mentioned by Maurice, these unit commands enabled the *strategos* to increase or decrease the depth of the files as needed, to widen or shorten the front, to prevent the enemy from outflanking the phalanx, to outflank the enemy on one or both flanks, to push through the enemy formation in the middle, and to face attacks from the flank and rear etc. These commands enabled the use of all of the tactical manoeuvres that we find in the treatises depicting the Hellenistic phalanx and also in Vegetius (3.19–20, 26: his seven battle formations, hollow wedge/pincer and wedge) even if Maurice does not state this.

Doubling the depth of the phalanx and attack

The standard method during pitched battles was to approach the enemy with a depth of four footmen who were deployed in open order, while the *merarchai*, each with two *mandatores* (heralds), two *campidoctores*, one *strator* (equerry), one *spatharios* (sword-bearer) and two *aquiliferi* (eagle-bearers, *ornithoforoi* = bird-bearers), all mounted, were in front of the phalanx to reconnoitre the ground and direct the formation. They were to stay in front until the enemy was close by, after which they assumed their places in their own *meros*. It is likely that both the *strategos* and *hypostrategos* with their own supernumeraries also stayed in front in like manner to direct the phalanx, after which they retreated behind the phalanx to direct the reserves from there. According to Maurice, the reasons for the use of the four-deep formation in open order for the approach were: 1) it made the phalanx look more formidable to the enemy; 2) the soldiers were more relaxed during the approach march; 3) it was dangerous to attempt to widen the frontage when the enemy was already near because it could take a long time to increase the width of the formation when it had already been deployed in close order formation; and 4) it was easy and quick to deepen the four-deep, open-order formation into an eight- or sixteen-deep array as required. The principal reason for deepening the phalanx was to prepare it for combat so that it would assume the more secure eight- or sixteen-deep array. The increased depth could also be used for a massed attack against the enemy with the idea of achieving a breakthrough. The infantry was not expected to march in full armour for longer than two miles away from its marching camp to fight a battle, ensuring that the soldiers would not become tired before combat. If the enemy delayed combat, the men were allowed to sit down and rest until the enemy brought its men forward.

The prescribed commands leave out the direction of entering the file, but in this case there was no need for this because the soldiers knew the direction, as they knew where their own *contubernium* and *lochos* were in the formation. The command for this was 'Enter!' ('*Intra!*'). This made the formation eight deep. If the commander wanted to make the array sixteen deep, he gave the command '*Intra!*' again. Maurice did not recommend deepening the array further, but if the commander still wanted to do this the command was 'Battle line into Battle line!' ('*Acia in acia!*'). This made the phalanx sixteen ranks deep. One may assume that '*Acia in acia!*' was once again preceded by the direction either 'Shieldwards!' or 'Spearwards!'. Maurice considered the depth of four men as the minimum and sixteen men as the maximum needed for anything.[36]

The tactical purposes for doubling the depth were either to equal the depth of the enemy formation or to try to provide additional mass for the formation, or parts of the formation, for it to push through the enemy line. Pushing through the enemy formation with a deeper formation meant primarily a pushing and shoving action with shields that the ancient Greeks called *othismos*. This meant that the front lines of both sides had advanced so close as to be shield against shield. In this situation the men crouched and pushed and the men behind did the same until one side crumbled under the pressure. In Maurice's opinion depths beyond sixteen did not add further mass to this action. The deepening of the array could also be used to open the formation in width. The experienced officers certainly knew this too.

When the phalanx had then been readied for combat by deepening it where this was required, the commander took the following steps when the phalanx was about two- to three-bowshots from the enemy. He gave the order 'Unite!' ('*Iunge!*') and the files moved towards the centre of the unit until their shields were rim-to-rim, after which the ranks moved forward until these were almost 'glued' to each other. In other words, the phalanx assumed close order (*pyknosis*).[37] In battle formation, the *mere* thus assumed close order while also closing in towards the centre of the phalanx, the navel, where the flag of the *strategos* was. The *mere* were required to maintain small intervals of 100- to 200-feet between them in order to be able to operate independently while supporting each other. The closing of the formation could be done while advancing or while staying in place. The *Strategikon* instructed the use of the moving *foulkon* in situations in which the front rank soldiers did not wear mail armour and shin-guards and the archers were about to open fire. In practice, the *foulkon* was obviously also used when the men were fully armoured. The command for this was 'Form *Foulkon!*' ('*Ad fulcum!*').

If after the assumption of the close order the battleline had become uneven because some of the men had advanced faster than the others, the commander was to straighten out the front by giving the order 'Arrange the front!' ('*Dirige frontem!*') so that the men would make the front even.

If the enemy force consisted of cavalry, then Maurice's *Strategikon* expected that the entire Roman *skoutatoi* phalanx, regardless of the amount of armour carried, assumed the sloping *foulkon* meant for use against cavalry (shields roughly rim-to-rim in width) and waited for the enemy to attack it in this formation while the archers, slingers, dart-throwers and javelineers used their missiles against the approaching cavalry. The combination of the missiles and *foulkon* bristling with spear points was expected to be enough to halt the enemy cavalry, with the enemy cavalry either fleeing before contact or halting in disorder just in front of the *foulkon* so that the spearmen could then attack it with the expectation of victory. The pursuit of the defeated foe was left to the cavalry, who were not to pursue too far from their infantry support. The infantry phalanx was expected to follow the cavalry. For additional details, see Chapter 7.6–7 and analysis of the *epikampios opisthia* formation.

On the basis of the narrative sources and Maurice's advice to sometimes deploy the Roman phalanx in difficult, rough terrain, it is clear that the Roman infantry sometimes used also the regular close-order *pyknosis* and the shields rim-to-boss in width *testudo/foulkon* in such a manner that they waited for the enemy cavalry or infantry to attack, and if in such cases the Romans used the rim-to-boss *testudo*, they usually knelt when waiting for the approach of the enemy and then rose to use their spears/javelins and swords when the enemy came close. If the Romans had left their advantageous position in such situations by advancing, they would have lost the advantage of terrain. By staying in place, they forced the enemy to break the cohesion of their formation during the attack. However, since Maurice paid most attention to the use of the infantry against infantry in such a manner that the Romans advanced against the enemy, it is clear that Maurice himself favoured attack over defence in pitched battles, because this increased the boldness of the soldiers. Julius Caesar would have approved.

When the enemy infantry was at a distance of about one bowshot and the soldiers had been deployed either in *pyknosis* or *foulkon* order, the *strategos* ordered the *mandator* (herald) to shout '*Ready!*' ('*Parati!*'), after which another shouted 'Help!' ('*Adiuta!*'), which the soldiers answered by shouting in unison 'O God!' ('*Deus!*'). This was the war-cry of the Roman army and the sign that battle had begun. The light-armed started shooting arrows overhead while the *skoutatoi* advanced towards the enemy. When the *skoutatoi* were at dart and javelin distance, they were expected to throw these if they had them. This required the men to place their spears on the ground and the opening of the formation in depth so that the men could throw their missiles. This was presumably done with the front rank taking a few steps forward, placing their spears on the ground and throwing their javelins and darts with power, followed by the ranks behind so that in the end the men had returned to their original unit order. It is possible that the front ranks could have maintained their unit orders (either *pyknosis* or *foulkon*) throughout, for example when only the rear ranks had javelins and darts (see below) so that only the rear rank *skoutatoi* actually threw their missiles behind the close order bulwarks in front of them, but Maurice does not include this alternative – regardless, one may only make the educated guess that this system could have been used when decided in advance.

If the *skoutatoi* did not possess javelins or darts, they advanced straightaway towards the enemy so that the front rankers threw their *kontaria* as if these were javelins and then drew their *spathia*-swords for close combat, while the men behind kept their heads covered with shields and used their *kontaria* in support. This does not necessarily mean that all men were in the *foulkon* order, but that all men naturally held their shields above their heads when the enemy used missiles. The best proof of this (*STR* 12.B.16.83–6) is the use of the shield-roof by the middle ranks regardless of the tightness of their array when the phalanx assumed the *amfistomos falagx* (double-front, *orbis*) unit formation as an emergency measure against outflanking, plus Onasander's (20.1) comment regarding the first-century Roman attack formation (at that time using the cylindrical rectangular *scutum*, usually rim-to-rim). According to Onasander, the Roman infantry attacked so that the front rankers used shields as tall as their body during the attack while all rear rankers held their shields above the head when inside bow range. Notably, Maurice's instructions signified a return to the old style *pilum* and *gladius* tactics of the earlier era, because during the reign of Justinian the *skoutatoi* fought mostly like the *hoplitai* of ancient times by primarily using spears in melee.[38]

The men posted at the front – especially when they did not wear armour or greaves – were to protect themselves with their shields until they came to grips with the enemy (just like in Onasander) so that the enemy could not hurt them with arrows. This instruction clearly referred in particular to situations in which the front-rankers threw their darts, javelins or *kontaria*-spears which required opening the formation for the throwing of the missiles. It is clear that in such situations the men who were using the advancing *foulkon* order had to open the shieldwall by tilting their shields sideways to obtain the ability to move independently – the danger was less pronounced when the men used the *pyknosis*. When the soldier did this, there existed the danger that he could expose his legs and torso to arrows, hence Maurice's cautionary words directed in particular towards those who did not wear mail armour or greaves. In other words, Maurice warned against incaution when the men in the first rank threw their missiles.

The throwing of the ca. 3.74m cavalry *kontarion* in infantry battle was obviously not quite as easy as the throwing of the *pilum* had been (it was because of this that Maurice required the *skoutatoi* to be equipped with shorter spears when fighting in difficult terrain), but obviously still possible. It is because of this that I have speculated in the *Military History of Late Rome, Volume 7* that it is possible that in practice Maurice sometimes varied the length of the spears between the *skoutatoi* ranks, as we find in the close-to-period texts of *Peri strategikes* and *Excerpts of Polyaenus*, with those posted in the front rank having shorter spears to enable them to throw them further with greater effect.[39]

According to the *Excerpts of Polyaenus* (18.8), the first, second and third rank men were to have spearheads proportionate to their position in the phalanx so that their spearheads presented an even front against the enemy. The fourth rankers and others behind them were to hold their spears upright so they could fill any emerging gaps in the phalanx or turn towards the rear and form up the *amfistomos falagx* in situations in which the enemy threatened the rear. There is no definite proof for the use of spears of different lengths by different ranks outside the *Peri strategikes* and *Excerpts of Polyaenus* for this period, excepting the following three pieces of circumstantial evidence: 1) the use of the *kontaria* as thrown weapons by the front rank in the *Strategikon* would work better if their spears were shorter than 3.74m and we have to remember that Maurice (*STR* 12.B.20.8) himself referred to the existence of *kontaria* of different lengths in the context of fighting in difficult terrain (cavalry *kontarion* usually employed by the *skoutatoi* and the shorter *kontarion* used in difficult terrain); 2) the fact that the spears are shorter than the Macedonian *sarissa*-pike in the *Excerpts of Polyaenus* and also in the *Peri strategikes* suggest period relevance; and 3) the Nydam bog deposits which appear to include captured Roman military equipment show that the third-century Romans used at least two different spear lengths with shaft lengths of eight feet (ca. 2.4m) and eleven and a half feet (ca. 3.5m) and when one takes into account the contraction of wood over time and the addition of the spear point to the length, it is probable that the actual lengths had been close to the equivalent Roman standard lengths which were ca. 2.5 to 3.74m.[40] The third-century usage of the two different spear lengths imply that the Romans varied the length of their spears even during that period and possibly also before, with the further implication that the Romans are likely to have continued to use spears of differing lengths throughout their history, giving the commanders the option to vary the armament within the ranks of the phalanx as stated by the *Peri strategikes* and *Excerpts of Polyaenus*. It is unlikely to be a coincidence that we also find different spear lengths in later military treatises dating from the tenth century. The Roman generals appear to have had the option of varying the weaponry according to their own preferences. This, however, is still speculation on my part.

Maurice noted that in battle the entire line moved at one signal only up to the moment of contact with the enemy. After that it was not necessary for the entire line to move at one signal, because the terrain and the situation did not always permit this, not to mention the fact that the size and length of the battle line could prevent this. Furthermore, the movements of the enemy were not necessarily uniform. It was possible that a deeper enemy unit was able to push a shallower Roman unit back, with the result that the other Roman units were required to assist it. Similarly, if the enemy was able to surround one Roman unit, the other Roman units could form a double front to assist it. It was because of this that the phalanx was divided into *mere* and *tagmata*. These were not to wait for orders but were required to be able to operate independently on their own initiative. The following analysis of the orders in the *Strategikon* gives us some indication of how the separate units were expected to operate on their own.

Dividing or thinning the files

If the commander wanted to divide the sixteen-man deep phalanx, he first stated the direction *Ad conto* ('Spearwards') or *Ad scuto* ('shieldwards') and then gave the command 'By eights march out!' ('*Ad octo, exi*'), after which every other man stepped out of line, which obviously reduced the depth of the files while widening the front. If the commander wanted to reduce the depth more while widening the formation, he first gave the direction and then the command 'March out!' ('*Exi!*'), so that the depth was reduced to four men.[41] The principal aim of the dividing of the line was to extend the frontage of the line. This had two possible purposes: 1) it could be necessary to make the width of the phalanx equal to the width of the enemy line so that the enemy would not outflank the Romans; and 2) the width could be increased to outflank the enemy on one or both sides – in fact the *Excerpts of Polyaenus* (17.2) included the very same manoeuvre for the double-outflanking of the enemy array. The *Strategikon* wisely warned against thinning the line too much to ensure that the enemy would not be able to push through. The second unmentioned consequence of this manoeuvre was that it could also be used in the adoption of close order (see the diagram opposite), because it resulted in a formation denser in width. After this it was easy to form close order just by having the ranks move towards the front. Maurice does not mention this, but it is obvious that the experienced officers knew this method of tightening the formation.

DIVIDING/THINNING OF THE PHALANX AND CLOSE ORDER
(in the example shown with three files of 16 ranks)

The thinning/dividing of the 16 ranks deep phalanx when the phalanx was in open order (*araiosis*) and its use for the assumption of the close order formation (*pyknosis*) for combat.

The dividing could also be performed towards the left with the command *"Ad scuto, ad octo, exi!"*

If the commander wanted to divide the eight ranks deep phalanx which was already in the close order, then the unit had to be opened up in width first with the command "*Largia ad ambas partes*!" This was then followed up with the command (if towards right) "*Ad conto exi*!".

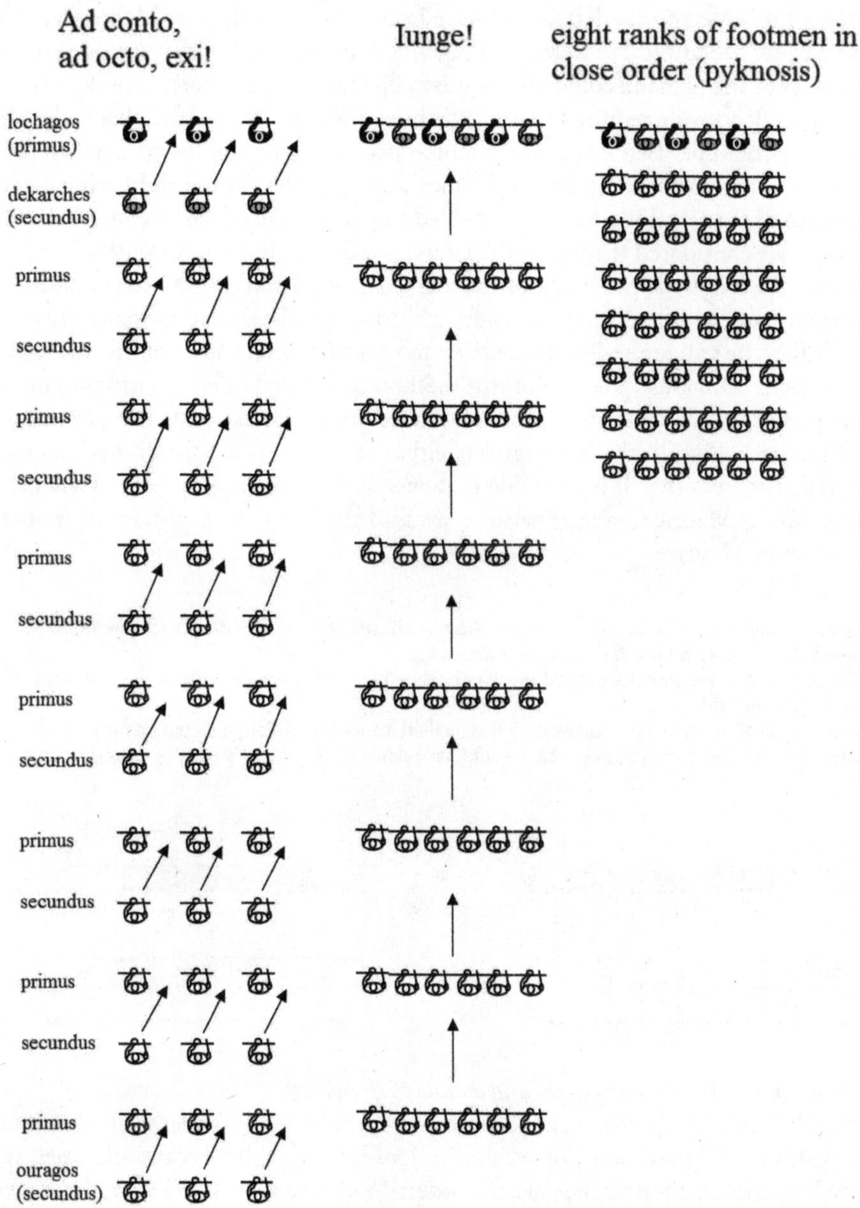

Extending the line and the outflanking

The commander was expected to adjust the width of his phalanx either to outflank the enemy or to avoid being outflanked by them, or to obtain more favourable ground, or to pass a defile. This was done by extending the line to the left and/or right. The extension of the line to the right was done by first commanding 'Spearward face!' ('*Ad conto clina!*') so that each soldier remained in place and turned to the right. The men were then marched in column formation to the designated place with the command to 'Move/March!' ('*Move!*'). Once they reached the correct place, the commander gave the order 'Return!' ('*Redi!*') and the soldiers resumed their original facing/front. The same set of orders to the left would have been: 'Shieldward, face! March!'. ('*Ad scuto clina, move!*') followed by 'Return' ('*Redi!*').[42] The use of this sideways movement of the units of the phalanx could also require adjustments elsewhere. It is clear that the Roman officers were required to use their own initiative in combat. This widening of the front could be used either for double-outflanking or single-outflanking. In the latter case, the preferred side to outflank the enemy was on the right wing, because that side of the attacking formation lacked the protection of the shield.

Vegetius considered the use of the lateral phalanx formation for outflanking risky, because the length of the line and the uneven ground endangered the cohesion and continuity of the line during the approach. Maurice clearly did not share this view. Regardless, he still reminded the *strategos* to ensure the continuity of the line right up to contact.[43] Obviously, the use of this method demanded at least parity in numbers. The period sources do not give us any instances of its use in battle for the reign of Maurice, which in this case reflects either the lack of opportunity for its use or imperfect sources (e.g. it is probable that versions of this array were used against the Slavs during Maurice's reign) because we find this very same system in use before the reign of Maurice.

The extending of the line in column formation to outflank the enemy either with one flank (*hyperkerasis*) or with both flanks (*hyperfalaggesis*)

- The extending of the formation could involve the thinning of the line or the use of the reserves to fill up the resulting intervals.
- It is probable that when the Romans used this method for the outflanking that the infantry attack was usually preceded by cavalry attack which would have removed the enemy's cavalry wing.

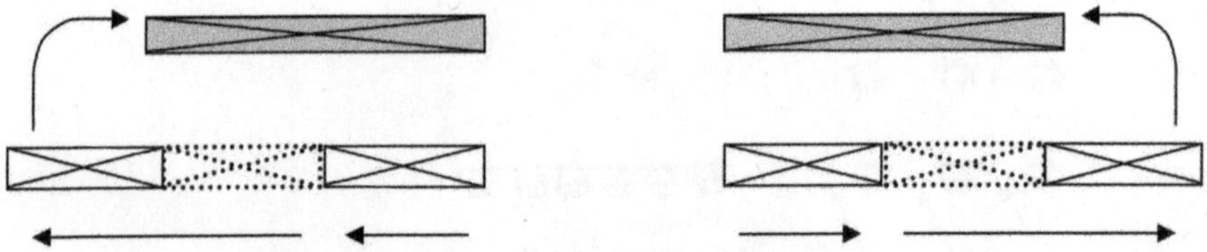

Extending the line on both flanks and double outflanking

If the phalanx had already assumed close order and the commander wanted to extend the width of the battle line, or wanted to loosen it in width because the men were crowding upon each other, he gave the order: 'Widen to both sides!' ('*Largia ad ambas partes!*'). This manoeuvre could be executed while marching or while at the halt, and could be done by a single *meros* or by the whole line. This order was actually necessary in circumstances in which the men were crowding upon each other, which is the natural biological reaction to humans when frightened. The intervals between the files of soldiers had to be sufficient for the men to be able to fight with swords. In other words, the minimum width for the infantry file was the *foulkon* formation (ca. 62.46cm). As already noted, the other use of this manoeuvre was to outflank the enemy on both flanks. This was obviously practicable only when the Romans outnumbered their enemy, because it was not recommended to thin out the line too much or to overextend it. It was also impossible to perform this manoeuvre if the enemy was already close, and it was because of this that Aelian and Arrian had recommended the use of light infantry and cavalry for extending the front because this did not disturb the phalanx. Maurice's solution to the same problem was obviously the use of the reserves or the posting of light infantry on the flanks if the enemy was already close by.[44]

The wheeling and its uses

According to Maurice when the commander wanted to wheel the battle formation towards the right or the left, for example because there was some emergency arising from that direction, the commands to this were 'Wheel to the right!' ('*Depone dextra!*') or 'Wheel to the left!' ('*Depone senestra!*').[45] Each of the *tagmata* performed the wheel one *tagma* at a time, so that the entire battle line assumed the new facing. The *tagmata* clearly had small intervals between them. It should be noted that even if in military theory wheels were always done as 90-degree quarter wheels, in practice the degree of the wheel depended on the need. Obviously, in actual pitched battles the reserves, caltrops and light infantry could also be used to counter the encirclement attempts by the cavalry.[46]

The use of a wheel was an emergency measure in the *Strategikon*, but it had also other practical uses, both in combat and before. The wheel was necessary for the columns to become lines and vice versa, and was required when one wanted to form the hollow square/oblong, *epikampios* arrays, hollow wedge, and wedge formations. Wheeling was also required when one outflanked the enemy and when one opened the formation (*antistomos difalaggia*) to allow enemy cavalry, elephants and chariots to pass through to a place between the two phalanxes. The *antistomos difalaggia* (the opening of the formation) could also be used against a deep infantry formation like a column or wedge so that those who opened the array could then attack the enemy

Wheeling the *tagma* to the left

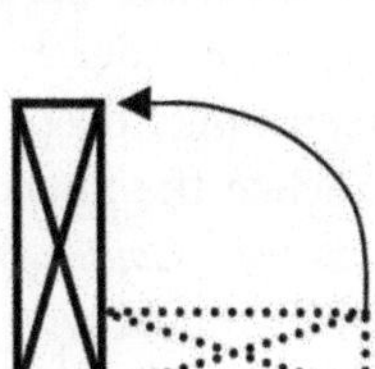

Wheeling the *tagma* to the right

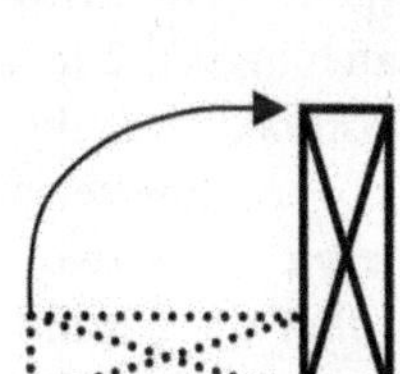

from the flanks and behind. It is because of this that we find in the *Excerpts of Polyaenus* (32.3) the incident in which Agesilaos 'opened the intervals' of the Spartan phalanx (means the wheeling of the units inwards into the *antistomos difallagia* formation to widen the intervals) so that the Thebans could run through it, after which the Spartans attacked them from behind. This instruction was still relevant during the Late-Roman and 'Byzantine-Roman' periods. In short, even if Maurice fails to name these uses for the wheel, it is clear that these uses were well-known to all period military commanders and one may assume that wheels were also used for these purposes whenever the commanders so decided.

Therefore, it is very likely that the Romans continued to employ all of the above battle formations and unit tactics, even when not specifically mentioned by Maurice in the *Strategikon*. The case for the use of the *peplegmene/serra* ('saw') formation is less clear, because Maurice does not specifically mention it even if he mentions the infantry wedge formation, but one may assume that officers who also read other military treatises, as expected by Maurice, retained knowledge of this tactic too – it was easy enough to form just by posting infantry wedges (which Maurice names) side-by-side. The saw-array was basically the infantry equivalent of the cavalry tactic of using *koursores* as skirmishers.

In short, it is probable that the Romans used the saw-array during the last years of the Late-Roman era if the situation required and the commander was well-educated, while it is also equally clear that Maurice considered this tactic unnecessary. The use of light infantry deployed in irregular *drouggoi* formations in front of the lateral phalanx the *Strategikon's* close relative of the saw formation. Maurice mentions the use of light-infantry *drouggoi* in front of the lateral phalanx only in the context of fighting in difficult terrain, which means that he once again thought it advisable to use the lateral phalanx in the regular way against enemy infantry – and his view was the official combat doctrine after his reign. The individual commanders could obviously decide to do otherwise, but if they did so we have no evidence for this in the sources. Therefore, it is safest to assume that, excepting some rare cases, combat methods were significantly streamlined and simplified by Maurice so that anything that was in any way complicated was discarded as unnecessary.

Facing attack from behind

Just as ancient Greek combat doctrine, Roman doctrine took into account the possibility that the scouts had failed to detect the enemy movements correctly so that the enemy approached from the rear. It is therefore not surprising to find the same alternative also in the *Strategikon*. The enemy could either approach only from behind or it could approach the Romans from both directions, front and rear, simultaneously.

If the enemy approached only from behind, Maurice instructed the soldiers to use countermarching if the formation had not yet assumed close order. In the example, Maurice assumed that the Roman phalanx was sixteen ranks deep, but one may assume that the same set of commands was true also when the phalanx was four-, eight-, or even thirty-two deep. The command for this was '*Countermarch!*' ('*Muta locum!*') so that the battle formation stood in place while the *lochagoi* (file leaders) turned and passed through the depth of the *acies* to the rear while the rest of the files followed them. In other words, the phalanx remained in place while it changed its facing towards the rear.

As already noted, Maurice simplified the practice by abandoning the Macedonian- and Laconian-versions of the counter-march and the separate counter-marches performed by file or by rank. The benefit of this was that in stressful situations there was only the one counter-march manoeuvre. If the phalanx had already assumed the *pyknosis* order (or the *foulkon*) and there was no time to open the formation with the command '*Largia ad ambas partes!*', the commander gave the order 'About turn"' ('*Transforma!*') and each soldier stayed in his own position, but turned around towards the rear so that the former file-closers (*ouragoi*) became the new *lochagoi*. It is probable that the about-turn was always performed towards one direction because the command leaves out the direction (left or right), but it should still be kept in mind that even if it was performed towards different directions by different soldiers it did not really disturb the formation because the about turn was performed by individuals staying in their own position.

If the enemy approached simultaneously from both front and rear, or the wagon train was not following, the Romans responded by forming a double phalanx (*amfistomos difalaggia*) if there was enough time for this. If the phalanx was sixteen deep (meaning that the array was already in close order) and the enemy was already very close, then the command was 'Divide in the middle, Form double phalanx!' ('*Medii partitis, Ad difallangiam!*'). In this case, the first eight men halted, the other eight men faced about and moved 300 paces towards the rear to form the double phalanx. If the phalanx was four- or eight-deep (means that the unit order was probably not yet fully closed), the command was '*Primi* halt, *secundi* into double phalanx march out!' ('*Primi state, secundi ad difallangiam exite!*'). In this case, the *secundi* led by the *dekarchai* made an about turn and marched through the intervals to the rear to form the second phalanx. If the enemy force approaching from the rear was stronger, then the command was '*Secundi state, primi ad difallangiam exite!*'. The second phalanx was always formed 300 paces behind the first so that the enemy arrows would not disturb the other phalanx. The command to revert these orders was 'Return!' ('*Reverte!*').

As can be seen, Maurice's double phalanx was primarily the version in which both halves consisted of the same units that had been divided. Maurice fails to specify what happened to the light-armed behind the phalanx and to the cavalry and infantry reserves, but it is obvious that these were either marched to the flanks or they marched through the intervals of the rear phalanx to assume a place between the two phalanxes. The light-armed were obviously divided between the two phalanxes, so that one half was behind the front phalanx and the other half behind the rear phalanx. When the reserves, possibly together with some of the light-armed men, were sent to the flanks, the end result was one of the versions of the hollow oblong. When posted between the phalanxes on the flanks, the light-armed troops and the infantry and cavalry reserves were in their turn protected by the presence of the phalanxes, because it was possible to retreat into the interval between the phalanxes, where any enemy intrusion would be met with crossfire from both sides. In fact, Syrianus Magister specifically states that the enemy was unlikely to advance into the open space because of this threat. If the cavalry wings also withdrew into the interval and dismounted, then the array was de facto a hollow oblong.[47]

If the enemy approached simultaneously from both the front and rear and there was not enough time to form the double phalanx, the commander was to use the double-front (*amfistomos falagx*). The command for this was the '*Undique servate!*' ['All directions face!']. The front half of the phalanx continued to face the front while the rear half turned about and faced the rear. The men in the middle remained in place and lifted their shields to form a shield-roof to protect their heads. In practice. the flanks could also face outwards, so that the entire formation faced all directions – like its Latin counterpart *orbis* implies.[48] This was an emergency measure when the

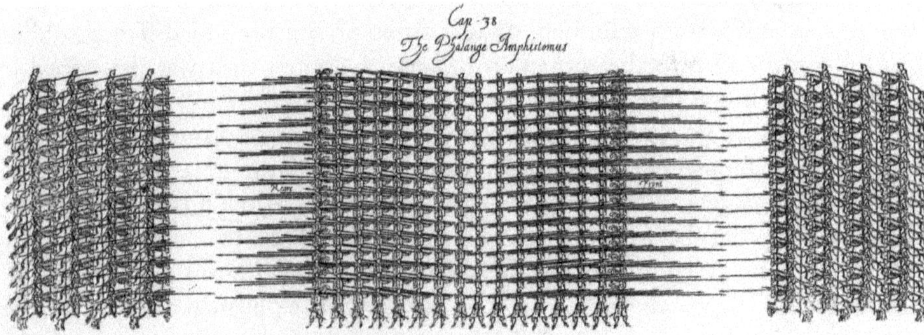

above: 17th century interpretation of the amfistomos phalanx

falagx amfistomos
(*Codex Burnley*)

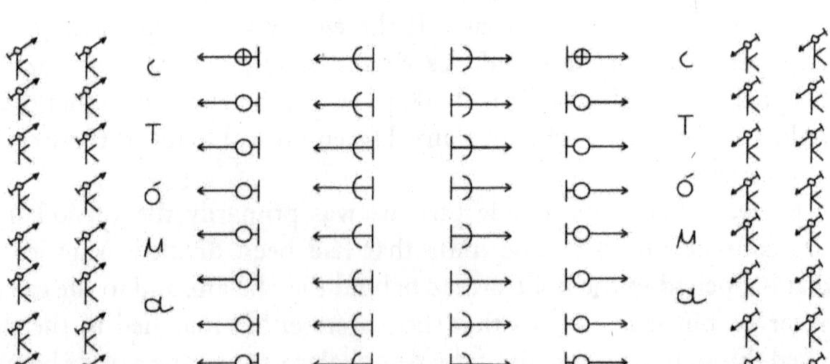

Amfistomos falagx

The double-fronted phalanx was used against outflanking cavalry in the text of Aelian (Matthew ed. Aelian 38; *Byzantine interpolation*. Devine ed.38.1-39.5, Dain ed. C1-D5), but obviously it was also used against outflanking infantry.

Note that this version of the *amfistomos falagx* has the *psiloi* (in this image the foot archers) forming up the inner ranks. It is very likely that in practice the foot archers entered the *amfistomos falagx* at the same time as it was formed up so that the rear half of the *skoutatoi* / *hoplitai* advanced a short distance to the rear to make room for the *psiloi* in the middle.

enemy attacked the Romans suddenly from both the front and back. It is probable that in such situations the light infantry either fought a separate battle outside the phalanx as best they could, or that the *amfistomos falagx* tried to admit them inside their formation by having the rear portion move further in order to create space between the two halves.

The sources suggest that the ability of Roman infantry to face threats from the rear or simultaneously from the front and rear varied greatly according to the quality of the force, so there are examples like the Battle of Mursa in 351 (see Appendix 2) and the Battle of Scalae Veteres/Cellas Vatari in 537 that demonstrate a superb ability to withstand outflanking attacks, while there are examples like the Battle of Thannuris where the infantry simply collapsed after the cavalry had fled. This was a reflection of the quality of the infantry forces, which as already discussed varied greatly during so long a period as we are dealing with here. Regardless, one can say as a general observation that the quality of Roman infantry forces remained relatively high throughout, and periods in which the quality of the infantry was such that it was unable to face threats from several directions simultaneously were not many.[49]

As already noted, according to the *Strategikon*, the *amfistomos falagx* could also be used to bring help to other sections of the battle line that had been surrounded by the enemy. The side of the formation which faced the enemy undoubtedly attacked it. This undoubtedly also means that in some cases the unit was further divided into a double phalanx, whose rear half was marched to help the threatened section.[50] The image on the previous page shows how the *Tactics of Aelian*, printed in London in 1616, interpreted the shape of the *amfistomos falagx*. Note that the equipment depicted dates from the seventeenth century. The other image is drawn after *Codex Burnley*.

In addition, there appear to be some significant tactical omissions in the *Strategikon* which we find in Vegetius, *Hermeneia/Definitiones* and *Peri strategikes*: namely, Maurice does not mention the use of the crescent (*menoeides*) in the context of infantry use for outflanking, the use of the oblique formation (*loxê*) for outflanking, the sending of the right wing forward so that cavalry and light-infantry spearheaded the outflanking attack, the sending of both wings forward to outflank the enemy, the convex (*kurtê/kyrtê*), the hollow wedge (*koilembolos*), the *antistomos difalaggia*, and the sending of light-armed forces in front of the phalanx when it was deployed on open, level terrain.[51] Most of these questions have already been analysed above, but a few additional comments as a summary of the most important points are still in order.

One may assume that the omission of the crescent in the context of infantry warfare was a purposeful and sound decision by Maurice because it did not really add anything to the standard use of wheeling the flanks inwards to outflank the enemy.

The omission of the oblique formation is not surprising because there is no evidence that this was used in practice during the Late Roman period, but this should not surprise us because the use of the oblique formation has been rare throughout history. This, however, does not mean that the Late Romans could not have used the array, because the training scheme certainly would have enabled this (one half of the large wedge is an oblique formation), on top of which our sources are so defective that it is safer to assume that the use of the oblique formation remained a viable alternative at least until the reign of Maurice and possibly beyond.

The same is true for the hollow wedge formation, both as a unit manoeuvre (the use of the wedge implies that its counter tactic was also known) and as a battle formation (it was certainly used in this capacity in ambushes from two sides when circumstances allowed). Fuller discussion of the hollow wedge (*koilembolos*) and *antistomos difalaggia* as battle formations follow later.

As regards sending the flanks in advance of the centre, on a closer look Maurice actually accepted this as he allowed the *mere* to operate independently of each other. It is therefore possible that when the *mere* came close to the enemy the flanks they advanced faster than the centre to outflank the enemy. In short, it is clear that the *Strategikon* allowed for sending one flank or both flanks in advance of the centre in the same manner as we find in Vegetius. This was certainly the case with the use of the so-called *epikampios emprosthia* array (forward-angled formation), which Maurice does not specifically mention in the *Strategikon,* but which was also used during his reign. It is very probable that Maurice's '*taxis alle*' (see below) is actually one of the battle formations which could be used to form the *epikampios emprosthia.* This is by no means surprising, because Maurice considered the *epikampios opisthia* (rearward-angled formation) as the other basic formation besides the *plagia falagx* (lateral phalanx).

Just like the omission of the crescent, the omission of the convex (*kyrte*) is very understandable. The convex did not really add anything of great value in comparison with the wedge, or the *epikampios opisthia*, or the use of the mixed formation, or just by having the different *mere* advance at slightly different speeds in the last stages of the advance.

The omission of the use of light-infantry in front of the phalanx in regular battle may be purposeful because Maurice did not see this to be useful in regular battle, or it may just be an unintentional omission because Maurice includes the use of light-infantry in front of the phalanx when the army was deployed in difficult terrain. However, I would suggest that the omission is intentional, as Maurice clearly envisaged the phalanx to operate differently in open terrain – there, its principal strength was the use of the *skoutatoi* phalanxes against the enemy and not the use of the light infantry.

The lateral phalanx in difficult terrain[52]

Roman combat doctrine took into account the demands of fighting in wooded, rough, mountainous or otherwise difficult terrain. This type of terrain required the use of the marching formation for the combat because it was not always possible to spread the array as a lateral phalanx. The best description of the fighting tactics employed by the Romans in difficult terrain comes from the pen of Maurice and the following account pays particular attention to this because Maurice based his treatise on earlier, no longer extant, military treatises and drill manuals. Maurice describes how the Romans fought when they were forced to deploy their phalanxes/divisions (*mere*) in column formation and it is clear that such tactics had always been used by the Romans when the terrain required the use of the column formation. In such cases, the infantry columns were basically forced to be on the defensive while most of the fighting was performed by light-infantry *psiloi* or cavalry, depending on location. When possible the Romans used the *epikampios opisthia* with a rear guard or hollow oblong/square array in wooded, hilly or mountainous terrain, while it was also possible for the Romans to place two marching columns/phalanxes back to back as a double front or double phalanx (*amfistomos falagx* or *difalaggia*), or to place the columns one after another as a single long column of phalanxes (*falagx orthia*) when terrain did not allow wider formation. It was obviously also possible to adopt the double-fronted (*amfistomos falagx*) array, even when marching as a single column, by having the flanks face out.

Fighting while marching in column was principally used against those enemies that inhabited areas that were considered difficult to traverse. During the early part of the Late Roman period, these foes consisted of such diverse enemies as the various Germanic tribes, the Picts, the tribes of the Caucasus Range, the Isaurians and the Moors who inhabited mountains. The Georgians and Armenians also inhabited mountainous regions, but both of these nations usually preferred to fight in terrain that suited their excellent cavalry forces, with the result that difficult terrain played a lesser role in the wars against them. The same was obviously true of the Sasanians, who usually exploited the difficult terrain only as a place to post their mounted archers for combat. After the great migrations brought the Germanic tribes into Roman terrain, there were instances in which these exploited the woods and other difficult terrain, but this was not the typical mode of fighting for them. During the Late-Roman period the Germanic tribes usually preferred to fight in open, unobstructed terrain, and resorted to the use of swamps, forests, field fortifications (meant usually an abatis of felled trees called *concides*), and mountains only when fighting defensive wars. During the sixth- and seventh-centuries the principal foes demanding the use of tactics developed for fighting in difficult terrain were the Slavs and Antes.

The *Strategikon* provides us with the necessary details of how the Romans fought in difficult terrain that the other sources fail to deliver, but we can be certain that these very same methods had always been used when traversing difficult terrain, the sole possible difference being the demand in the *Strategikon* for the shield-bearing (*skoutatoi*) infantry to use larger shields in such terrain. The *De rebus bellicis* has the exact opposite instruction. Its author wanted the soldiers to carry the smaller shield in such conditions.

Fighting in difficult terrain (wooded areas, rough or hilly terrain, narrow passes, mountain ranges) required a lightly-equipped army without wagons and which did not have many horsemen. Thanks to the demanding circumstances, the men were required to carry only the most essential supplies and equipment. The *Strategikon* expected that the 'heavy infantry' shield-bearers (*skoutatoi, hoplitai*) did not carry any heavy armament such as helmets, long cavalry spears, and mail coats with them. In such cases, the *Strategikon* instructed the 'heavy infantry' *skoutatoi* (shield-bearers) to carry only 'larger' shields, swords, and spears shorter than the cavalry spear (i.e. javelins). The same practices had already been followed during the Republican and early-imperial period, but the instructions in the *De rebus bellicis* (9–11, 15, 19.1) show that

Diagrams in the *Strategikon* 12.B.22.101-12

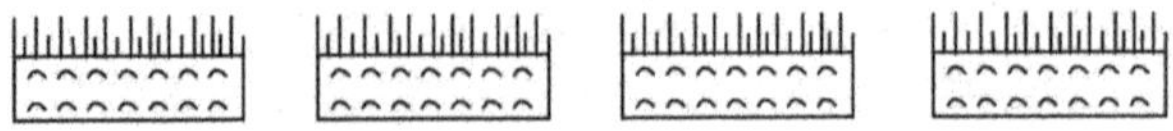

The impressive looking lateral phalanx used in open terrain.

Left: *taxis orthia* (column) *monofalaggia* was used in narrow passes or when there was only one road. If there was a need to use this array, the combat doctrine required the sending of the light-armed in advance to occupy the heights on both sides of the narrow passes.
Right: *amfistomos orthia falagx / orthia monofallagia* facing all directions (my reconstruction on the basis of the diagrams in the *Strategikon*). There is at least one instance in which the Romans purposefully entered enemy controlled pass. This took place in 373 when the Romans under Theodosius purposefully entered a pass in the *amfistomos orthia falagx/orbis/rotundus* order and defeated the Moors opposing them. This was a very risky move. Note, however, that the Romans could also exploit valleys and passes place their army in a valley purposefully if the enemy outnumbered them, the idea being to negate the enemy's numerical advantage (see e.g. Syvänne, *MHLR 3*, 78).

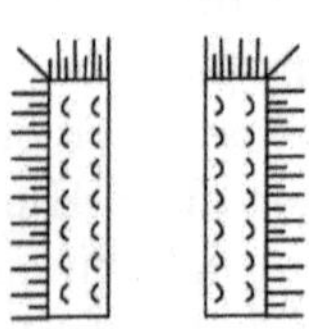

Taxis orthia difalaggia (double column phalanx) was considered an essential formation for use in thickly wooded terrain because it provided a place of refuge and rallying point for the *psiloi* (light infantry) and *kaballarioi* (cavalry) while also protecting the *touldos* (baggage train) in the middle.

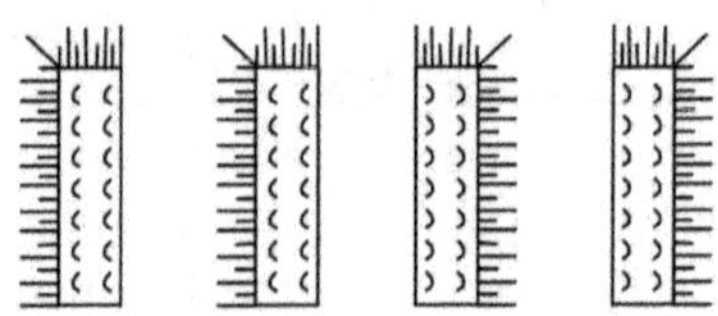

Taxis orthia tetrafalaggia (four column phalanxes) was considered the standard array to be adopted when the Romans had very large force of infantry in a situation in which the terrain allowed its use. This formation enabled the Romans to advance faster than in the double column and eased the wheeling of the formation into lateral phalanx.

Roman military thinkers did not have a consensus view of how to equip the soldiers for difficult terrain. *De rebus bellicis* recommended the use of the *thoracomachus* (i.e. *subarmalis*, padded under-armour), small round shields (*clipeocentrus, parma, modicus clipeus*), boots, iron greaves, helmet, *gladius*-sword, *lanceae* (javelins, throwable lances) and apparently also *plumbata* darts (*DRB* 10–11) in icy regions, because the woods were impenetrable if the soldiers used large shields. The recommended reform means that there were others who thought otherwise. The *Strategikon* (12.B.20.8–10) agreed partially with the *De rebus bellicis* in the matter of the need to use small shields in wooded terrain, because Maurice instructed the light-infantry *psiloi* to be equipped with smaller and lighter shields, bows, short spears (*Bêruttai*, i.e. pl. *veruta*, sing. *ver(r)utum*, 15 inch/0.37cm iron shank with a 3.5 feet/1.05m long shaft), short Moorish javelins, and some metal darts (*martzobarbouloi*, i.e. pl. *martiobarbuli*, sing. *martiobarbulus*, 'barbs of Mars', also known as *mattiobarbulus* and *plumbata*) for difficult terrain.

In short, it is clear that both Maurice and the author of the *De rebus bellicis* agreed about the need for the same type of equipment in wooded terrain, but that Maurice did not share the same view about the equipment worn by the so-called shield-bearers. Maurice wanted them to use larger shields without helmets and greaves, while still being ready for close-order fighting in the open. One may therefore draw the conclusion that the equipment worn by the soldiers in difficult terrain varied from one time-period and place to another during the Late Roman period, the only unifying principle being that most of the military theorists – a view most likely also shared by most practitioners – preferred lighter gear in wooded terrain. Notably, the *Strategikon* demanded that all footmen were to possess axes, which were carried by the pack animals. This was obviously necessary in wooded terrain and one may expect that the same instruction had been in effect as long as the Romans had used fortified marching camps.

Depending on the size of the army, the *Strategikon* expected that the lightly-equipped shield-bearer infantry *(skoutatoi)* was deployed in two, three or four marching columns/phalanxes (*mere*), each two, three, or four ranks deep. In extremely narrow places it was allowed to use only one marching column. The use of marching columns was obviously age-old, and was used by the Romans well before the so-called Late Roman era. The marching columns were placed about a stone's throw apart from each other. This formation was used until the Romans reached open terrain. When this happened the marching columns made a quarter turn and formed the lateral phalanx formation.

If there was cavalry present and/or a baggage train (consisting of mules and horses), these were posted so that the baggage train was behind the phalanxes and the cavalry behind the baggage train. It is of note that Maurice did not require the cavalry to be present[53] in an army that proceeded into difficult terrain, even if he clearly expected this to be the norm on the basis of the rest of the instructions – the reader should just remove all references to cavalry in the text and accompanying diagrams to see how Maurice envisaged the pure infantry army to look. The Romans posted a separate rear guard to protect the extreme rear, which consisted of a detachment of shield-bearing infantry (*skoutatoi*) and light infantry (*psiloi*). They also posted a separate protective screen, a vanguard, about a mile (ca. 1.48km) ahead of the main body. The vanguard consisted usually of light infantry (*psiloi*) and a small unit of cavalry. When the terrain was relatively trouble free, the cavalry proceeded in front of the *psiloi*, but in truly difficult and densely forested terrain the light infantry advanced in front. The flanks of the formation were protected by *drouggoi* (irregular groupings) of *psiloi*, who were supported by small numbers of cavalry when needed. It is uncertain if the small numbers of cavalry reinforcements always accompanied the *drouggoi* of *psiloi*, or were dispatched only when needed from the cavalry units that were posted behind the baggage train. The latter is probably likelier. The *psiloi* in question consisted of *drouggoi* (irregular groups) of four or five men: three or four of these were to be armed with javelins and shields, so that there was only one archer in such a group. Most of the archers were kept with the phalanxes because they were considered useful in rough, steep, narrow, and open terrain, whereas the javelin throwers were considered particularly useful in thickly wooded areas – as noted above this view was shared by the *De rebus bellicis*. The *drouggoi* were deployed for their own safety so close to the main body that they could still hear its trumpets and bugles, and the *drouggoi* were deployed one after the other so that they could protect each other's back. The light infantry together with their supporting units of cavalry were required to be ready for possible enemy stratagems (e.g. trees sawed partway through) and ambushes in difficult terrain.

The above describes the standard version of fighting in difficult terrain (*Strategikon* 12.B.20), but on the basis of the *Strategikon* (9.4, esp. 9.4.14–6) it is clear that it was sometimes necessary for the commanders to send forward a detachment of either infantry (*pezoi*) or cavalry (*kaballarioi*) to seize wooded terrain and/or narrow locations in advance of the main force. This means that the advance guard could also include shield-bearing 'heavy infantry' (*skoutatoi, hoplitai*), either in light gear meant for difficult terrain or even in their regular gear. However, the inclusion of the *skoutatoi* among the vanguard can also be found in the basic instructions of the *Strategikon* (12.B.20.54ff.) concerning combat in difficult terrain. The *Strategikon* instructed the general: 1) to deploy the four divisions so that the divisions were manoeuvred to face the enemy from whichever direction it came; 2) to use the phalanxes in regular close order when this was possible while the *psiloi* and cavalry encircled the enemy; 3) to deploy the battle lines (*akiai*, i.e. *acies* rather than files) in greater depth and width (i.e. the files and ranks had greater intervals in between) in more forbidding terrain so that these could pass through (note that Modares deployed his infantry phalanx precisely in loose order when he attacked downhill in 379);[54] and 4) if the terrain was considered too difficult even for this, then the main body halted while the *psiloi* with the support of a few *skoutatoi* (shield-bearers) and *kaballarioi* (cavalry) were sent against the enemy. I included the *skoutatoi* among the vanguards in the diagrams of the *Military History of Late Rome* vols. 6–8 for the sake of completeness, but I have in the accompanying diagrams on pp.325–6 left those out because the inclusion of the shield-bearers in the vanguard was restricted to special instances. Notably, if there was a need, the men attached to the vanguard were also required to cut down the trees and clear the route. The vanguards could therefore also include engineers and other specialists. Furthermore, it should be kept in mind that the vanguards always included officers/engineers whose duty it was to locate and measure the marching camp for the following night, not to mention the fact that the general was always required to post scouts and spies at greater distances away from the marching formation. These are not included in the accompanying diagrams.

The main striking force of this marching array therefore consisted of the *drouggoi* of *psiloi* and the small numbers of cavalry supporting them that were deployed outside and around the phalanx structure. In combat situations, when the leading *drouggos* encountered enemy resistance, the following *drouggoi* (throngs) were expected to provide help while attacking the enemy's rear if possible. The main objective of the *drouggoi* was always to seize the higher ground and get above the enemy. The men of these *drouggoi* were clearly the crème-de-la-crème of the Roman infantry. They were expected to be able to operate independently of each other without any outside supervision and without the psychological and physical support given by the close vicinity of fellow soldiers in rank-and-file formation.

marching formations in difficult terrain

The diagrams depicting two or four divisions (*mere*) as marching columns are author's reconstructions. The symbols are based on those used by the *Strategikon* and later military treatises.

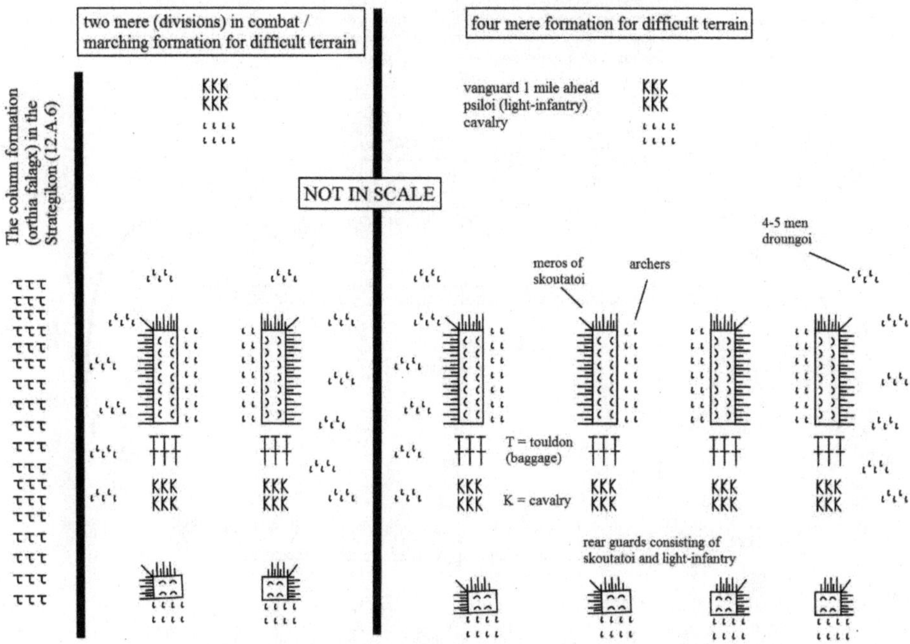

The *Excerpts of Polyaenus* (19.2), which may date from this era, includes an interesting variant borrowed from the original text of Polyaenus (3.9.48). In the example given, the famous Athenian condottiere Iphicrates/Ifikrates (*Ifikratês*) had looted, with his men, a great amount of booty in Laconia and was retreating from enemy territory. The local garrison commander gave chase. Ifikrates responded by posting the *hoplitai* in front of the booty, while he deployed the light-armed (*koufoi*, presumably javelineers, can also mean the peltast javelineers) and light-infantry (*psiloi*, presumably archers) here and there in small groups (i.e. as *drouggoi/globi* even if the *Excerpts of Polyaenus* fail to use this term) against the pursuing Laconians. When the light-infantry '*drouggoi*' then attacked the Laconians both in front and flank so that the Laconians were facing attacks from the light-armed, Ifikrates encircled them, presumably with his hoplites, from both sides and attacked them from behind and killed them all. These details suggest that the Romans had copied the use of the small irregular groups of light-armed from the Greeks and also show how effective these small groups could be against enemy hoplites in such terrain that enabled them to skirmish freely. This instruction has clearly been chosen from the original text of Polyaenus precisely because it fit perfectly the tactics used during the Late-Roman and 'Byzantine-Roman' periods.

The Romans also faced guerrilla warfare by small bands of guerrillas/bandits in difficult terrain. In the usual circumstances it was the duty of the civilian police forces

The likely deployment pattern in the open terrain for four divisions

The light infantry *drouggoi* together with cavalry seek to outflank the enemy while the vanguard prevents the enemy from making a surprise attack while the four phalanxes and cavalry deploy for combat. While this happened the vanguard retreated back to the main formation. The cavalry that had followed behind the baggage train (*touldon* = T) was probably divided into four detachments so that the cavalry units on the flanks were sent to support the outflanking *drouggoi* while the remaining two in the middle advanced to the wings of the main battle formation to protect its flanks. It is probable that the baggage train of each *meros* was posted behind their respective *meros* while the four infantry detachments of the rear guard formed up in such a manner that they could serve as reserves for the wings and centre. If each *meros* was about 6,000 strong, then the width of each of these was 1,500 men if these were deployed four ranks deep (in this example three ranks of *skoutatoi* and one rank of *toxotai*).

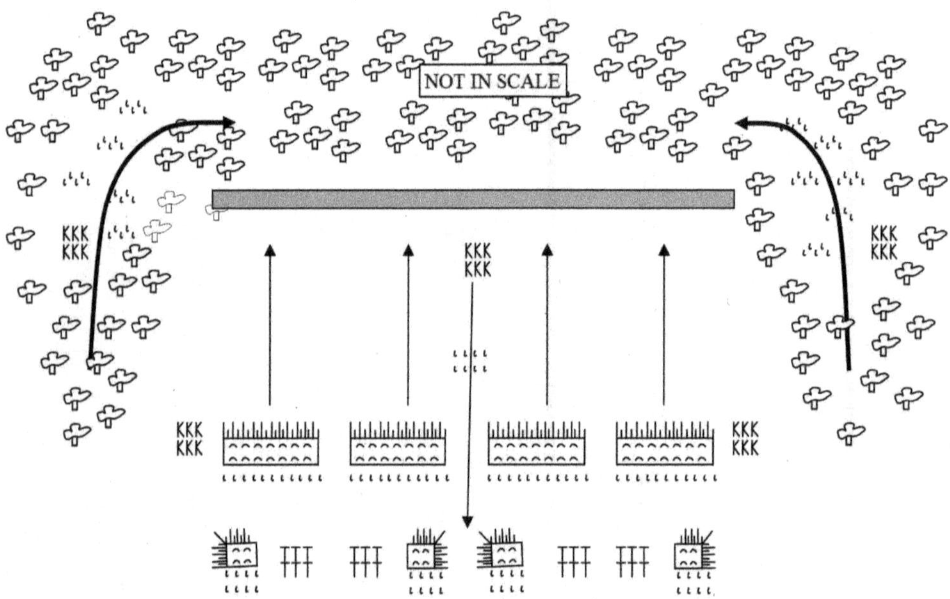

and paramilitary forces to deal with such problems, but when the barbarians invaded in strength, as happened for instance in the quoted example below, it required the intervention of the regular army, which was ill-suited to such warfare. The typical enemies exercising this type of guerrilla warfare against the Romans were the Germanic tribes across the Rhine and Danube and the tribes inhabiting wooded and/or mountainous/hilly areas (e.g. Isaurians, Tzani, Jews, Samaritans, Caucasians), and then in the sixth- and seventh-centuries the Slavs and Antes. In addition to this there was a special class of bandits within the Roman Empire who were called '*Bacaudae*'. These were native, freedom-loving bandits who had abandoned civilization and chosen to live in forests, woods and mountains. The *Bacaudae* consisted of local peasants, mountaineers, former soldiers, members of the lesser gentry, and younger sons of the nobility. As natives with local connections up to the very top of Roman society, it was very difficult to surprise and subdue them. This phenomenon lasted from the third- until the sixth-century, and some of these bandits actually became autonomous or independent nations, the best examples of this being the *Bacaudae* of the Alps, Armorica and Tarraconensis. The reason why these bandits were able to gain autonomy and temporary independence was the collapse of order in Gaul during the fifth century.

The typical Roman response to large scale banditry/guerrilla warfare was to use the regular forces in the conventional manner, so that these were posted on the level, open terrain in sufficient numbers so that the raiders were forced to seek a place of refuge among the forests, woods and mountains. This was often enough for both the Romans and the enemy to conclude a peace, but if the Romans had enough determination and resolve to face the problems that the difficult terrain caused, the Romans advanced into these places of refuge despite the heavy casualties that this caused. The latter option obviously brought a more permanent peace than the former, on top of which it usually gave the Romans the chance of adding the enemy forces into their own army.[55]

However, it was sometimes necessary or preferable to employ other bandits/robbers of barbarian origin against other barbarians because they were familiar with such tactics. This was the approach adopted by Julian in Gaul in 358, and it was also the approach the Romans found useful later, the best examples of this being the use of the Slavs and Antae against Slavs in the Balkans in the sixth century and the employment of the mercenary warlord/bandit Mundus and the Slavic commander Chilbudius.[56] The best counter measure against a bandit/robber was to employ another bandit/robber against him. This is the same method as we find the Americans using in the nineteenth century against the bank- and train-robbers, and which we find Putin using in Chechnya. The attached quote from Zosimus shows how the small scale operations in the woods and wilds were conducted in practice:

> Caesar [*means the Caesar Julian in 358*] by this time perceiving that the Barbarians dared not again engage him, but were intent on secret excursions, and rapine, by which they did great damage to the country, scarcely knew how to act, until at length he invented a stratagem to confound the Barbarians. There was a man of extraordinary stature, and of courage proportioned to his size. Being by nation a Barbarian, and accustomed to plunder with the others, he had thought proper to leave his own country and go into Gallia Celtica, which was subject to the Romans. While he was residing at Treves, which is the largest city of all the nations beyond the Alps, and saw the Barbarians from beyond the Rhine [*in this instance the invading barbarians were the Germanic tribe of the Quadi, but on the basis of Ammianus it is probable that this is a mistake for the Alamanni*] ravaging cities on this side of the river, and committing depredations everywhere without opposition, (which was before Julian was made Caesar), he resolved in himself to defend those towns. As he did not have authority and permission to do this, he at first went alone into the thickest part of the forest, and waited there until the Barbarians made their incursions. In the night, when they lay intoxicated and asleep, he fell on them and slew them in great numbers, bringing their heads and shewing them to the people of the town [*presumably Treves*]. This he practised continually to such an extent that he frightened the Barbarians, who though unable to guess what was happening, yet were well aware of the losses they had sustained, their army diminishing daily. When some other robbers joined this man and their number had increased to a considerable body, Charietto (which was the name of the man who first used this kind of ambuscade against the Barbarians) came to Caesar, and told him the whole story, which only a few had known before. Caesar was at this time

unable to restrain the nocturnal and clandestine incursions of the Barbarians, because they robbed in small, scattered parties, and when day appeared, not one of them was visible, all hiding themselves in the forests and subsisting on what they gained from robbery. Considering therefore the difficulty of subduing such an enemy, he determined to oppose these robbers, not with an army of soldiers, but with men of similar description.

For this reason, he sent Charietto and his band, adding to them many of the Salii [*means the federate allies the Salii/Salian Franks*], against the plundering Quadi, who though they lived on what they stole, yet were probably less expert in the art of robbing than these men who had studied it. In the day he guarded the open fields, and killed all that escaped his robbers. He did this for a long time, until the Quadi were reduced to such extremities, and to so small a number, that they and their general surrendered to Caesar, who had a taken great number of prisoners in the former excursions and engagements, and among the rest the son of their king, who was taken by Charietto. From this cause, when so lamentably petitioned for peace, Caesar demanded their chiefs as hostages, and required the king's son to be one of them; the general, or king, broke out into a most pathetic complaint, and declared with tears in his eyes that his son was one that had been lost… Caesar… shewed him his son who had been nobly entertained; but told him that he would retain the youth as a hostage as well as other of the chiefs whom he had in possession. He condescended, however, to make peace with them… Caesar, after he had thus settled affairs, added the Salii, the Quadi, and many of the inhabitants of Batavia to his legions, which we still retain in our service. [*Zosimus therefore states that there were still units of Salii, Quadi and others in the East Roman armed forces in the sixth century.*]

Zosimus 3.6.4–7.7, tr. by J. Davis (or anon?), 1814, 71–2 with my comments and some changes and corrections in Italics added inside parentheses.

The above quote also shows one of the principal means of unorthodox warfare as practiced by the Romans, namely the targeting of family members of high-ranking enemies for use as hostages. Charietto repeated the same tactic later and we find the Romans using it later, usually with great success. The capturing of hostages failed only once, when employed against the Lombards in the late-sixth century.[57]

9.5. Outflanking formations:[58] Double outflanking: *epikampios emprosthia* (forward-angled), wings advancing before centre (fourth and fifth formations of Vegetius, variants of the *epikampios emprosthia*), *koilembolos* (hollow wedge), *antistomos difalaggia* (inward facing double phalanx), *menoeides* (crescent), *hyperfalaggesis* (double-outflanking) Single outflanking: *hyperkerasis* (outflanking), a wing advancing before the rest of the phalanx, and *loxe* (oblique attack)

The true crescent (*menoeides*), the *koilembolos* (usually a unit formation used to counter a wedge, but it was also used as a battle formation in ambushes against a column formation), the *hyperfalaggesis* and *epikampios emprosthia* (forward-angled formation), the *antistomos difalaggia* (a unit formation used against cavalry, elephants or chariots, but which was also used as a battle formation in ambushes), and the wings advancing before centre (the fourth and fifth formations of Vegetius) formations were used to outflank enemy formations on both flanks. There were some significant differences between these. The question of ambushes as an outflanking formation will be discussed separately in a chapter devoted to ambushes and surprise attacks.

Firstly, it is clear that as grand tactical battle formations the hollow wedge (*koilembolos*) and inward-facing double phalanx (*antistomos difalaggia/difalangia*) were used only in ambushes mainly against enemy armies marching in column formation between the Roman wings, because both were otherwise too vulnerable to attacks against their flanks. Therefore, these are not analysed here as true pitched-battle formations.

Secondly, in the true crescent the line remained continuous and curved while in the *epikampios emprosthia* (together with the wings advancing first) and *hypefalaggesis*, the angle of the wings to the line was either 90 degrees to begin with (*epikampios* and its variant wings advancing first) or it varied according to the degree of wheeling required by the situation (*hyperfalaggesis*). If the *hyperfalaggesis* was used only to subject the enemy to missile attacks from both sides, then the angle of the wings was usually probably about 45 degrees, but if the outflanking was performed against the enemy flanks then the wings were wheeled by about 90 degrees and then further to attack the enemy's rear if this was possible. In other words, the first difference between the crescent and the other double outflanking arrays was probably only the angle in which the wings were posted to take on the enemy, and the possible separation of the wings from the centre in the *epikampios* array.

The third of the differences concerns the time when battle formations were assumed. When deployed as true battle formations, the crescent proper and the *epikampios emprosthia* (with its variants) were deployed as such from the start with the aim of outflanking the enemy. In contrast, the outflanking of the enemy in the *hyperalaggesis* formation took place only when the phalanx had advanced very close to the enemy or into contact with the enemy. The disadvantage with the crescent and *epikampios emprosthia* formations was therefore that their use immediately betrayed the battle plan to the enemy, allowing the enemy to adjust his formation accordingly. The counter tactic against these (and the *koilembolos*) was the use of the so-called triple phalanx, in which the flank phalanxes were sent against the wings of the crescent or *epikampios emprosthia* and the centre (one or two *mere*) was held back as a reserve, or the use of the oblique formation against one flank. It should be noted that the triple phalanx in this case is basically just the same as Vegetius's fourth and fifth formations and therefore basically only another name for the *epikampios emprosthia* formation, the only difference being that in this 'triple phalanx' the commander could choose to overcome the enemy's wings by sending superior numbers against them.[59] In comparison, the advantage with the *hyperfalaggesis* tactic was that the enemy could counter it only by having an equally-long line or by having adequate reserves behind their own phalanx. In other words, it left the enemy less time for counter measures, not to mention the fact that after the wings were wheeled inwards its final shape could still be the *epikampios emprosthia*. In other words, it was safer to use the *hyperfalaggesis* for the achievement of the *epikampios emprosthia* formation. However, it should be noted that the use of the so-called 'Another Formation' ('*Taxis allê*') also removed the problems present in the use of the *epikampios emprosthia*. It should also be kept in mind that it is possible that the crescent and double-outflanking formations were sometimes confused with each other by the narrative sources, even if there was no such confusion present in the theoretical treatises. Regardless of their inherent problems, the infantry crescent continued to be considered relevant at least until the reign of Justinian when Syrianus Magister still included it in his treatise, but we should not forget the fact that these continued to be included in the copies and adaptations of the earlier treatises so that knowledge of these battle formations persisted until the fall of Constantinople in 1453. These treatises included all of the Hellenistic battle formations and unit orders. It was always up to the commanders to decide how to exploit the information they learnt from the books. The *hyperfalaggesis* and *epikampios emprosthia* were both retained in the *Strategikon* of Maurice, the former clearly and the latter as the end result of the use of the '*taxis allê*' formation.

As I discussed in my doctoral dissertation *The Age of Hippotoxotai*, the form and use of *epikampios emprosthia* is problematic, because it appears to have had two different versions, both of which could be called *epikampios emprosthia*. The first version is the one in which the wings were in contact with the lateral phalanx behind, and the second version is the one in which the wings were sent forward in various different ways, as we find for example in the fourth and fifth tactics of Vegetius or in the triple phalanx. However, we know that it was used during the Late Roman period, but so that it is named only by one source, Agathias, in the context of the Battle of Casulinus River in 554 (*MHLR 6*, 346ff.). According to Agathias's account, the Romans encircled the attacking Frankish infantry wedge by using the *epikampios emprosthia* formation while the infantry centre also opened to admit the enemy infantry wedge inside so that its force would fall on empty space. In my learned opinion, this *epikampios emprosthia* is actually likely to be a variant of the 'Another Formation' rather than the *epikampios emprosthia* proper and represents the use of Hannibal's Cannae tactic – for which see the analysis of the '*taxis allê*' ('Another Formation').

However, we also possess other, less definite evidence for the use of this array. The Romans used the '*bicornis*' formation at the Battle of Brotomagus in 356 and it is clear that this meant either the *epikampios emprosthia* or crescent array (*MHLR 1*, xvii, 343), or possibly even the hollow wedge. Priscus also used an array in which the wings separated from the phalanx, allowing him to surround the Avars in 599 (*MHLR 7*, 312–3), which clearly suggests the use of the *epikampios emprosthia*/ Vegetius's fourth and fifth formations. The image opposite shows how the shape of the *epikampios emprosthia* array was envisaged in the '*Byzantine Interpolation of Aelian*' and how the crescent was envisaged in the text of Asclepiodotus (called *koile* by him). For the other interpretations of the battle formations in the '*Byzantine Interpolation of Aelian*' see Chapter 9.1–2.

The oblique formations towards right or left, sending one wing in advance of the phalanx, and the *hyperkerasis* were meant to outflank the enemy on one flank. The preferred side for outflanking was the right side, because the attacker's right side lacked the protection of the shield, while outflanking on the right also enabled the

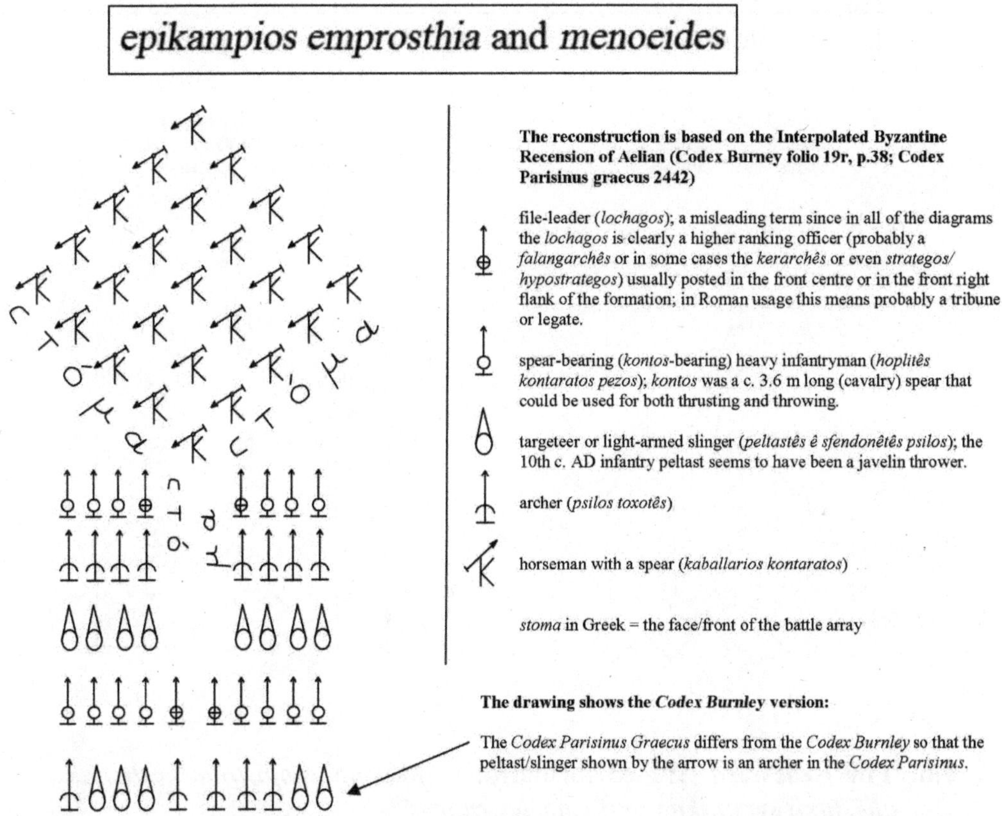

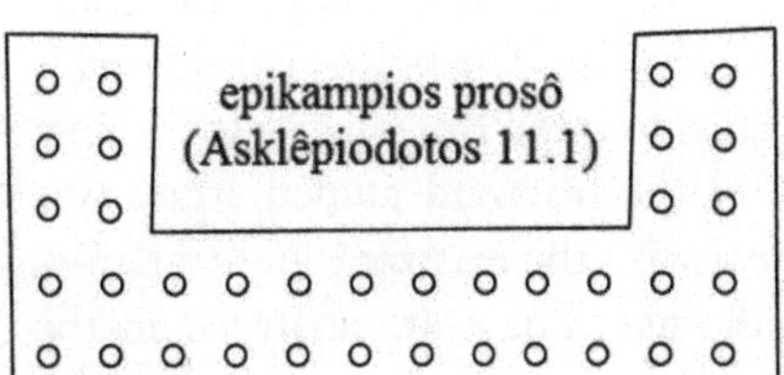

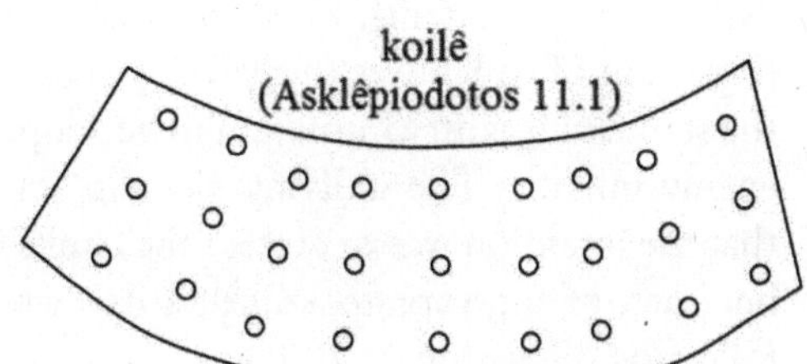

mounted archers to shoot towards their left. Excluding the oblique formations, the other two were actually usually employed when one was using the lateral phalanx, so these should not be considered battle formations proper unless of course one of the wings was sent forward too soon. As already discussed, the knowledge and know-how for the oblique attack was retained in the Roman military, but we do not possess any definite unequivocal evidence for the use of the oblique attack formation for the Late Roman period. This may be due to poor survival of material or it may result from the conscious decision by multiple commanders that it was not the right solution to the situation they faced.

koilembolos and difalaggia as infantry ambushing formations

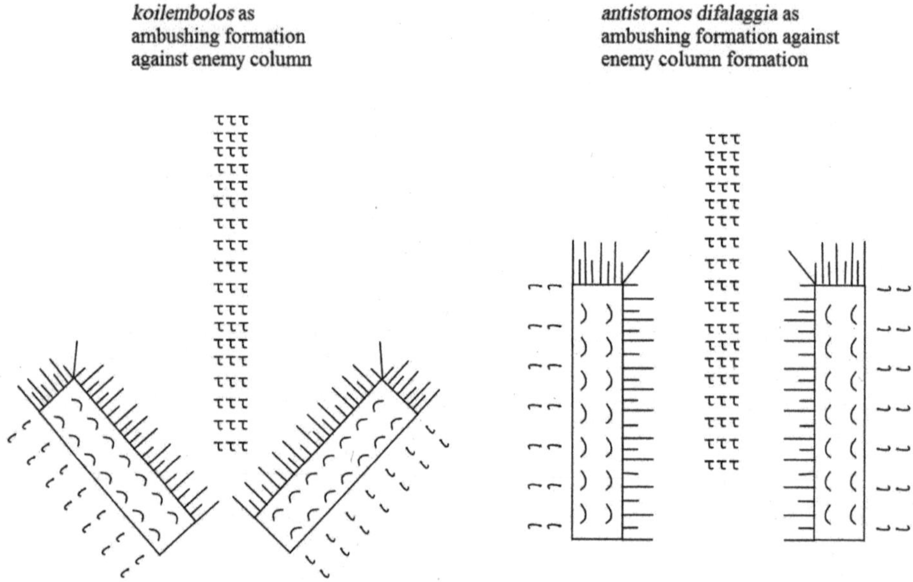

9.6. The Rearward Angled formations: *epikampios opisthia, kyrte, taxis allê, taxis symmiktos* and *embolos/cuneus*[60]

The Romans used several rearward angled formations (*epikampios opisthia/epikampios koile*, with its variants, *kyrte, taxis allê, taxis symmiktos, embolos/cuneus*), the most important of which was the *epikampios opisthia* because the *Strategikon* considered it the standard infantry formation to be adopted by dismounted Roman cavalry against enemy infantry. The unifying element for each of the rearward-angled arrays was that the intention was to protect the flanks and rear with the rearward-angled flanks, but each of the variants still had its own specific use. These are analyzed in the following discussion.

We possess several sources for the use of the rearward-angled formation (*epikampios opisthia*) for the Late Roman period. The most important of these are the military manuals the *Strategikon*, the *Taktikon* of Urbicius, the *Peri Strategias*, the *Byzantine Interpolation of Aelian* (with the reproduced versions of Aelian), the *Epitoma rei militaris* of Vegetius and the *Apparatus Bellicus*. The information provided by these is confirmed by the narrative sources which include examples of this battle array in actual use (see especially *MHLR Vol.7*). According to the *Strategikon*, the *epikampios opisthia* was meant for use against cavalry when the Roman force consisted of a mixed force of infantry and cavalry, or a partially dismounted cavalry force, with or without wagons. The array is already known from the Hellenistic military treatises so it was not a new invention.

The formation got its name from the infantry flanks, which were bent backwards. In conjunction with the cavalry reserves these features provided extra protection for the flanks and rear of the army. It was not as secure as the hollow square, but its flanks were still more secure than were those in the traditional lateral phalanx (*plagia falagx*). The main advantage of the array was that the army could concentrate on the use of a frontal attack or on receiving the enemy cavalry attack. There were several ways of forming this formation, of which the easiest was probably the wheeling of the flanks of the lateral phalanx backwards, but this was certainly not the only way to manoeuvre the units – as can be seen for example from one of the battles fought between Priscus and the Avars in 599.[61] The formation was also adopted by Roman cavalry as an emergency measure, dismounting some of the men when the enemy cavalry force had the advantage of numbers.

The Epikampios Opisthia *in the* Strategikon

The *Strategikon* provides us with the best evidence for the use of this array during the Late Roman period. The depth of the phalanx and the exact deployment pattern were reflections of the strength and composition of the army. If the Romans had more cavalry than infantry, the cavalry files could assume the depth of eight or even ten men, while the infantry was deployed only four to five men deep. If the Romans had more infantry than cavalry, then the infantry were posted in eight- to ten-man deep files while the cavalry had only four to five ranks of horsemen. According to the *Strategikon*, the best balance between the forces was to possess one-third cavalry and two-thirds infantry, but even one-fourth of cavalry was acceptable. The treatise also instructed that when the Romans were to fight against a powerful nation like Persia, the files of footmen were to consist of eight heavy-armed men and two light-armed, and the cavalry files of eight or ten men. The idea behind this was to make the array more secure against enemy breakthroughs during the period when the Romans were still getting used to the enemy. This requirement was no longer in effect after the men had become used to the enemy, so the relatively-shallow depth of four heavy footmen and one light-armed was regarded sufficient because the cavalry forces behind the infantry provided adequate protection. According to the *Strategikon*, the flank guards were to consist of ca. 1,000 to 1,200 men on both sides, and the rear guards, in square- or in wedge-formation were to consist of ca. 500 heavy- and light-armed men (*tagma*) indicating the small unit version for the wedge, but see below.

These are slightly stronger than in the formation provided by the *Apparatus Bellicus* (discussed in greater detail below), but on the basis of the other descriptions of this array in the other sources it is clear that the flanks were usually far more powerful. It was usually a rearward-angled formation which would become a hollow square if the frontal phalanx assumed double phalanx formation, after which the rear half marched to the rear, or when it was used as a marching formation in such a manner that the baggage train, infantry and cavalry followed the *epikampios opisthia* so that the bottom of the hollow square/oblong array followed behind the baggage train as a rear guard. It is therefore probable that the actual size of the flank guards varied according to the size of the army and the needs of the situation. The diagram of the array in the *Strategikon* suggests the same. It implies a larger formation. It

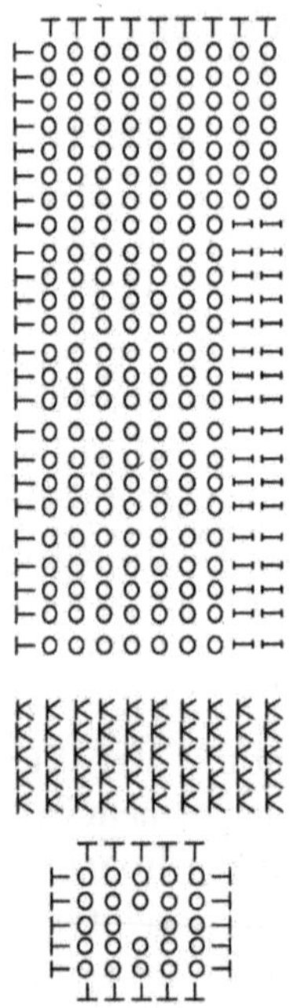

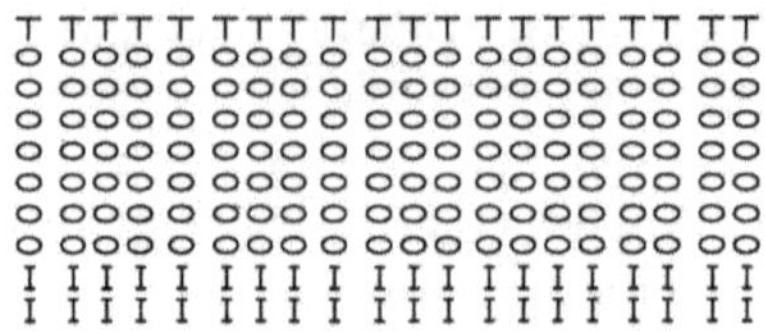

Epikampios opisthia (*STR* 12.A.7)

Epikampios Opisthia (Str. 12.A.7).

This diagram leaves out the cavalry reserves behind the centre described by the text. This diagram clearly describes only the principle and is not necessarily the original. The battle array could also contain baggage train and *carroballistae*. This would have given the formation an all around defensive capability not shown here. The centre could also be significantly wider so that the actual appearance could have been very close to the one depicted by the *Apparatus Bellicus*.

T Commander, front-rank infantryman, file leader, *dekarches*
O Heavy-armed infantryman (hoplites), shield bearer (*skoutatos*), armed man
I Light-armed infantryman, javelin thrower, archer
K Cavalryman

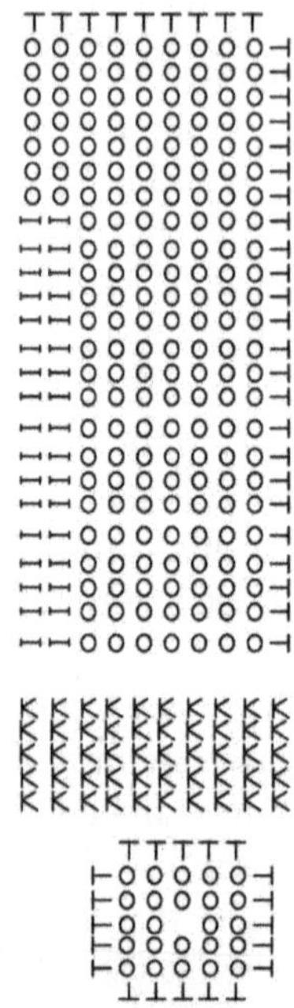

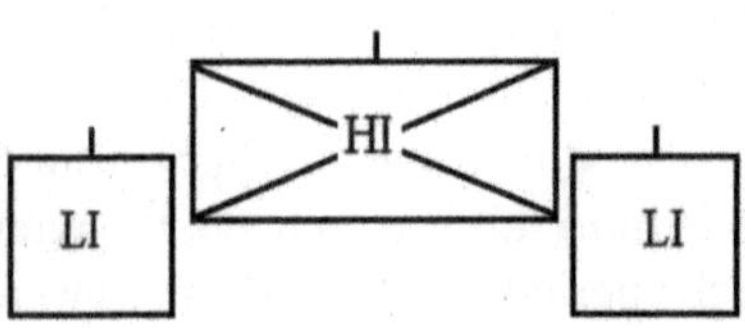

Below left: a variant of the *epikampios opisthia* in which the light infantry (LI) formed the wings and the heavy infantry (HI) the centre. The name of the array was *hypotaxis*. Arrian (Teubner ed. 26.7) equates this *hypotaxis* with *epikampios*. The latest edition of Aelian by Matthew (30) includes also a variant of this same array in which (despite its name), the light infantry was posted slightly in front of the heavy infantry making the resulting formation a variant of the *epikampios emprosthia*. The late Romans certainly used this variant too because e.g. Narses posted his light infantry on the flanks at the battle of Taginae in 552 – although it is likely that the end result of Narses's array in that battle would have been called in military theory as *hyperfalaggesis* or wrongly as *menoiedes*.

does not include the cavalry posted behind the phalanx mentioned in the text, the baggage train or the reserves, but it is clear that this array usually had those making the formation the equivalent of the hollow square/oblong when the baggage train followed. It should also be noted that the *epikampios opisthia* was considered the principal combat formation that the cavalry was to use when it needed to dismount against enemy in difficult terrain, or when it regrouped after retreat to fight against pursuing cavalry. The early-sixth century *Taktikon* of Urbicius includes a variant in which the rearward-placed wings consisted of light infantry *psiloi*. The idea behind this variant was to prevent the outflanking of the Roman array with missiles and light forces so that the heavy-armed could concentrate on winning the frontal combat.[62]

The shape of the wings in the *epikampios opisthia* array is indicative of the use of a large infantry army with large numbers of light-infantry, because the shape depicted by the diagram in the *Strategikon* could have been most easily formed from the large lateral phalanx by wheeling the wings inwards towards the rear: and vice versa, the large lateral phalanx could easily be formed out of the *epikampios opisthia* just by having the wings wheel outwards towards the front. This, however, is only my educated guess based on the shape and structure of the different formations, because Maurice does not describe the procedure of how the array was formed out of the lateral phalanx – obviously he did not need to because the Romans could assume the formation from the start.

It is probable that in such instances the cavalry reserves from the flanks (as *koursores*) and the *bucellarii* of the *strategos* (as *defensores*) were formed behind the centre to serve as the cavalry behind the centre of the phalanx – which is missing from the diagram in the *Strategikon*. These horsemen would have been used as the middle *meros* in a three-*mere* cavalry formation that could be posted in front of the infantry when needed, in one of the versions of the cavalry formation with infantry support. Note also that the same array is readily divisible into the mixed array (*taxis symmiktos*), meaning that the intervals between the four *mere* could be widened to admit the flank cavalry *mere* and the centre cavalry *meros* into the phalanx structure, while the flank guards and infantry would be used to strengthen the wing units formed up as flank guards. See the analysis of the mixed formation a few pages after this. Even the smaller version of the lateral phalanx could be used in like manner to form the *epikampios opisthia*, but in this case the commander would have used the infantry reserves of the flanks as rear guards. The attached diagrams overleaf show simplified versions of the manoeuvres, because in practice the baggage train would also have been required to perform the manoeuvres of their own at the same time. However, note that the phalanx manoeuvres performed by armies led by Comentiolus and Priscus prove that there were also other ways to form the *epikampios opisthia*, and in fact these instances serve only as concrete pieces of evidence of how the *epikampios opisthia* and hollow square were formed from the lateral phalanx.

We know on the basis of the *Byzantine Interpolation of Aelian* (Devine 45.3–46.2, Dain J3-K2) that the array could also be used against infantry. According to the *Byzantine Interpolation of Aelian*, the *epikampios opisthia* was the ideal combat formation when the enemy used the *epikampios emprosthia* (forward angled half-square), while the ideal combat formation against the *epikampios opisthia* was the *kyrte* (convex). However, in the *Strategikon* and in the *Apparatus Bellicus*, which is clearly based on texts older than the *Strategikon*, the *epikampios opisthia* was used only when the enemy employed cavalry. It is also clear that the triple phalanx (strengthened *epikampios emprosthia*) was actually better against the *epikampios emprosthia* than the *epikampios opisthia*, because the latter formation allowed for outflanking.

According to the *Strategikon*, the best way to use the formation was to fool the enemy into believing that it faced only cavalry. This could be achieved by posting a thin screen of cavalry in front of the formation to hide the presence of the infantry behind it. When the enemy cavalry was at a distance of four- to five-bowshots away, the cavalry screen was withdrawn through the opened infantry files, after which the withdrawn cavalry formed the front of the cavalry held in reserve behind the phalanx. The *Peri strategikes* (36) of Syrianos contains a similar stratagem. In this text the Romans also posted their cavalry in front of the infantry phalanx and withdrew them only at the last moment to the flanks of the infantry. In the *Strategikon*, the

One of the deployment patterns for the large lateral phalanx into *epikampios opisthia* formation and back (simplified)

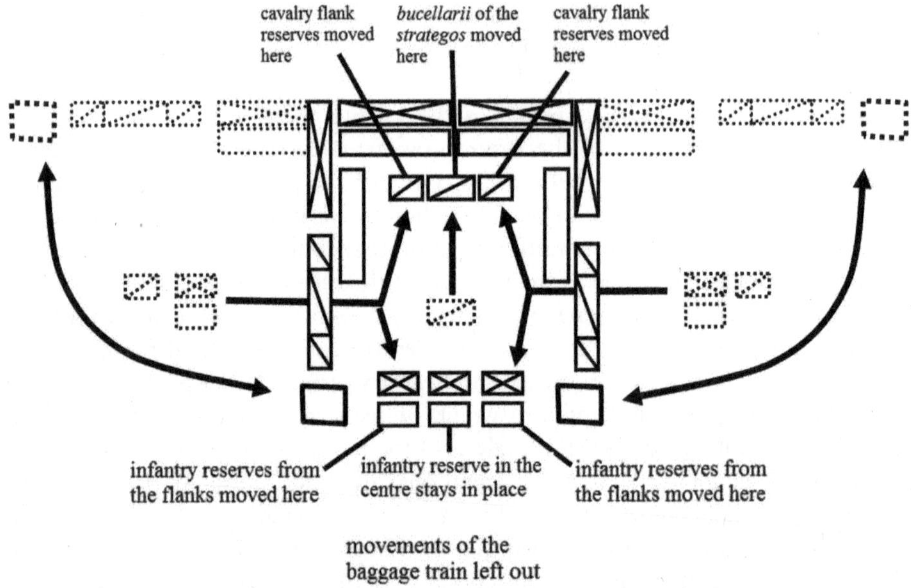

one of the possible ways how the small lateral phalanx could be deployed as *epikampios opisthia* (simplified)

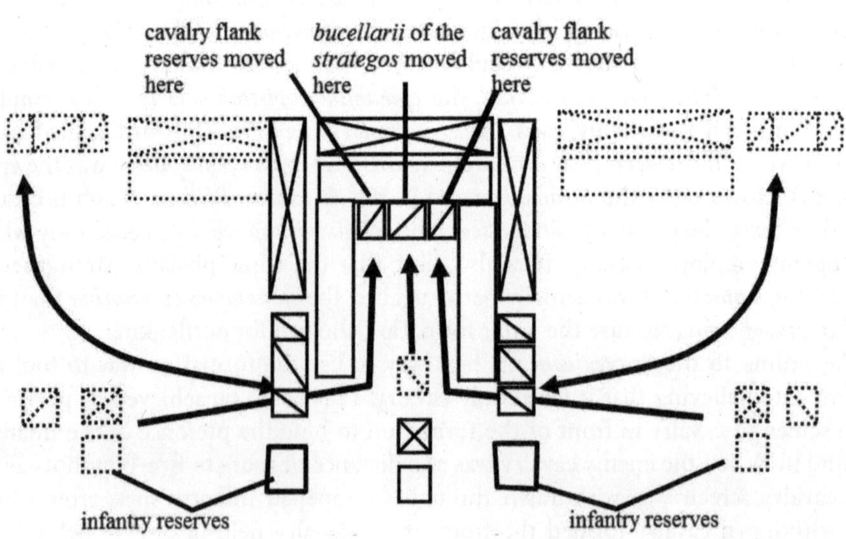

array was made even more secure by posting infantry in square or wedge behind the wings, and between them and the infantry additional cavalry reserves.

When the enemy cavalry then charged towards the Romans, the footmen, whether dismounted cavalry or infantry proper, assumed the kneeling type of *foulkon* to oppose them while the light-armed and cavalry used bows. This was usually enough to stop the enemy attack. When this happened, the infantry assumed the deep formation so that the Roman cavalry could charge through the intervals between the units and pursue the enemy. The first half of each cavalry *akia* (meaning both the file and *acies*) in the cavalry formation pursued as *koursores*, while the second half of each *akia* followed them as *defensores*. It is clear that the word *akia* has to be interpreted in this case as a single cavalry *acies* consisting of separate units of *koursores* and *defensores*. The use of half of each *acies* implies that both halves of the *akia/acies* consisted of the same number of men, but one should not take 'half' literally in this case because the preface to book six in the *Strategikon* makes it clear that the standard pursuing formation used by he period cavalry was based on the Italian Drill Formation. In other words, each cavalry *meros* consisted of the *koursores* posted on the flanks of the *meros* and of the *defensores* posted in the centre, which means that the first half of the pursuing *acies* consisted of a third of the horsemen of the *acies*. In the cavalry section Maurice provides different versions regarding the proportion of the *koursores* and *defensores* and the text states that the ideal number of *koursores* was to be a third of the *meros,* which means that the first half of the *acies* would not have consisted of half of the *acies*, but the diagrams give us a different picture. The diagram of the *meros* of the *foideratoi* has six *banda* of *koursores* to ten *banda* of *defensores,* while the *mere* of the first line of the battle formation has six *banda* of *koursores* to eleven *banda* of *defensores*. However, the *mere* of the second line turn the tables by having six *banda* of *koursores* to every five *banda* of *defensores*. The only formation that would have always divided the *acies* into two equal halves would have been to use the Alan Drill Formation (*Strategikon* 6.2) in which every other *moira* consisted of either of *koursores* or *defensores*, but this was one of the advanced forms of drill so the *strategos* could not expect all of his cavalry forces to be drilled to that standard.

In sum, it is clear that we should not interpret the text too literally. Maurice simply meant that the cavalry units designated as *koursores* pursued, while the units that were designated as *defensores* protected them. The use of the *akia/acies* (a battle line and file) in the *Strategikon* suggests that Maurice used some other earlier military treatise(s) as his sources. The description of the cavalry tactic is also very similar to Arrian's *Ektaxis kata Alanôn* (ca.135 AD), which shows nicely how old the tactical concept was.[63] The flank guard cavalry *banda* could also gallop after the pursuers and get close behind the cavalry flanks to act as flank guards. The rear of the pursuing cavalry was protected by the advancing *epikampios opisthia*.[64]

It was not considered safe for the cavalry to pursue further than three or five bowshots away from the formation. The aim was to secure the perfect functioning of the combined-arms concept so that each of these arms, cavalry and infantry, was always supporting each other very closely. The pursuing cavalry was never allowed to go further than was safe. As long as both arms of service, cavalry and infantry, functioned together the enemy cavalry had no chances of defeating this type of force.

epikampios opisthia in the Column of Marcus Aurelius, the text of Asclepiodotus and in a late sixth century battle

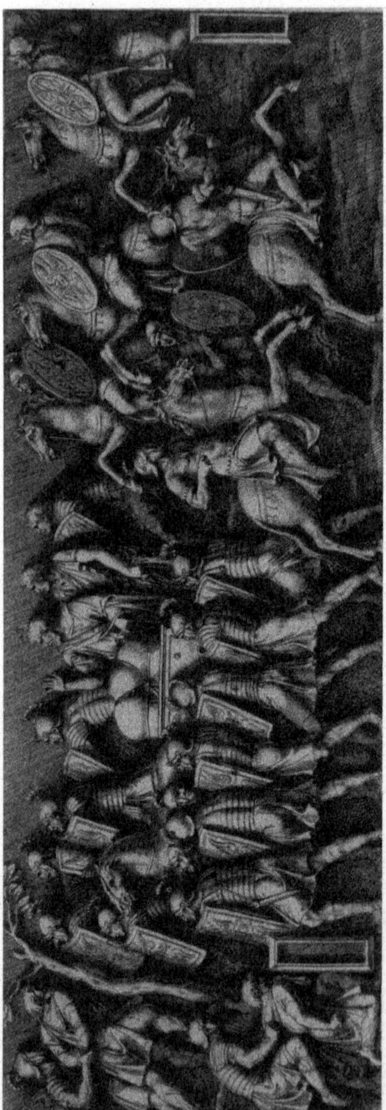

Epikampios opisthia formation in the Column of Marcus Aurelius. It is probable that the auxiliary soldier between horsemen represents the rear portion of the almost square formation similar to the one used by Julius Caesar in Gaul. 17th century drawing.

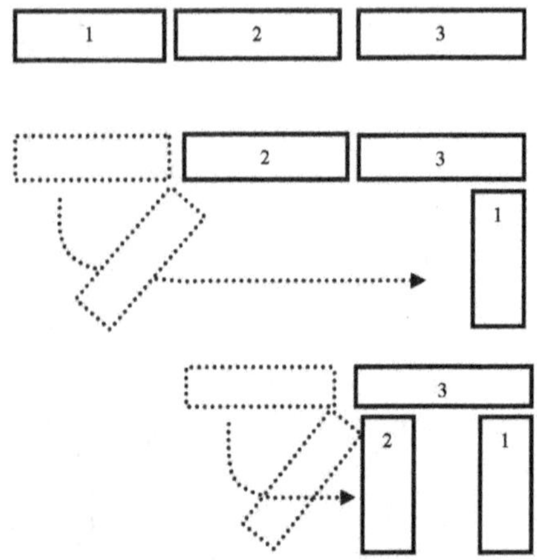

Above: A simplified version of the forming of the *epikampios opisthia* in a combat situation. This is the manner in which the *strategos* Priscus arrayed his army when he faced the Avar cavalry in front of him at the fourth battle he fought against them in 599. For further details, see Syvänne, *MHLR 7*, 314-5.

Below: An illustration of the *epikampios opisthia* in the military treatise of Asclepiodotus. The fact that the treatise of Asclepiodotus was copied and distributed also during this era makes it clear that its information also retained its relevance.

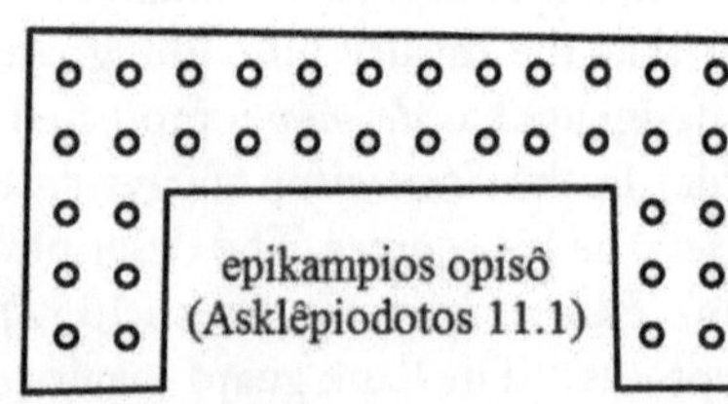

If the enemy counter-charged and both lines of the cavalry were unable to check its advance, the horsemen were to retreat behind the infantry. The infantry in its turn came out of the deep formation, filled the gaps in the line and stood ready to receive the enemy charge in the *foulkon* formation. The cavalry rear guards retreated to the flanks of the infantry. If the enemy horsemen then advanced against the rear of the Roman formation, the main phalanx was divided into a double phalanx and the light armed were divided between them. When this took place, the rear portion of the infantry phalanx moved through the light infantry and cavalry posted behind them, after which one half of the light infantry joined themselves to the rear of the heavy-armed infantry. The end result of this manoeuvre was that the array became a hollow square/oblong in shape. This was obviously a complicated manoeuvre and it shows nicely the very high quality of the Roman forces. It should be kept in mind that this battle formation and all the manoeuvres it entailed were considered to be one of the main combat formations that the *Strategikon* expected the soldiers to be able to do. The *epikampios opisthia* was not as safe as the hollow square/oblong in open terrain, but it had one advantage over the hollow oblong/square which was that it pitted more men into combat than the former. Furthermore, as we have seen, the *epikampios opisthia* could also be reformed as a hollow oblong/square when necessary. In short, the *epikampios opisthia* was a multipurpose formation that could be used by both cavalry and infantry, which gave them protection against flank attacks and enabled them to advance into difficult terrain with dismounted cavalry. The smaller than usual distance given for the pursuit shows that the main asset of the formation was the combined-arms approach to combat. The *Strategikon* expected that the cavalry and the infantry cooperated and supported each other constantly.

The Epikampios opisthia *in Vegetius*

Vegetius's basic marching formation in the *Epitoma rei militaris* (3.6) is a version of the *epikampios opisthia* formation, even if it is not immediately apparent without further analysis. According to Vegetius, the marching order was as follows: First, the cavalry (*equites*); then the foot soldiers (*pedites*); the baggage train (*impedimenta*); the pack-saddle horses/pack-horses (*sagmarii*); the soldiers' slaves/servants (*calones*): and the vehicles (*vehicula*, wagons and carts) were gathered in the middle, with the *expediti* (unimpeded, unencumbered, lightly-equipped) *pedites* (infantry) and *equites* (cavalry) bringing up the extreme rear. The *impedimenta* (baggage train) was to be protected on both sides by an equal number of armed men ('*armatorum manu*'). When one remembers that the *pedites* preceded the *impedimenta*, the latter of which was protected on both sides by the armed men, it is clear that the formation was *epikampios opisthia*.

The *impedimenta* was also arranged into units to make its handling easier, the idea being that it should be clearly separated from the defenders (*propugnatores*, front-fighters). The Romans therefore chose from among the *calones* officers who were called *galearii*. These *galearii* were then placed in charge of not more than 200 pack-horses (*sagmarii*) and young slaves/servants, and each such group was given a *vexillum* to ease their identification. This allows the making of further deductions. Since the *impedimenta* consisted of the pack-horses guided by the *calones* under the *galearii*, and it was the *impedimenta* that was placed between the infantry wings, it is clear that the pack horses were placed nearer the soldiers, while the vehicles (*vehicula*) were clearly behind the *epikampios opisthia* but in front of the *expediti pedites* and *equites*. The pack-horses (or mules or camels) would have carried spare-arms, food (e.g. dried fruits, jerky, hard tack, nuts) and drink (water, beer, or wine). The likeliest place for the wagons and carts is therefore behind the rear of the *epikampios opisthia*,

Epikampios opisthia of Vegetius

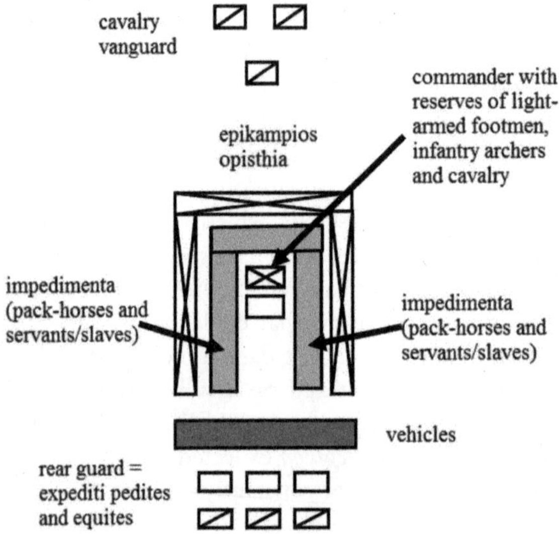

The reconstruction of the formation adopted by the cavalry vanguard, the interior of the *epikampios opisthia* and the formation of the rear guard are hypothetical educated guesses based on the text of Vegetius.

meaning that these formed a defensive barrier for the rear of the formation, while the extreme rear was then brought up by the *expediti pedites* and *equites*. One may also assume that the pack-horses, vehicles and reserves were organized in similar manner in Maurice's *epikampios opisthia*, so one can use this information to complement the diagram and text in the *Strategikon*. This means that the array was in practice a version of the hollow square, as is Maurice's version when one takes into account the existence of the baggage train and reserves. This is basically the same formation that we find Julius Caesar using at least twice during the *Gallic War* and which was called 'almost a hollow square' (*paene agmen quadratum*) in the *Gallic War* (8.8), and as we have seen in the illustration on p.338 was also used by Marcus Aurelius during the Marcomannic Wars.[65]

If the Romans suspected that the enemy would approach from one side, it was to be strengthened with select cavalry, lightly-armed (*levis armatura*) troops, and infantry archers, but if the enemy was expected to threaten all sides, then the commander was to be ready to send help to all sides (i.e. the commander was to send reserves (*subsidia*) wherever needed). This means that the commander had *subsidia* of select cavalry, lightly-armed footmen and infantry archers inside the *epikampios opisthia* that he could use as his reserves. The general was to be wary of enemy cavalry, in particular in open fields, while the opposite was true in wooded, mountainous and swampy terrain where he was to fear enemy infantry. In short, even if the *epikampios opisthia* was primarily a marching formation in Vegetius, it was also a battle formation if the Romans faced an enemy force during a march.

The Epikampios Opisthia *in the* Apparatus Bellicus[66]

As noted, it is clear that the *Strategikon* had used some earlier treatise as its source, and we possess a somewhat similar description of the same array which has been borrowed from a much earlier treatise in the tenth-century military compilation *Apparatus Bellicus*. As will be made clear below, on the basis of the contents this text contains an earlier version of the *epikampios opisthia*.

The author of the *Apparatus Bellicus* uses the Hellenistic ideal army as his basis, so the army he describes consisted of 16,384 footmen and 4,096 horsemen. The hoplites were divided into 32 *pentakosiarchiai* each with 512 hoplites. Of these, 30 *pentakosiarchiai* were posted in line with small intervals (a width of five horsemen) between them. The remaining two were posted on the flanks towards the rear, arrayed as a *plagia falagx*. The author leaves out the 8,192 light infantry *psiloi*. This may be explained in three different ways: 1) it is an omission; 2) the author expected the hoplites to be multipurpose soldiers who also possessed bows, as Vegetius and Syrianos state and Procopius's description of the Roman forces in Italy confirm;[67] and 3) the author expected that the light infantry just worked around the formation as the situation dictated and did not consider it necessary describe how this was done because it was the responsibility of their commanders. I would consider the second option (multipurpose hoplites) as the likeliest, because the material has clearly been borrowed from a treatise which dates from the period when many of the Late-Roman practices were simultaneously in use with the traditional types of Roman cavalry (*kontaratoi*, *logchoforoi*, and *hippotoxotai*, see below) and when the Romans had already adopted the marching camp which is preserved in the *Strategikon, Peri Strategias/Strategikes* and *De re militari*.[68]

The cavalry was similarly divided into 32 units called *hippilarchiai*, each consisting of 128 horsemen. Of these, 30 were posted behind the infantry phalanx and the remaining two behind the two infantry wing units deployed towards the rear. The term *hippilarchia* probably results from a confusion in which the 128 horsemen *epilarchia* and 512 horsemen *hipparchia* are combined as one term.[69] In my opinion the author's use of the rhombus to describe the subdivisions of *hippilarchia* in its turn results from the fact that he or his source(s) did not understand that the 128-horseman *epilarchia/epeilarchia* was usually deployed as a rhombus, which in its turn consisted of two 64-horseman *ilai* (sing. *ilê*) which were deployed as wedges. Therefore, the author uses this term, the rhombus, to describe also the smaller subdivisions (64- and 21-horseman) because all of these were part of the rhombus. In sum, in the basic variant each of the 32 cavalry units was deployed as a 128-horseman rhombus.

The *Apparatus Bellicus* does not say why the units consisted of 128 horsemen and why these were deployed as rhomboids. In my opinion the likeliest reason for this was that the 128-horseman rhomboids could operate independently both towards the rear and front. If the commander decided to send these to the front in their entirety, this was easy to do by ordering the hoplites to tighten their array, shields rim-to-boss, or by having them double the depth. The width of the rhombus was 15 horses (1, 3, 5, 7, 9, 11, 13, 15, 15, 13, 11, 9, 7, 5, 3, 1) so this was easy to do. Each of these rhomboids then consisted of two 64-horseman *ilai* which were initially deployed as wedges back-to-back. The 64-horseman *ilai*/wedges in their turn consisted of three

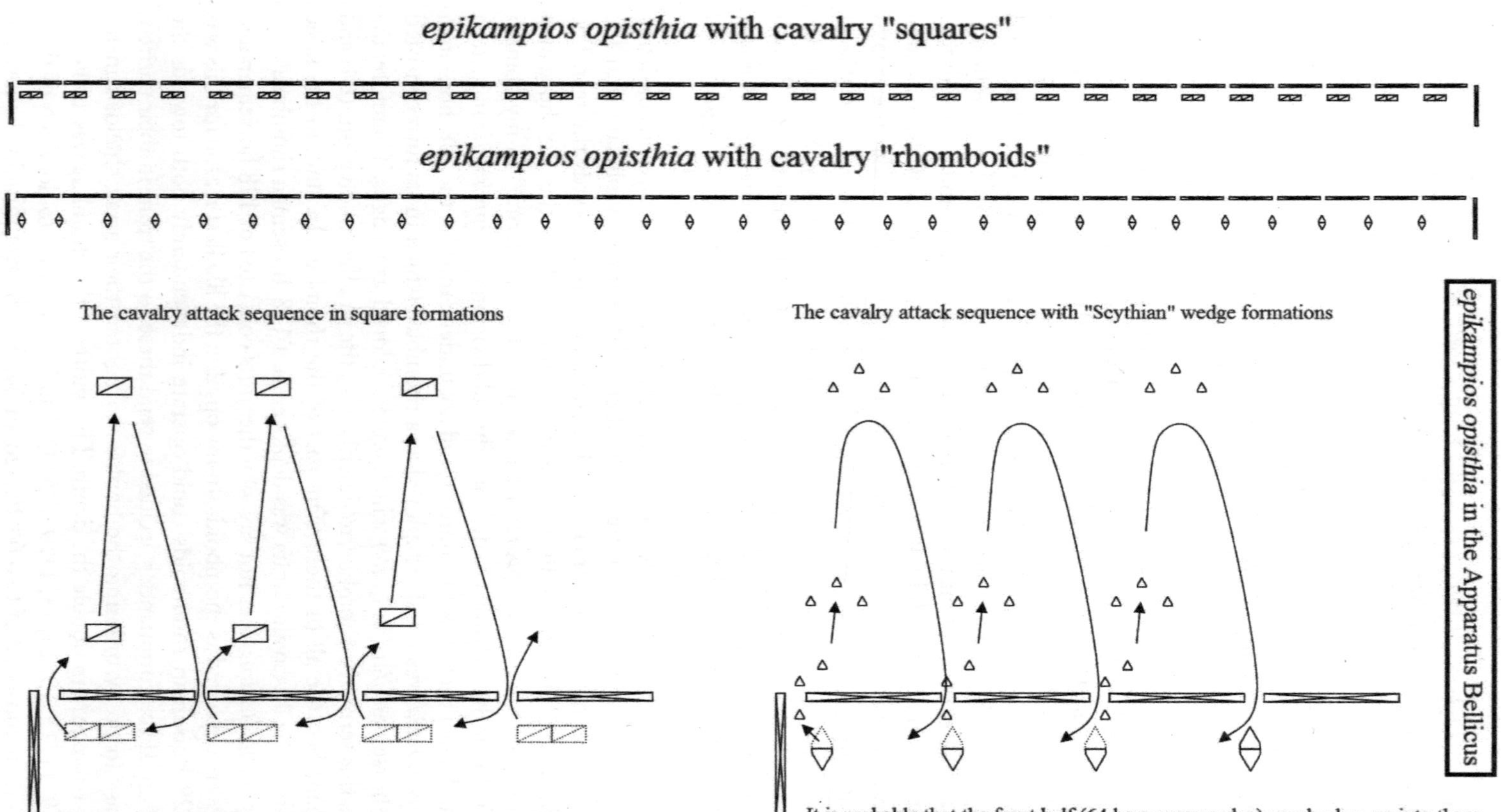

The cavalry squares were probably posted just behind the infantry units so that it was easier for the left wing of the formation to pass through the left interval as a four horsemen wide column. It is clear that the return through the right interval had to be timed so that the cavalry units did not collide with the cavalry posted to the right of them.

It is probable that the front half (64 horsemen wedge) was broken up into three wedges immediately when it was sent to fight in front of the phalanx because the intervals between the infantry units was five horses wide. The rear half of the rhomboid could be used as a reserve for the front half. Note that when the cavalry units were deployed as rhomboids the rear half could also be used as a 64 horsemen wedge towards enemies approaching from the rear.

21-horseman wedges (wrongly called rhomboids by the author but clearly part of the rhombus) and one *ilarchos* (commander of the *ilê*). In addition to this, the author notes that the 128 horsemen could be deployed in square formations.

The author gives us two different versions of the armament of the cavalry forces. In the square formation, the 64-horseman *ilê* consisted of 32 *kontaratoi* (i.e. *kontoforoi, kontos*-bearers), 8 *spathatoi* (*spathaforoi, spatha*-bearers), 8 *rhiptaristai* (javelin-throwers), and 16 *hippotoxotai* (horse-archers). The terms *spathatoi* and *rhiptaristai* date the text to the tenth century, but both of these terms were just updated versions of the names of the Roman cavalry already in use during Arrian's time. According to Arrian, the Roman cavalry consisted of the *kontoforoi* and *logchoforoi* (lance-bearers, spears both thrown and used in melee together with the *spatha* sword), the latter of which can therefore be equated with the *rhiptaristai* and *spathatoi* of the *Apparatus Bellicus*. In sum, the 64-horseman unit consisted of 32 *kontaratoi*, 16 *logchoforoi* and 16 *hippotoxotai*.

In the second version, the armament of the three small 21-horseman wedges (*embolos, trigônikos*) of the 64-horsemen *ilê* consisted of three different types of forces. The one posted on the right in combat formation consisted of the *kontaratoi* (*kontos*-bearers) and the one on the left of the *hippotoxotai*. The author fails to state how the middle wedge was armed, but we can make the educated guess that this wedge consisted of *logchoforoi* (*rhiptaristai/spathatoi*). It is possible that this variance in equipment resulted from the use of two different sources, or from the fact that Roman cavalry was taught how to use each of these types of weapons at least since the days of Hadrian, so Roman horsemen could vary their gear according to the situation. Regardless, it is clear that the description of the equipment comes from an earlier source which is likely to be representative of the Roman cavalry forces in general from the early-second century onwards, which also means that it was in all probability used also during the Late Roman period.

When the enemy cavalry forces approached the Roman formation and were presumably brought to a halt in front of the Roman hoplites, the unknown author provides two alternative responses, one for the 'square formation' ('*tetragonon ilên*') and another for the 'rhombus formation' ('*rhomboeidê* = *rhombus*).[70] In Hellenic military theory the square formation meant either an actual square in shape or an oblong, depending on the width and depth of the cavalry unit. The 'square' formations could therefore be nine in width and three in depth, or eight in width and four in depth, or ten in width and five in depth, or twenty in width and ten in depth and so forth depending on the size of the unit. In this case, the size of the unit was 128, which means that it was probably divided into two 64-horseman units, each of which was deployed sixteen in width and four in depth, or eight in width and eight in depth, the former being likelier because it enabled the formation to pass through the Roman infantry in a column formation. The rhombus in its turn consisted of two 64-man wedges which were deployed as ranks (1, 3, 5, 7, 9, 11, 13, 15) and files posted back to back. In combat, the front half of the 128-man group advanced in front of the infantry phalanx through the interval on the left as *koursores*, while the remaining half remained behind and also advanced through the left interval when ordered to act as *defensores* (called *ekdikêtai* in the *Apparatus Bellicus*) for those who had already sallied out.[71]

If the Roman cavalry advanced in front of the hoplites in square formation, it did so presumably as a column which was then deployed as a line of sixteen files with a depth of four ranks (this in turn consisted of two 32-horseman units deployed eight by four) because this conforms to the unit sizes given. In this case, the front consisted of two ranks of *kontaratoi*, behind which was one rank of *logchoforoi* (*rhiptaristai/ spathatoi*) and one rank of *hippotoxotai*. If the enemy then regrouped and counter attacked, the author expected that the rear of the retreating Roman square would consist of the mounted archers, but if the enemy got close to the retreating Romans the author expected that the rear would consist of the *kontos*-bearers. Since it is clear that wheeling the formation would have been too slow and cumbersome a manoeuvre in such cases, it is clear that the author envisaged a looser cavalry formation in which it was possible to pass between the cavalry files. In sum, when the enemy cavalry regrouped, the last rank in the Roman formation, the *hippotoxotai*, continued to move forward through the intervals until it reached the front, after which all the horsemen made an about turn and started retreating while the Roman mounted archers used the Parthian shot against the pursuers. If the enemy got near the Roman formation, the *kontaratoi* and *logchoforoi* slowed down their retreat while the *hippotoxotai* continued the retreat with the previous speed so that the *kontaratoi* then became the last rank facing the pursuers and the *hippotoxotai* the front rank. If the second 64-horseman *ilê* (the right flank) had already advanced in front of the hoplites, then it acted as reserves (*defensores/ ekdikêtai*) for the left *ilê*. The 64-horseman *ilai* always sortied out of the formation by using the left interval in front of them and returned back to the formation by using the right interval so that they would not collide with each other during the combat. The turning was done towards the right so that the soldiers always maintained their shields towards the enemy – when the *kontaratoi* had their backs towards the enemy, their shields were raised above their heads and moved to cover the back as described by Arrian in his *Technê Taktika* (43.2, the *touloutegon* manoeuvre for the *kontoforoi*).

The cavalry formations in the *Apparatus Bellicus* and in Syrianos's *Peri strategias* (17) both suggest that there was far more variety in the tactical orders of the cavalry troops than there was after Maurice produced his *Strategikon*. The *Strategikon* became the Bible of the East Roman forces until the tenth century, so there was presumably far less variety after that. The *Strategikon* favoured silence in attack and close order with the trot or canter in attack, but the descriptions of Belisarius's operations suggest that his forces shouted and galloped and hence probably used a looser formation, similar to the one implied by the *Apparatus Bellicus*. The cavalry commanders and theorists never found a consensus on this issue, so we find them arguing about the relative benefits of the close order vis-à-vis looser order with the gallop until the end of the nineteenth century.

If the front *ilê* of the rhombus advanced in front of the hoplites and then retreated, this was called a Scythian retreat. As noted by Constantin Zuckerman (1994, 386–7), the *Strategikon* (4.2–3, 6.1, 11.2) also includes references to Scythian manoeuvres. In fact, Maurice (6.1) specifically mentions that in the past Roman cavalry had performed Scythian drills every March. It is therefore quite clear that both authors had access to some earlier treatise which included these.

In the *Apparatus Bellicus*, the front 64-horseman *ilê* advanced through the left interval between the hoplites, presumably having already divided itself into three 21-horseman wedges and one *ilarchos* because the intervals between the hoplites were five-horses wide, making it easier to pass through if they did so. The 21-horseman wedge was deployed six ranks deep (1, 2, 3, 4, 5, 6) meaning that these did not have files and ranks. This wedge could therefore have passed through the hoplites just by having the two flanking rear rank men pass through the infantry later than the rest or just by tightening the infantry formation very slightly, the former being likelier. The three 21-man wedges formed with the *hippotoxotai* posted on the left flank, the *logchoforoi* (*rhiptaristai/spathatoi*) in the middle, and the *kontaratoi* on the right flank. When the enemy regrouped, the one closest to the enemy (clearly the middle wedge consisting of the *logchoforoi*) was instructed to flee and draw the enemy in pursuit, while the other two turned either left or right so that when the enemy pursued the middle wedge, the other two could attack the enemy from behind. The second 64-horseman *ilê*, the rear of the rhombus, could also act as reserves, as it did for the square formation, and the retreat was also performed by using the right interval between the hoplites so that the cavalry units did not collide.

It is probable that the *epikampios opisthia* formation described by the *Apparatus Bellicus* is earlier than the one in the *Strategikon* because the one depicted by the *Strategikon* had added security features missing from the former. However, it is quite possible that the cavalry troopers in the version depicted by the *Strategikon* were to be used in like manner, so that sometimes the troopers assumed the 'square array' and at other times the 'wedge array' (presumably to be identified with Maurice's *drouggos*). Unfortunately, it is impossible to provide exact dating for the array and its tactics because its descriptions would fit any period from the third-century (or even from the second-century) onward until about the mid-sixth century. No definite examples of the use of the array depicted by the *Apparatus Bellicus* are known, but it is still clear that it was used. The paucity of information results merely from the poor survival of detailed sources for certain periods of time, which include most of the third-century, parts of the fourth-century, practically the entire fifth-century and parts of the sixth- and seventh-centuries. For example, it is possible that Aetius deployed his infantry phalanx (which formed the left wing of the allied army that as a whole was deployed as a mixed formation with infantry wings and cavalry centre) in this manner in the famous battle of the Catalaunian plains because he posted his cavalry initially behind the infantry and he needed extra protection for both of his infantry flanks. Similarly, it is possible (but note that there is no definite evidence for this) that Arnegisclus deployed his army against the Huns in this manner at the Battle of the River Utus. For these battles, see the *Military History of Late Rome Vol.4* (425–457).

Kyrte, *the convex formation*

Ammianus Marcellinus's description of the Battle of Maranga on 23 June 363 proves that Julian the Apostate arrayed his army in the convex formation called in Greek military theory *kyrte*. This array is not included in Vegetius's *Epitoma Rei*

Militaris, Urbicius's *Taktikon*, Syrianus's *Peri strategikes* or Maurice's *Strategikon*. The only period military treatises to mention this array are the so-called *Definitiones/ Hermeneia* (military vocabulary closely related to the texts of Aelian and Arrian before the sixth century) and the so-called *Byzantine Interpolation of Aelian*, if its original version is dated to the sixth century. However, there is every reason to believe that the original text of Aelian also included the same information as the so-called *Byzantine Interpolation*, hence the inclusion of this very same material in the latest edition of Aelian's treatise by Christopher Matthew. Similarly, it is clear that the military treatises dating from the fourth- to fifth-centuries that Dain called *Corpus Élianique* included the same material. It is also possible that the *kyrte*/convex hides behind some of the references in the sources to the use of the wedge or crescent. The reason for this suspicion is that we know that Julian actually used the convex rather than the crescent at the Battle of Maranga only because Ammianus likened his crescent with the wedge. The convex array can also be found in the earlier Hellenistic treatises, such as that by Asclepiodotus, which were copied and re-circulated during this time with the implication that the formation retained its relevance.[72]

According to the text of Aelian, the *kyrte* was particularly useful for engaging the *epikampios opisthia* array, but in the accompanying diagram of the *Codex Burnley* it is actually pitted against the *epikampios emprosthia* (forward-angled half-square). This means that there is a mistake either in the text, or in the diagram, or that both are actually true because the referral to the *kyrte* (Devine ed. 46.2; Dain ed. K2; Matthew 46) follows immediately after the referrals to the *epikampios emprosthia* (Devine ed. 45.3; Dain ed. J3; Matthew 45) and *epikampios opisthia* (Devine ed. 46.1; Dain ed. K1; Matthew 46). I would suggest that the *kyrte* was useful against both.

Neither the text nor the diagram of Aelian explains why the *kyrte* was effective in each of these instances. The statement that the shape of the array hid the actual strength of the army from the enemy does not explain the truth, because the same was also true of the *epikampios opisthia*.[73] However, it is still easy to understand what the likely reasons were for the effectiveness of this formation. The first of these is that the round shape exposed only the centre of the Roman formation to the enemy and it was possible to post the best men there. The second would have been that if the enemy wanted to engage the Romans who were deployed in a round shape, they had to break their own formation in a situation in which the Romans were perfectly arrayed while the enemy broke its order. Furthermore, every time any enemy unit would then turn to face the round shape of the Romans, the enemy unit would expose its flank to the Romans who were further back in the convex array. This is the manner in which Julian appears to have used his convex formation against the Persians. For further details, see *Military History of Late Rome, Vol.1* (102–3).

The use of the round-shaped *kyrte* required very well trained and motivated soldiers. Therefore, it is no wonder that we do not hear of any other battle besides the Battle of Maranga in which the Romans could have used this array during the Late Roman period. This, however, is an argument from the silence of the sources, and the survival of sources for this era is very uneven. Regardless, it is still clear that there were such periods during the long Late Roman period when the Romans were unable to use this array in practice. This was certainly true for the reign of Anastasius,

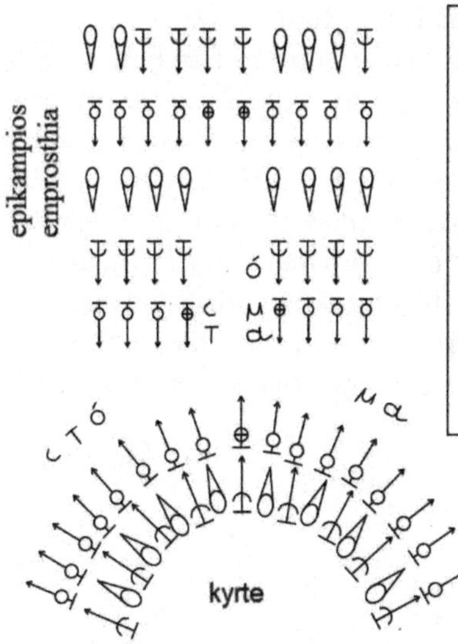

Kyrte* vs. *Epikampios Opisthia* in Aelian (drawn after *Codex Burnley*) and Asclepiodotus's *kyrte

In the text of Aelian the *kyrte/kurte* (convex) was used against the *epikampios opisthia,* but in the accompanying diagram in the *Codex Burnley* the *kyrte* is actually pitted against the *epikampios emprosthia* (forward-angled half-square). It is clear that there is something amiss either with the text or diagram. My educated guess is that both alternatives should be accepted for the reasons I state in the text.

The *kyrte* was particularly useful because it forced the enemy to break up its array if it wanted to engage also other Roman units besides those posted in the centre and if the enemy did so its units exposed themselves to flank attack by the Romans posted further to the rear.

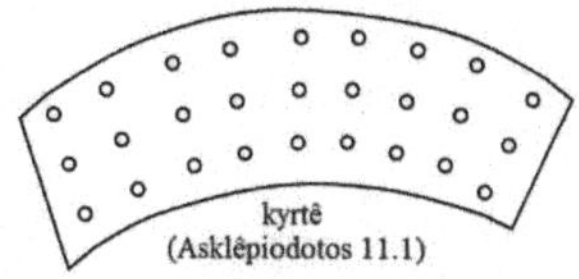

when Urbicius wrote his military treatises, and in the immediate aftermath during the reigns of Justin I and early reign of Justinian I the Great. It was during that era that the Romans generally had very little faith in their infantry forces. However, it is still probable that the Roman infantry was still at least sometimes drilled to use this array because the Romans continued to produce copies of Aelian's treatise together with the diagrams. It is also clear that the Romans continued to train to use the round-shaped formations in combat, because the round tortoise array and the crescent also required this, hence it is probable that they trained to use the convex too. Regardless, it is very likely that the array was no longer used from the reign of Maurice onwards because he simplified the general training scheme and tactics of the Roman armed forces.

Taxis allê *('Another' Infantry Formation), see Appendix 3*

The 12th book of the *Strategikon* includes a unique battle formation, which Maurice calls 'Another' formation (*'taxis allê'*), which is not specifically included in any of the other extant military treatises. According to the *Strategikon*, this formation was used against infantry when the Roman infantry outnumbered their cavalry.[74]

Unfortunately, the extant text of the *Strategikon* does not explain how this formation was used. On the basis of the location of the cavalry wings in the array, my educated guess is that the cavalry wings were used to fool the enemy to post their own cavalry wings opposite them. When the enemy then did this, the Romans withdrew the cavalry wings at the right moment with the result that the enemy cavalry suddenly faced a deep formation of infantry. The Roman cavalry wings could perform this in two different manners: 1) the cavalry could be sent outwards to outflank the surprised enemy cavalry, which then faced the Roman infantry in front of them while being threatened from the flank by the Roman cavalry; and 2) the Roman cavalry could be withdrawn behind the infantry columns to the place where the cavalry was in the *epikampios opisthia* formation. If the Roman cavalry was trying to outflank the enemy, the surprised enemy could then attempt to face both threats simultaneously or it could try to withdraw and then regroup. If the Romans withdrew their cavalry wings, then the enemy cavalry could try to attack the Roman infantry formation (this was very unlikely to succeed) while being threatened by the Roman cavalry if it made an about turn. Or they could try to follow the retreating Roman cavalry with the result that the Roman infantry column could subject them to a missile attack from the flank. In a sense, I would therefore categorize this rearward angled array to be in truth an outflanking formation in which the array was initially an *epikampios opisthia*, but which ended up being an *epikampios emprosthia* that outflanked the enemy formation.

Strategikon 12.A.4: Another Formation (*taxis allē*)

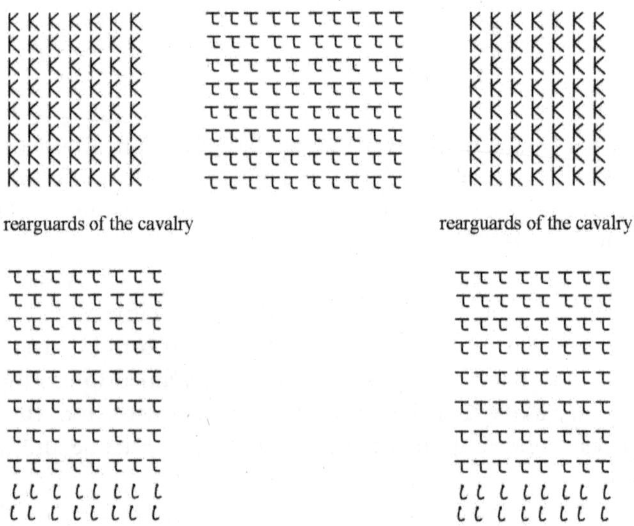

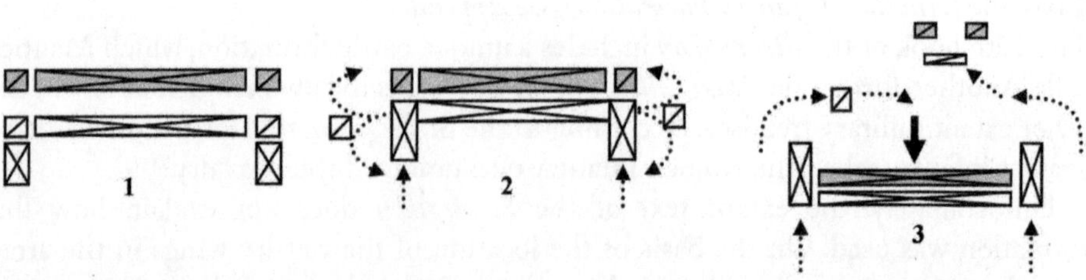

1) Cannae Theory: Another Formation = Cannae 216 B.C.? **2)** The two alternative methods of manoeuvring the cavalry while the infantry flanks moved forward: **A)** The retreat behind the wings into the position of the rear guards; **B)** The encircling manoeuvre against the enemy cavalry wings. The use of the convex infantry centre at the Battle of Cannae left out. Delbrück (1. 317) has suggested that the horseshoe shape (convex, crescent or wedge?) of the centre at the Battle of Cannae resulted from the natural tendency of a long line to bend as it moved forward making the final formation accidental, but it may just as well have been intentional. **3)** The final phase of the battle. The combination of the infantry wings and cavalry has routed the enemy cavalry, which is pursued by part of the cavalry. The enemy centre has pushed itself into the trap, which is closed by marching the infantry wings forward and then by turning them. Notably, the Cannae tactic can also be found in the *PE* 19.4.

As I noted in 2004, it is very likely that this array had actually been used by Hannibal at Cannae 216 BC and probably also by Narses at the Battle of Casulinus River in 554.[75] There is also every reason to believe that the battle array that Hannibal used at Cannae was also included in a tactical military treatise which was still extant in the sixth century and could have served as inspiration for the *Strategikon*. A good candidate would be Frontinus's no longer extant *De re militari*, because Frontinus also included the Battle of Cannae in its sister treatise the *Stratagemata* (2.3.7).

***The Mixed Formation* (taxis symmiktos)[76]**

According to the *Strategikon*, the mixed formation was used against cavalry when the Roman cavalry force was of the same size or even smaller than the infantry force. The array was formed by posting infantry and cavalry next to each other. On the basis of the diagram (see image below) the array did not have any reserves. This meant that the *strategos* was in the middle of the formation, just like Belisarius was at Callinicum. The diagram also leaves out the *hypostrategos*. This may imply that the *hypostrategos* could have been posted somewhere behind and that he had with him some reserves. Furthermore, if the battle arrays of John Troglita during the sixth century are interpreted as mixed arrays (the evidence is uncertain) it is possible that this array could also possess some reserves. However, most of the extant evidence points to the use of a single line without any reserves that could have been used in actual combat. The only reserves that such arrays appear to have had in practice consisted of men posted behind, in front of, or inside a marching camp or with the wagon train when it was following. These guards served merely as the last place of refuge for the forces posted in the actual combat formation.

The *Strategikon* fails to state the reason for mixing the cavalry with the infantry, but it is clear that the intention was to counter the enemy cavalry with the compact infantry formation while enabling the Roman cavalry to begin its charge immediately

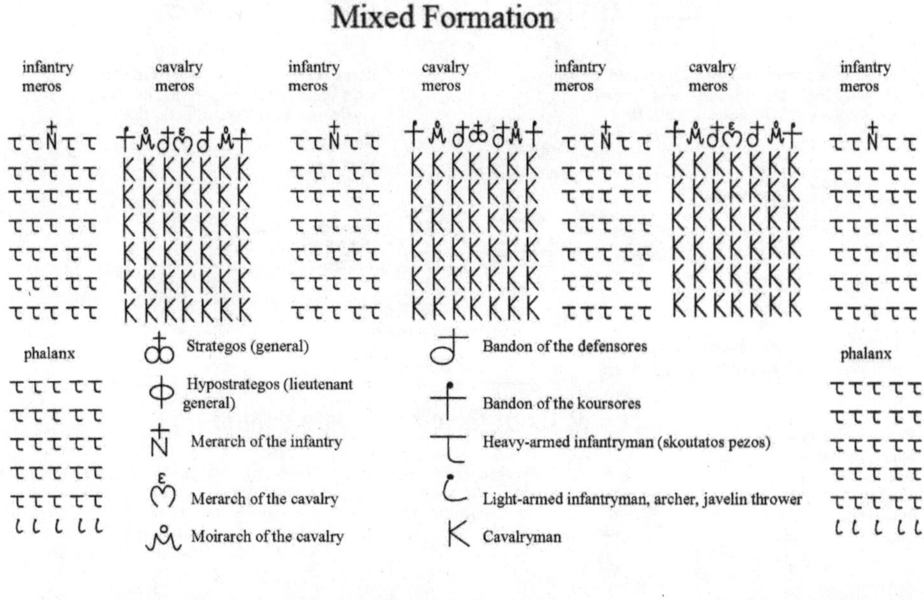

after the enemy cavalry had started to retreat, so that its retreat could be turned into flight. The formation had two weaknesses: 1) it lacked reserves; and 2) it required perfect cooperation between cavalry and infantry. If the enemy forced the Roman cavalry into retreat and it collided with the Roman infantry, or the morale of the Roman infantry collapsed when they saw their own cavalry in retreat, it was possible that the structure of the formation (no reserves) contributed to the Roman defeat. It is quite probable that this was one of the reasons for the Roman defeat at Callinicum in 531. The mixing of cavalry and infantry in the same single line formation also increased the danger of the cavalry becoming separated from its infantry support. The pursuing cavalry force lacked reserves if it had advanced too far from the infantry, with the result that it could become isolated while also leaving the infantry line behind them horribly vulnerable – the wide intervals in the infantry formation made it easy for the enemy to penetrate the array once the cavalry had left its places in the battle formation. Therefore, it is not at all surprising that there are not that many examples of its use.

Belisarius employed a very unique version of the mixed formation (I class it mixed because it had cavalry in the centre) at the Battle of Dara in 530. The array was

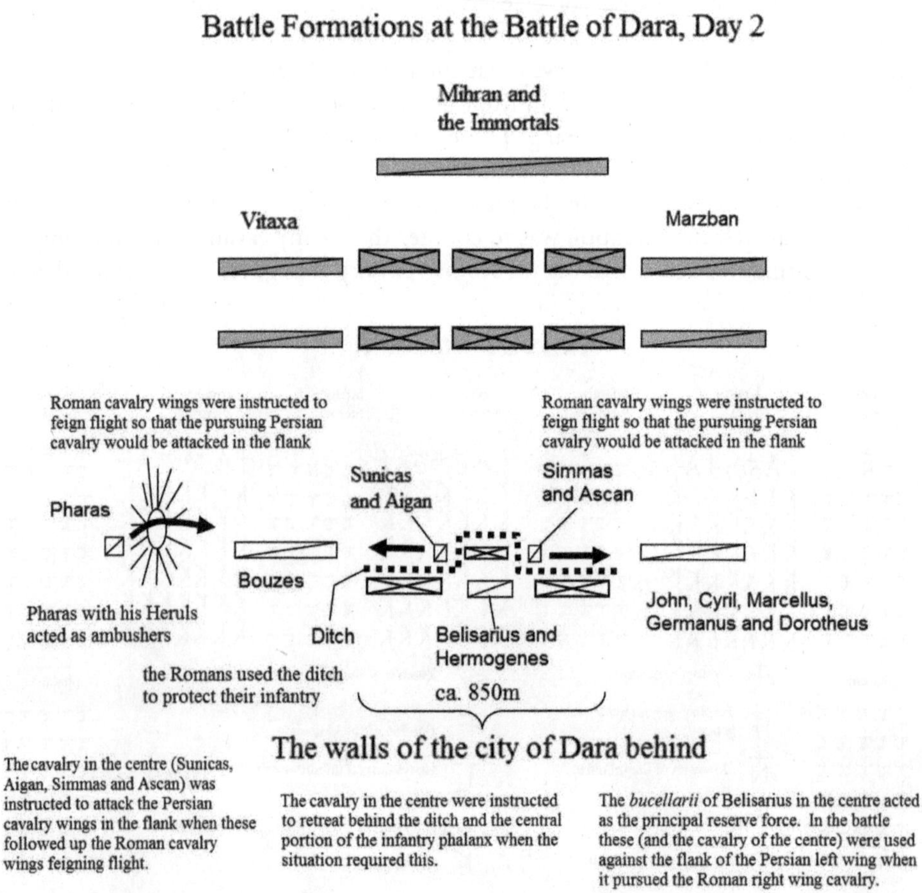

improvised to take into account the numerical superiority of the enemy, the Persians, while seeking to protect the unreliable Roman infantry from the enemy. For a fuller analysis of the battle see Syvänne (2004, 461–2; *MHLR 6*, 56–63).

The mixed formation of cavalry with infantry was an ancient one – for example the Achaemenid Persians, Seleucids, and the Numidians had used it – but it is not known when the Romans started to use it. The earliest example that I know of is the instance of Labienus using a mixed formation of Numidians (cavalry wings with mixed centre) against Caesar at the Battle of Ruspina 47 BC. The Romans were certainly employing the mixed formation by the end of the third century AD because we find them using this battle formation at the Battle of Turin in 312. In this battle the enemies of Constantine the Great attempted to crush his formation with a massive formation of cataphract cavalry (probably arrayed as a massive wedge or as a deep phalanx). In fact, the mixed array is mentioned in the third-century novel *Aethiopica* by Heliodorus. In the fictional text, Heliodorus has the Persians using the mixed formation in which the cataphracts formed the centre and the infantry the wings. It is quite clear that he has copied an actual battle formation used by the Persians for his novel and that the Romans copied this array from the Persians because the Romans employed their cataphract cavalry in like manner in the above-mentioned Battle of Turin. The failure of this array at that particular battle did not discourage the Romans from using it because we find them using it again at the Battle of the Catalaunian Fields in 451 (victory) and again at Callinicum in 531 (defeat) and Fihl in 634/5 (defeat resulting from the wrong kind of use; it was used against a mixed force of infantry and cavalry while it was meant to be used against cavalry). In short, the battle array had a mixed combat record. Most importantly, the mixed formation was included in the *Strategikon*, which suggests that it was considered a useful combat formation. However, it required highly-trained and motivated soldiers and the right circumstances to function as expected.

Wedge (cuneus, embolos)[77]

The Romans used three variants of the infantry wedge (*cuneus, kounion/kouna, embolos*): 1) the small unit wedge (ca. 200–300 strong wedge, or ca. 400–500-man wedge; which could be formed from a single *tagma* or cohort) without any hollow space inside which looked like Δ (delta); 2) the divisional wedge with empty space between the phalanxes which looked like Λ (lambda); and 3) a battle formation which looked like Λ (lambda). In short, the former two were unit formations while the third version was the actual battle formation for the entire army.

The small unit and divisional wedges could both be used as *globi* separately from the main phalanx, for example as flank guards for the lateral phalanx, or rear guards of the *epikampios opisthia*, or in front of the main phalanx to break up the charge of the enemy cavalry, or as a variant form of the *peplegemene/serra* formation. Besides this the divisional wedge could also be included in the main battle formation in the following ways: 1) the most common way was to form the wedge in the centre of the lateral phalanx so that it could be used to break the enemy's centre (according to Vegetius this required the use of reserves, but this is an exaggeration because we know from the other treatises that the two middle *mere* could be united in the navel and formed into a wedge or a single *meros* could do assume the wedge shape on its own); 2) the wedge shape could also be used by the flanking *mere* to break through the enemy formation; and 3) the battle formation could adopt the use of several wedges simultaneously, called either the *peplegmene* (woven) or *serra* (saw) formation in Graeco-Roman terminology. We do not have any definite evidence for the Roman use of the saw formation for this period, even if the Romans could be deployed as several wedges for marching and therefore knew the array well, but we possess definite evidence for its use by the Alamanni at the Battle of Argentorate in 357. In that battle each of the Alamanni divisions was deployed as a wedge.

Roman military theory also knew the wedge as an actual battle formation in which the entire battle line was formed up as a lambda-shaped Λ wedge. We do not possess any definite evidence for its use by the Romans for the entire battle formation, but the military treatises make it clear that the Romans also trained for its use in this manner. The only uncontroversial period-evidence for the use of the Λ-wedge comes from the Battle of Casulinus River in 553, fought between the Romans led by Narses and the Franks led by Butilinus.[78] In this battle it was the Franks who adopted the wedge in order to break through the Roman lines, which Narses countered by using the Cannae tactics known as the *taxis allê*. The use of the wedge by the Franks was not surprising, because both the Germans and Romans used it and the Franks were inheritors of both military systems, so that there were even complete Roman units with their native gear among the Frankish armies.

9.7. Four-sided/all-facing formations: hollow square/oblong (*plinthion*, *plaision*), several hollow squares/oblongs side-by-side, wagon laager (*carrago*, *karagos*), double phalanx (*difalaggia*) and circle (*orbis*)[79]

The hollow square/oblong, double phalanx, several hollow squares/oblongs side-by-side, wagon laager and double phalanx were actual combat formations, but the circle (*orbis*) was actually a unit formation, the equivalent of the *amfistomos falagx* (double-front, which actually faced all directions) in Greek theory.[80] Some of the authors may have confused the circle and hollow square with each other, because the corners obviously had the habit of becoming rounder if these were attacked by the enemy, but the proper name was the hollow square even in such cases. In Persian and Muslim usage and in naval warfare the circle was an actual defensive combat formation, but this was not the case for the Romans, excepting perhaps in the case of the use of the wagon laager as will be discussed below.

The principal all around defensive combat and marching formations used by the Romans therefore consisted of: 1) the double phalanx, with the intervals being protected by the reserves, light infantry and cavalry; 2) the hollow square/oblong; 3) several hollow squares/oblongs side-by-side; and 4) the wagon laager. The use of the wagon laager appears to have been a fourth- and fifth-century addition to the Roman repertoire, meaning that it was used from the ca. 366 until the 530s as one of the principal combat formations. Valentinian appears to have been the person who introduced this combat formation into the Roman army – although one cannot entirely preclude its earlier adoption by the Romans, for example in the third century when the Romans faced Goths employing the wagon laager. The use of several hollow squares/oblongs (each approximately 2,000- or 6,000-strong) side-by-side also appears to have been invented by Valentinian, which the Romans appear to have continued to use in the fifth century but not in the sixth.

The question of the double phalanx has already been discussed,[81] so the following analysis concentrates on the use of the hollow square/oblong proper, several hollow squares/oblongs side-by-side, and on the use of the wagon laager. The Romans had employed the hollow square/oblong as one of their basic marching and fighting formations probably as long as they had organized armed forces. It is because of this that we see the Romans using it during the campaigns of Julius Caesar (*BG* 8.8), Marcus Crassus (Battle of Carrhae in 53 BC), Tiberius (Battle of Klis in AD 9), Varus (Battle of Teutoburg Forest in AD 9) and Germanicus, so that it had become the principal marching- and combat-formation by the third century AD. We see a greater variety of battle formations in the fourth century until the 360s, but from the reign of Valentinian onwards the principal battle formations were the single large hollow square/oblong for the entire army, or the line of 2,000- or 6,000-strong hollow squares/oblongs deployed side-by-side or in depth when there were reserves.

Hollow square (plinthion) *and hollow oblong* (plaision)

The following four illustrations from the *Byzantine Interpolation of Aelian* (*Codex Burnley*) and Asclepiodotus (*tetragônon* = square; *tetrapleuron* = oblong) show how the hollow square and oblong were presented in the earliest versions of the military treatises. However, as will be discussed below, we can arrive at greater accuracy for the shape of the hollow square/oblong formation on the basis of a comparison between period evidence and the tenth-century *Syntaxis Armatorum Quaddrata*, which is included in the later version of the *Byzantine Interpolation of Aelian*, and on the basis of the tenth-century treatises *Sylloge Tacticorum*, *Praecepta Militaria*, and *De castrametatione* (*Peri katastaseôs aplêktou* = PKA).

The principal advantage of the hollow square/oblong as a marching- and battle-formation was that it provided an all-around defensive perimeter for the army. The enemy could not outflank it and subject the Romans to flank and rear attacks. Its primary importance lay in the psychological side of the soldiers. The soldiers would not automatically think that noises and commotion in the rear blocked their route of retreat, so lowering their chances of survival. In other words, there were fewer chances that the animal instinct of fight-and-flight would to turn into the latter option. The knowledge of the presence of the comrades-in-arms behind made the men in the front readier to fight.

The standard combat tactics with the hollow square/oblong consisted of several different variants. Firstly, it was possible to use the hollow oblong/square as if it was a regular lateral phalanx placed between two cavalry wings, with a separate vanguard of cavalry *prokoursatores* preceding it. In this case it provided a safe base for the cavalry to operate and place to withdraw if necessary. The cavalry vanguard obviously sought to surprise or ambush the enemy, and if this was impossible then it withdrew inside the square to serve as reserve force. The infantry was then used against the enemy in regular manner so that its front part fought as a phalanx while the Roman cavalry wings attempted to defeat the enemy's cavalry wings. If the enemy force consisted of cavalry, then the battle progressed very much like we find in the description of the *epikampios opisthia*, but with the difference that this formation was even safer against flank and rear attacks than was the case with the latter. If the enemy force included infantry, then the Romans probably attempted to outflank and defeat it in the manner described below.

Secondly, most of the cavalry could be placed in front to engage the enemy in advance of the infantry force. In such cases, the vanguard of the army consisted of the so-called *prokoursatores* (front-runners), who were placed in front to reconnoitre and to engage careless enemies. The cavalry followed it in regular formation. Unlike in the tenth-century, in the Late-Roman period this array would not have had cavalry rear-guards because the infantry square/oblong served as such. The cavalry was then used against the enemy in the regular manner. If the enemy force included infantry, the reaction to this depended on the time period. In the third-, fourth- and fifth-

plinthion and *plaision* in the *Byzantine Interpolation of Aelian* and Asclepiodotus

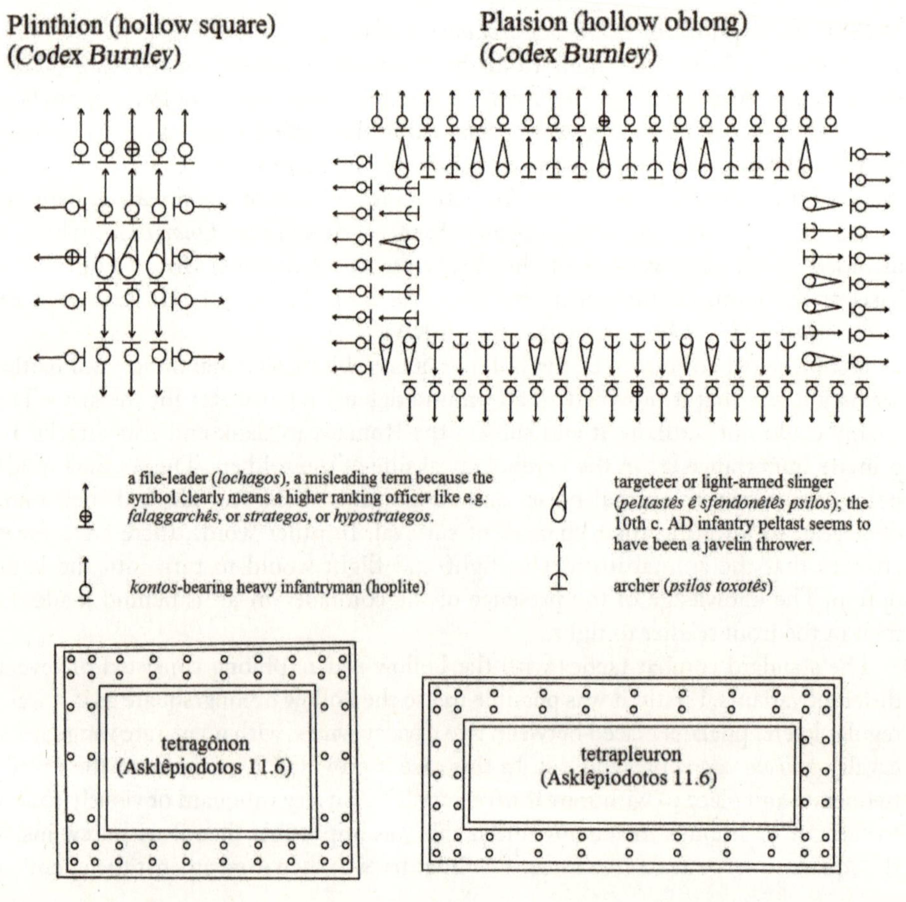

centuries, the Romans were in the habit of using their *catafractarii/clibanarii* against the enemy infantry, in the sixth-century the cavalry would have withdrawn inside the square/oblong, while in the early-seventh century they would once again have engaged the enemy infantry with their cavalry, as happened several times in the course of the wars of Heraclius.[82]

If the enemy force included infantry, there were several different ways for the next stages of the battle. Firstly, when the Roman cavalry had withdrawn inside the square/oblong and the front section of the infantry square/oblong had come face-to-face with the enemy infantry formation, it was possible for the Roman cavalry to sally out from one or two sides of the square/oblong against the enemy's flanks. The light infantry, especially the javelineers, could be used in support of this attack because these were considered particularly valuable in operations that required speed against infantry flanks. If the cavalry and light-infantry failed to achieve their goal, then the heavy-armed infantry moved to the attack, which meant the advance of one flank of the square so that the rest of the forces (the other three flanks, light-armed and cavalry) were used in its support as required by the situation. This is how we find the Romans operating already during the reign of Augustus in AD 9 when Tiberius attacked the fortress of Klis with a hollow square. The other three sides of the square/oblong protected the flanks and rear of the attacking side. We find similar instructions also in the tenth-century *Praecepta Militaria* (2.1–18) and Ouranos's *Taktika* (e.g. 56.12–3). Ouranos adds that if the enemy also employed hollow square/oblong then the light-armed were not to attempt to attack the enemy's flanks but were to support the front of the square/oblong against the enemy infantry. This is obviously common sense and one can expect that a similar procedure was also followed during the Late-Roman period because it would have been futile to attempt to break up the enemy's heavy infantry square with light-infantry flank attacks.

The hollow square/oblong could also be used as a secure base as if it was a walled city from which to launch attacks with cavalry so that the cavalry advanced from there against enemy cavalry and retreated inside it if pressed hard. This could be continued as long as was necessary, which could even mean days because most of the casualties among the enemy cavalry would have been caused by the Roman cavalry rather than by the infantry because the enemy cavalry was unlikely to charge straight into the Roman infantry wall of spears when the Roman cavalry threatened their flanks. We possess evidence for the use of the hollow square in this manner for the Armenian theatre of war in the early 370s:

> ...And the multitude of the legion, that is to say the shield-bearers of the Greek forces and likewise the shield-bearers of Armenia, supported the Armenian host enclosed within their shields, like a fortified city [standing] in the rear. Whenever the Persian army began to crowd the Greek army or the contingent of the Armenian spearmen, these [turned] to the Greek shield-bearing legion or to the shield-bearers of Armenia as though entering into a fortress, and sought respite. But when, barely having caught their breath, they sallied forth from there and attacked, they killed and struck off the heads of countless Persian warriors fallen before them. ... But when the Persian army gained over them

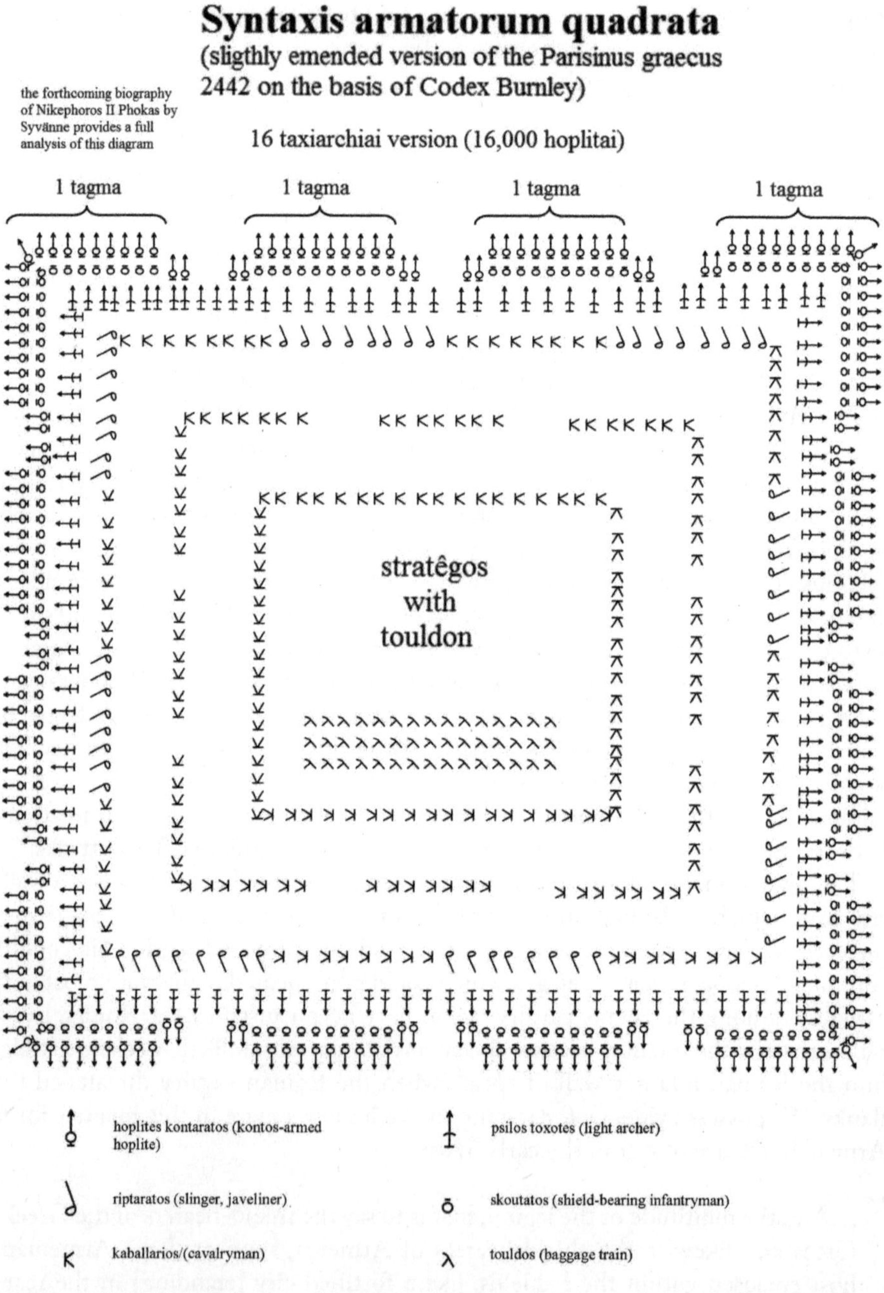

to some degree, they turned as to an impregnable fortress to the shield-bearing forces of the legion, and these, opening their shields, let them in and held them enclosed. And on that day, the Greek army with its stratelat, Terēnt, and the Armenian contingent with its sparapet, Mušel, struck down and destroyed the Persian host. And Šapuh king of Persia with a few men fled from the battlefield. *Pawstos (Faustos) of Buzand*, 5.5, translation by Garsoïan p.195.

As I noted in my doctoral dissertation back in 2004 (*Age of Hippotoxotai*) and *MHLR, Volume 2* (18–20, 143–7), the battle descriptions and number of legions (twelve) employed by Terentius correspond exactly with the tenth-century description of the smaller version of the hollow square formation which had twelve *taxiarchiai* in a hollow square. It is therefore now the time to elaborate the structure and shape of the Roman hollow square and oblong in greater detail.

It is clear that the Late Romans, and in all probability also their ancestors, used the hollow square in the same manner as the 'Byzantine Romans' in the tenth century. It was used by combined armies (consisting of both cavalry and infantry), pure infantry armies and also by pure cavalry armies when they dismounted to fight on foot. On the basis of later and period evidence (see the quote below), the two principal variants of the hollow square consisted of either 12 or 16 divisions which the Late Romans called *legiones* or *mere* and which the tenth-century Romans called *taxiarchiai* (sing. *taxiarchia*).[83] The organization of the different *mere/legiones/taxiarchiai* within this array could be varied according to the type of equipment worn. There was only one constant, which was the placing of the heavy infantry (*hoplitai, skoutatoi, aspidoforoi, gravis armatura*) or the shield-bearing dismounted cavalry at the outer edges, but when the Romans retreated before an enemy using cavalry forces, they usually placed foot or dismounted archers on the outside.[84]

The light infantry was usually placed inside the hollow square/oblong behind the heavy infantry, as shown in the attached illustration taken from the *Byzantine Interpolation of Aelian* (the tenth-century *Syntaxis armatorum quadrata,* similar to the *ST*), but since both Vegetius (3.14) and Syrianus (*PST* 16.40–53) had versions of units in which there could be light-armed placed between the heavy-armed front and rear, one may assume that the units were sometimes arrayed so that the each

SMALL HOLLOW SQUARE
Squares with not enough horsemen to form five *parataxeis* per side were to be formed as four *mere/taxeis/parataxeis* per side which does correspond with the number of gates in the 12 taxiarchy formation. (*PKA* 6, 8.24ff.)

LARGE HOLLOW SQUARE
16 taxiarchies with 8,200 horseman divided into 24 units/*taxeis* (six per side) on addition to which were the imperial reserves (1,000 men) with the emperor. If there were more cavalry, as there were expected to be, these were to be distributed to the sides according to this model (*PKA* 1, 81.-21). According to the *Praecepta Militaria* (2.15), one was to place 12 *parataxeis* into the 12 intervals, but in addition to this figure there appears to have been the reserves and scouts (2.3-8, 2.11). Essentially the same info can also be found from Ouranos' *Taktika.* It should be noted that unlike the *PKA,* these treatises consider each of the corners to be one taxiarchy with the result that the previous researchers have not noted the variant with 12 taxiarchies.

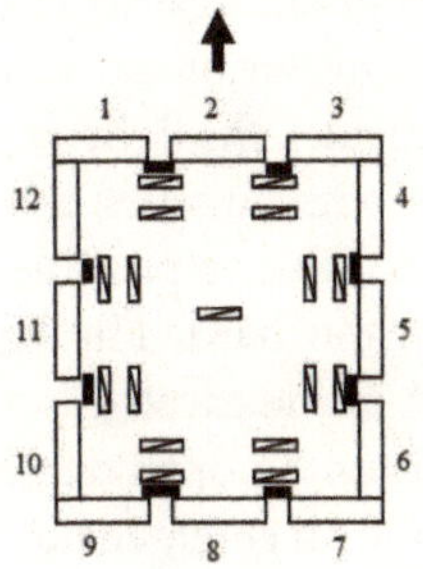

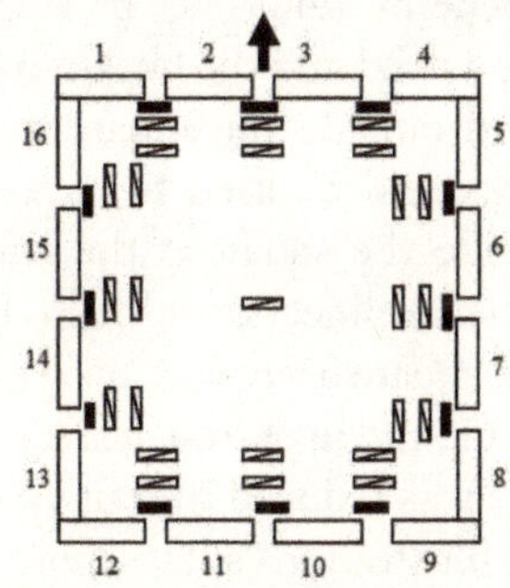

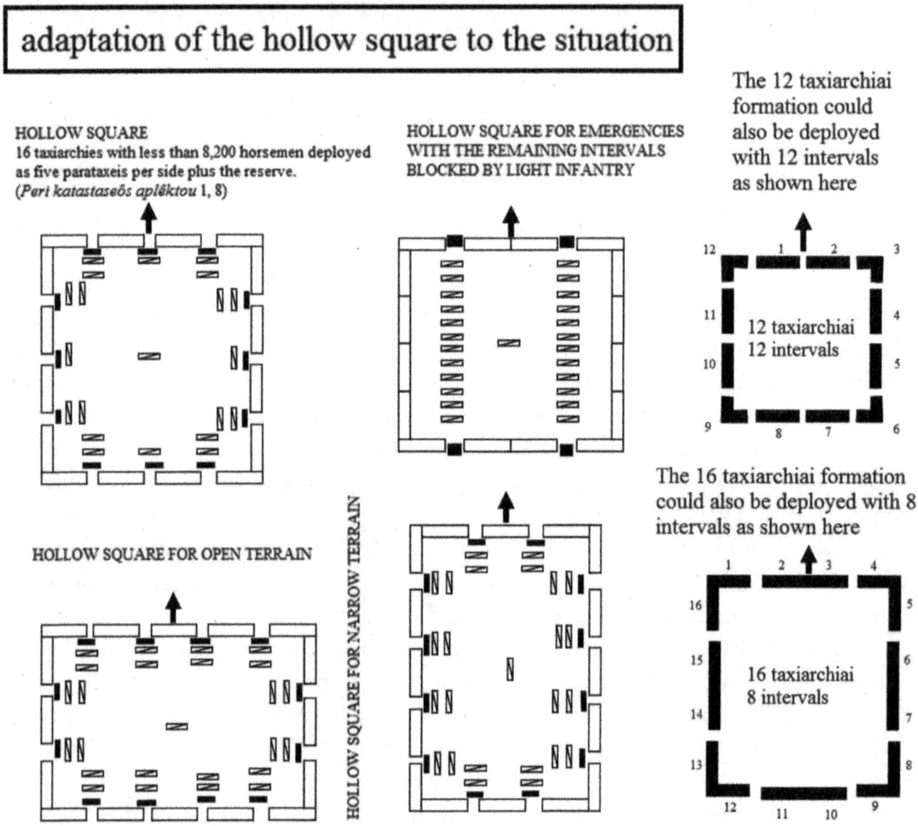

unit could face both forwards and backwards if necessary, as in the latter-half of the tenth century treatises *Praecepta militaria, De castrametatione* and Ouranos's *Taktika* – this provided protection for each unit if the enemy managed to penetrate inside the square. However, during this period the standard method of using the hollow square/oblong was to place the light-armed behind the heavy-armed. The expectation was that the light-armed would be able to fight and survive without assistance from the heavy-armed. The extra light infantry were placed behind each of the intervals of the square/oblong, with the javelineers forming the front ranks and the archers and slingers the rear. Their primary mission was to protect the intervals against enemy attacks so that the enemy would not be able to penetrate the interior of the square/oblong. If there was a need and the terrain was suitable (i.e. difficult), the light-armed could also be placed outside the square to protect the array with missiles against harassment or they could be used to harass the enemy. It was always possible for them to retreat inside the square if the enemy pressed them too hard. The light-armed, especially the javelineers, could also be used in support of the cavalry.

The shape of the square array was always adapted to the terrain. In open terrain it assumed a wider shape and in narrow it assumed a narrower shape. In emergencies, the number of intervals was reduced to four and the light-armed and cavalry were used to cover these. The illustrations above show the different variants. It is probable that

Battle of Iatrus in 598

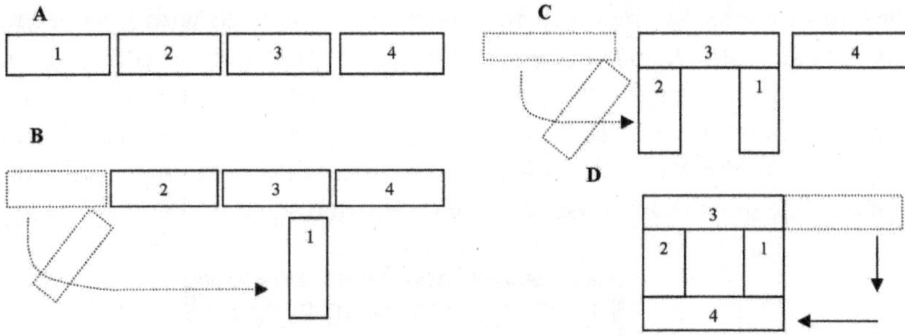

even if the Late Roman figure of 12 legions under Terentius corresponded roughly with the numbers given in the tenth-century hollow square for a 12,000-footman army, in practice during both periods the size was always adapted to the situation, as was also expected in the tenth-century treatises for both infantry and cavalry (e.g. *PM* 4.14, 4.18, 4.20), where if there were fewer footmen the line was made thinner and when there were more footmen the divisions (*mere, legiones, taxiarchiai*) were made wider while retaining the same number of intervals allowed in the larger, sixteen-*taxiarchiai* variant. The reason for this conclusion is that we know the Romans used smaller and larger armies and it is unlikely to be a mistake that Maurice allowed the use of unspecified numbers for the building of different sized *mere* in the large lateral phalanx array. He did not give any upper limit, because in infantry warfare the size of the division depended on the size of the infantry force.

The period military treatises fail to describe the manoeuvres required to form the hollow square and oblong formations. The basic manoeuvre was obviously wheeling. If the enemy did not threaten the Romans, the procedure could be performed in many different ways at their leisure, but how was it done in emergencies? Fortunately, we possess a description of how this was done at the Battle of Iatrus in 598 in Theophylact's texts. The attached simplified diagram explains the different steps required. The forming of the rear could also be performed by wheeling, but I have here followed the manner in which the Europeans formed it in the seventeenth century.

Hollow squares/oblongs side-by-side ('regimental squares')

As already noted a number of times, Valentinian introduced the use of several hollow oblongs and squares simultaneously so that the numbers of men in these appear to have varied between 2,000 and 6,000. The concept of using several hollow squares/oblongs side-by-side resembles the use of the divisional wedges side-by-side as a marching/combat formation,[85] and should therefore be seen as a further development of the same concept – in other words the use of formations that improved the survivability of the Roman units when these operated with wide intervals. In combat, these were used as if they were regular divisions of the lateral phalanx that used the *amfistomos falagx* unit order, but obviously with the difference that all of these units also had light-armed men posted inside their formation. This made each hollow oblong/square completely an independent separate unit that could fight against enemies with each of its sides, while also providing support for any neighbouring division in the same array, and this these units could do with greater security than was the case with the normal *meros*-divisions of the lateral phalanx. In other words, one can consider this to have been an improvement over the earlier practice, which went out of use by the sixth-century only to reappear as a new, modern era invention made first by the Russians in their wars against the Ottoman Turks in the 18th-century. The most famous use of these regimental squares obviously took place when

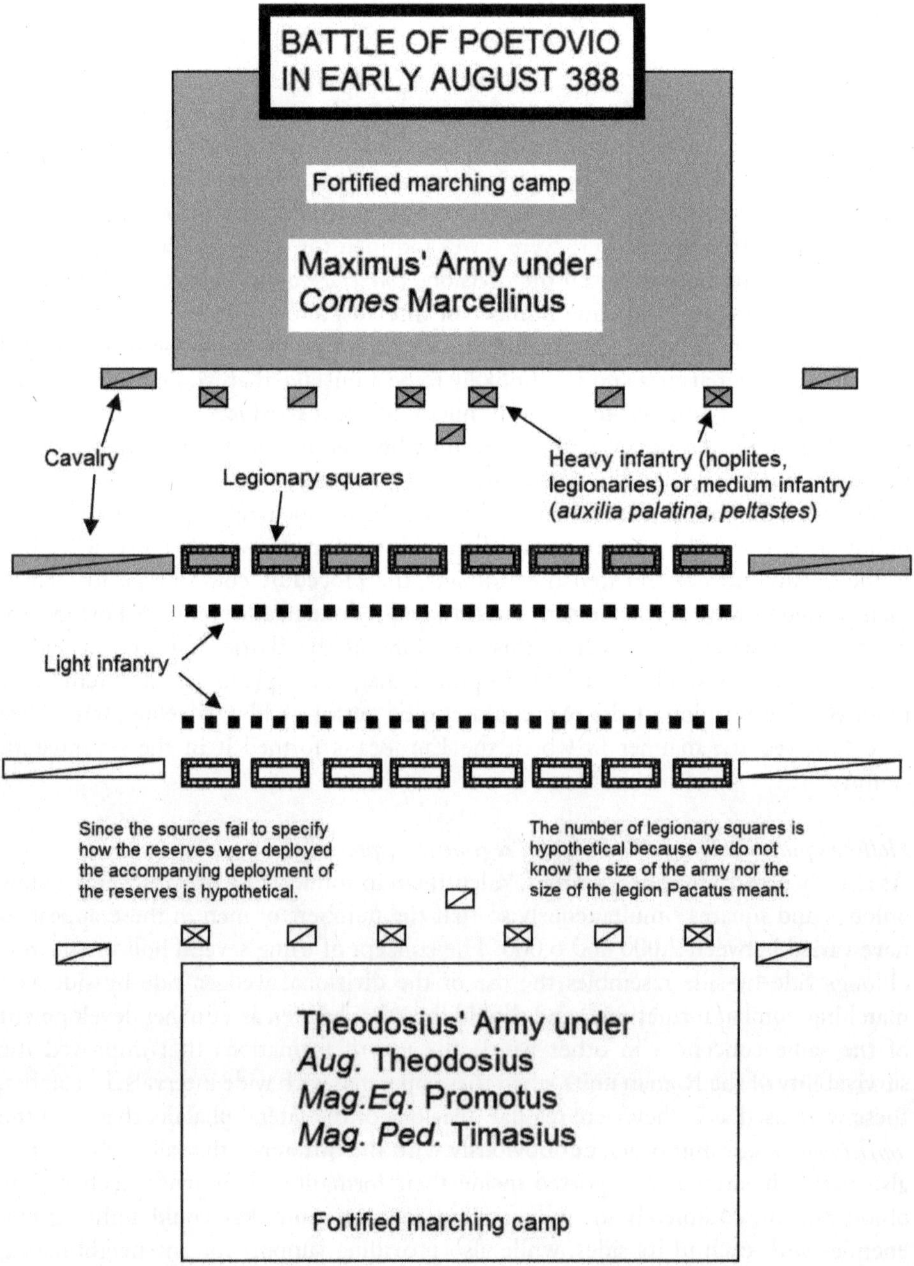

Wellington's redcoats defeated the French cavalry at the Battle of Waterloo in 1815. In short, the invention made by Valentinian was well-ahead of its time and forgotten soon after, even by the Romans. The drawing on the previous page shows the likely shape of the battle formations when both sides employed hollow oblongs side-by-side at the Battle of Poetovio in 388. For further details, see Syvänne (*MHLR 2*, 245–50: See also Syvänne, *MHLR 2*, 156–7).

Wagon laager (carrago, karagos)

The second of the innovations of Valentinian, namely the copying of the use of the wagon laager (*carrago, karagos*) from the Goths and others who used when serving in the Roman army, was also an important new development because now the Romans had a real moving fortress in their arsenal which had wagons, anti-cavalry machines and cart/wagon mounted artillery pieces placed as a defensive outer layer. The wagon laager had a longer lease of life than the use of the hollow squares/oblongs side-by-side because it did not require men as well-trained as the former. In fact, the wagon laager appears to have been one of the most important marching and combat methods for the period from ca. 366 until the 530s. However, even if the array was already the best-protected battle formation used by the Romans, the lowering standard of the infantry required the addition of even more protective features during the reign of Anastasius. Firstly, the oxen drawing the wagons had been demonstrated to be a liability if the enemy cavalry got a chance to shoot arrows at them. This was corrected by adding protective coverings to the wagons and with the strict instructions to hobble or tie the oxen which we find in the *Strategikon*. In addition to this, Urbicius, the *magister militum per Orientem* of Anastasius, introduced his own invention to make it even more secure. He added a form of 'Spanish-riders' well outside the wagon laager to serve as an extra security feature.

We do not know the exact ways in which the Romans used the wagon laager. Therefore, it is possible that they deployed and used it like the Hussites of the fifteenth century, which I speculated to have been the manner in which the Goths used it in the *MHLR, Volume 1* (82–3). The Hussites used both rounded- and rectangular-formations, so when one remembers that Vegetius (3.8) still allowed marching camps to be round in shape, it is possible that the Romans did indeed continue to use this shape at least sometimes when they used wagon laager. The following diagrams, drawn after Jan Durdík, show how the Hussites built their wagon laagers and it is entirely plausible that the Romans used the same methods.

However, I would still suggest that the Romans usually used their wagon laager in the same manner as the Russians did in the late-seventeenth century, where the wagons were drawn on all four sides of the marching hollow square or oblong so that the wagons and *carroballistae* formed a real fortress for the infantry inside, while the cavalry would have been deployed outside during the march.[86] If the Romans then encountered terrain that prevented this, they redeployed the wagons and infantry in such a manner that it was possible to march through the obstacle.

As noted, Urbicius considered the period infantry forces to be too poorly-trained to employ the unit manoeuvres he described in his *Taktika* and too poorly trained to resist the impetus of an enemy cavalry charge. His solution to this was that the

The Hussite marching order according to Jan Durdík, 158.

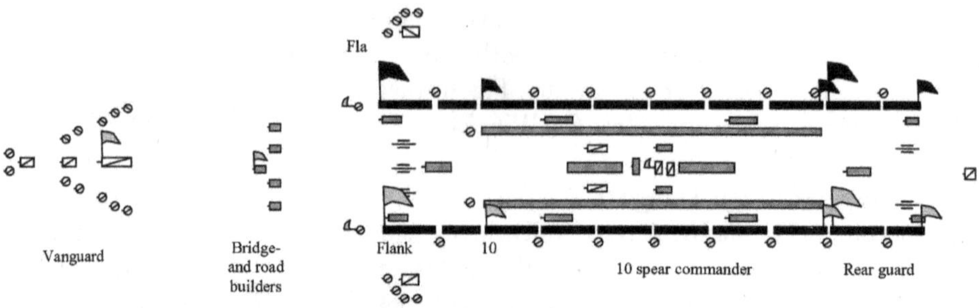

The closing of the wagons by the flanks and rear: Durdík, 161.

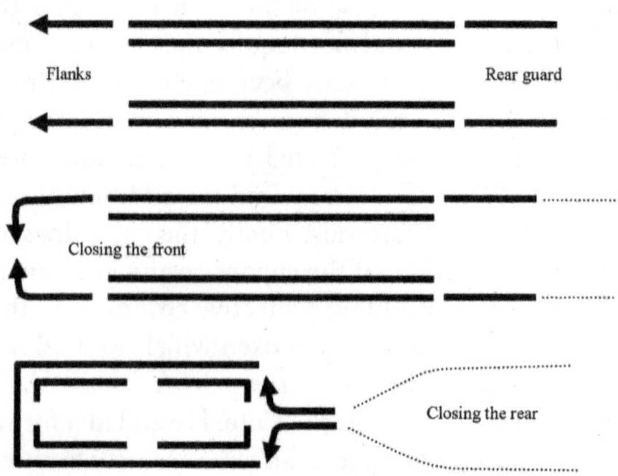

The closing of the wagons according to the lay of the land: Durdík, 161.

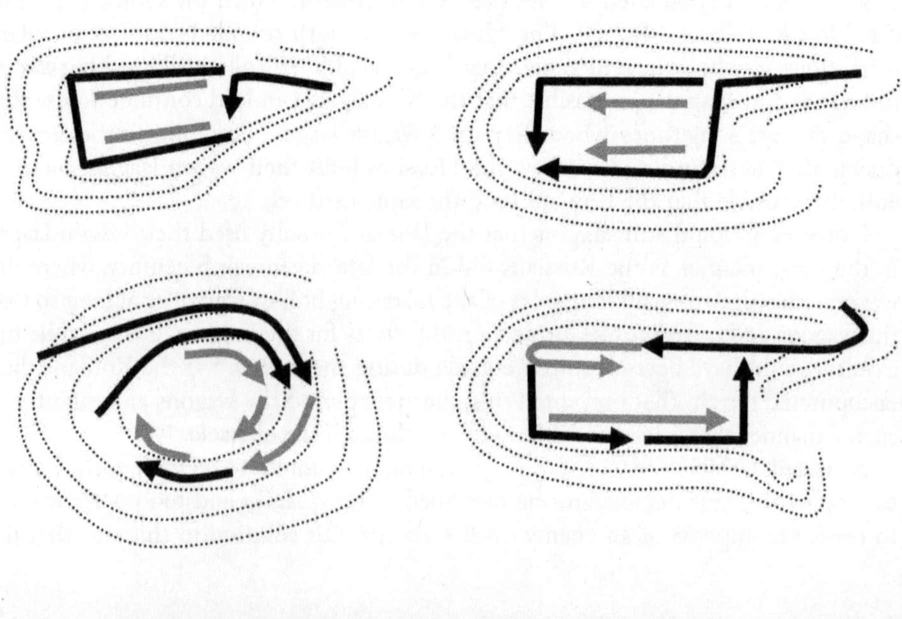

infantry was to use only one combat formation, the hollow square with wagons, *ballistae* carts and his new anti-cavalry device, which was also to be used during marching and encamping. As I noted in the *MHLR, Volume 5*, this minimized the manoeuvres that the infantry had to make and gave every soldier the illusion of safety because his rear and flanks were always protected.

Urbicius's wagon laager had the extra protective feature of a 'fence' (the new invention built according to his instructions) built outside the infantry square at such a distance that enemy arrows would be unable to reach the soldiers while they were arraying *ballistae* in movable wagons (*carroballistae, ballistroforoi hamaxai*) on the outer edges of the square to shoot at the attacking enemy horses. According to Urbicius, the *ballistae* could shoot three times as far as bows, so this worked well. He foresaw a situation in which the Roman arrows would annihilate the foremost barbarians just in front of the fence so that the horses and men would pile up there. Some of the fallen enemy horses would also become stuck in the nails of the fence. The barbarians behind them would further collide with the obstacle while trampling their fallen comrades. According to Urbicius, if the enemy dismounted and attempted to dig up the poles of the fence, the Roman soldiers could move forward to kill them with their spears if the *ballistae* had not yet done that. The expectation was that when the enemy saw their front ranks annihilated that they would be so demoralized that they would flee. When this happened, the Romans were to pick up the fence, place it back on the pack animals, and then pursue the enemy, and repeat the above if the enemy appeared.

We unfortunately do not know the appearance of the device invented by Urbicius because his drawings have been lost. We only know that each *decuria* of soldiers took three poles fitted with nails from a pack animal carrying 30 such poles for a

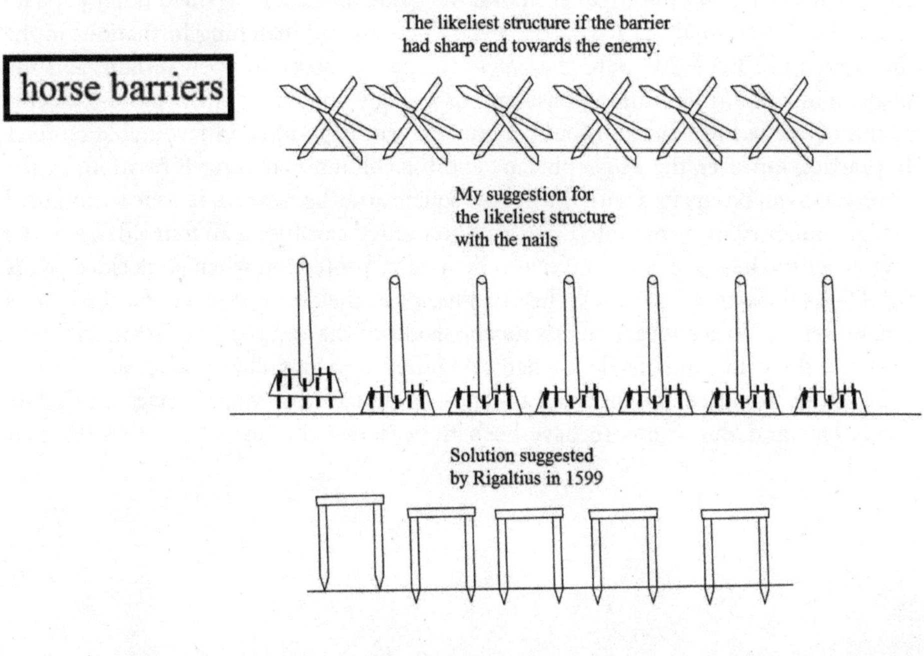

group of ten *decuriae* and then drove these into the ground outside bowshot range when they learnt of the approach of the enemy. Urbicius likened the end result to a fence which had nails protruding in such a manner that the fallen horses became stuck in these. Unsurprisingly, Urbicius thought that his fence was far better than the barbarian manner of using a circle of wagons for protection, because the latter was easy to destroy by fire if the attackers showed some initiative. With this he refers to the defeat suffered by Hypatius at Acrae in 514 which appears to have been caused by the use of fire arrows against the wagons and oxen. However, it should be noted that Urbicius is slightly inaccurate in his criticism of the wagon laager, because his version with the *ballistae* carts and wagons was still a version of the same. The attached illustrations show some possible shapes for the anti-cavalry fence of Urbicius.

The improving quality of the Roman infantry during the 540s and 550s resulted in the abandonment of the wagon laager as a marching and combat formation, with the result that it is no longer included as a combat formation in the texts of Syrianus Magister and Maurice.[87]

Hollow square/oblong, marching camp and double phalanx in the Strategikon

Maurice no longer considered the hollow square or the wagon laager to be of relevance as battle formations. He restricted even the use of the hollow square and oblong only to situations in which the army was forced to retreat. The following list shows how Maurice envisaged the use of the hollow square/oblong or double phalanx in different circumstances: infantry *tetragonos/plinthion* or *difalaggia* during retreat (*STR* 7.B.11.45–52); infantry rear guards in the *epikampios opisthia* formation deployed almost as a hollow square *plinthion* (*STR* 12.A.7.21, 12.A.7.66); a four-sided formation (*tetragonon*) used as a marching camp (*STR* 12.B.22, esp. 12.B.22.52–8); and infantry used in hunting to surround the game inside a four-sided hollow square (*plinthion tetrapleuron*) (*STR* 12.D.138–9). The standard marching formations in the *Strategikon* (12.B.19–20) were the *plagia* (lateral) phalanx in open terrain, and the phalanx in column formation consisting of a single column (narrow passes), double column (wooded terrain) and four columns (a very large force in favourable terrain). In practice, however, the *plagia* phalanx and the column marching-formations of the *Strategikon* can be equated with the hollow square array because these were surrounded by light-infantry, or dismounted cavalry, or mounted cavalry on all four sides, and the wagons of the baggage train could also be used as protection when so decided (*STR* 9.3.87–9.4.65 with 9.3.87–9.4). The other uses that the *Strategikon* had for the hollow square were: 1) to act as rear guards for the flanks of the *epikampios opisthia*; and 2) to surround the wild game inside it when using hunting as a military exercise.

In short, Maurice did not see the hollow square nor wagon laager as battle formations and this seems to have been the case for the rest of the Late Roman period.

9.8. Regular Cavalry Formations with Infantry Support

There also existed a mixed formation that consisted of the regular cavalry formations of one or two lines and their infantry support in front of a camp or city. The battle lines were to be posted an arrowshot apart from each other. These formations have already been dealt with in Chapter 6.7. The role of the infantry in these formations was minimal. They were only to provide a place of refuge for the cavalry. In other words, in such cases the infantry gave the cavalry a chance of regrouping, or at least they gave the cavalry a chance of withdrawing without chaos inside the fortifications behind them.[88]

Chapter Ten

The Infantry Battle

10.1. The Deployment of the Army

The ideal pre-battle procedures for the infantry army were the same as for the cavalry army. The army was required to use a separate marching camp closer to the enemy, adequate amount of provisions were to be at hand, the soldiers and horses were to be rested before battle, a council of officers was to be held, and common prayers and speeches by the general and officers were to be delivered to the men. The men were prepared physically, emotionally, and psychologically for combat. The soldiers and general obviously still felt the stress of the forthcoming battle, but so did their enemies. We find descriptions of both from the sources too.[1] When the army was arrayed for battle, the Romans posted some *banda* of cavalry nearer to the enemy to prevent hostile reconnoitring or harassment of the infantry while it was in the process of deploying. Then the rest of the army was marched out of the camp in the order of heavy infantry first, followed by the wagons. What happened next depended on the chosen battle formation. The infantry army was not expected to march more than two miles (c. 3km) from the camp so that it would not become tired through marching.

10.2. The Role of Generalship and Officers[2]

Before the battle the role of the general was of paramount importance, but his ability to make the right choices depended on the availability of accurate information about the enemy. In other words, the availability of information provided by the spies, scouts and patrols dictated what the commander could achieve with his personal skills and army. It was the duty of the commander to ensure the most favourable circumstances for the battle. He had to choose the right time and place for the battle, and also the right combat formation and tactics to use in that specific situation. Ideally, he was able to choose the right type of terrain for the battle, together with the wind direction and the sun shining from behind the Roman army. In the case of the infantry phalanx, this usually meant open and level terrain if the baggage train followed, but when the Romans had inadequate numbers of cavalry while the enemy had an abundance of cavalry the ideal location for the battle was wooded, rough and hilly terrain. If the enemy possessed a numerical superiority and the Romans still decided to engage them, they could use the terrain or manmade structures to negate this. The typical solution in such circumstances was to post the army behind fortifications, for example as Belisarius did at the Battle of Dara in 530 and the Siege of Rome in 537, or in a pass that protected the flanks, or resting the army against a river, lake, swamp or mountain to protect one flank. On the basis of the narrative sources and the tenth-century *De velitatione bellica*, we know that in such situations the shield-bearing infantry was deployed in the pass as a phalanx or double phalanx so that they blocked the route, while the men armed with slings, bows and javelins occupied the elevated (or woody) sections. For example, in 399 the Pamphylian paramilitary forces under Valentinus of Selga defeated the Goths of Tribigild by using an infantry phalanx to block the route while the slingers held the heights (Syvänne, *MHLR 3*, 92–3). This stands as a good example of the effectiveness of local citizen militias in the right conditions.

In open terrain it was possible to advance against the enemy if it consisted of infantry as this bolstered the morale of the advancing men, but if the enemy force consisted of cavalry, then the Romans usually waited for the enemy to attack. If the Romans had an advantageous higher position or had posted their army in difficult terrain because the enemy consisted of cavalry, then the Romans obviously did not leave their advantageous terrain but waited for the enemy to attack. Each arm of service had its own role to play in the infantry battle.

This was the ideal, but in practice the Late-Roman commanders were not always able to force their will on their subordinates or on their soldiers, and these forced the commanders to act against their wishes. It was also possible that the spies, scouts and patrols had not performed their duties well so the Romans were surprised, as happened for example at the Battle of Adrianople in 378. However, in most cases the spies, scouts and patrols acted as expected and the Roman commanders performed their duties at least adequately. Similarly, we possess instances in which the commander in charge was a rash fool – once again the Battle of Adrianople stands as a good example of this. The successes that Roman arms usually had in battles during the fourth and sixth centuries prove that the most advantageous plan was most often the one chosen.

The Late Roman period also witnessed jealousy and rivalries between officers, which sometimes required imperial intervention in the form of posting a *magister officiorum* in the theatre of operations, or the placing of *scribones* on the general's staff (served also as a security measure against the commander), or the giving of an imperial letter confirming the supreme position of the commander. In the worst case scenario these rivalries resulted in violence and murders. However, in most cases the subordinates still obeyed their superiors – although during the first-half of the sixth-century this often require imperial intervention or the presence of large numbers of personal *bucellarii* of the commander, as was the case with Belisarius. Furthermore, we should not stress too much the importance of the jealousies and rivalries between the officers and the resulting frictions, because these are ever-present in any army or organization. The generals were humans with their own ambitions and personalities.

More important than this was actually the influence of the massive corruption in both the higher and lower echelons of the Roman armed forces that started during the reigns of Valentinian and Valens. This increased friction between military factions and lowered the general quality of the military leadership and armed forces. After this there were periods during which the general standard of Roman generalship was really poor, largely because corruption resulted in the broken hierarchy of the leadership.

After the fall of Stilicho most of the West Roman commanders were utter failures. The only exceptions were Constantius III, Aetius, Marcellinus, Avitus, Majorian, Aegidius and Ricimer, but their abilities could not save the Empire when the meagre resources were wasted by incompetent commanders and especially when the command hierarchy broke down, so that, for example, Marcellinus and Aegidius were de facto independent rulers of their own domains, not to mention the position of Ricimer as kingmaker (he appointed a number of puppet emperors). The position of the general therefore depended not on the competence but on the perceived loyalty of the appointee towards his patron. The incompetence led to military failures.

The general competence of the East Roman commanders during the fifth century, in particular between the years 457 and 518, also left a lot to be desired. After the death of Theodosius I in 395, the military hierarchy had broken down so that for most of the time the emperors were not free to appoint and sack commanders as they wished but were entirely reliant on the different power blocks within the military establishment, which were often in violent competition with each other. During the fifth century only Theodosius II and Anastasius I had the authority to walk over their generals – in the latter case, however, this resulted in the appointment of the incompetent nephews of Anastasius to the highest positions. It is largely because of this that we see incompetent leaders on the top ladders of the East Roman military establishment. These incompetent leaders include for example Aspar, Zeno, Basiliscus and Illus. The only exceptions to this rule were Anthemius (while still a general in the East), Marcian (when he was emperor), Sabinianus the Great, the Ostrogothic king Theoderic the Great (but often in revolt against Rome), John the Scythian, John the Hunchback, Romanus, Celer, the rebel Vitalianus, and Marinus, but when able commanders were placed in command of some of the major armies their successes were more than counterbalanced by the massive failures caused by poor leaders, the most serious of which was obviously the destruction of the large amphibious landing force that Basiliscus led against the Vandals in 468.

In contrast, the general standard of Roman commanders during the reigns of Justinian I, Justin II, Tiberius II and Maurice was relatively high – obviously there were rash and poor commanders among them but most of the defeats were caused by imperial-level decisions rather than by tactical-level decisions. The victories achieved during the reign of Justinian bespeak of the improved standards of generalship. It was during those years that Belisarius, Narses and Germanus achieved great victories with numerically inferior forces. The general standard of generalship reached its apogee during the reign of Maurice who required his officers to be literate and well-educated. Petrus (brother of Maurice), Comentiolus, Droctulfus, Romanus, Priscus and Phillippicus were all very competent military commanders.

The next period during which most of the East Roman commanders were utterly incompetent took place towards the end of the Late Roman period, between the years 602 and 641. During those last dark years of the Late-Roman era the only commanders to demonstrate any real ability were the emperor Heraclius and his brother Theodorus, and from the former generation of generals the elderly Priscus and Philippicus.

The variable quality of generalship was the result of two things: 1) there did not exist any military academy in the modern sense of the word where the officers would have received a well-rounded education of military theories and history, meaning that it was up to the emperor (or the power behind the throne) to decide what standards he used in the appointment of high ranking officers; and 2) rising through the ranks was not necessarily the result of ability, but rather the result of a patron-client relationship. This meant that the commander was not necessarily in any position to choose the best infantry formation or tactics to the situation at hand because he simply did not know of its existence. It depended on the level of education he had received, and in order to be well-educated in the military theory of the Graeco-Roman times one had to be well-versed in military history and theoretical works dealing with the battle formations. It is because of this that the emperor Julian, who did not have any military background, was able to choose the best combat formations and siege tactics for each situation during his years in charge of the Roman armies: the narrative histories and military treatises had less to say about logistics so it is not surprising to find that Julian, who had learnt his military trade solely from books, failed in this respect when he did not listen to the professional soldiers who had had experience of this side of the military life, although they were probably being less well-educated in the military formations and siege equipment.

When the Romans were commanded by competent generals and the spies and scouts had performed their duties well, the Romans could expect to win thanks to the fact that their armed forces consisted of professional soldiers, with all arms of service included so that they could adapt their tactics according to the enemy and situation. The fact that Roman society was technologically the most sophisticated on earth enabled them to also use military engineering to their advantage. When led by competent commanders the Romans could overcome the enemy by exploiting their weaknesses.

The role of the *strategos* in an infantry battle varied greatly. It depended on the battle formation and on the personal preference of the general. The recommended place for him was in the middle, so that messages would reach him quickly, but he was often in charge of the wings or cavalry. The general could influence the course of the battle with: his use of the reserves; by showing a personal example in the forefront; by leading a section of the front by standing just behind it; or by using the standard stratagem of claiming to the troops on one flank that the Romans on the other wing had already been successful.

The position where the general stood was important because the direct impact of the general was usually felt only in the direct vicinity of his person. It was because of this that the army had a command hierarchy and officers in charge of the units. The general could increase his impact by moving about behind the army and by sending commands via flags, trumpets, horns, couriers, and heralds, or in person with verbal commands. The other officers used trumpets, horns, verbal commands, flags, and hand signals for the same purposes. The distances between the different sections of the battle line and the terrain sometimes influenced the ability of the general to observe what was happening. This was particularly true when the army consisted of huge numbers of men, as was the case when hundreds-of-thousands of men fought at the Battle of the Catalaunian Fields in 451. In that battle, Aetius was entirely unaware of what had happened on the right wing because the length of the battle formation was tens of kilometres. This battle, however, was an exception. In most cases the commanders could observe the entire battlefield with their own eyes if they managed to find a higher locale nearby (there could also be a raised rostra prepared by the Roman soldiers for this purpose) or were mounted on an elephant like the Persian generals. The fact that the generals were mounted also helped them to observe the course of the infantry battle because they could post themselves behind the infantry.

The role of the other commanders was more restricted. The commander of the infantry, the *hypostrategos*, usually led the whole infantry line only until the moment of the contact with the enemy, after this he acted just like any other wing commander in that he was in charge of only the centre of the lateral phalanx. It is probable that he was usually in the centre of the other battle formations too, but it is less clear where he was when the Romans used the mixed formation which was organized differently. The other wing commanders (left and right wings) and the commanders of the *mere* and *tagmata* were also expected to be able to operate independently without waiting for instructions from the *strategos* if the situation required this.[3] It is therefore clear that Roman infantry combat doctrine expected that the general standard of the commanders of the wings (left, centre, right), *mere* and *tagmata* to be very high. Since the Romans were usually successful in their infantry battles, or at least able to retreat safely, it is clear that the general standard of the subordinate commanders remained relatively high for the most of the period (excluding the late-fifth and early-sixth centuries), even during those times when the *strategoi* consisted of incompetent men who had achieved their position through corruption. This is not surprising, because these lower-ranking officers usually consisted of the men who had risen through the ranks so they could be expected to know at least the basics.

The battle formation was held together by the mutual trust that the officers and men had towards each other. The general, wing commanders, and commanders of the *mere* and *tagmata* had to trust each other so that they did not fear to be abandoned on the battlefield. The officers also had to have the trust of their subordinates. The soldiers and officers in the *tagmata* and members of the tent groups had to be able to trust in each other. Graeco-Roman combat doctrine recognized the importance of mutual comradeship in the usage of the tent group, and in the usage of soldiers from their native area in their own units. The two most important factors influencing the ability of the *strategos* to lead his men were his reputation as general and his reputation as paymaster. Successful and generous generals known for their largesse could expect their soldiers to fight well while a commander who showed signs of being greedy quickly lost his soldiers' respect.[4] The ability of the *strategos* to command the army depended also on the actions of the emperor. If the regular payment of salaries was in arrears (which happened often during the reign of Justinian), or the emperor lowered the salary (Maurice), or interfered with the distribution of war booty (Justinian and Maurice), the morale of the army suffered.

In normal circumstances the fighting was done during the spring, summer and early autumn, so if the emperor ordered the men to continue the campaign during the winter problems could ensue. This is what happened when Maurice ordered the men to winter north of the Danube, with the result that he was overthrown in a military mutiny. The same was not true for all the Late Roman period, because there were also instances in which the soldiers willingly fought during the winter months.

The commanders of the wings, *mere* and *tagmata* reacted to a situation by using the unit manoeuvres already discussed above. The *strategos* and *hypostrategos* could extend the formation on both sides to outflank the enemy or to prevent outflanking, or they could send light-armed or reserves to the flanks for the same purposes. The left- and right-flank commanders could prevent outflanking or attempt to outflank the enemy, for example by using a combination of cavalry wings, cavalry reserves, light infantry and heavy infantry reserves. The left- and right-wing commanders and commanders of the flank *mere* could also widen the width of the formation by dividing the phalanx, by extending the line, and by using the light-armed or reserves for this, followed by the use of wheeling. They could also deepen the formation for a breakthrough with doubling or by using reserves, or they could form a wedge to break an enemy cavalry attack, or for the purpose of pushing through the enemy formation, or they could form a hollow wedge/pincer/scissors against an enemy wedge. The commanders of the wings, *mere* and *tagmata* could also open their formation to allow the enemy cavalry, infantry, elephants or chariots through so that these could then be engaged with an *antistomos phalanx* and *drouggoi* of light infantry.

If the enemy approached from the rear, the *strategos* could counter-march the army, or use double phalanx or *amfistomos falagx* to face the enemy from the rear. The individual commanders could also use the *amfistomos falagx* array for the support of the units next to them. Further, commanders could also employ independently-operating *globi* (*drungi, drouggoi*) in front or on the flanks of the battle formation, either to break up the enemy formation or for outflanking. If the fighting took place in difficult terrain, the use of these was *de rigeour*, but their use could also be advantageous even in battles that took place in the open terrain. In certain circumstances the commanders could also decide to use the saw (*serra*) formation, but its use appears to have been abandoned by the time Maurice wrote his *Strategikon*. It was particularly important for the officers of the cavalry to ensure that their cavalry forces would not become separated from their infantry support during fighting or during a pursuit of the enemy. In combined-arms combat it was important to ensure that the infantry and cavalry cooperated smoothly on the battlefield, because when it did not, as happened at the Battle of Yarmuk in 634, the results could be disastrous for the Romans.

If the initial attack or defensive measures did not result in the flight of the enemy, so the battle became prolonged with the result that the battle lines withdrew a short distance from each other, it was possible for the *strategos* and his subordinates to reorganize their lines. A pause like this enabled the *strategos* to regain command of the whole formation from the divisional commanders so that he could try something else. A good example of this phenomenon is the Battle of Mons Lactarius in 552 (*MHLR 6*, 344–6), which actually had several pauses because it lasted for two days.

10.3. The Role of Cavalry in the Infantry Battle[5]

The military treatises in the Hellenistic tradition give us a good overall picture of the use of cavalry in infantry battles. Asclepiodotus summarises this aptly as follows:

> Now the cavalry, like the light infantry, take their positions according to the demands of battle, and especially is this true of skirmishers; for these are the most useful to draw first blood, to provoke the enemy to battle, to break their ranks, to repulse the horse, be the first to occupy points of advantage, carry such positions as the enemy have already occupied, reconnoitre terrain that looks suspicious, lay ambuscades, and in general to open and support the struggle; for by their swift manoeuvring they render many valuable services in battle.
>
> Asclepiodotus, 7.1. Tr. by C.H. and W.A. Oldfather.

The role of the cavalry in an infantry battle varied according to the type of formation used and according to the battle plan of the commander. The cavalry could be used for hiding the presence of infantry from the cavalry based enemy or for pre-battle skirmishes. In some formations, for example the *taxis allê*, it is clear that it was used for outflanking the enemy and pursuit, while in other formations, such as the lateral phalanx, the *epikampios opisthia* and the *emprosthia* formations, their role depended on the battle plan and composition of the cavalry force. In the *epikampios* arrays, the cavalry was primarily used for the protection of the infantry formation and for pursuit, and only secondarily for offensive purposes. In the lateral phalanx, the cavalry could be used either defensively for the protection of the flanks and rear or offensively for outflanking. If the array included *cataphractarii/clibanarii* these could also be used for breaking the enemy infantry formation, but their use was no longer seen as advantageous by Maurice, who did not recommend the use of large numbers of cavalry in an infantry battle – the cavalry only protected the flanks against the enemy cavalry and performed the pursuit. In the *Strategikon* the final outflanking of the enemy infantry formation was always performed by the Roman infantry, even when the cavalry had already outflanked the enemy by forcing the enemy cavalry away from the field with the best *banda* posted on the extreme flanks. However, preferences changed once again during the reign of Heraclius because he and his commanders used the armoured cavalry for frontal charges against enemy infantry.

The encircling cavalry and mounted archers could be used to pin down the enemy infantry by arrows and by threatening to charge, which in its turn enabled the Roman infantry to perform its outflanking movement unhindered. The best examples of the use of horse archers to pin down an enemy are the Battles of Taginae and Casulinus River. Maurice recognized the general truth that frontal charges against a well-motivated heavy infantry in close order formation were unlikely to succeed, but this view was not held by all Late-Roman military thinkers, and it is also clear that there were several instances in which the *clibanarii* cavalry did indeed break up the opposing infantry formation.[6] Regardless, it is still clear that Maurice's approach to the use of the cavalry was the safer option. If the Romans used the mixed formation, then the infantry served only as a base of operations for the cavalry and the Roman cavalry was used only for frontal charges against the enemy cavalry: if the enemy held its ground, the cavalry retreated into the intervals between the infantry *mere*.

Nonetheless, the narrative sources suggest that the role of cavalry was often decisive or at least very important in infantry battles. The reason for this was the fact that the greater mobility of the cavalry often enabled them to outflank the infantry formation which usually resulted in a collapse of morale among the outflanked enemy. Only the best trained and well-motivated infantry could hold its ground when encircled. The posting of the best *banda* at the outer edges of the array in the *Strategikon*, the use of light-cavalry on the outer edges in the *Epitoma rei militaris* and *Peri strategikes*, and the use of the 500-strong flank guards in the *De scientia politica dialogus*, reflected the importance of archery in cavalry warfare. Archery, whether by foot archers or by mounted archers, was considered an effective countermeasure against a cavalry charge.

Regardless of the battle formation, the cavalry performed its pursuit of the enemy with the units assigned as *koursores* performing the pursuit proper while the units designated as *defensores* protected them. When the Romans used the combined-arms system, the pursuit of the enemy cavalry was not performed with the same vigour as when the Romans used a pure cavalry army, because in the combined-arms formation it was dangerous for the cavalry to be separated from its infantry support by more than three- to five-bowshots.

In the lateral phalanx the cavalry reserves posted just outside the wagons were deployed in a very open formation so that they could perform wheeling and other manoeuvres with ease. The reserves were expected to protect the rear and flanks of the formation and to support the cavalry wings, acting as flank guards for the front line and its cavalry wings.[7] It is probable that the cavalry reserves were deployed in like manner, in open order with the other infantry formations so that they could be used for the protection of the flanks and rear with greater ease.

If the cavalry was forced to retreat, it was to reform itself in the space between the phalanx and the wagons when these were using the lateral phalanx, and presumably also when it they were using the *taxis allê* and *epikampios* formations. And if they still could not hold out, they were expected to dismount and defend themselves on foot. If the cavalry dismounted, it meant the adoption of the *epikampios opisthia* formation, or square formation if the infantry at the same time assumed the two-phalanx formation. If the Romans used the hollow square/oblong formation then the cavalry retreated inside the square/oblong, and if the Romans used the mixed formation, the expectation was that the cavalry would regroup between the infantry *mere*.[8] In the case of the 'regimental' and 'divisional' hollow squares/oblongs deployed side-by-side, the only evidence for the place where the cavalry retreated to consists solely of the Battle of Poetovio in 388 and the Battles of Stilicho, but we see Roman cavalry retreating only at the Battle of Pollentia in 402.[9] The evidence such as we have suggests that when the infantry was deployed side-by-side as hollow squares/oblongs that the cavalry usually retreated behind these in the same manner as if these infantry squares/oblongs were normal *mere* of the lateral phalanx.

10.4. The Role of the Light Infantry *psiloi/levis armaturae*[10]

The Late Romans used the light infantry in battles just as it always had. Aelian summarised it as follows:

> The javelin-men, archers, and all other skirmishers are useful for opening an engagement, for drawing the enemy out of formation, for breaking their arms and equipment, and wounding and killing them from a distance, for breaking up their formations, for driving off their cavalry, for occupying commanding positions more quickly, for forcing those who had occupied them beforehand to retreat, for reconnoitring suspect positions, and for preparing ambushes. Aside from all this, they are also used as skirmishers in front, for tactical co-operation, as skirmishers in the rear, and generally for rapid movements and attacks from a distance, and thus achieve many and significant results in fighting.
>
> Aelian 17.1. Tr. by A. M. Devine p.50. Note the similarity with the functions of cavalry in the ancient treatises.

The principal asset of the light infantry was its manoeuvrability and flexibility on the battlefield. Before the battle the *psiloi/levis armaturae* could be used to occupy key strategic positions, and they could be used for pre-battle skirmishing – in other words, the light-armed could be used to disorder the enemy before the actual engagement took place. In difficult terrain the *psiloi* were used for the protection of the main line and for outflanking the enemy, and if the terrain was really forbidding the *globi/drouggoi of psiloi* actually served as the main striking force for the entire army while the rest of the forces remained behind.

In regular battles in open terrain the light-armed could be placed between the heavy infantry files (*entaxis*) or in front of the heavy infantry units (*prostaxis*, a sort of saw formation), or in the rear position at an angle behind the flanks of the phalanx (*hypotaxis*, a sort of *epikampios opisthia*) to counter an enemy cavalry attack on the flanks, or behind the phalanx (*epitaxis*). On the basis of the *Strategikon*, the javelineers and dart-throwers were usually deployed behind the heavy infantry or on the flanks of the phalanx (and not in the middle), but the *Peri strategikes* (16.40ff.) suggested that the javelineers (equipped either with the shorter *doration*-spear or *akontion*-javelin) could also be posted among the inner files when the outer edges of the phalanx consisted of men carrying the regular long *dory*-spear. The difference between these is not as great as it would seem at first glance. Maurice meant the javelineers, which belonged to the light infantry class (*psiloi*) while Syrianus Magister referred to *aspidoforoi* (i.e. *skoutatoi*) class who were equipped either with *doratia* or *akontia* instead of the regular *dorata* (cavalry *kontaria* of Maurice). According to the *Strategikon*, the slingers were usually placed only on the flanks, but it is clear that they were sometimes places in front to skirmish or behind the other light-armed troops.

On the basis of Vegetius and Syrianus Magister, it is clear that the Romans also possessed multipurpose heavy-armed men during the Later Roman period who could be used simultaneously as infantry archers and regular heavy infantry. We possess definite evidence for their use for the reign of Justinian, which proves that

Roman heavy infantry achieved a very high level of sophistication after the 540s. This type of heavy infantry combined both light-infantry foot archers and heavy-infantry spearmen (or javelin men if equipped as such) within the same unit. The dismounted cavalry could also be employed in this manner, as foot archers and spearmen. This was actually a more effective means of countering enemy cavalry with archery than the posting of foot-archer files between the heavy infantry files (*entaxis*), because with this system the entire front rank of the multipurpose footmen or dismounted cavalry could shoot straight at the enemy and then switch to spears when the enemy was near. It was also better than the posting of light infantry in front of the phalanx to break the enemy cavalry charge, because this system did not disorder the phalanx in any way – which could happen if the light infantry retreated either through the intervals between the *mere* or through the intervals between the files if the phalanx was in the *pyknosis* order.

The usual place for the light infantry throughout the Late Roman period was the rear position (*epitaxis*), which the Romans used against both cavalry and infantry. This was the standard combat method in particular when the enemy army consisted of heavy infantry, because arrows and slingshots were less useful against heavy infantry. In other words, there was no point in placing archers and slingers in front of the phalanx or within it. This, however, does not mean that the Romans would not have attempted to use foot archers against heavy infantry as well. For example, Constantine the Great used a prolonged period of skirmishing with foot archers in the Battles of Cibalae in 316 and Campus Ardiensis in 317.[11] In both battles the Licinians occupied higher advantageous ground, so Constantine placed his archers in front with the idea of weakening the enemy first with missiles. The arrows undoubtedly had some effect on the enemy, but they did not cause the defeat of the enemy infantry force. Regardless, in the right circumstances it was possible to defeat heavy infantry and dismounted cavalry with a barrage of arrows.[12]

On the other hand, it was entirely possible for the foot archers to cause an enemy cavalry charge to fail when they were supported by heavy infantry reserves, or were in rough terrain, and/or used caltrops. Syrianus even claimed that caltrops, arrows, slingshots and javelins were all that were needed for the protection of the flanks against cavalry. In contrast, if the outflanking force was composed of infantry, then one needed heavy infantry to oppose it.[13]

When the foot archers or slingers were placed on the flanks, so that they could also be wheeled forward and inwards so that the end result looked like a crescent without being a crescent proper, it was possible to place the enemy into a crossfire from both flanks. This is what happened for example at the Battle of Taginae in 552, where Narses used either the light- or multipurpose-infantry in an outflanking formation against the charging Gothic cavalry. The use of bows and slings was particularly effective from the flanking position against cavalry because even the armoured horses made large vulnerable targets.

When used for frontal softening fire, the infantry archers were also more deadly than their cavalry counterparts because it was possible to pack more footmen with more powerful bows into a smaller space than was possible with the horsemen – the speed of the horse did not add enough power to the shot to compensate for this. The foot archers could also aim their bows better than the men on horseback, and thanks to their stiffer bows they could also outshoot them in both range and speed. And as if this was not enough, the infantry archers could use arrow-guides and darts, which doubled the range of their shots and made their attacks less easy to see. The maximum range with the combination of arrow-guide and dart was about 680 metres. The enemy was also unable to reuse the darts. It is therefore not surprising to find the recommendation to use foot archers against cavalry in the texts of military theorists. The only real advantage of mounted archers over infantry was their greater mobility. The horsemen were more difficult targets if they used the fast canter or gallop. The foot archers were therefore required to anticipate the speed and direction of the horsemen and adjust that with the effects of the wind, but the use of large cavalry formations and the size of the horses still made them vulnerable targets.

In infantry combat the Romans used the light infantry primarily for softening up the enemy formation with a succession of missile attacks prior to any hand-to-hand combat. The aim was to cause enough casualties (wounded and killed) that the enemy formation would become disordered and the men demoralized so that the enemy would flee at the first sight of the advancing line, or would at least be weakened enough that the combination of thrown spears/javelins and the advancing men with swords would finish the job. The level of disorder and demoralization among the enemy obviously depended on the quality and amount of protective equipment worn and the level of training of that army. If the enemy consisted of well-armed and armoured disciplined infantry using a well-ordered formation most of the arrow- and slingshot-hits were likely to cause only less severe wounds rather than fatalities. In fact, the stones or lead balls of the slingers were actually more effective against well-armoured enemies than the arrows because these caused at least concussions and bruises and were quite often deadly if these hit the helmet. At close range the thrown heavy javelins such as *pila/spiculum* would render the enemy shields useless and could also penetrate body armour. The lead darts in their turn were more effective against those parts of the body which did not have armour.

The effectiveness of all archers, infantry included, depended upon the terrain and weather. Terrain, walls offering cover, and higher ground increased the effectiveness. Rain, moist weather, winter and contrary winds decreased the effectiveness.

The light infantry was vulnerable in close-quarters combat against heavy cavalry and heavy infantry unless they were protected by difficult broken terrain, city walls, field fortifications, wagons or heavier forces behind or next to them. However, it was possible to use the light infantry as if it were heavy infantry when they consisted of the javelineers and dart-throwers. These units could form a shield-wall and fight as if they were regular *skoutatoi*. This could work quite well against cavalry because the adoption of close order formation was usually sufficient to deter enemy cavalry from approaching too close, but all obviously depended upon the morale of the men on both sides. In most cases the pseudo-*skoutatoi* were obviously at a disadvantage when they faced real heavy infantry because they usually did not carry as large shields and as much armour as the regular heavy infantry. However, the amount of armour was not the sole decisive factor in combat. Morale and *esprit de corps* were at least as important, if not more so. It is therefore clear that the light infantry could also play a decisive role in battle, but their effectiveness depended primarily upon the terrain.

The fact that the light infantry were required to be able to operate in small units as *drouggoi* without the moral support of comrades close by and without the support offered by a hierarchy of officers means that the men belonging to these units had to be the crème-de-la-crème of the Roman army. Their mobile style of fighting in loose formations required an ability to improvise and excellent fighting skills with bows, javelins and swords as individual fighters. These fighting skills could also be very important if the enemy made its way through the intervals of the heavy infantry *mere* or outflanked it. In such instances the light-armed acted as reserves for the heavy-armed.

10.5. The Role of the Heavy Infantry (*skoutatoi, hoplitai, aspidoforoi, gravis armatura*)[14]

When the combined-arms system was used in open terrain, the final outcome of the battle depended on the heavy infantry. If the heavy infantry collapsed, the battle was as good as lost even when the Roman cavalry managed to defeat the enemy cavalry forces, but if it held its ground it was impossible for the enemies to defeat the Romans even when they had defeated the Roman cavalry forces or their outlying light infantry. Furthermore, the presence of the heavy infantry was usually needed when the cavalry had forced the enemy cavalry into flight and the enemy still had to face infantry forces and/or a fortified camp. It was the duty of the heavy infantry to defeat the enemy's heavy infantry and it was also usually the duty of the heavy infantry to storm the enemy camp by using the *foulkon/testudo* formation while the archers, slingers and javelineers protected their approach.

When the Roman battle formation advanced towards the enemy, the commanders and their assistants mounted on horses moved in front of the formation in order to be able to observe the terrain and the enemy until they were about to start fighting. If there was a need they adjusted the formations. When the Roman infantry phalanx waited in place, the straightening and adjusting of the line was obviously done in advance or while the enemy was approaching. If the Romans used the lateral phalanx and marched towards the enemy, then the initial formation of the files was only four deep because it was less tiring to the soldiers and made the line look more impressive to the enemy – the commanders adjusted the depth and width as the situation required. If the *taxis allê* advanced towards the enemy, it may have followed a similar initial approach of initially using a shallower depth. However, if the Romans used the hollow square/oblong (or these side-by-side), *epikampios opisthia*, wedge, convex, crescent, or *epikampios emprosthia*, it is probable that the Romans formed these well in advance.

In the lateral phalanx the *mere* had intervals of ca.30–60m, so they were clearly separated but still able to provide support to each other. In the case of the other formations, the intervals between the *mere* obviously varied according to the formation and tactics used. In the case of the lateral phalanx, the *mere* were required to coordinate their movements with the centre where the standard of the general was posted. It is also probable that there were very small intervals (perhaps about 1–2m), not mentioned in the *Strategikon*, between the *tagmata* to separate them from each other so that these could also operate independently as their training allowed. This would have been the place where the supernumeries (i.e. officers, standard bearers, trumpeters and bodyguards) assumed their posts. It is probable that the widths remained roughly the same, even when the units assumed the *foulkon/testudo* formation because the *Strategikon* expected the intervals between the *mere* to remain constant and the tightening of the array was always made towards the centre of the *meros*. This means that after the *meros* or units within it had assumed the *foulkon* order, the *tagmata* and *mere* often adjusted their places to follow the standard requirements, but it is equally possible that in practice the assumption of the *testudo* order by some units while other units did not assume it resulted in the widening of the intervals between those, which then had implications for the final stage leading up to the melee phase. It is possible that in some cases those units that had assumed the *testudo* spread out during the final attack phase in order to facilitate the use of their swords. As regards the intervals between the units in all of the situations described above, it was expected that the depth of the heavy infantry formation together with the presence of the light infantry and the small intervals of only 30- to 60-metres between the *mere* were enough to protect the intervals against any possible enemy attempts to penetrate through the intervals.

Roman combat doctrine expected that the armoured men in *pyknosis* order and the unarmoured men in the *foulkon/testudo* order (the rim-to-boss order gave each soldier the protection of two shields) were able to withstand enemy missile attacks so that they would reach the enemy formation in full combat readiness. The narrative sources prove this to be true. In the case of the *testudo* order, the arrows could obviously nail two shields together, but they were thin enough to be broken in two so this was not a problem. Javelins were another matter, but all combat systems were still compromises that sacrificed some of the advantages in return for something else. If the enemy used javelins with shanks that bent on impact, such as the Franks (the *spiculum, angon* and harpoon javelins both acted in this manner), then the Romans in *testudo* order simply retained the shield wall while fighting in melee.

Roman combat doctrine did not expect the heavy infantry to march long distances in full armour before battle, no more than two miles (ca.3km) from the camp. Longer marches could have exhausted them before combat. If the enemy delayed the battle, the soldiers were to sit down and rest, and in hot weather they were allowed to take off their helmets and were given water from the wagons but not wine. It was important that the soldiers were rested before the battle, because there existed the possibility that the combat could become prolonged. However, in practice the Romans did sometimes use forced marches before engaging the enemy in combat. This usually happened when the commander wanted to surprise the enemy with the speed of his march. A good example of this is the extraordinarily long march of the army of Theodosius I from Emona to Aquileia in one day in 388. His army marched 55 miles (88km) in a single day and then stormed the city walls and captured the usurper Maximus. The risk paid off handsomely.[15]

The army was also expected to follow standard pre-battle procedures in which the religious personnel (in pagan times pagan priests, soothsayers etc. and in Christian times Christian priests) encouraged the soldiers with religious ceremonies, and the *strategos* and officers encouraged the men with speeches and promises. When the infantry army had been deployed and was ready to begin battle manoeuvres the herald shouted: 'Silence! Follow orders! Worry not! Keep your rank! Follow the flag! Do not abandon the flag and pursue the enemy!'[16] Thereafter followed the commands that the higher ranking officers thought necessary, for which see Chapter 8.4.

When the heavy infantry reached a distance of two- to three-bowshots from the enemy, it assumed the *pyknosis* order. After this, when the men approached even closer to the enemy the soldiers would have started to feel the psychological stress of the forthcoming combat, with the result that some of them would have started advancing faster than others to get it over with while others would have crowded upon each other to seek the psychological comfort that this offered. The officers were required to prevent both with the necessary commands, because both of these would have broken the cohesion of the array. The men were usually so well drilled that they obeyed these commands, but obviously there are also examples in which the soldiers charged wildly in disorder or massed together. There are also instances in which single men or groups of men separated themselves from their formations in anger, but these instances occurred only after prolonged periods of stalemates, and they broke the enemy's will to resist. The role of individuals, small groups and the personal initiative of officers and commanders could play a decisive role when the battle became prolonged. It is clear that discipline was usually maintained during the initial approach and that the instances in which the Roman infantry charged wildly without orders were rare indeed.[17]

The final phase of the infantry attack began at a distance of about a bow-shot from the enemy. At this stage the soldiers assumed the *foulkon/testudo* order if the men did not wear armour or when the officers in charge thought it appropriate. Then followed the commands: '*Parati!, Adiuta!* ', which the soldiers answered with '*Deus!*'. The attack began immediately after this, with the archers firing arrows over the heads of the heavy infantry while the heavy infantry phalanx advanced until they came to the range of thrown weapons. If the men had darts or other lighter missile weapons, they threw these. As already discussed, it is probable that the formation opened up in depth to enable the men to throw their weapons. It is probable that the men used the overhead throwing technique with both darts and javelins, even if this diminished the range of the darts, because this was the safer technique to use when there were men in front of the thrower. The throwers probably placed their spears on the ground when they threw darts and javelins. Following this the *skoutatoi* would have tightened the array again for the final approach.

When the phalanx was fighting against infantry and reached the range of thrown spears, they threw their spears as if these were javelins and advanced into melee with swords in hand while still protecting themselves with their shields. If the men used the *spiculum*-heavy javelin or shorter spears, the javelins had greater impact on the enemy before the melee phase with swords, but it is still clear that even the longer *kontaria* had an impact on the enemy when thrown with a forward step by a trained soldier. The use of the sword indicated to the enemy that the attacking side had the will to engage them in brutal hand-to-hand combat. The throwing of the spears or javelins had two purposes: 1) it disordered the enemy ranks just before the melee; and 2) the use of swords made the soldiers more aggressive because they had to advance closer to reach the enemy. According to Julius Africanus (1.1.80–81, the ancient Roman *pilum* (heavy javelin) attack had achieved one kill for every ten thrown javelins and it appears likely that similar results could also be achieved by Late Roman soldiers. Additionally, the volley of spears would have wounded some enemies, while also rendering some of the enemy shields useless. It should be noted that the enemies like the Franks who also threw their spears, javelins or axes could achieve similar results against the Romans, which ensured that both sides faced each other with a similar amount of confusion. As regards the shields rendered useless by missiles, this did not necessarily result in immediate problems if the men had retained their *pyknosis* or *synaspismos/foulkon* order. If the battle became prolonged it was possible to replace the damaged shields during lulls in the battle (Syvänne, *MHLR* 6, 346). It was also possible to replace other lost equipment during such lulls. It was because of this that the Romans carried extra equipment among the baggage.

It is probable that the Romans timed their missile attacks (archery and slingshots), whether by infantry or cavalry, so that the enemy faced suppression fire at the exact moment that these were most needed by the advancing Roman infantry phalanx. For example, it is quite probable that the first volley of arrows helped the Roman units to assume the *foulkon* order (if this was needed) and start its march towards the enemy and that one of the succeeding volleys of arrows was timed so that the heavy-armed could prepare to throw their darts and javelins while the enemy was forced to cover themselves against this shower of arrows. It is also possible that one of the missile volleys was timed to take place just before the Roman infantry came within javelin range. If the Romans had managed to place their army advantageously so that they had the sun, wind and dust blowing from behind them into the face of the enemy, the effectiveness of these archery volleys increased proportionally because the enemy could not see the showers of arrows well, while their own archery was hindered by the contrary wind and inability to properly target the enemy if the sun shone straight in their eyes. In high winds in dry sandy terrain, the approaching Roman army was actually partially hidden (although so were the enemies from the Romans) because when the heavy-armed moved forward, they produced clouds of dust that moved faster than the marching Romans.

The men behind the first rank were required to support the front rank with their spears while holding their shields above their heads against enemy missiles – this shows nicely that the use of a shield roof did not always mean the adoption of the tightest version of the *foulkon/testudo* in which the width of the file was only ca. 62cm. The second rank therefore supported the first by advancing close enough to give support by spear thrusts over the shoulders of the first or by throwing the spear if this was the right thing to do. The use of spear thrusts by the second rank was potentially a very useful technique, because the attention of the enemy soldiers would be occupied by the front-rank men. It is also likely that some of the men posted behind threw their spears overhead in a higher trajectory to attack the enemy

The attached stylized drawing of the first century AD stele at Mainz shows how the Roman legionaries of the first century were already using light equipment in conjunction with the shield roof in attack.

In late Roman context the second ranker would also have pointed his spear towards the enemy by using the overarm grip so that he could use either the throw or thrust.

Left: image from the *Ilias Ambrosiana*, dated turn of the fifth or fifth century. It depicts the Trojans attacking a city. Drawing by Mai (1819).

One could protect the head also by raising the shield in the manner depicted in this image. It was not always necessary to form a shield roof especially when the men wore helmets and armour. Obviously the men depicted in this image are unarmoured and without helmets which means that Maurice would have recommended the use of the real *foulkon*.

The fully cylindrical and more flat versions of the rectangular shield appear to have remained in use throughout the late Roman period because these continue to be depicted in the works of art and because we find referrals to the use of the large shields in the *Strategikon* which Leo the Wise (*Taktika* 6.21) equated with the *thyreos*-shield. Leo described *thyreos* as large, oblong and curved/cylindrical. According to Leo, the *skoutatoi* were equipped with the *thyreos* in situations in which it was necessary. At other times the *skoutatoi* used the *skoutarion*-shield, which was presumably the round hoplite type of *aspis*. These two shields had different grips. The *thyreos* had a single grip for the hand in the middle while the *aspis* had one strap for the forearm and another for the hand.

from a blind spot. The danger of this happening to the Romans was recognized in Maurice's requirement for the rear rankers to protect their heads with their shields.[18] Obviously the use of shields for the protection of the head still meant that the larger shields were interlocked in depth (when deployed as a roof), if not merely raised up as depicted in the accompanying drawing (this would also have protected the head from all missiles excepting those that dropped from directly above) when close to the enemy, so it would have been a rare thing for the *skoutatoi* to throw their spears if they followed the instruction of Maurice.

Sometimes the Roman *skoutatoi* fought like the hoplites of old times and did not throw their spears/javelins. This option had obviously always been possible even when the Romans still used the *pilum-gladius* technique because even Vegetius (1.20) referred to the fighting stance with the right foot in front when using the *pila* for thrusting. The evidence suggests that this was actually the preferred fighting style at least from the late-fifth century until the reign of Maurice, because we find the *aspidoforoi* fighting in this manner in the *De scientia politica dialogus* which dates from the reign of Justinian. The *Taktika* of Urbicius and the *Peri strategikes* of Syrianus confirm this conclusion. One may also assume that the same tactics had been used in open terrain well before this, because Vegetius sought to reform the Roman infantry in such a manner that they would use the thrown *spiculum* and *spatha* combination at close range. This suggests that at that time the typical close-range tactic was the use of spears for thrusting in phalanx formations. Why else suggest the return to the old ways as a way of reforming the army? This, however, does not mean that the Romans would have abandoned the javelin even at that stage – it was just that its use was probably restricted to fighting in difficult terrain. It is therefore very likely that in open terrain before the reign of Maurice the Roman heavy infantry usually fought like the ancient hoplites. This means that in open terrain the Roman shield-bearers usually fought by using spear-thrusts by several ranks simultaneously, and that it was very difficult for the enemy to break the Roman formation frontally. The odds were that the Roman phalanx would win all frontal combats. When the Romans used the version depicted by Syrianus, a brave enemy could attempt to get past the spears by ducking below, but when he did this he first had to get past the spear of the front ranker and subsequently became vulnerable to a shield-bash with the spiked shield. When the Roman phalangites used the *pyknosis* order, the front ranker had even greater freedom to operate against anyone who attempted to duck below the spearheads.

If the attacking heavy infantry formation faced cavalry, as could have happened on the flanks (e.g. with an outflanking lateral phalanx or *taxis allê*), or along the entire front (if the enemy had been fooled to believe that they faced cavalry), or even in the centre if the enemy used mixed formation,[19] then they retained their spears or javelins in all instances so that they fought as spearmen in the *pyknosis* or *synaspismos* order, or in the kneeling *foulkon* order. When the footmen remained upright and advanced towards the cavalry the closeness of the order together with the spear wall usually resulted in immediate victory because the enemy horses and riders were usually unwilling to have themselves killed on a wall of spears. If the Romans had fooled the enemy cavalry into the belief that they faced only cavalry, then the Roman infantry waited for their attack in place by using the defensive *foulkon* order with the front rank kneeling. This unit order forced the footmen to remain in place while it created an insurmountable obstacle bristling with spears to the enemy cavalry as long as Roman morale held. This was usually enough to cause the enemy cavalry to flee, especially as the Roman formation was also supported by foot archers, slingers and javelineers.

From the point of view of morale, the numbers mattered because only large bodies of heavy infantry usually produced sufficient confidence among the men to face an enemy cavalry charge. There are instances in which small numbers of Roman footmen assumed a defensive formation against enemy cavalry, but typically their resistance proved short-lived.[20] And even when the Romans had large numbers of footmen, the Romans did not necessarily have the will to stay in place and resist an impetuous enemy cavalry charge at gallop. The Roman infantry had to consist of well-motivated and -drilled men, and when neither of these was true, for example when the army consisted of recent recruits, then the infantry needed wagon forts and other devices to withstand the enemy cavalry attack, and the narrative sources prove that not even these were enough during the reign of Anastasius to protect the inexperienced infantry against the cavalry of the rebel Vitalianus. However, excluding the period when the majority of the Late Roman infantry consisted of poor quality forces (late-fifth century until the 530s), the Roman infantry could expect to defeat any enemy cavalry force there existed.

If the Romans waited for the enemy infantry to attack, the combat procedure was the same as when they were advancing. The Romans started peppering the enemy with arrows at bowshot range and then employed slings, darts and short javelins at the right distance. If the enemy still continued its advance and the Roman morale had remained intact, then, depending on the time period and place, there were two possible procedures: 1) when the Romans used the thrown-spear tactic, the front rank threw their spears when the enemy came within the range and engaged them with swords, while the rear ranks used their spears in support; or 2) the Romans waited in place and engaged them with a spear wall of four ranks. Both of these systems had a fair chance of causing the enemy to flee because their lines would have been disordered by the rough terrain and Roman missiles, with the result that when the Romans suddenly attacked them with a perfectly aligned formation it was likely that the enemy would simply flee. Obviously in real life this did not take place every time. A well-motivated enemy would have continued to fight, even when their lines were in disorder.

At the melee stage of the battle, when the soldiers had thrown their spears or javelins against the enemy footmen, the ability of the soldiers to use their swords depended on the battle order they used. The *testudo/foulkon* order required the men to maintain their shield wall in such a manner that the swordsmen had less freedom to fight as individuals, while the *pyknosis* order (rim-to-rim order) gave the individual swordsmen greater freedom for fencing, shield-bashing and martial-arts manoeuvres (e.g. kicks to the shield, knee, feet or shin – or even wrestling moves if he had lost his weapon etc.) as individuals.[21] It is also clear that the Romans sometimes spread the *testudo/foulkon* order when the men came within throwing distance of javelins and spears. When the men ran wildly against the enemy, it is clear that their unit order became looser. The first-century military theorist Onasander (29.1–2) describes this as follows: 'the men to spread out in attack while shouting, sometimes on the run, while waving their swords above their heads with the polished swords and spear points flashing in the sun'. Vegetius (1.9) even recommended that the attack was to be done by jumping up and down to scare the enemy. This, however, was not the standard combat doctrine during the Late Roman period – rather, an attack like that was seen as a sign of undisciplined soldiers. However, Maurice still recognized that this could happen during the final stage of the attack because he instructed the first line to protect themselves until they came to blows with the enemy. The spreading out of the formation and the wild attack on the run while jumping up and down could expose the men to enemy arrows and javelins, especially if they did not have coats of mail or greaves. This in its turn suggests that there could in practice be wide intervals between the *tagmata* when some of these assumed the *foulkon/testudo* order and did not tighten the formation enough towards the centre of the *meros*.

However, the narrative sources still prove that there were situations in which the infantry or dismounted cavalry were ordered not to use spears but were ordered to attack on the run with swords in hand. The attacks with swords were used in confined spaces where the use of the long weapon was a hindrance. Therefore, we see this type of attack being used only when fighting in confined spaces (streets, walls, ravines, woods, forests). The three best examples of this are: 1) the attack downhill against the Gothic encampment by the sword-and-shield equipped Roman infantry deployed in loose order because of the woods in 379; 2) the sending of the *bucellarii* of Belisarius and Mundus inside the hippodrome to butcher the population during the Nika Revolt in 532; and 3) the attack on the run with swords against the Goths in the Vivarium during the Siege of Rome in 537.[22]

In the melee stage, the soldiers benefitted greatly from the large shield, adequate armour and the support of their neighbours, because they faced several enemies simultaneously (at least three in the front rank and one to two from the second rank). The soldier was also vulnerable to attacks from the sides and from blind spots. The use of the smaller shield, the oval or round *scutum*, or the hoplite-style *aspis* (or even the hexagonal and octagonal shields that were still used in the fourth century) obviously gave the soldiers less protection during both the approach stage and the melee stage – hence the instructions in the *Strategikon* to use the 'larger' shield if unarmoured and the warning not to expose the body when attacking. However, the use of the smaller shield was not without its advantages. The use of the smaller shields gave the individuals fighters more freedom to move as individuals (i.e. for fencing, shield-bashing and martial arts manoeuvres) than the larger shield at the cost of losing the extra protection offered by the large shield.

Despite the fact that the Roman *skoutatoi* infantry formed the backbone of the infantry phalanx, it would have been quite helpless against cavalry without the contribution of the light-armed and cavalry – the sole exception to this rule are the multipurpose spearmen and archers that we find in the texts of Vegetius and Syrianus. They needed the missiles of the foot archers and slingers and the mobility of the cavalry to win the battle. The foot archers were needed in particular to counter enemy cavalry. Furthermore, it was also preferable to possess specialist units of club-bearers or mace-bearers against enemies that fielded cataphract cavalry. The heavy infantry could only form the solid base around which the other arms of service operated. Otherwise the heavy infantry would have met a repetition of the Battle of Carrhae in 53 BC. The heavy infantry spearmen could attack enemy cavalry only when the latter had carelessly approached close enough to be attacked by the spearmen of the heavy infantry phalanx.[23]

When this happened, for example because the enemy cavalry had charged against the infantry phalanx with the predicable result of having been forced to stop just in front of the shield wall bristling with spears, the Roman spearmen were in a position to charge against the static cavalry force in front of them. In such situations, the spearman would have preferred to approach the horseman from the right side of the horse when the horseman had lost his spear and used his sword, because the horseman had difficulty in reaching an opponent on foot on his left hand side. The stationary horseman was also vulnerable on his right side because it was difficult to protect that side with a shield. In practice the soldiers obviously usually lacked the luxury of choice and fought as the situation dictated. The reach of the standard 3.74m spear allowed the targeting of the horse and rider up to the face, but it is probable that the principal target of the attacks was still the horse. If the Roman soldier had thrown his spear or it had been broken by the cataphracted enemy horse, the soldier undoubtedly targeted his sword attacks mainly at the horse, and against the legs and the torso of the horseman. In contrast, the rider with his spear could target the head and upper body, while with the sword he could target the head, shoulders, and neck. He could also urge his horse forward to trample the footman. The physical limitations set the limits to what could be achieved with different types of attack.

However, the *skoutatoi* were still the ones that usually decided the infantry battles, because heavy infantry was needed against other heavy infantry, whether it meant outflanking or the breaking of the enemy formation frontally. The other arms of service were less well-suited to this. The contribution of the heavy infantry for the success of the army was absolutely essential. In the right circumstances it could even defeat all the other arms of service simultaneously, as the Persian wars between the Greeks and Achaemenid Persians nicely show.

If the enemy retreated thanks to the successful use of outflanking or by breaking of the enemy formation with a frontal attack, Roman combat doctrine expected that the pursuit was performed by the cavalry and/or light infantry. The role of the heavy infantry was to reform its lines and follow the pursuers as quickly as they could while still maintaining their formation. The combat doctrine expected that the pillage of the enemy valuables was performed by separate units designated for this purpose, so that the heavy infantry and other arms of service could concentrate on the fighting and pursuit of the enemy. This ideal was not always achieved, the most famous example of this being the pillage of the Vandal marching camp after the Battle of Tricamarum in 533. In that battle, Belisarius lost all control of his army when they started looting for booty.

10.6. Fighting Techniques at Distance and in Melee[24]

Even if the appearance of the Roman military gear and fashion was often quite fancy, it still took into account practicability. The best example of this is the instructions of the *Strategikon* drawn mostly from earlier treatises. Maurice recommended the use of simple loose Gothic tunics or short tunics split in the middle because these allowed freedom of movement, as did the simple mantlets when worn. He also recommended the use of Gothic shoes rather than boots, the reason being that these gave the men a better 'feel' of the ground than the boots. The use of shoes shortened the reaction time. This means that the footmen in their shoes were better adapted to fighting on foot than the dismounted horsemen in their boots. The short military haircut ensured that visibility was not hindered by hair in combat situations.[25]

The fighting techniques employed by the soldiers obviously depended on the equipment used and training received. The long-distance fighting was conducted by the archers and slingers. The archers employed almost exclusively the so-called Roman shooting technique, usually without the shield, until the reign of Maurice who was the first to require that the archers train to use both the Roman and Persian shooting style while carrying the shield. For details, see Chapter 2.8–9. The Romans were trained to shoot both accurate sniping shots and volleys directed against units/areas. The same was obviously true of the slingers or staff-slingers and they could deliver both accurate throws plus attacks directed against a unit. Therefore, the archers and slingers were prepared for all of the things required from them. In pitched battle the throwing of darts and javelins was usually aimed at the general area of the enemy unit, but, obviously, in situations in which the men using these weapons were in the front ranks they could attempt to hit a specific person or horse in front of them.

The different types of melee weapons also required different fighting techniques. After having thrown his spear or javelin, the heavy infantry front-ranker relied on his sword. In that situation his ability to fight as an individual swordsman depended on the unit order. If the soldiers were deployed in the psychologically demanding open order which did not provide the fighter with the comfort of having a comrade standing right next to him, the swordsman was free to use the entire repertoire allowed by his training and personal ability, but if the swordsman was deployed in a unit using the *pyknosis* order the expectation was that he would fight as a part of the unit while the second ranker would support him with his spear, but with the likelihood that the *pyknosis* order still allowed the swordsman to attack and retreat as an individual. If the unit of the swordsman was arrayed in the *foulkon/testudo/synaspismos* order, he was expected to fight as a part of the unit by using his sword above the shield so that the cohesion of the shieldwall was maintained throughout the fighting. The same realities were also true when the *skoutatoi* fought with their spears: 1) in open order the spearman fought as an individual; 2) in the *pyknosis* order he fought as a member of his unit while still retaining the freedom to charge forward from the unit to deliver a thrust if the opportunity for that presented itself; and 3) in the *synaspismos* order he was expected to fight as a member of his unit.

The spear thrusts had four basic techniques. Let us begin with the less common methods. If the spearman wielded the spear with two hands, for example because he was a horseman who had dismounted, he could use 'fencing' techniques such as thrusts, circular parries, parries from below or from above, beats, glides and lunges with the left leg in front, and takedowns by pushing the spear between the legs of the enemy. The two-handed fighting techniques with the spear had been honed to perfection by this time and were used by the hunters and the gladiators. The Isaurians also appear to have employed a special fighting technique in which the fighter used two Isaurian javelins (*duo akontia Isaurika*) simultaneously, one in each hand, so that the javelins were thrust at the enemy while the fighter turned from side to side. This was undoubtedly a special technique that only some gladiators, martial artists and Isaurians could employ. Notably, there were also martial artists who used this technique while mounted. The circular movements and the javelins were therefore used for simultaneous parries and attacks. Neither of these techniques, the two-handed use of the spear or the use of two javelins/spears, were particularly good in pitched battles because their users were very vulnerable to enemy missiles. They were better suited to duels between martial artists. Therefore, it is not surprising that both of these techniques were very rare in pitched battles.[26]

The regularly-armed footman used the spear or javelin with a shield. The four basic thrusting techniques with the spear were: an overhead thrust, a reverse thrust, a low thrust, and an underarm thrust. See the attached illustration. The last three versions were interchangeable: the spear could be raised and back from the low position to

The four basic thrusting techniques with a spear

Note that all of these techniques could also be performed with the right foot foward as recommended by Vegetius (1.20).

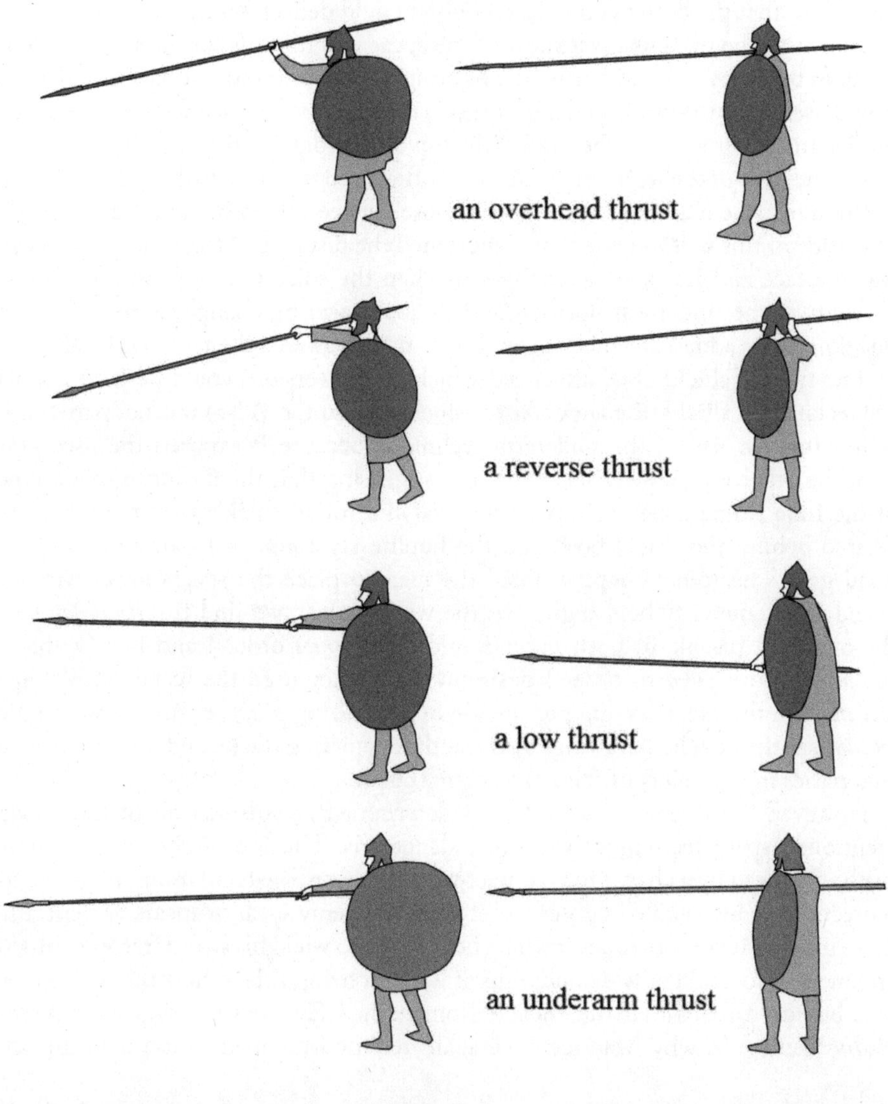

the underarm position and from it to the reverse position easily in the *pyknosis* order. All of these variants were therefore usable in the *araiôsis* and *pyknosis* orders, but in the *synaspismos* and advancing *foulkon/testudo* orders, the use of the underarm thrust was rare because when the Romans tightened the array the spears were held above the shields, which left three different alternatives for the Late Roman *hoplitai* in the *testudo* order: the overhead, reverse, and underarm thrusts. In pitched battles the principal target area for the spear thrusts was the face and the secondary target areas were the throat, right arm and armpit, and the left shin; any areas left unprotected even for the blink of an eye. If the first spear thrust was well-timed and aimed, it could cause a fatal wound immediately, but it was more typical that the ultimately-weaker opponent succumbed only as a result of several thrusts (it should be remembered that enemies also placed their best fighters in the front rank), that in conjunction with physical the exhaustion caused by close-quarters fighting gradually weakened him enough so that the better conditioned fighter could deliver a well-aimed and -timed fatal thrust. The principal weakness of the spear was its wooden shaft, which could break or be cut by a sword during the fighting. If the spear did not have a butt-spike, it was useless, but when it had a butt spike, it could also be used against those lying on the ground or even on the instep (the top of the foot, i.e. the metatarsus).

On the basis of the works of art the most often used variant was the overhead thrust position because it allowed the user to choose between throwing and thrusting. The downside of this method was that it shortened the distance of the thrust if the target was the face and head, as it usually was when the soldiers fought in a shieldwall. In contrast the underarm position, which could also be changed into the reverse position easily, added distance to the thrusts made towards the face and head.

The type of shield also influenced which of the versions could be used because the rectangular shield (the large *thureos*-shield of Leo the Wise) was not particularly well-suited for use of the underarm technique because it exposed the user more than the other techniques, hence it is not surprising that the favourite shield types of the Late Roman spearmen were the oval and round shields with the hand-grip located behind the shield boss, and the hoplite-style *aspis* with an arm-band and hand-grip. The round shape enabled the men to place the spears lower with the shield simultaneously held higher. In the works of art, we find the *skoutatoi* using the overhead (usable in both *pyknosis* and *synaspismos* orders) and low (usable in the *araiôsis* and *pyknosis* orders) positions when they used the rectangular/oblong *scutum* with the spear/javelin. For the use of the pairing *pilum/hasta* and rectangular *scutum*, see the attached drawings by the author depicting the second-century Roman legionaries in the reliefs of the Tropeaeum Traiani.

However, if the enemy consisted of determined swordsmen all of the above-mentioned spear techniques were very dangerous. The use of the thrown spear/javelin followed by a charge into contact with the sword-in-hand using the shield for protection enabled skilled fighter to get past the enemy spear or spears by deflecting those into the inside range, after which he was free to wield his sword freely against an opponent who could only defend himself without being able to hurt the swordsman. It is because of this that the ancient Romans had favoured the aggressive *pilum-gladius* tactic and why Maurice once again required his men to use it in infantry

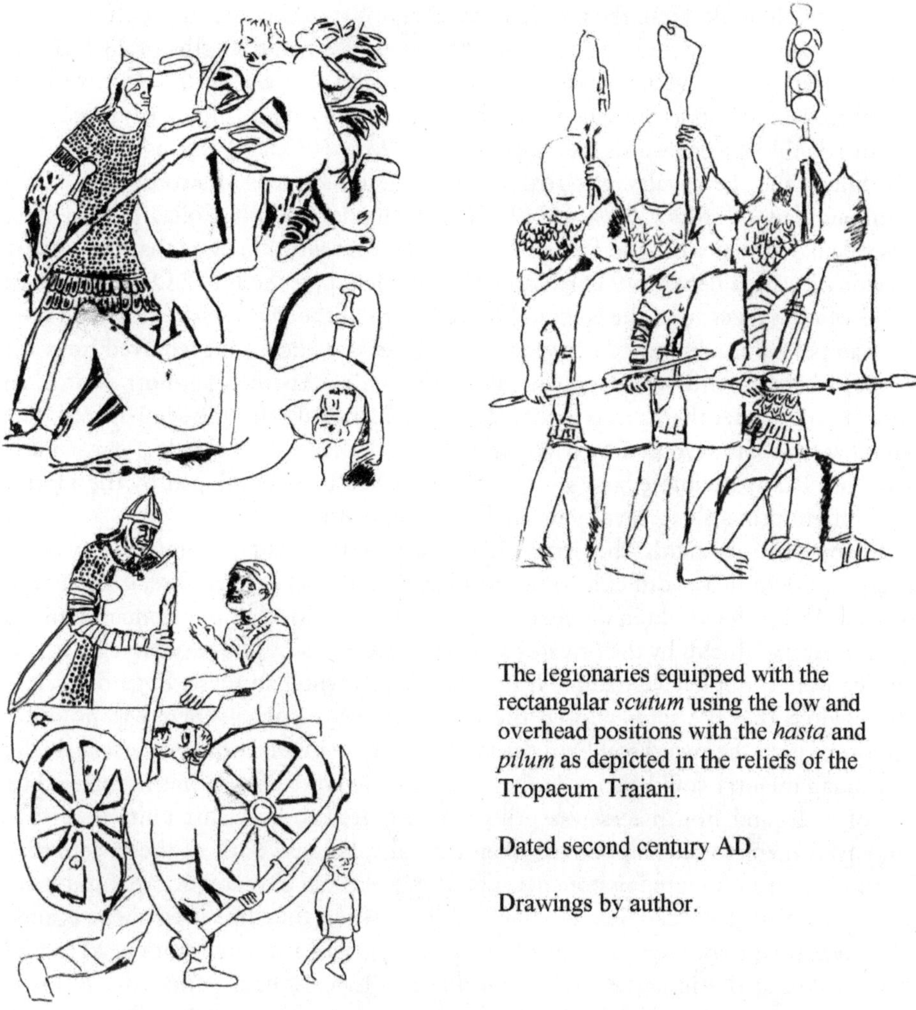

The legionaries equipped with the rectangular *scutum* using the low and overhead positions with the *hasta* and *pilum* as depicted in the reliefs of the Tropaeum Traiani.

Dated second century AD.

Drawings by author.

battles. In this situation the spearman had no other alternative than to drop his spear, and draw his sword or dagger while seeking to protect himself with his shield.[27] It is therefore clear that Maurice's recommendation to use the spear for throwing at close range was very practical. The spear as a thrusting weapon was only useful when facing cavalry. Regardless of this general reality, the narrative sources still offer us several instances in which spears were successfully used as thrusting weapons against enemies fighting on foot.[28]

Late Roman infantry carried two types of swords, the double-edged longsword *spatha* and the double-edged short sword *semi-spatha*. The principal weapon of choice was the *spatha* (*spathion*), which was a cut and thrust, double-edged straight long sword. On the basis of archaeological finds, modern archaeologists have concluded that the blade length of the *spatha* varied between ca. 63.5 to 91.5cm, and that there do not exist examples of the *semi-spatha*. I would suggest that this is a mirage which results from the problems in categorization, because the archaeologists class short swords (including the *gladii*) to encompass blade lengths of 23.1–69cm. There clearly exists overlap. I would therefore suggest that we use the tenth-century sword measurements as our guide.[29] In the tenth century, heavy infantry was required to have double-edged swords measuring four *spithamai* (93.6cm) in length, including the hilt, while the cavalry was required to have double-edged swords that had a minimum length of four *spithamai* (93.6cm) including the hilt. Kolias T. Taxiarches has calculated that what the *Sylloge Tacticorum* meant with this was that the cavalry swords measured between 93.6cm and 110cm, including the hilt.[30] On the basis of this I would suggest that the standard blade length of the *spatha/spathion* of the Late Roman period usually varied between about 74cm and 92cm (hilt removed from the figure) so that swords shorter than this should be re-classified as short swords. In sum, it is clear that there exists archaeological evidence also for the semi-*spatha*. The Herulian *spathion* favoured by Maurice (*STR* 12.B.4.1–2) was clearly a particularly long and handy double-edged sword because it had been developed by the Heruls who fought primarily as cavalry in the East Roman army.

By modern standards the *spatha/spathion* therefore had an unwieldy weight/length ratio, so it was difficult to use in parries and blocks. Therefore, it needed to be paired with the shield in defence. It was because of this that the Romans banned the carrying of shields by the private *bucellarii* inside cities.[31] The swordsmen without shields were extremely vulnerable if they faced men who had these. Regardless, it is still obvious that the soldiers used the *spatha* defensively when there was no other option, or used the sword scabbard or cloak or other piece of cloth as an ersatz shield.

Roman infantry could also use other very specific types of weapons in melee. The use of clubs and iron maces was primarily restricted to specialist units that were employed against the cataphracted enemy cavalry, because both of these were very suitable against armoured opponents. The Romans could also use pickaxes and hand axes in combat if needed, just as they had done when they had used them against men wearing the *crupellarii* (a type of gladiator who was armoured from head to toe) armour during the imperial period. During the Late Roman period the Romans appear not to have used pickaxes or axes in this manner against heavily armoured infantry, but we still possess evidence for the use of the *bibennis* (double-edged battle-axe) in a pitched battle against the Moors at the Battle of Antonia Casta in 547 (Syvänne, *MHLR 6*, 275). On that occasion, it was apparently used primarily for the breaking of the defences of the Moorish marching camp rather than as a martial tool needed against the Moors.

The *spatha* was primarily used for cuts and slashing, but it was also readily usable as a thrusting weapon, especially those versions in which the blade tapered towards the point. The most effective way of using the *spatha* was to employ the edge of the sword from the middle of the blade to the tip of the point for cuts. The principal target areas for these cutting techniques were the head, the shoulders, the neck, the right arm and the left leg (when placed in front). In cavalry battles the *spatha* was used almost exclusive as a cutting weapon, but in infantry combat the *spatha* could be used with relative ease as a thrusting and stabbing weapon. This is obvious from the fact that Vegetius recommended the use of the thrusting technique in infantry battles, and the descriptions of the battles show this to be the case. The principal target areas for the thrusts and stabs were the face, the throat, the right hand, arm and armpit, the left leg or the leg positioned in front, the instep (the top of the foot), and the sides of the body – basically any part of the body left unprotected by armour. It was also possible to penetrate the armour with a powerful thrust, especially if the armour consisted of chain mail or ersatz armour like the *subarmalis*. The same was true also for the swords cuts – a perfectly executed cut could cut through the helmet and armour. This was obviously less common when infantry fought against infantry than when the swordsman rode a horse, but was still possible. Stabs or cuts to the head or face, including the eyes, were always effective, whereas hits to the arms or legs had the effect of gradually weakening the enemy for the killing strike. Most of the head wounds from cuts would have been to the left side of the head, because the sword was held in the right hand.

As already noted, the Late Romans also employed the short sword, the *semi-spatha*, in combat. Its use was rare at this time and undoubtedly restricted only to battles between footmen, because the short sword was not effective when fighting against cavalry which required a longer reach. The continued use of the short sword is not only supported by Vegetius (and the existence of swords shorter than the *spatha* in the archaeological record), but also by narrative sources and art sources. During this era the short sword and dagger was often called a *machaira*. The short sword was at its best in situations in which the length of the sword could be a hindrance. It was because of this that the Romans had favoured the use of the *gladius* with the *scutum*. One may assume that the Late Romans usually restricted the use of the *semi-spatha* to infantry combat, and to situations in which they used, for example, the *testudo* order against an enemy *testudo* or *pyknosis* order. The short sword was used primarily for thrusts at close quarters, just like its ancient counterpart the *gladius*.[32]

On the basis of the extant works of art it is clear that most of the fighting guards (the stance and position adopted by the fighter in readiness to fight) had the left foot in forward position when the soldier had a shield. However, we find Vegetius (1.20) instructing the soldiers to place their right foot forward when fighting with the *pilum* or *gladius*. In his opinion the soldiers placed their left foot in front only when throwing javelins, but this is just a view held by him and by the martial arts masters that he used as his sources. Vegetius also disapproved of the use of the slash and cut, and favoured the use of the thrust because it was more likely to cause a deadly wound than the cut. He (1.12) also quite correctly noted that when one used the cut it exposed the right arm and side. This does not mean that the Romans would not have employed the cut with the left foot in front. Vegetius's instructions just presented one school of thought.

The right-foot-forward stance was also adopted by soldiers when they used the advancing *testudo* and all men knelt. In this kneeling position all men were required to place their right foot in front. This means that the right-foot-forward was also used as a variant for different kinds of guards in combat. This is not surprising, as it is obvious that the soldiers needed to be trained in the use of different guards/fighting stances for different situations involving different strikes, cuts, slashes, and thrusts. Real life combat situations were simply too varied for the use of a single stance.

However, the two basic versions of the fighting stance which allowed the use of the different guards were the following: when one used the shield, the positioning of the left foot in front was better as a defensive posture because it placed the defender's body further away from the enemy; the placing of the right foot and right side in front was the more aggressive posture as this enabled the soldier to deliver faster sword, javelin or spear attacks, while it also gave the swordsman better control over the blade because this method did not involve as much turning movement of the body when using the sword. When the right foot was in front and the formation was not tighter than the *pyknosis* order, the fencer could use the lunge for both cuts and thrusts. It was this in particular that made the attacks faster to deliver, at the cost of sacrificing some safety. The different martial arts masters and individuals had different personal preferences, so it is not surprising to find that the martial artists never formed a consensus view of what were the best guards to use with the combination of the sword and shield, with the result that we find Renaissance-era masters suggesting various different guards that had either the right or the left foot in front.

The important point here is that the evidence for the use of different fighting techniques proves that the Romans were constantly seeking to hone their fighting techniques to perfection, so that their soldiers would be trained to use their weapons in the most efficient manner allowed by human physics and the equipment used. Roman combat doctrine also took into account the fact that the fighting methods of different peoples varied, so that it was necessary to familiarize the soldiers with these before engaging the enemy in pitched battle to ensure that they would not be unfamiliar with these and thereby surprised. However, in pitched battles the importance of a familiarity with the enemy's martial arts style or weapons was not as pronounced as it was in single combat – the closeness of the formations and the use of shields negated most of the advantages and disadvantages connected with such familiarity.[33]

The unit order obviously influenced the guards that the soldiers could assume. The *pyknosis* and sloping *foulkon* against cavalry (when the men rose up) both allowed the use of all of the guards because the shields were placed roughly rim-to-rim so that the soldiers could still place their swords, javelins or spears between the shields as they did when using the sloping *foulkon/testudo* against cavalry. However, the rim-to-boss order, the *synaspismos*/advancing *foulkon*, was far more restrictive.[34] In fact, I would suggest that when the Romans used the *synaspismos* order in combat that they used solely those fighting guards that placed the left foot in front, because the main intention of this unit order was to exploit the protective cover of the shieldwall and restrict the freedom of the individual to lunge forward from the formation. Furthermore, the use of this unit order often resulted in the two opposing formations advancing shield against shield, with the result that the two formations fought by using the mass-shoving action called *othismos* in Greek. In this situation each soldier pressed his left shoulder against the shield and shoved so that their left foot was in front while the right was used for pushing from behind.[35]

Obviously, the type of sword used in conjunction with the unit order also influenced the fighting guard. It is probable that when the Romans employed the *araiôsis* or *pyknosis* order and the short sword, they leaned forward and used their shield offensively for shield bashing while employing primarily underarm thrusts. However, when the men employed the *spatha*, the combatants would have preferred a more upright stance which allowed them to exploit their longer sword for longer thrusts and cuts. The rim-to-boss order obviously restricted the use of the sword to taking place primarily from above the shieldwall, which actually limited the power of the thrusts or cuts in comparison with attacks that had the full power of the moving body behind them, as would have happened when the man lunged forward. It was because of this that the rim-to-boss order could result in the *othismos*-shoving match – the men could not kill their opponents as effectively as in the other unit orders, so it was possible to advance to the shield-to-shield distance more easily.

Regardless of the unit order, all swordfights were a matter of feints, blows and counterblows, while the helmet, armour and shield acted as defences. Such fights did

The drawings by author depict the swordfighting techniques of the 2nd century Romans, but these are still representative of the late Roman fencing too. Drawn after the reliefs of the Tropaeum Traiana.

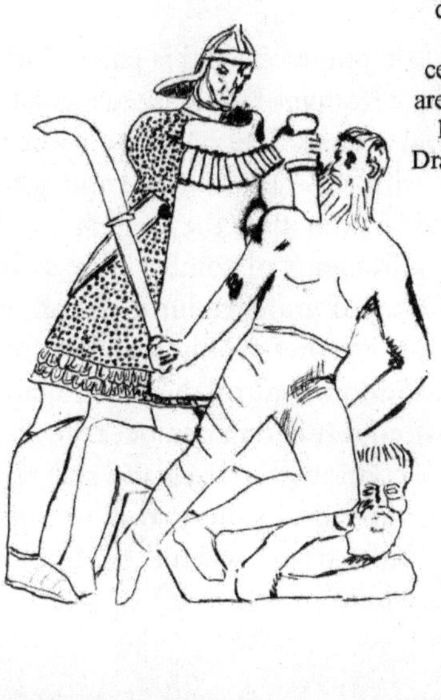

not last long, because the human physique set limits to the length of the fighting. It was because of this that the fighters often paced their attacks so that they could concentrate on delivering killing strikes with terrible force when the opportunity for that arose. See the attached drawings drawn after the reliefs of the Tropaeum Traiani (dated second century AD) for some examples of how the Romans used their swords in combat.

It is impossible to give all the details of the fencing and sword fighting techniques in the context of this study. The reader should also remember that there was and is no safe way of attacking. Any attack with a shafted weapon or sword opened one up to counter attacks or ripostes. It is because of this that in fencing the fencers usually tries to create openings by feints, by applying pressure on the blade or by using beats, glides along the enemy blade, by using invitation and so forth.[36] Direct attacks are also possible when one senses that his opponent is inattentive, tired or careless. The ability to see this is acquired through practice. It is because of this that Maurice stressed the importance of having well-rested men in pitched battles. Well-rested men have better senses. A skilful soldier could use a variety of feints and invitations with his weapon-and-shield combination to lure the enemy into a premature attack or premature defensive move against an attack that would therefore not take place, which opened the enemy to his own true attack. With the shield, the above-mentioned fencing techniques meant, for example, the pretended opening of the defensive posture, the use of the edge or rim or the embossment of the shield to bash into the enemy's shield (or weapon), or applied pressure against the top or bottom of the shield or towards the left (the right side of the enemy shield) to create an opening. The use of the aggressive parry with the shield could also be used in like manner. Once again the unit order set limits to this. It is probable that when one used the rim-to-boss unit order that the aim was to perform the shield pushing in unison so that it had the tendency of becoming a pushing and shoving match. In actual combat the ability of the soldiers to survive depended on his personal talents, the amount of training and experience, amount of protective gear, the support their comrades gave, the situation, and their luck. The melees were short and brutal and psychologically demanding.

The requirement to kill other human beings at close range was and is particularly difficult. It was partially because of this that the ancient Romans familiarized civilians with the sight of blood and gore in the gladiatorial arenas. The narrative sources prove that Roman soldiers usually had no problems with the sight of blood and gore, which means that the methods the Late Romans used even after they had stopped using gladiatorial contests were sufficient to the requirements of combat. The sight of the killing of wild animals in the arenas, the killing of animals during hunting, and the sight of animals being killed by butchers or by other family members for food were clearly adequate replacements for the abandonment of the real pagan-style gladiatorial games. The fright of combat also often resulted in combat rage and 'overkill' of the enemy, which meant that the soldiers continued to hack and stab the enemy to ensure that he was really dead. There was and is obviously also a sound reason for this, because even if the enemy had received a deadly wound he did not necessarily die instantly but continued to fight for a while.

The ability of the soldier to survive depended also on the behaviour and combat performance of his comrades. If the battle formation had become too spread out it was possible for the enemy to penetrate it and subject individual Roman soldiers to flank attacks. If the soldiers had crowded upon each other through fear, the soldiers lost their ability to use their weapons and they became a helpless mass easily killed by enemies.[37] The narrative sources also prove that there were instances in which individuals disregarded Roman combat doctrine and advanced in front of their battle formation. When Roman soldiers did this, they obviously lost the support of their comrades while subjecting themselves to being punished according to military law.[38] However, in the instances reported by the narrative sources, the soldiers who did this appear to have usually been expert martial artists who did this when the battle had already become prolonged, aiming to encourage the men with their personal example.

When the Romans fought in difficult terrain in irregular *drouggoi*, they obviously did this at the cost of losing the support of the close-order formation. In these fights the individual fighting skills and the help provided by the members of the same tent group were obviously at their extreme. As already noted, the light-infantry that fought in woods and rugged terrain were the crème-de-la-crème of the Roman army. These men had to be self-reliant and able to work together. They had to be able to fight as individual martial artists in situations which can only be described as a long series of duels fought between individual fighters.

At really close range, the Romans soldiers used hand-to-hand fighting techniques borrowed from boxing, wrestling and *pankration*, and whatever weapons they had at hand like daggers, knives or head-butts with helmets – it was because of this that Syrianus recommended the use of spiked helmets. All of their enemies had their own martial arts techniques for situations like this too. Real life hand-to-hand combat obviously favoured otherwise forbidden dirty tricks, such as kicks to the knee, stamps to the instep (the top of the foot), choke holds, eye-gouging, joint locks, head butts, biting, punches or open-hand techniques or 'karate chops' to the Adam's apple or neck (jugular vein), etc.[39]

10.7. The Role of the Foreign, Native, and Ethnic Contingents

The Roman armed forces included also ethnic units (as regular units, allies and *foederati*) that used their own equipment and tactics. These included units consisting of Moors, Arabs, Nubians, Tzani, Armenians, Quadi, Georgians, Sarmatians, Alans, Isaurians, Goths, Franks, Vandals, Burgundians, Heruls, Slavs and Antae. Of these at least the Nubians, Tzani, Moors, Lazicans/Georgians, Slavs, Antae, Arabs and the Germanic tribes provided footmen besides cavalry: the Nubians provided superb foot archers; the Tzani heavy infantry spearmen; the Moors provided mainly lightly-equipped javelineers; the Georgians both heavy and light infantry; the Slavs and Antae mainly lightly-equipped infantry using javelins, shields, and axes, together with some bowmen that used their own Slavic archery technique and wooden bows with poisoned arrows; the Arabs provided mainly foot archers; the Germanic tribes contributed mainly heavy infantry spearmen, while the Franks also provided men armed with the *Francisca*-axe or javelin; the Isaurians added lightly-equipped spearmen, slingers and javelineers that fought both in phalanxes and as irregular light infantry. Typically, these units fought under their own leaders, especially those ethnic units of *foederati* that had entered the Roman Empire in the early-fifth century and became successor states of the West Roman Empire.

This means that the Roman infantry phalanx could include several different types of troops using their own peculiar sets of equipment and fighting styles. The *Strategikon* (11.2.85–89) acknowledged this possibility by stating that the ethnic infantry units were to be deployed according to their customary manner in the front line of the *epikampios opisthia* formation. In fact, we find the Romans and Lazicans using the *epikampios opisthia* formation together in this precise manner at the Battle of the Hippis River in 550. The Romans had always used auxiliary infantry units in this manner to reinforce their legions.

The ethnic heavy infantry units that fought by using the rank-and-file phalanx structure in combat were all easily added to the phalanx structure used by the Romans. These units included for example the heavy infantry provided by the Germanic tribes (which provided both true heavy infantry and dismounted cavalry for infantry battles), Tzani 'hoplites', Georgian 'hoplites', and the dismounted Sabiri Huns. However, even if the ethnic heavy infantry contingents could be attached to the regular Roman formation, the general could not expect them to be able to perform the same combat manoeuvres as the Roman regulars. It was because of this that Maurice placed them in the front line of the *epikampios opisthia* and Narses in the centre of his phalanx formations. All that the Romans could expect from the foreign ethnic heavy infantry units was that they would maintain their place where assigned and then move forward from that position if necessary. They could not be expected to perform anything more complex than that. Regardless, the Romans still considered the contribution of the Germanic tribesmen in particular very valuable. They were bold, fearless and undaunted thanks to their warrior ethos. They stood where ordered and attacked boldly when called to do so. In addition to this, the Germanic tribes that entered the Roman Empire during the early-fifth century sometimes fought as separate independent tribal armies under their own leaders or kings (both had official Roman titles[40]) without any Roman soldiers at all (excepting those officers and diplomats that were accompanying the army as envoys and representatives of the emperor); in effect, these tribes were officially Roman armies. The best examples of this are the campaigns that the Visigoths fought in Gaul and Spain during the first-half of the fifth-century, the campaigns of Alanic units throughout the fifth-century, and the conquest of Italy by the Ostrogoths led by Theoderic the Great in 489–93. In all of these instances the Germanic armies were officially Roman armies.[41]

The same was true of the ethnic units of light infantry, because their use in the phalanx battle formations was more flexible than the units that were needed solely for the heavy infantry phalanx. Their tactics allowed greater flexibility and freedom of movement and a less disciplined organization, because they simply adapted their movements to the movements of the heavy-armed. The lightly-equipped specialist ethnic units included the Slavs, Antae, Arabs, Moors, Isaurians, Nubians, Lazicans/Georgians and probably some of the other peoples of the Caucasus. There were also units that were called Pseudo-Isaurians who were equipped in light gear and armed with short spears.[42]

In contrast to the infantry, the use of ethnic cavalry units in their native gear did not pose any similar problems for arraying the army in combat formation. The cavalry units had always fought by using different unit orders, tactics, equipment and formations on the flanks of the infantry phalanx. It is because of this that we find Julius Caesar employing only Gallic or Germanic cavalry instead of native cavalry consisting of equestrians. The only possible problem that these ethnic cavalry units posed for the infantry phalanx concerned their possible influence on the morale of the infantry if they fled.

In the right circumstances, which means the inclusion of limited numbers of ethnic units alongside the Roman regulars, the foreign ethnic contingents added both numbers and flexibility to the Roman forces which an able general such as Narses could exploit against enemies. When employed in their native lands against their fellows, they could also cause divisions among the enemy, not to mention the fact that they also knew their own native territory better than the Romans. The foreign ethnic units were also valuable as inexpensive 'cannon fodder', in fact too much so for the West Romans during the latter half of the fifth century, because the neglect of their own native forces cost them the Empire.

10.8. The Face of the Infantry Battle[43]

If one would like to generalize something that cannot really be generalized, as each battle had its unique characteristics, one can still say that the most typical infantry battle formations of the Late Roman era were: the lateral phalanx and hollow square/oblong throughout the era; side-by-side deployed hollow squares/oblongs from about 366 until about the 470s; and the *epikampios opisthia* in the sixth century. Of these, the most commonly-used formations were the lateral phalanx and the hollow square/oblong. The other battle formations saw far less use in battles.

Common Elements in Infantry Battles

The common elements in the Late-Roman period consisted of the following features. The most common of the similarities was the often-decisive role of the cavalry. The typical infantry battle of the period started with the cavalry encounter: 1) the Roman and enemy cavalry posted on the wings of their respective armies would engage each other in combat; 2) the Roman cavalry could fool the enemy cavalry force into the false belief that it faced only cavalry, after which it withdrew behind the Roman infantry – with the result the enemy cavalry attacked the Roman infantry formation; and 3) the enemy cavalry attacked the Roman infantry formation without any prior combat between the Roman and enemy cavalry. The actual results of these encounters between Roman infantry and enemy cavalry depended on the tactics, terrain and situation.

The most efficient way to decide the battle was to outflank the enemy formation, which was usually performed with the assistance of cavalry. The appearance of the enemy on the flank or rear quite often panicked the soldiers as their often-imaginary route of retreat was suddenly cut off – survival instinct then set in and caused the men to act in a manner that endangered their survival. This is the reason why the cavalry was so often the decisive arm of service in the battles. Their greater mobility allowed them to outflank the enemy formation, which then affected the morale of the enemy infantry. If the Roman cavalry posted on the wings of the lateral phalanx was successful against the enemy cavalry posted on the wings of its infantry, they usually pursued the defeated enemy cavalry up to their camp or elsewhere, while some of the cavalry units (usually the *clibanarii* posted next to the Roman infantry phalanx) attacked the now-exposed flanks and rear of the enemy infantry formation while the Roman infantry extended its flanks to attack the enemy infantry. The enemies of Rome were rarely able to withstand an outflanking manoeuvre, whereas the Roman infantry was usually able to withstand it by forming the hollow square/oblong, double phalanx (*amfistomos difalaggia*), or double-front (*amfistomos falagx*).

If the Romans managed to surround the enemy cavalry or managed to force it against some obstacle, they usually annihilated the enemy force. However, it was far more typical for cavalry battles to take place in open and unobstructed terrain so that when the combined Roman force of infantry and cavalry defeated the enemy cavalry, most of the enemy horsemen were still able to survive by fleeing to fight another day. In some cases, they fled only as far as their camp and infantry support, while in other instances they fled further while leaving behind their encampment with its servants and infantry. If the enemy cavalry fled to their marching camp, it was possible that this was enough to force the Romans to abandon their pursuit, but it was far more usual that the Romans then also captured the camp, forcing the enemy to continue its flight. The high-quality of the Roman infantry usually enabled them to capture the enemy marching camps too.

The Roman infantry phalanx could also defeat the enemy by breaking through its infantry formation by using a deeper unit order, by outflanking with individual units, by using the wedge, by the use of counter tactics against the enemy attack (e.g. hollow wedge/scissors against wedge, or *antistomos difalaggia* or *amfistomos falagx* against an enemy breakthrough), or by using independently-moving *globi/drouggoi*. If the fighting was in difficult terrain, the principal attacking forces were the light-infantry *drouggoi* which were expected to outflank the enemy. If the enemy infantry force collapsed and retreated to their camp, the pursuit was to be performed primarily by the cavalry with the Roman infantry following so that it then participated in the storming of the enemy camp. If the enemy infantry held its ground, the battle became prolonged and would be decided by the will of each side to fight – or by luck.

In practice, however, there were as many variations as there were battles and battle formations. There were also pure infantry battles, ambushes and night attacks. If the general did not use the lateral phalanx which could be manoeuvred into other formations, and opted instead to use some other specific battle formation from the start, this decision restricted the general's tactical choices and affected the outcome. It was also possible for the cavalry to win the battle outright without any significant effort by the infantry, or for the Romans to defeat the enemy cavalry by a concerted effort of the cavalry and infantry, followed by a pursuit of the defeated enemy cavalry to their infantry support and/or marching camp – in most cases the enemy infantry reserve collapsed as soon as their cavalry reached them, but there were also instances in which the enemy infantry reserves protected their cavalry successfully against the Roman cavalry and infantry with the result that the Romans failed to achieve complete victory over the enemy. The opposite was also true. The enemy could inflict a defeat on the Roman cavalry with the result that they fled back to their infantry forces. In most cases the Roman infantry held its ground even after this and protected their defeated cavalry forces, but during the period when the quality of the Roman infantry forces was poor (late-fifth century until the 530s) it usually failed to do this. There always existed the danger that the defeat of the Roman cavalry would also demoralize the infantry. It was because of this that the *De scientia politica dialogus* (*Peri politikes epistemes* 4.49–52) stressed the mutual contribution of both arms of service and the importance of cavalry training because it often decided the outcome, and the *Strategikon* recommended that one was not to use large numbers of cavalry against infantry.[44]

Land battles could also include other phenomena. The chosen site of the battle (fighting from within city walls, in front of the city, use of a valley, river, bridge, difficult terrain etc.) influenced the course and outcome of the fighting. Typically, the Romans used their fleet to transport supplies and men, but sometimes the fleet had also an important role in river crossings in which it could be necessary for defeating enemy opposition. The navy could also influence the outcome of the infantry battle with its naval artillery and archers, as happened for example at the Battle of Phasis. The fleet could also land infantry behind enemy lines, as happened at the Battle of Trachea in 550.

As a generalization, one can state that the professional Roman infantry soldiers had a clear advantage against all enemies during the Late Roman period, so the Romans could expect to win the typical infantry battles that lasted for some hours or even longer, the only exception to the general rule being the dark years at the turn of the sixth century. The Roman infantry was better conditioned, drilled and equipped than any of the opposing infantries. Before the arrival of the Muslims, the only enemies that came close to matching their abilities were the Franks, Saxons, the *Murmillones* infantry of Persia (descendants of Roman fugitives), and the Daylami mountaineers of Persia. The lightly-equipped Moors, Slavs and tribes of the Caucasus were no match for Roman heavy infantry in pitched battle. The only infantry forces of the Late Roman period that were a real match for the Romans were the Muslim infantry, and in fact they were more than a match for the Romans because they defeated the Roman infantry in every encounter. This, however, appears not to have been the fault of the infantry but the fault of the Roman commanders who employed their infantry forces in such a manner that their defeat was all but certain.

As regards the encounters between Roman infantry and enemy cavalry armies, the results varied according to the quality of the Roman infantry and situation. Excluding the low point of the infantry at the turn of the sixth century, the Roman infantry was expected to hold its ground against any enemy cavalry force. Obviously, Roman infantry did suffer some defeats even when its general quality was good, just like there were occasions in which the Roman infantry held its ground against enemy cavalry even when its general quality was low. As noted the actual results of these encounters between the Roman infantry and enemy cavalry depended on the tactics, terrain and situation.

The Influence of Enemies on Tactics

The tactics used obviously depended also upon the type of enemy the Romans faced. The enemies which usually employed cavalry forces in front of their infantry (e.g. the Goths, Persians, Moors and Arabs) usually resulted in cavalry encounters before the opposing infantry could meet each other. There were also enemies that used only cavalry for combat (e.g. steppe nomads like the Huns, Bulgars and Avars, Sarmatians, Quadi, Alans, and sometimes the Goths, Arabs and Persians), with the result that the Romans faced dismounted enemies only when they advanced up to their marching camps. In these instances the battles could be decided by the initial cavalry encounter, or by a successful defence conducted by the infantry phalanx, which included the use of a specific unit order (*pyknosis, synaspismos, testudo/foulkon, globi* or wedge posted in front of the phalanx, *antistomos difalaggia,* or *amfistomos falagx*) to negate the effects of the cavalry charge, and by the use of a suitable battle formation (e.g. outflanking the enemy with wings, lateral phalanx posted on difficult terrain, *taxis allê, epikampios emprosthia, epikampios opisthia*, or hollow square/oblong). If the Romans employed the *epikampios opisthia*, double phalanx, or hollow square/oblong formations against steppe nomads it was possible that the battle would become very prolonged, even lasting for days, because the nomadic cavalry was unable to break the Roman formation when it consisted of well-drilled and motivated soldiers, which then resulted in the Roman infantry seeking to move from one supply depot to another with the nomads following and harassing them with arrows. Battles like this could continue until one or the other had exhausted either arrows or supplies, or either's morale collapsed.

In contrast, the enemies that fielded primarily infantry forces (such as the Saxons, Franks, Slavs, or other Romans in civil wars) increased the relative importance of the Roman infantry forces vis-à-vis their cavalry. The presence of infantry was always required for true pitched battles when the majority of the enemy force consisted of footmen. As already discussed, the outcome of the infantry combat could vary from one *tagma* and *meros* to another so that it was necessary for the officers and soldiers to be able to improvise on the spur of the moment, as they had been trained to do. The terrain could be uneven and the movements of the enemy were not necessarily uniform. If one *tagma* or *meros* in the phalanx was deployed in shallow formation so that the enemy was able to push it backwards, the other *tagmata* and *mere* that were deeper were required to help them. If the enemy had managed to surround a Roman unit, the other units could form the *amfistomos falagx* (double-front) and assist it with the side that faced the enemy. The *strategos* could influence the outcome of such combat with his personal presence and with the reserves he commanded. The infantry battles were usually decided by the quality and number of men and the quality of generalship, so that the outflanking of the enemy on one or both flanks, breakthroughs in the middle, and ambushes were the manoeuvres that decided the infantry battles.

The Phalanx as a Battle Formation

The criticism levelled by some modern-day military thinkers and historians against the phalanx are not based on facts. They are correct to point out the fact that all of the soldiers in the phalanx were subject to the immediate stress of the battle, but period evidence suggests that when the soldiers were well-drilled and motivated this did not cause any problems for them.[45] The depth of the phalanx provided internal reserves for the battle and the men were not unduly influenced by the stress of combat. The soldiers of the phalanx were expected to be able to assume the double phalanx, *antistomos difalaggia*, wedge, and hollow-wedge formations in the middle of combat. This shows nicely that stress did not influence the ability of the soldiers to manoeuvre and fight. The quality of the soldiers was expected to be very high. Obviously when this was not the case, as happened at the turn of the sixth century, it is possible to think that the criticism of the modern historians would have some evidence to back it up, but this is once again a mirage. The problem was not caused by the phalanx formation, but by the very poor quality of the infantry forces which could not hold their ground against cavalry (Goths and Bulgars), even with the help of the hollow square formation or wagon laager. However, the poor quality of the infantry forces was an anomaly in the history of the Roman armed forces that lasted from the late-fifth century until the 530s. The usual state of matters was that the Roman infantry was so well-drilled and motivated that it could perform all of the drilling-ground manoeuvres in the heat of the battle. The Late Roman phalanx had its own internal reserves.

The Numbers in Infantry Battles

The Late-Roman infantry battles could be really small or massive in scale. For example, it was possible that a pitched battle involved only two *auxilia palatina* units or one to two legions without any cavalry support (i.e. anything from ca. 1,000 or 2,000 men[46] up to 6,000 or 12,000 men when the 'legions' in the sources meant *mere*-divisions) while at the other end of the spectrum were the spectacularly-large battles involving huge numbers of infantry and cavalry (and naval forces in some cases) fought between Constantine the Great and Licinius in 324 and the 'Battle of the Nations' known as the Battle of the Catalaunian Fields in 451. These, however, were the extremes. Usually the major pitched battles were fought by combined armies which consisted only of tens of thousands of men or at most of about 120,000 men (plus the non-combatants). The armies that had fewer men were usually insufficient for major battles involving cavalry-based enemies, as the defeats suffered by the 10,000–15,000 men armies against the steppe nomads prove.

The Length of the Infantry Battle

Depending on the situation, infantry battles had varying lengths:

1) It was possible that either the enemy or Romans fled without any real fight, which typically happened if one side had managed to surprise the enemy with an ambush or stratagem;

2) It was possible that the different stages of the Roman advance towards the enemy (the silent advance of the well-ordered Roman line with possible parade ground manoeuvres that demonstrated the superiority of the Roman drill, tall men in front with spiked helmets, polished armour and weapons, successive missile attacks, the use of the *barritus* or other war cry) demoralized the inexperienced enemy sufficiently so that it fled almost immediately when the Romans came to the range of thrown spears/heavy javelins, especially when the Roman wings were threatening to outflank them;
3) The typical infantry battle was not decided fast. According to Vegetius (3.9), the typical infantry battle lasted two-three hours. The length of the mock battle in the *De scientia politica dialogus* (*Peri politikês epistêmês,* 4.18) suggests that its author expected infantry battles to last three-four hours. When one remembers that Maurice expected the infantry battle to progress in such a manner that the different *mere* and *tagmata* had to adjust their actions to the varying situations, it is clear that the Late-Roman military theorists expected the battles to last on average two-four hours, and that they also included either periodical localized lulls in fighting or lulls along the entire front.

The three minute rounds in boxing or martial arts matches are based on the ability of the humans to fight uninterrupted without exhaustion. The maximum length of time that well-conditioned athletes, such as the Roman soldiers, can fight without breaks was about fifteen to twenty minutes. After this the men were too exhausted to do anything. It was because of this that the period soldiers tried to pace their attacks just like the modern day boxers and martial artists. The pacing of attacks means that the soldiers did not constantly seek to kill the enemy but concentrated more on defence. This, however, does not mean that the men would have become passive so that they just held their shields in front, unless of course they were stunned and unable to parry. While they retained their full senses they acted like modern-day boxers when they are pacing their attacks.

Enemy attack could result in an immediate counter attack. Training produced instinctive reactions and it was these that played themselves out on the battlefield. Pacing was important because the men who became exhausted and thereby inattentive were likely to be killed if their enemies were not similarly exhausted. In short, the lulls in combat would have resulted from the physical impossibility of fighting in melee for more than fifteen to twenty minutes at most.

The lull in fighting usually meant that the two sides withdrew a short distance often just outside the distance of javelins to take a breath, but it was also possible that one side held its ground while the other retreated.[47] In some cases, the retreat could be longer. This happened when one side gained a significant advantage and defeated the opposing side with the result that it fled in panic for a couple of hundred metres and sought safety among woods, on a hill, or in a camp and then renewed the fight from that safer position.[48] It was also possible for the two sides to rest while still within javelin distance, but this was obviously done at the cost of not resting well enough because one still had to pay attention to the incoming missiles from the other side. This is what happened for example at the Battle of Mons Lactarius in 552. See the quote describing this battle below. The same battle shows nicely that sporadic fighting could also take place during the lulls, with some men throwing or shooting missiles (arrows, darts, javelins or stones), while some bold individuals could also advance from the battle-line to show their personal bravado and give an example to others. Sometimes their example actually encouraged the entire line to charge, with the result that they won the battle.

If one side had held its ground while the other retreated, it lifted the morale of the advancing side while it disheartened the side that retreated. When this took place for long enough, it resulted in the demoralization of the side retreating with the result that it fled.

On the basis of Maurice's text, we know that the Romans could reorganize their units even in the middle of fighting, but it is clear that the Romans and their enemies exploited these lulls in combat by making further adjustments to their formations. Units adjusted to the prevailing situations both during the fighting and during the lulls in fighting, so these varied from one section of the front to another in the course of the battle. This means that while there was fighting in one sector of the front, there was not necessarily fighting taking place in another sector because there was a lull in the engagement in that sector. During this sort of fighting some of the engagements would have taken place: from distance with both sides using missiles; while in some other sector the units would have packed together and fought by using the *othismos*; while in yet other sectors the men would have engaged each other in melee using spears and swords; while in further sectors some units may have outflanked the opposing units, or vice-versa the enemy had outflanked them; while in some other sectors the units on both sides just rested.

In the *othismos* one could not gain any extra benefit for the push by adding new ranks behind the sixteenth rank, as confirmed by the *Strategikon*. One may therefore assume that when the Roman commanders used a sixteen-ranks-deep array these units were always able to withstand enemy's *othismos* regardless of their depth while being able to repulse the enemy's shallower formations off the field.

During fighting that lasted for up to four hours, the amount of rest and food received prior to the battle could become a decisive factor, hence Maurice's recommendations for not tiring the soldiers too much. The enemies tried to exploit this to their advantage, because they knew that the Romans always (unless the general had ordered the soldiers to eat their meals earlier) ate at midday. When enemies timed their attack to take place just before or during this, they prevented the Romans from preserving enough energy for a long battle. If the soldiers on one side had better physical condition (physical training, adequate rest and energy from food intake), it was possible for them to defeat the less well-conditioned fighters in melee when the soldiers of the opposing side started to falter. On the basis of the narrative sources, Rome's enemies did not usually derive any special advantage from the above because the professional Roman soldiers were usually in such physical shape that they easily outlasted their enemies in combat.

Another thing that could be very decisive in a battle lasting for hours was the number of casualties suffered by the elite soldiers. Maurice quite correctly notes in the context of siege warfare, that in a sortie and hand-to-hand fighting, it was typical for the best soldiers to be killed or wounded with the result that the rest would be discouraged. The same was true for a normal melee. It is clear that in a battle lasting for hours the first four front ranks which consisted of the better fighters suffered most of the casualties in the melee, but that was also true of the enemy.[49] In situations like this, the deciding factor was how well the morale of the rest held when they saw the best being brutally killed and the enemy advancing forward into the vacated places. If it held, the breach in the formation was quickly plugged by the rear ranks advancing to the vacated place, while the forward ranks attacked the flanks of the enemy advancing into the formation.[50] The presence of the corpse obviously made footing precarious for the man replacing him, just as it did for the enemy if he intended to step into the vacated spot. When the battle became prolonged the battlefield also became littered with mangled corpses, used weapons, and horse carcasses. This affected the morale of both armies, but in particular the side with the greater number of casualties.[51]

However, it was also possible that the men panicked when they saw the enemy breaching their formation. This could happen for example when the soldiers in the rear consisted largely of green recruits because the better men had already been killed. It was because of this that deeper formations were safer. The sixteen-rank-formation had four ranks of better fighters in front, while the eight deep had two ranks of better men in front and the four rank array had only one better man in front. The shallower formation exposed the inexperienced green recruits to the stress of combat sooner than the deeper formations. Therefore, the use of a deeper formation by both sides increased the likelihood of a prolonged battle because the deeper units typically collapsed more slowly than the shallower formations. Whatever the depth of the formation, one side would eventually lose their will to continue the battle when they witnessed the killing of their best soldiers, plus/or they had gradually lost ground by retreating. It was then that the winners began their pursuit and the losing side would suffer most of its casualties. When the rout began it most often started from the rear, which is why the Romans posted the more experienced file-closers (*ouragoi*) there. Their task was to force the men forward and prevent the flight of those unwilling to continue the fight.

The 19th-century French savant Ardant du Picq described the rout of one side as follows:

> If the weapons were nearly alike, preserving ranks and thereby breaking down, driving back and confusing the ranks of the enemy, was to conquer. The man in disordered, broken lines, no longer felt himself supported, but vulnerable everywhere, and he fled. It is true that it is hardly possible to break hostile lines without doing the same with one's own. But the one who breaks through first, has been able to do so only by making the foe fall back before his blows, by killing and wounding. He has thereby raised his courage and that of his neighbour. He knows, he sees where he is marching; whilst the adversary overtaken as a consequence of the retreat or the fall of the troops that were flanking him, is surprised. He sees himself exposed on the flank. He falls back on a line with the rank in the rear in order to regain support. But the lines in the

> rear give way to the retreat of the first. If the withdrawal has a certain duration, terror comes as a result of the blows which drive back and mow down the first line. If, to make room for those pushed back, the last lines turn their backs, there is small chance that they will face the front again. Space has tempted them. They will not return to the fight. Then, by that natural instinct of the soldier to worry, to assure himself of his supports, the contagion of flight spreads from the last ranks to the first. The first closely engaged, has been held to the fight in the meantime, under pain of immediate death. There is no need to explain what follows; it is butchery. … demoralization and flight began in the rear ranks.
>
> Picq, 113–6, tr. by Greely & Cotton.

It was possible that the battle became extremely prolonged and lasted for days or even months. The former could result from the fact that the two sides were evenly matched both in morale and size in a situation in which neither side was able to achieve a significant tactical advantage (e.g. outflanking) against the other. The latter, extremely prolonged, battles usually involved the exploitation of city walls, rivers, wadis, field fortifications etc. in defence by one party so that the aim was to reduce the enemy more through a war of attrition than with a real pitched battle. Good examples of this are the multiple skirmishes and battles in front of the city of Rome in 537 and the prolonged Battle of Yarmuk in 634, which actually consisted of several battles similar to the fighting in front of Rome. For an analysis of these, see Syvänne, *MHLR, vols. 6 and 8.*

Some Examples of Pitched Battles from the Narrative Sources

The Battle of Antonia Castra fought between the Romans and Moors in 547 shows simultaneously the different elements of pitched battles, the different stages of combat between cavalry and infantry which did not last long, followed by the pursuit up to the Moorish camp where the fighting then became prolonged, but so that the entire battle appears not to have lasted longer than the expected maximum of three to four hours.[52] The battle began with the advance of the Moorish cavalry against the Roman phalanx formation. The Roman infantry phalanx met the enemy attack with a perfectly aligned formation and showers of arrows, with the result that the Moorish light cavalry fled. The Roman cavalry *koursores* pursued, but the Moors regrouped and counter attacked (the Moors had performed a feigned flight) and forced the Roman *koursores* into retreat. The Roman *defensores* and reserves under John Troglita came to the rescue and forced the Moors to flee to their enemy infantry and marching camp. After this the Romans brought their infantry to the scene and attempted to storm the enemy camp. At this stage both sides regrouped and fought so that both had cavalry on their wings while the Roman infantry in the middle attempted to take the enemy camp. This resulted in a prolonged infantry battle with lulls in fighting:

> Now the grim and brooding war god crouched over the ranks of foemen and forced them, driven as they were by fear, to retreat toward the safety of their own camp. Our soldiers began to cut down the Moors among their own camels

> and, carried by their own momentum, to break through the besieged trenches of the enemy. And this was where the greatest clash occurred with fierce fighting on every side. The brave Romans and the Moors … rushed at one another all at once. Brother did not recognize brother then, covered as they all were with thick dust. … The frontal assault mixed unit with unit and made them indistinguishable in the grim combat, as rank after rank rushed forward with serried weapons. In the confusion of battle, the very density of the men scarcely permitted the struggling units to move their hands. Chest was crushed upon chest, shield on shield, and they dashed the necks of the brave men near them with the tops of their helmets. [*This describes the* othismos *stage. Note also the use of the helmet spikes as the* PST *suggested!*] Now, without warning, one of the enemy would be cruelly wounded, stabbed by weapons he could not even see. The dead bodies on top of one another in a thick heap. A thousand men fell dead, blood poured over sand and its gathering stream, a ruddy current, was trampled by the men's feet as their footsteps were traced in the wet red mud. … They ran forward and, observing the Romans making their way into the camp among the camels, launched a savage counterattack. [*This shows how success varied between different parts of the line.*] They pressed their entangled enemy hard, now hurling their spears at close range, now striking with unsheathed blades. Then they reformed their ranks and advanced beyond their own walls in an attempt to drive our brave companies off. It was not only their blades that inflicted horrible wounds as they flew onward, but now stones and firebrands fell, … Our soldiers held their positions with all their might and, with faith in the valour of their unconquered leader, they pressed in on the enemy camp. And yet the enemy, in a fit of vicious anger, struggled on and hurled another attack at our men, even managing to drive them back and regain their trenches. [*This shows how there were lulls in the fighting.*] But our general John, … renewed their spirit … Their (Moors) troops gave way in fear and turned their backs in flight. The Romans followed, driving them before them. … the general … exhorted them to break through the trenches, … you can hope for spoil once this enemy is annihilated, … we will gladly grant the entire camp to our men as the price for their toils. … So he spoke and, making for the camp, was the first to strike one of the huge camels with his sword. … All at once the camp was thrown into confusion, as the animal barriers, cut by a single stroke, lay open. … They broke through the lines, slaughtering, plundering and lashing out. … Neither age nor harmless sex softened their hearts.
>
> Corippus, 5.350–468, tr. Shea, 140–144 with my comments inside square brackets in italics.

Corippus obviously embellished the material with poetical licence on the basis of what he had heard from participants, but we can still consider the account to be an accurate description of the actual event because Corippus's audience included participants of the battle. The battle clearly shows how there could be lulls in the fighting and how bloody and brutal the fighting was at close quarters. It also shows how the battle could become a shoving match, which the Greeks called *othismos.*

Thanks to the physical restraints, the shoving phase did not last long. During this short period of pushing, stabbing, wrestling and shoving, the better-equipped and -armoured Romans had an edge over the lightly-equipped Moors. The battle also demonstrates how the presence of the marching camp helped the Moors in that they could retreat there and regroup. On the other side, the presence of the general and reserves steadied the Roman line so that each of the Moorish counter attacks failed. The material also shows that the Moors fought with dogged determination. It required the personal interference of the *strategos* John to tip the scales. He promised the contents of the enemy camp as booty and then spearheaded the attack in person.

The Battle of Ad Salices in AD 376 gives us an example of a prolonged battle which ended in a stalemate. The Romans were both outnumbered and tired thanks to a sleepless night so it is not surprising that they lacked the will to continue the battle the next day.

> … Richomeres by common consent took command, and was joined by Profutus and Trajanus, who were encamped near the town of Salices. Not far from there a countless mass of the barbarians had arranged their numerous wagons in the form of a circle, …the Romans, on learning this, also spent the night without sleep, … Therefore the light of day had hardly appeared, when the trumpets on both sides sounded the call to take up arms, and the barbarians, after taking oath together, … tried to reach the height, in order that from there by a swift rush down the slope like so many rollers they might overwhelm all before them. On seeing this, our soldiers hastened each to his own *manipulus*, where they stood fast without moving about or leaving the ranks and rushing forward. So, when both armies after advancing cautiously remained unmoved, the opposing warriors stared at each other with savage and sidelong glances. The Romans in unison sounded their war-cry, as usual rising from a low to a louder tone, of which the national name is *barritus*, and thus roused themselves to mighty strength [*the barritus of Germanic origins was later abandoned and we do not find it in the sixth-century military treatises*]. But the barbarians sounded the glories of their forefathers with wild shouts, … And now, after attacking each other from a distance with javelins and other missiles, they came together menacingly for a hand-to-hand conflict; the shields were fixed side to side in the form of a tortoise shed, and they stood foot to foot. The barbarians, who are always alert and nimble, threw at our men huge clubs, hardened in the fire, and ran their swords through the breasts of those who showed most resistance; thus they broke through the left wing. When this gave way, a strong troops of reserves bravely hastened to their aid from near at hand, and rallied them when death already sat upon their necks. Then the battle grew hot and the slaughter was great; all the more active rushed into the thick of the fray and met their death from the arrows that flew like hail, or from the swords. Those who fled were pursued on this side and on that by troops of cavalry, who with mighty strength slashed at their heads and backs; and likewise on both sides by foot soldiers, hamstringing those who were in the toils of fear and had fallen. And while the whole battlefield was covered with corpses, some were lying among them

> who were mortally wounded, … some were smitten with a bullet from a sling or pierced with arrows tipped with iron; the heads of others were split through mid-forehead and crown with swords and hung down on both shoulders, a most horrible sight. But not yet wearied by the obstinate struggle, both sides continued to assail each other without a decision, and felt no diminution at all of their native hardihood, … But at last day gave way to evening and ended the murderous contest, and withdrawing in disorder wherever each one could, all the survivors returned in sorrow to their tents.
>
> Ammianus 31.7.5–1, tr. by Rolfe, 427–35 with comments inside square brackets in italics and changes.

The prolonged Battle of Mons Lactarius in 552 stands as a good example of a prolonged battle which lasted for two days and also of the use of the dismounted cavalry as infantry. The battle began with the Goths launching a surprise attack on horseback against the Roman cavalry who did not have time to organize their lines, after which both sides dismounted so that the battle became an infantry battle.

> … thus from both sides they charged their nearest opponents with great fury, … Now the battle began early in the morning, and Teïas, easily recognized by all, stood with only few followers at the head of the phalanx, holding his shield before him and thrusting forward his spear. And when the Romans saw him, thinking that, if he himself should fall, the battle would be instantly decided in their favour, all those who laid claim to valour concentrated on him – and there was a great number of them – and they all directed their spears at him, some thrusting and others hurling them. He himself meanwhile, covered with his shield, received all their spears in it, and by sudden charges he slew a large number. And whenever he saw that his shield was filled with spears fixed in it, he would hand over this to one of his guards and then take another himself. And he continued fighting in this manner for the third part of the day, and at the end of that time his shield had twelve spears stuck in it and he was no longer able to move it where he wished and repel his assailants. So he eagerly called one of his bodyguards without leaving his post so much as a finger's breadth nor giving ground nor allowing the enemy to advance, nor even turning round … nor, in fact, did he even turn sidewise … he stood there, shield in hand, killing with his right hand and parrying with his left and calling out the name of the bodyguard. And the guard was now at his side with the shield, and Teïas immediately sought to take this in exchange for the one weighted down with spears. But while he was doing so his chest became exposed for a brief instant of time, and it chanced that at that moment he was hit by a javelin and died instantly from the wound. Then some of the Romans raised his head aloft on a pole … But not even then did the Goths abandon the struggle, but they kept fighting till night …when it began to grow dark, the two armies separated and passed the night on the battle-field in their equipment. And on the following day they arose at dawn, and arraying themselves again in the same manner fought till nightfall, neither army retreating before the other nor being routed

> nor giving ground, though large numbers were being slain on both sides … But finally the barbarians sent to Narses some of their notables … they came to terms …
>
> Procopius, *Wars*, 8.21–36, tr. by Dewing, 413–419.

The battle narrative demonstrates nicely how the dismounted cavalry could fight using pikes, spear thrusts and thrown javelins together with the swords. In this case it is probable that the dismounted cavalry also included *clibanarii* cavalry without shields, so that they held their pikes (*sarisae, conti*) with two hands. The use of the two-handed grip enabled them to use their pikes as if they were fencing, with parries, glides etc. This means that the dismounted units that had *conti/sarisae* employed slightly different individual fighting methods than the infantry proper.

The Sight of Blood

The most difficult thing for most humans to face in combat was and is fighting with bladed weapons, face-to-face with another human being. It was therefore particularly important to possess men who were used to the sight of blood and gore, the noise of the battle, and the general mayhem. As already noted, the ancient Romans had used gladiatorial games for this purpose and this practice continued until the reign of Theodosius I and even after, as we find the prisoners of war still engaging in the killing of wild animals during the reign of Maurice. It was also because of this that Vegetius recommended the recruitment of butchers and stag- and boar-hunters. They were used to the killing and seeing of blood. The fact that most of the period humans had either butchered animals, or had seen those butchered by others, helped to alleviate the problem. The public executions of criminals and prisoners of war also desensitized the people to the sight of killing. The use of large scale hunts of wild animals as a form of military training not only trained the soldiers to perform their military manoeuvres, but also helped the men to get used to the killing of animals. The officers also needed to be bloodied, and it was here that the greatest problem lay. The upper levels of society did not necessarily take part in the killing and preparations of animals for human consumption, so they were out of touch with the realities of life, just like are today's urban 'snowflake' populations.[53] Procopius gives us an example of this phenomenon. The *magister militum* Areobindus, who was not acquainted with the sight of killing, fled when he saw his soldiers being killed in battle. This was just too much for his faint heart to bear.[54]

The Endgame

Roman combat doctrine expected that soldiers would concentrate on the fighting and pursuit of the enemy to ensure that the enemy would be not be able to regroup. The plundering of enemy possessions was forbidden during the battle and pursuit. It was the duty of chosen persons and units to pillage the enemy so that the soldiers could concentrate on the task at hand. This ideal was not always achieved, and there are instances in which the commanders lost control of their army when it began looting enemy possessions. The pursuit of the defeated enemy was primarily the duty of the cavalry, but when the enemy sought a place of refuge in their fortified marching camp or in some other fortified place, it was primarily the duty of the following infantry to capture that place. The effectiveness of the pursuit varied greatly from one battle to another because the circumstances (battle formation, composition of the troops, personal preferences of the commander etc.) all influenced that. Maurice and many other commanders considered it imperative to press the pursuit hard so that the enemy would not be able to regroup, but there were other commanders who either preferred to allow the enemy to flee to minimize the risks after victory or were otherwise unwilling to pursue the defeated enemy too closely because the defeated enemy outnumbered them. The composition and nationality of the defeated force also influenced the manner in which the pursuit was conducted. If the defeated enemy cavalry consisted of the 'Scythian' nomads, the pursuit was not pressed home as hard as it was against other enemies because the nomads were known for their use of feigned flights and ambushes.[55]

Roman combat doctrine also expected that when the defeated enemy were able to reach a place of refuge behind fortifications or on a hill etc., that the pursuing Romans would attack them or at least cut off supply lines, or else agree to an advantageous treaty with the defeated enemy. However, it was also possible for the Romans to allow the enemy to continue their flight from their place of refuge in order to minimize risk – this enabled the Roman commander to claim victory while it also enabled the soldiers to retain their recently-obtained booty without any further risk. Sometimes it was also impossible to pursue the defeated enemy force effectively because of lack of adequate numbers of cavalry or because the enemy managed to flee to an inaccessible place. As already noted, there were also instances in which the entire enemy force was annihilated in the battle, which was the typical consequence in situations in which the Romans had encircled the enemy or had forced them against a terrain obstacle. The pursuit of the enemy and the looting of their property were not the only considerations that the general had to take into account in the aftermath of the battle. It was at least as important to attend to the wounded, to reward those who had distinguished themselves, to bury the dead, to distribute the booty fairly and to punish those who deserved it. The upkeep of morale depended upon it, and this was also recognized.[56]

If the Romans had suffered a defeat, they followed similar procedures as they did with the cavalry army. The defeated Romans were expected to retreat inside their camp or city or to some easily defended location like hills, woods or mountains. It was possible to prepare for such an eventuality in the manner described by Syrianus Magister, namely by posting a cavalry force a few miles behind the army to threaten the enemy if the Romans were forced to retreat. One may assume that this system was also used in practice for two reasons: firstly, we find the other instructions of Syrianus Magister in use during the reign of Justinian the Great; and secondly, the separate cavalry force behind the battle line could be interpreted to be the equivalent of the rear guard which followed the Roman army in the marching formations of Maurice.

Roman combat doctrine did not recommend the use of the same defeated soldiers for pitched battles within the next few days. It was necessary to raise morale first with successful skirmishes and stratagems. Regardless, there is one major difference in the conditions that a defeated Roman infantry force faced in comparison with a defeated Roman cavalry army. If the Romans had an infantry army and the enemy army consisted of cavalry, they could easily use the square and double-phalanx formations in retreat, but if the enemy army consisted of a combined force of infantry and cavalry then the infantry-based army faced serious problems because the enemy infantry could attack the retreating Roman infantry on equal terms. The standard method adopted by the Romans in such cases was to cover the retreat of the infantry by the cavalry, which could also include the above-mentioned separate force left behind for this purpose. This meant that the defeated infantry force retreated as they could while the horsemen manoeuvred behind them by threatening the enemy from flank or rear, or by making counter-attacks, charges, and retreats as required. The Romans could also throw caltrops behind them to block the route, or they could throw valuables behind them to break up the pursuing enemy formation (this was a stratagem).[57]

The Casualties, Wounded and Medical Attention

The number of casualties varied from one battle and situation to another, so it is impossible to make any other generalizations about it except the following two: if the enemy consisted of cavalry it usually suffered relatively light casualties when the Roman combined force defeated it, unless the Romans had managed to outflank it or had forced the enemy against a terrain obstacle; most of the casualties in battles were usually suffered by the losing side when it fled in panic while the enemy pressed its pursuit home.

The number of casualties also depended upon the medical services. It was possible to evacuate the wounded while the fighting was still going on, with comrades standing next to the wounded man shielding him while the man behind grasped him and dragged him away from danger. After this the man in the next rank did the same under the shield roof until the wounded was taken to the rear of the phalanx,[58] where the medics could take them further behind the lines to receive medical attention, similar to the case with cavalry battles. The Romans and their enemies obviously also exploited the lulls in combat not only to replace broken or used weapons and shields, but also to remove the exhausted, wounded or dead from the formation. However, when the Romans reorganized their formation, they sought to keep it as intact as possible. They replaced only those fighters from the front ranks that were truly unable to continue to fight. It was always preferable to retain the best soldiers in the front of the formation.

The Roman military medical services with doctors and hospices were the best before the twentieth century and the arrival of antibiotics. The armies were accompanied by combat medics, medical orderlies, and trained physicians that provided first aid and after-care. However, modern historians have not found a consensus opinion regarding the exact details and quality of the medical services. Some historians have suggested that the Romans knew about the sterilization of surgical instruments and used rudimentary anaesthesia, but this is not accepted by all. However, there are some things that we know for certain. The Romans used medicinal herbs to fight infections and it is known that some of these were actually antibiotics. Furthermore, it is known that the Roman soldiers were given antidotes against poisonous arrows as a precautionary step before the battle, for example when fighting against the Slavs. The level of medical care obviously varied. When victorious, the Romans were in a position to give full medical attention to the wounded, but in defeat this was not possible. We learn from the texts of Procopius that the Romans could perform remarkably sophisticated surgical operations. Procopius obviously preserved these instances because the survival of the wounded was so remarkable, but these examples are still valuable because they show what the physicians and surgeons could achieve. The principal cause of death among the wounded was severe blood loss, which the Romans did not know how to correct.[59]

Chapter Eleven

Late Roman Infantry Tactics

Late Roman infantry tactics were flexible and varied. The Roman infantry was trained to fight in all terrains and weathers. The Romans possessed the most varied battle tactics in existence so that they trained their infantry to use vast numbers of different unit orders (*araiôsis, pyknosis, synaspismos*, several different versions of the *testudo/foulkon* unit order for different situations, and the *drouggos/globus*) that prepared their infantry units to face any kind of situation they could be expected to face on the battlefields. These units were taught to operate independently of each other while providing support for each other. The units could assume a mindboggling number of unit formations: different kinds of wedges (*cuneus, embolos*), hollow wedge/pincers/shears (*koilembolos, forceps, forfex*), *drouggos/globus, antistomos difalaggia, amfistomos falagx* – and possibly also crescent, convex, saw (*serra, peplegmene*) – the sixteen-deep unit for the breakthrough, and the *epikampios* unit formations. The Romans also used a vast variety of different kinds of battle formations: lateral phalanx, double phalanx, triple phalanx, quadruple phalanx, mixed formation, *taxis allê, epikampios opisthia, epikampios emprosthia*, crescent, convex, hollow square, hollow oblong, hollow squares /oblongs side-by-side, wagon laager, attack in open order, and battle formations meant for use in difficult terrain. The employment of field artillery and naval forces in support of their land forces made the Roman armed forces unique to their era.

The mindboggling variety of different battle formations, unit orders, unit manoeuvres, weapons systems and multipurpose soldiers gave the Roman general an unparalleled ability to adapt his army to the prevailing circumstances. The general could choose to use his army in long-range combat by using a great variety of weapons to soften up the enemy starting with (field, mural and naval) artillery (darts, fire-darts, spears, stones, fire-bombs), bows with arrow-guides, and ending up with the lead darts, javelins and spears. The Roman infantry was taught to use a great variety of fighting techniques with a great variety of weapons such as swords, axes, maces, spears, pikes, bare hands and shields. The actual fighting method depended upon the type of unit, weaponry, and the chosen unit order and tactics. The favoured melee phase tactic of the infantry in open terrain varied, the Romans sometimes preferring to use their spears for melee in infantry combat while at other times the favourite doctrine was the throwing of the spear or javelin and then the use of the sword. However, in difficult terrain the favourite tactic in infantry combat was always to use the spear or javelin as a thrown weapon, while against cavalry the principal method was always to use the spear or javelin for thrusting, even if it could also be thrown.

The importance of the infantry vis-à-vis the cavalry varied in the course of the Late Roman period. The Late Romans showed a preference towards the use of the combined force tactics from the late-third century until the fifth-century, but at the turn of the sixth-century the preference had tilted to the side of cavalry because the quality of the infantry forces had plummeted as a result of the loss of so many infantry armies in the latter half of the fifth century and early sixth century (e.g. the loss of the fleet with infantry during the failed amphibious operation to take Carthage in 468, and the destruction of several infantry field armies from the 480s until the 520s). The infantry regained its importance from the 540s onwards, so that during the reign of Maurice it was assigned very important tasks. However, the loss of several armies during the period between 602 and 622 resulted once again in the neglect of the use of infantry in the proper manner during the reign of Heraclius. The Romans paid a heavy price for this neglect, because the combined Muslim armies exploited this oversight to the hilt.

The ability of the Roman infantry to defeat the enemy depended not only on the amount of training they had received and the combat methods they used, but also on the abilities of their commanders. A commander well-versed in military theory and history (i.e. a well-read and educated commander) knew how to use the infantry to best effect, while a cavalry general like Belisarius did not really know how to use infantry effectively. And then there were the numerous commanders who did not know how to use infantry or cavalry at all because they had received their positions through personal contacts, or because of their family background, or because of their perceived loyalty to the emperor, or because the commander was the emperor himself in a situation in which he had insufficient knowledge of military matters (e.g. Heraclius). These questions are analysed in detail in the *Military History of Late Rome* series, volumes 1–8, and further developments will be analysed in equal detail in the forthcoming book series *Military History of Byzantine Rome*, meaning that the current author will be responsible for volumes covering the years 641–1204 and Juho Wilskman the years after this.

Chapter Twelve

Other Fighting Methods on Land

12.1. Military Trickery and Unorthodox methods[1]

The Romans used a number of other ways of waging war than the fighting of pitched battles. These methods included the use of diplomacy (e.g. the divide and rule approach); the trade embargo; the bypassing of commercial competitors (direct trade with India, China and Africa; the stealing of commercial secrets such as silk); the finding of new trade routes through exploration (at least during the reign of Justinian the Great); bribery; assassination; the use of hostages; kidnapping; germ warfare (food, drink, air); the poisoning of armies and horses (food, fodder, water and drink); the denial of supplies; the spreading of disinformation (e.g. false deserters, letters allowed to fall into enemy hands etc.); the killing of civilians to create terror; the targeting of families to force husbands to change sides or leave their current operations unfinished in order to protect their families; the support of usurpers; and the use of stratagems (these also included the already-mentioned methods). These are not analysed here because the topic of this book is tactics. Readers interested in these are advised to consult any of my other studies mentioned in the bibliography or which can be found at the academia.edu website. The following discussion concentrates on the use of ambushes, surprise attacks, night attacks and guerrilla warfare. These sometimes included the use of other stratagems so their use receives some analysis here, but the concept of the stratagem included far more than the mere exploitation of the enemy's gullibility for some tactical purpose, hence for a fuller discussion I would advise the reader to consult the bibliography. Underhand methods are analyzed separately here because in most cases these defy a clear division into standard cavalry or infantry tactics.

The underhand methods used to overcome the enemy can be divided into two categories: 1) those that depended upon the simple exploitation of an opportunity provided by the enemy; and 2) those that resulted from the explicit manipulation of circumstances so that one could achieve a clear advantage over the enemy. The first mentioned consisted of surprise attacks, night attacks and ambushes when the enemy army was marching or encamped carelessly and so forth. In other words, the enemy provided a suitable opportunity for the Romans by its own actions. The second consisted of means used to manipulate the enemy in such a manner that it started to act carelessly or was forced to commit a mistake so that the Romans could surprise or ambush them. These methods included the use of visual deception (the use of a small camp and fewer flags to create the illusion of small numbers etc.); the use of false deserters and false peace negotiations; cutting off the supply routes together with a denial of supplies to force the enemy into retreat while simultaneously forcing the enemy to keep its forces united with ambushes; the blocking of certain routes with fortifications or armies so that the enemy would be forced to use another route; giving the impression of a large force to induce the enemy to leave (extra flags, open formation, extra-large marching camp with extra lights etc.; feigned flight etc. The following discussion analyses some of these questions in greater detail.

12.2. Surprise attacks[2]

The better Greek and Roman commanders had always recognized the fact that the use of all kinds of underhand methods was the most cost effective way of fighting wars. The Late Roman period was no exception. This was particularly relevant when the enemy outnumbered the Romans. They needed to even out the odds against them. The best way to achieve this was the use of guerrilla warfare, skirmishing, ambushes, surprise attacks (e.g. when the enemy marched carelessly, was crossing a river, encamped incautiously, guards were asleep or drunk or bribed etc.), stratagems and night attacks. However, the use of these was always preferable even in those circumstances in which the Romans outnumbered the enemy. It was the sign of a rash commander to seek a pitched battle without first trying to defeat the enemy through other means, but there were obviously still plenty of such rash commanders during the Late Roman period, just as there are in every period encompassing so long a period of time as is being discussed here. There were also generals who agreed to meet the enemy in battle on some chosen battlefield, or allowed the enemy to cross a river unmolested, or who decided that a duel would decide the outcome of a battle.[3] This kind of behaviour, however, was not the preferred way of fighting among the military theorists of this era and it was also not the way that Roman combat doctrine expected the generals and soldiers to behave. The general preference for the use of stratagems, surprise attacks and ambushes can also be detected in the reproduction of old collections of stratagems (Frontinus and Polyaenus) and their inclusion among the military treatises.

Maurice aptly stated that it was always preferable to attack the enemy without suffering any injury oneself. Experience had shown that surprise attacks were very effective against forces of equal strength, and even against vastly superior forces. Such attacks were to be meticulously planned and swiftly carried out. The best period-analysis of the different types of surprise attacks can be found in Maurice's *Strategikon*, who clearly drew his categorization from works of history, collections of stratagems and period experiences. The narrative sources confirm their general accuracy for this period. Maurice and Syrianus both omit the use of the fleet in surprise attacks, but the narrative sources prove that the Romans did indeed use the navy to disembark men behind enemy lines or elsewhere unexpected, just as they could use swimmers and cavalry able to swim across rivers. The ability to cross rivers, lakes and seas gave the Romans a clear advantage in this respect. They could disembark an army unexpectedly in some location where the enemy did not expect it. See *MHLR*, vols.1–8. The following list is from Maurice.

1) The *strategos* was required to be ready to exploit any opportunity and pretext possible so that he could attack the enemy before they were able to get organized. One can include in this category surprise attacks against all enemies. However, surprise attacks against the Slavic settlements along the Danube and other rivers were a category of their own which received special attention from Maurice (see Chapter 3.4). In the *Strategikon* the preferred arm of service for these attacks was cavalry, but mixed armies and infantry armies could also be used.
2) The general could also fool the enemy by welcoming his envoys and give them generous terms after which the ambassadors were dispatched back. After this, the *strategos* followed them and attacked the enemy when it was unprepared.
3) It was also possible to send envoys offering favourable terms to the enemy and then launch the surprise attack.
4) Yet another stratagem was to send fake envoys to the enemy camp to obtain information about its defences, after which the commander made a night attack against the camp on a moonlit night two – three hours before daybreak. This type of attack required archers. The question of night attacks against enemy camps will be dealt with in greater detail below.
5) When scouts reported that the enemy was marching in disorder, it was possible to launch a surprise attack against them and inflict serious damage on the enemy.
6) It was also possible to hide an ambush (this question will be discussed in greater detail below).
7) Some generals used the feigned withdrawal from the theatre of operations to lull the enemy into a false sense of security, after which they turned back and surprised the enemy.
8) It was also possible to drive herds of animals in front of the army so that the enemy would then gather the loose animals as booty, so that when the enemy was scattered and disordered by this they could be surprised.
9) The general could bridge a river with wooden crossbeams or with a pontoon bridge, after which he would build towers at both ends to protect the bridge. When the campaign was conducted by the book, the Romans carried with them bridge building materials (boats, timber, barrels, leather bags) for this purpose, but obviously there are also instances in which the Romans failed to follow their own military doctrine. Notably, the successor kingdoms of Western Rome, the Goths and the Franks, both used wooden towers to protect bridges against the Romans, just as the Romans did. The bridge could then be used for a safe crossing and returning, because the bridge could always be destroyed behind if the enemy pursued. Maurice always recommended the building of a bridge when a marching camp was located close to a river with the camp on the hostile side of the river. If there was a reverse in battle, it was possible to retreat into the camp rather than attempt to cross the river when the enemy pursued. Even if this was the more secure option and worked very well for example for the army of Priscus in 599, even when the only means of escape across the river were actually boats and ships and not bridge, it obviously did not mean that the fortified marching camp could always be used as a safe place of refuge when the soldiers had lost their morale. The Battle of Yarmuk in 634 is the best evidence of this.[4]

One can add to the ninth category some additional alternatives from the other sources together with the Roman counter-measures against the same.

The *Excerpts of Polyaenus* (47.6, 48) include the use of smoke screens (dry wood and green wood burned) to ease the crossing of rivers, as well as the use of smoke screens and fire behind the army to protect its withdrawal across the river.

If the enemy used bridges with towers, as the Franks and Goths did against the Romans in the sixth century, the Romans had the means to counter this.[5] A good example of this is the use of the fleet, improvised skiffs and infantry to open the supply route along the River Tiber to the city of Rome, which was besieged by Goths. The Goths had blocked the Tiber with a bridge that had towers at both ends with a metal chain posted in front of the bridge. Belisarius had to improvise because he lacked sufficient forces to engage the enemy in a pitched battle. He joined two skiffs together and had a tower built on it, and on the tower he placed a small boat filled with flammable substances. In addition to this Belisarius erected wooden parapets with holes in them to 200 dromons. These acted as a cover for the archers and sailors manning the dromons. After this he advanced against the bridge with the tower skiffs and 200 dromons, while his infantry marched along the right shore. When the Romans reached the iron chain in front of the bridge closing the river, they killed some of the guards with arrows, forcing the rest to flee. After this the Romans removed the chain and cleared the opposing shore with arrows so that they could bring the skiffs with the tower close to one of the towers of the Goths. Then they dropped the flaming boat on top of the tower. The entire Gothic garrison was engulfed in flames and killed. In similar situations, the Romans also resorted to other improvised tactical solutions. For example, just before the Battle of Casulinus River, the Romans burnt the tower of the Franks protecting the bridge by using a wagon filled with hay. Before the battle of Mons Lactarius, the Goths blocked the advance of the Roman army under Narses with a bridge and tower. Narses resorted to a counter blockade so that he captured the enemy ships of the Goths with a stratagem and then built a series of towers with *ballistae* along the riverbank to isolate the enemy. In short, the Romans were ready to improvise and they possessed the necessary technical expertise to do so.

Still another method to be added to the ninth category (in locations which could be used safely against the enemy) was the use of city walls, field fortifications and fortifications in general as safe bases from which to launch attacks against the enemy and into which one could retreat back safely. This was the favourite tactic of, for example, Belisarius who employed field fortifications at the Battle of Dara in 530 and the city walls of Rome in 537 for guerrilla campaign against the Goths.

holding riverline with artillery and towers

The method used by Narses at Mons Lactarius: Towers with the *ballistae* built to cover the length of the river. The probable manner of building was to use the *ballista(e)* of the previous tower and foot archers to provide covering fire for the building crew of the next tower.

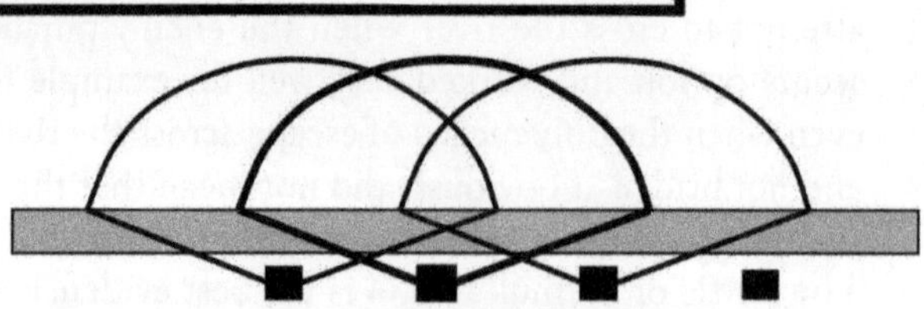

Schematic representation of the principal usages of the walls as part of tactics.

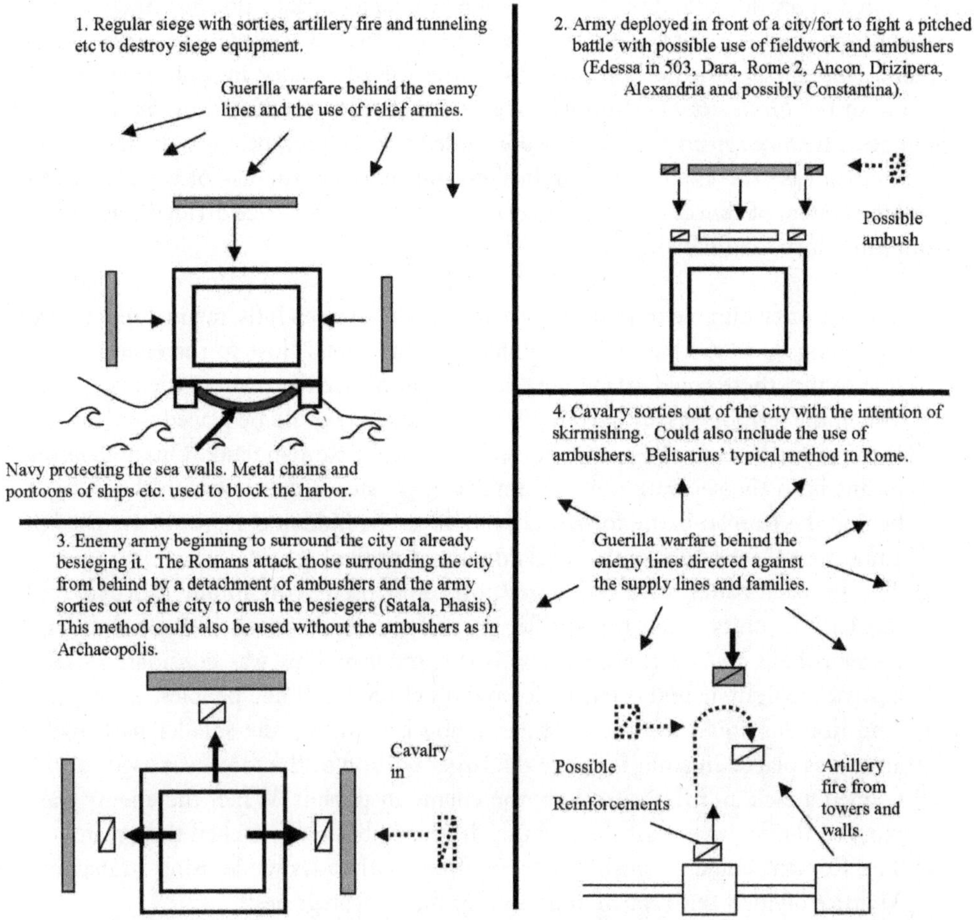

In *The Age of Hippotoxotai* (295–6), I summarised the use of walls in the manner as follows: 1) one could place the infantry on the walls and use the cavalry or infantry to make sorties out of the protective system of the walls or trench, supported by mural artillery; 2) one could use the army just outside the walls to fight a pitched battle; 3) one could use the whole army to make a sortie out of the city; 4) one could ambush the enemy force with part of the army, followed by a sortie of the rest of the army; 5) one could weaken the enemy through a regular siege, possibly punctuated by periodical sorties or limited pitched battles, that were sometimes followed by an all-out attack from within the walls; and 6) a defeat in battle could result in a defensive siege and a siege could proceed to a battle or involve a battle while the siege lasted.[6] When the fighting was done from inside the city walls, the commander could also bolster the number of his forces with civilian paramilitary forces. In some cases, these actually outperformed the professionals in combat because they were fighting on behalf of their families and personal possessions.

12.3. Ambushes

As noted in my doctoral dissertation, on the tactical level the ambushes and surprise attacks were above all carried out by the cavalry arm, even when infantry accompanied them. Unsurprisingly, the best analysis of the use of cavalry in ambushes can be found in the *Strategikon* (4) and its conclusions are also confirmed by the narrative sources. The *Strategikon* obviously also included a set of precautions to be taken into account in case the enemy used ambushes, starting with the use of scouts, patrols and the system of *koursores* and *defensores* with reserves.[7] Maurice divided the cavalry ambushes into the following types:

1) The use of favourable terrain (dense woods, valleys, steep hills, ravines, mountains extending up to the battle line) to hide the ambushers close to the enemy battle line so that these could attack the enemy's rear before the main battle began.
2) When the terrain was not favourable, the ambushers could be placed on the flank between the Roman and enemy lines, or behind the Roman flank. This apparently means both the separate units of ambushers posted either between the lines or behind the Roman battle formation, as well as the hidden *drouggos* or *drouggoi* of ambushers placed behind the outflankers and/or flank guards.
3) The Romans could place the larger portion of their army in ambush deployed regularly in units while the smaller part of the army attacked the enemy and feigned flight to lure the enemy into the ambush. This was considered useful against the light-haired (Germanic) and other undisciplined peoples.
4) The Romans could use the Scythian ambush in which the smaller part of the army was placed in ambush while the larger portion of the force was used for the feigned attack and flight to draw the enemy in pursuit. When the enemy then pursued the fugitives past the ambuscade, the ambushers attacked their rear.
5) The Romans could ambush the enemy from both sides while using stratagems. Maurice divided this type of ambush into further categories:[8]
   a) The Romans could dig a trench eight-to-ten feet deep, fifty-to-sixty feet wide for the distance required. The trench was to be covered with wood, hay and earth so that it looked natural. The trench was to possess some solid ground clearly marked through which the Romans could pass. The ambushers were to be concealed on both sides of the trench so that when the enemy pursued those who feigned flight they would fall into the trench. This ambush should therefore be separated from the trench that was posted between the cavalry lines and which can be considered as a variant version of this ambush/stratagem. According to Maurice, the Hephthalites had ambushed the Persian king Peroz in this manner. The Persians actually used the trenches in like manner against the Romans at the Battle of Thannuris in 528 and it is probable that the Romans then copied this stratagem from them.[9] Maurice considered this type of ambush uncertain because it required a lot of time and plenty of labourers, meaning that it was easy to discover through deserters and scouts. Therefore, Maurice suggested three additional alternatives.

b) The Romans could use the method that the Goths had used against the emperor Decius at Abrittus in 251, which consisted of the use of a feigned flight by cavalry to draw the pursuers into a swamp which had ambushers hidden on both sides.[10] There was obviously no need to build the swamp because it already existed.
c) The Romans could use iron caltrops tied to cords for ease of use. These were to be spread out to a depth of 100 feet along the entire width of the cavalry battle formation, but there were to be four-to-five clearly-marked passageways, each with a width of 300- or 400-feet. The Romans were to post hidden ambushes on both sides of the caltrops. The cavalry forces that performed the feigned attack and flight were required to remove the markers of the passageways when they passed through it and the ambushers were to launch their attack against the enemy's flanks when the latter's horses were crippled by the caltrops. According to Maurice, the necessary preparations could be performed by a small group of reliable soldiers even on the day of the battle or on the previous evening. This ensured the secrecy of the stratagem. Maurice quite correctly considered the use of caltrops to be the easiest to prepare and the likeliest to remain secret.
d) The Romans could also use so-called horse-breakers (this was an ancient stratagem, already used for example by the ancient Greeks) which were round pits (diameter of one foot and depth to three feet) with a sharp stake at the bottom dug here and there. The pits were to be dug in alternating rows about three feet apart in all directions so that these covered an area 150-feet-deep for the distance of entire battle formation, excepting the gaps that were to be left in it for the Romans to be able to pass through. In this version the Romans posted their first line about a mile in front of the pits and the second line two to three bowshots behind the pits, with the units were placed so that the frontline could retreat easily through the passageways. If the Romans arrayed their entire force behind the pits, then the battle line was to be located three bowshots behind them. In this case the passageways were to be kept smaller than in the other case. The Roman cavalry was to attack the enemy at the moment when their horses were starting to fall into the pits. According to Maurice, the digging of the pits could also be performed by a small group of soldiers even on the day of the battle or on the previous evening.

If the Romans then decided to employ ambushers, Maurice assigned one or two cavalry *banda* per wing, or more if the army was large. We find the same set of instructions also in the context of the actual cavalry formation (e.g. *STR* 3.13–4, 3.16), so it is clear that Maurice considered the use of ambushers to be essential in all circumstances. The ambushers were required to be good and their officers were to be courageous and intelligent – unsurprisingly in the *De militari scientia* the ambushers consisted of the *bucellarii* and *foederati*. The ambushers were required to time their attack so that if the enemy attempted to launch their attack first, the ambushers attacked them and prevented this. If the enemy did not attack, then the ambushers attacked the enemy in their sector, or raided the enemy's baggage train, or they attacked the enemy's rear or flanks. If the enemy had two battle lines or had ambushers of their own behind their battle line, the ambushers were to engage these. The ambushers were to time their attack against the enemy so that it did not take place too early or too late. In the former case, the enemy could defeat them before the armies fought, while in the latter their attack could take place too late to influence the outcome. The ambushers and the main body were to time their movements so that they moved at the same time, or alternatively so that the main force moved earlier to draw out the enemy's attention, but the ambushers were to time their attack just before the main body attacked to achieve maximum effect on the enemy. The attack by ambushers would have disordered the enemy, which the main body could then exploit. If the ground permitted, the double ambush was the best option – in particular when the Romans had a large cavalry army. In such instances it was even possible to divide the ambushers so that one portion of the ambushers was used to counter enemy assaults while the other half attacked them. If even one of the units of ambushers remained intact while the Roman main line was forced to retreat, then this unit could stay clear of the fugitives and attempt to attack the enemy's rear in the hope that the sight of this would rally the fleeing Romans.

Maurice quite correctly noted that the best unit order for the ambushers was the *drouggos*, because this allowed greatest mobility and speed while also being the easiest formation to conceal from the enemy. The irregularity of the order allowed the exploitation of the lay of the land for the concealment of the ambush. The ambushing force was to be deployed regularly, in rectangular rank-and-file units, only when it was medium to large in size; in other words, a regular cavalry army in formation. The narrative sources prove that Maurice's instructions regarding cavalry ambushes reflected the actual battlefield conditions that the Romans faced from the third-century to the seventh-century.

Even if the cavalry were ambushers par excellence, their usefulness against disciplined infantry was limited. It is therefore not surprising to learn that there are numerous examples of the use of infantry in ambushes. As already discussed in the context of infantry tactics, the two standard infantry ambushing methods were the use of the *antistomos difalaggia* and *koilembolos* formations against an enemy force between the Roman phalanxes – the best place for their use was a valley through which an unsuspecting enemy would be marching or where the enemy pursued Roman cavalry which was feigning flight. This is what probably happened when the combined East Roman army (included cavalry and infantry, or alternatively dismounted cavalry, used to occupy the heights) ambushed and isolated the Goths in some unnamed valley in the Balkans in 467. The Battle of Dara in 530 shows nicely that the cavalry could also be used as ambushers and for the feigning of flight, even when arrayed alongside the infantry phalanx. The narrative sources also prove that the infantry could be used for an ambush from only one side, as happened for example when Sabinianus ambushed the unsuspecting Goths with a combined force of infantry and cavalry in 479. If the terrain possessed woods or was otherwise broken, it was more typical for the Romans to use infantry (or dismounted cavalry) as an ambushing force, but the successful surprise attack by Aetius's cavalry against the Goths near the Snake Mountain in a location where the Goths were advancing as a column proves nicely that it was also possible to use cavalry effectively in the middle of a snaking valley against the frontal portion of the enemy column formation.

When the combined Roman force dispatched a separate force behind the enemy, the composition of the force depended on the terrain where it was to be used. At the Battle of Campus Ardiensis in January 317, Constantine the Great sent 5,000 horsemen to attack the enemy's rear, while at the Battle of Mount Bourgaon in 534 Narses dispatched 1,000 infantry to occupy the hills behind the Moors during the night. The infantry could also surprise the enemy with the speed of advance or by marching through the night. The enemy could also be lured into an artillery ambush with a feigned flight (usually performed by cavalry) – for example the Chersonites' feigned flight against the Sarmatians lured them into a killing zone where the Chersonites had placed a fortress of *ballistae* carts and ambushers in 286. The volleys of *ballista* bolts stopped the enemy while the ambushers surrounded them. The use of field artillery in this manner was an ancient Greek stratagem which can also be found in the *Excerpts of Polyaenus*. The enemy could also be lured through a stratagem into a position where they were completely helpless, so that the situation was just a one-sided butchery of helpless foes.

The narrative sources also prove that the Romans could use their navy as part of the ambush when the fighting was done close to a river or the sea, or the navy could be used for a surprise attack by landing men behind the enemy line – an alternative version was that some of the soldiers or cavalry swam and surprised the enemy. A good example of this is the ambush of the Greuthungi Goths by the army and river fleet of Promotus along the Danube River in 386.[11]

The use of ambushes was the sign of an able general. The contemporary or near-contemporary military treatises all described how these should or could be used in practice.[12] However, there exist differences between their approaches. Maurice detailed many variations in connection with cavalry tactics, while Vegetius and Syrianus Magister described only the use of feigned flight and surprise attacks (including night attacks). Syrianus did not consider the use of feigned flight to be quite as effective as it should have been. Syrianus Magister suspected that any intelligent person would understand that when a small force attacked a large army and then fled that the intention was to lure them into an ambush. It was because of this that he recommended the dropping of valuables so that the flight would become more credible. Some of the pursuers would then dismount to take the spoils while others would continue their pursuit, with the result that their formation would break up.

Syrianus is indeed correct to suspect the effectiveness of the feigned flight against a smart commander, but reality was different. The feigned flight with small forces worked again and again against gullible, rash cavalry commanders – of which there were plenty. Syrianus also noted quite correctly that it was especially important to place the ambush in the right place, so that the ambushers would not face the pursuers while they were still in formation, but not too late so the enemy had already defeated those sent to lure them. Maurice omitted this, but this is understandable because this was too obvious and self-evident to be included.

The preferential treatment given to the use of ambushes, stratagems and surprise attacks in the military treatises is a clear sign of the Roman aim to achieve military effectiveness with a minimum of effort and risk, which only rash idiotic commanders ignored.

12.4. Night attacks and attacks against camps[13]

One of the forms of surprise attack was the use of the night attack, which was used to gain possession of some important location (heights, river-crossing, bridge etc.), or for the purpose of launching a surprise attack against the enemy – its opposite was the use of the night to hide the flight (a stratagem that the Romans also used when necessary). This had two variants: the sending of ambushers behind the enemy, which has already been discussed, and the surprise attack against the enemy (means typically an attack against a marching camp) just before or at dawn. There were two ways to achieve this: firstly, if the enemy was encamping carelessly without adequate safety measures such as pickets and scouts, it was possible to surprise them without any preparations; and secondly, if the enemy was acting cautiously, it was possible to use different kinds of stratagems to lull the enemy into a false sense of security (e.g. by sending a false deserter followed by a feigned withdrawal; peace negotiations; by arraying the army in front of the camp for several days in succession without engaging to give an appearance of fear, as the Avars did against Priscus; leaving behind booty etc.)[14] and these the commander could learn from the collections of stratagems, from books of history, or by inventing a new stratagem.

Maurice recommended the use of both infantry and cavalry archers and javelin-throwers in the night attack, but narrative sources prove and Maurice's own text implies that there were instances in which the night attack was made by cavalry or infantry alone. Night attacks required archers in particular for spreading panic among the unsuspecting enemy. The night attacks were particularly effective against such peoples, whether on foot or on horseback, who did not fortify or set up their camps properly but pitched their tents all over the place and without protection (means mainly the nomads) or whose encampments were considered disorganized by the Romans (means the Persians), and against those who did not have sufficient numbers of archers and javelin-throwers and who preferred to engage in hand-to-hand fighting (means mainly Germanic peoples; they typically used wagon laagers, but the lack of archers still made them vulnerable). In sum, the use of the night attack was considered potentially effective against all but the best-prepared foe. The narrative sources prove that the Romans could surprise even the Arabs and Moors in the desert, but this demanded the use of local allies as scouts.[15]

Unsurprisingly, the narrative sources confirm the effectiveness of night attacks and surprise attacks. Most of the surprise attacks were directed against enemy camps, but even forts and cities could be targeted in like manner by a surprise attack. The Romans were expert users of surprise attacks, but they were also themselves surprised more than once. The ability to surprise the enemy obviously depended on many variables. The most important of these were the generalship shown before the attack, and the abilities of the spies and the scouts to find what the enemy was up to – plus the ever-present, ever-elusive luck. When the enemy was susceptible to being surprised, the next single most important factor influencing the outcome was the timing of the attack. One could arrive too early or too late. If too early, the enemy had a greater chance of detecting the presence of the enemy, and if too late then the enemy was already awake.

Roman combat doctrine included some very sound pieces of instructions for conducting a night attack, but there were still disagreements between those. Maurice recommended that the night attacks were to be carried out on a starry or moonlit night so that the soldiers would not be bumping into one another or getting lost during the approach. In contrast, Syrianus Magister favoured attacks on moonless nights, but still allowed attacks on starlit nights where the stars could be used to guide the army to its destiny – if the night was cloudy the guides led the men from the front with lanterns hanging from their spears (the lanterns had hide coverings on all four sides, but three of these were painted black and the one painted white was towards the rear). The former therefore favoured safety during marching, while the latter favoured the camouflage of darkness. However, all military thinkers favoured silence during the approach. The soldiers were marched in silence under the guidance of trustworthy men to the site. The attackers were to be unencumbered (*'expediti'*), carrying only what was necessary. The *Peri strategikes* adds that the leading soldiers were required to have iron soles in their shoes to protect their feet from the caltrops of the enemy – the use of caltrops was a typical defensive measure employed by the Persians, so the caution was very sound. According to Maurice, the Romans were to time their attack so that they reached a spot that was about one or two miles (c.1.5–3km) away from the enemy's camp when it was one- to two-hours before dawn. The soldiers were then hid and rested, and launched their attack against the enemy just before dawn.

When the Romans got close to the enemy, they deployed from columns into a line of battle, and, depending on the terrain, launched their attack from two or three sides. They were not to surround the enemy completely, but were required to leave open one side (the side which led back to their home country in the *PST*) for those who wanted to flee. If the Romans had a large army, the attackers blew one or two trumpets, but if the Romans had a small army then they were to blow several trumpets to give an impression of greater numbers. Syrianus adds that one could also employ men who spoke the enemy's language to confuse them further. It is possible that it was because of this that the *Strategikon* also recommended that the heralds be able to speak Persian. The commander was required to hold some of the men in reserve in case of reverses. If the Romans faced a fortified camp (the Persians and other Romans used such) it is probable that the Romans filled the trenches with material prepared for this in advance and used the sloping *testudo/foulkon* formation when necessary for the storming of the walls.

The surprise attack was to be conducted very carefully, especially if against cavalry, so that if the enemy had been warned about the approach and had lined up against the Romans, the Romans would be prepared to meet them in pitched battle if necessary. If the enemy force consisted of infantry, and the Romans used cavalry, then

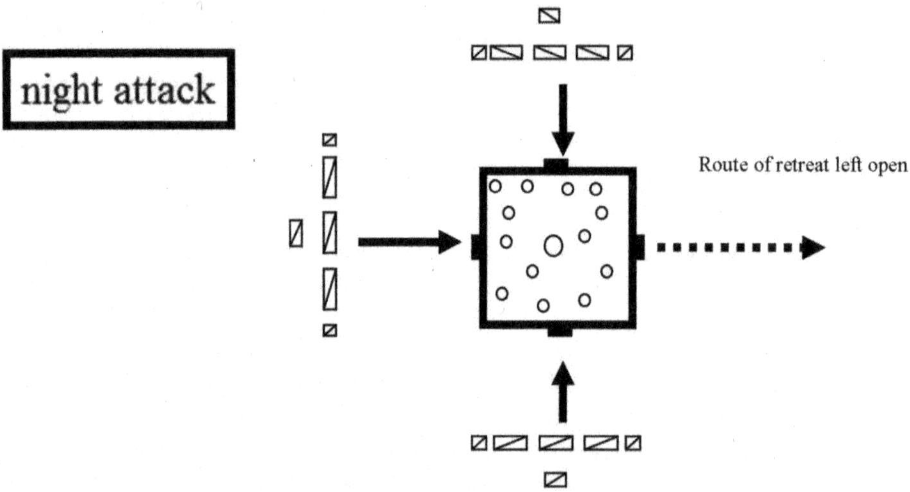

the expectation was that either the Romans would still be able to cause serious harm to the enemy infantry, or alternatively the Romans could retire without suffering any injury from the enemy because infantry could not pursue them. Syrianus (40) added that if the enemy had learned of the Roman approach and had posted an ambush, the Romans would counter this with a new ambush. The commander was always required to detail units for the capture of supplies and booty so that the fighters would be able to concentrate on the fighting. The same requirement held true for both night or daytime attacks against a fortified camp, or for surprise attacks against marching enemy troops or their baggage train.

12.5. A Summary of the Other Methods

The Romans used the other tactical methods depicted in this chapter either because the circumstances forced their adoption, or because the commander preferred to use them, or because circumstances (for example a gullible enemy commander, or the enemy made a mistake, etc.) allowed their use. The better Roman commanders were always ready to resort to the use of unorthodox warfare to overcome an enemy if there was a chance for this.

The Romans could remedy an inferiority of numbers by using against the enemy the city walls or the fortified marching camps, field fortifications, bridged rivers, field artillery, engineering skills, ambushes, stratagems and surprise attacks. If the enemy used these same methods against the Romans, the Romans had systems in place to detect this (spies and scouts) and possessed superior technical and siege skills to counter these, and also naval forces where this was needed to tilt the balance in their favour. In short, under an able commander the Romans could expect to be able to defeat any enemy, even when unorthodox methods were needed, thanks to their greater flexibility of organization, better engineering skills and greater variety of military methods.

Chapter Thirteen

Conclusions

We have seen that the Romans had a combat system that could be adapted to meet any foe imaginable. To put it simply, the Late Romans were ready to meet any challenge posed by any enemy in any terrain or environment. The Romans drilled their armies to fight in any conditions against any type of enemy force. The only real omission was the lack of a true military academy in the modern sense of the word which could be used to train and educate the higher ranking officers. Training was left in the hands of the individuals themselves and this was a major problem in some cases as the generals were clearly not up to their tasks even when their soldiers were. The typical result of this incompetence was that the commanders did not follow combat doctrine.

Roman combat doctrine expected that the Romans would always possess accurate information about the enemy so that the commander could plan his operations accordingly. The preferred way of fighting was to defeat the enemy either through hunger and thirst without having to take major risks, or with a surprise attack, night attack or ambush. It was only when these had been unsuccessful that Roman combat doctrine expected the commander to risk a battle with the enemy, and then only when the circumstances for that were favourable, and even then it was expected that the commander would fight the battle by employing stratagems, plus an ambush or ambushes.

Roman combat doctrine expected the general to take a wide variety of security measures during marching, before battle and during the battle so that the enemy could not surprise the Romans. It also took into account the relative weaknesses and strengths of each side and sought to pit Roman strengths against the enemy's weaknesses. The fact that the Romans possessed all possible varieties of troop types (*clibanarii*, *catafractarii*, regular heavy cavalry, light cavalry, heavy infantry, multipurpose infantry, light infantry, field artillery, naval support, marines, civilian paramilitary forces, allies, engineers, and architects) together with sophisticated unit orders, unit manoeuvres and battle formations, enabled them to do this when the general in command was up to his task.

Roman military theory and practice neglected none of the questions that could happen before, during and after the battle. The Romans took into account all of the different circumstances that the Roman armed forces could face, starting with the pre-battle procedures for different situations, things to do during the battles, the pursuit of the defeated foe, what to do if the enemy inflicted a defeat, and so forth. Their combat doctrine even considered what to do when the Romans faced an unfamiliar enemy.

There is no doubt that the Late Roman armed forces were the most versatile forces on earth. The reason for the fall of Western Rome did not rest with Roman combat tactics or with a faulty combat doctrine, but with the political decisions made by its political and military leadership. Similarly, it was not the fault of the Roman military doctrine that the Roman armed forces lost to the Muslims. It was the fault of the Roman military leadership which employed the Roman armies in a manner that was contrary to its combat doctrine. Excluding the turn of the sixth century, when the infantry forces were not up to their job, the Roman armed forces and their combat doctrines can be considered to have been the most efficient tactical systems in existence, because when used correctly these enabled competent commanders to use the well-tested and realistic combat systems to overcome any type of enemy. The key problem for the Romans was the incompetence of the commanders at key moments in history. This was the result of not possessing a military academy in the modern sense of the word, which would have ensured that the generals placed in command would have had adequate knowledge of military theory and history.

For the development of combat tactics after 641, see the forthcoming book series *Military History of Byzantine Rome.*

Appendices

Appendix 1: The Cavalry Battle in Mesopotamia in 422

Background

The purpose of Appendix 1 is to give the readers an example of one of the variants of Roman cavalry tactics. The chapter on cavalry contains another good example, the Battle of Tricamarum in December 533, but for other reconstructions of Late-Roman cavalry battles, the reader is advised to consult *MHLR* volumes 1–8. For additional analysis of the battle depicted here, see *MHLR 3*, (274–8) which is also the source for the material depicted here. For the commands used here, see the Chapters dealing with cavalry.

In 422 the Romans and Persians were negotiating to end the war that had started in the previous year. The Romans had dispatched to the scene the *Magister Officiorum* Helion (a sort of Prime and Foreign Minister at the same time) who arrived with instructions to conclude a peace.

Quote from Socrates

It is best to start the analysis with a quote from the original source, Socrates/Sokrates 7.20 (tr. Bohn Library pp.354–356 with some changes and comments added in parentheses):

> He [*Helion*] sent … as his deputy Maximinus, an eloquent man who was the assessor of Ardaburius, the commander-in-chief of the army [*strategos*], to make preliminary arrangements concerning the terms of peace. Maximinus, on coming into the presence of the Persian king, said he had been sent to him on this matter, not by the Roman emperor, who was ignorant of the state of things, and thoroughly condemned the war, but by his generals [*clearly a face saving lie*]. And when the sovereign of Persia would have gladly received the embassy, because his troops were suffering from want of provisions [*resulting from the time of year and the Roman guerrilla campaign*]; the corps among them which is distinguished by the name of the Immortals [*Athanatoi*], numbering about 10,000 of his bravest men, counselled the king not to listen to any overtures of peace, until they should have made an attack upon the Romans, who, they said, were now off-guard. The king, approving their advice, ordered the ambassador to be imprisoned and a guard set over him; and permitted the immortals to put their design upon the Romans into execution. They therefore, on arriving at the place appointed, divided themselves into two divisions [*tagmata*], with a view to surround some section of the Roman army. The Romans, observing

but one body of Persians approaching them [*a single line of cavalry which hid behind the Immortals on both flanks. The Persians would have placed the Immortals so close to the first line that the presence of the second was not visible to observers on the plain*], prepared themselves to receive it [*i.e. Ardaburius deployed a single cavalry line which was probably slightly longer than the Persian line, but his purpose was merely to fool the Persians as we shall see*], not having seen the other division because they had attacked suddenly. But just as the battle was about to start, Divine Providence so ordered it, that a Roman army [*stratos*] under the general [*stratelates*] Procopius appeared on the heights. He, perceiving that the comrades were in danger, attacked the Persians in the rear. [*It was possible to see the presence of the hidden enemy line from a higher locale*]. Thus were they, who but a little before had surrounded the Romans, themselves surrounded [*i.e. the Romans had used one of their standard cavalry tactics to fool the enemy into surrounding their only visible cavalry line so that the outflanking enemy wings could then be ambushed by the second Roman cavalry line which had been hidden behind a hill*], and in a short time utterly destroyed; and those who broke forth from their ambuscade [*ek tês enedras*], being next attacked by the Romans, were in like manner every one of them slain with darts [*kai toutous sympantas katêkontisan, the katakontizô comes from kata, akontizô, to kill with akontia-javelins*] [*This is actually a doublet of the former sentence and should rather be interpreted that the Romans followed their success by outflanking the main Persian line after they had destroyed the outflanking wings of Athanatoi, it seems probable that Socrates has misunderstood his source*]. In this way it became known to the Persians that the Immortals were mortals; Christ having executed this vengeance upon that people, because of their having shed the blood of so many of his pious worshippers. [*Theophanes's confused account, AM 5918, confirms the account of Socrates. He states that the Romans encircled the Persians and killed them all including the 10,000 Immortals together with their generals*]. The king of the Persians, on being informed of this defeat [*i.e. he had remained behind, either in a fortified camp with infantry and some cavalry or at Nisibis*], pretended to be ignorant of what had been done; ordering the embassy to be admitted he thus addressed Maximinus: 'I agree to the peace, not as yielding to the Romans, but to gratify you …'. Thus was that war concluded which had been undertaken on account of the suffering Christians in Persia …; and with it terminated the persecution which had been excited in Persia against the Christians.

Reconstruction of the battle

The above text demonstrates that the Persians used the peace negotiations as a stratagem in that they attempted to surprise the Romans in the midst of the negotiations. The Persian plan included two stratagems: 1) attack during peace negotiations; and 2) hiding of the second line as ambushers.

Socrates describes what happened next as a miracle, but this is clearly false. The Romans were clearly aware of the Persian plan to surprise them during negotiations and also of the presence of the ambushers. The hiding of the support line behind a hill was one of the standard Roman tactical ploys, the idea being to lure the enemy into posting his second line to the flanks for the purpose of outflanking the Romans. The Persians attempted the same by hiding the ambushers/second line directly behind the first, which was one of the standard methods also used by the Romans. In this case the Roman stratagem required the Romans to use only a single line, which must have been wider than the Persian line for it to be believable. The fact that the Romans were able to array their forces in readiness to meet the Persians means either that they had spies among the enemy army or that they had been warned by their scouts and patrols of the approach of the enemy. If the former is true, then they would have known of the second Persian ploy from the start, but if the latter is true then the presence of the hill behind Ardaburius's force would have enabled the Romans to see the presence of the *Athanatoi* behind the first Persian line, so that they were able to place their second line under Procopius in ambush behind the hill. It is impossible to know which of these is true, but we know that the Romans were clearly aware of the Persian plans and devised counter tactics against this. We also know that the Persians were unaware of the Roman ambush.

On the basis of analysis of multiple sources not included but which can be found in *MHLR 3*, it is probable that the commander of the Persian army in this battle was Mihr-Narseh and that he had 40,000 cavalry with him, the 10,000 *Athanatoi* (Immortals) included. The *strategos* Ardaburius appears to have had about 31,000-horsemen, which he would have arrayed in the first line. The enemy was unaware of the presence of the commander of the second line, *stratelates* Procopius, because he was bringing reinforcements to the scene. It is probable that he had about 15–20,000 horsemen at his disposal, which was more than enough to destroy the 10,000 Immortals from ambush, and this is precisely what happened. Socrates's account proves that the Roman ploy worked like a dream and the Persians were soundly defeated and his account is confirmed by Theophanes (AM 5918).

When the first Roman line approached the enemy, the *strategos* Ardaburius would have given the orders: '*Ad latus stringe, ad decarchas, ad pentarchas!*', followed by the command '*Iunge!*' when about to reach the range of bowshot. When the entire Roman first line was then closed from the rear, the mounted archers opened fire, and the *strategos* gave the command '*Percute!*' at which time the *dekarchai* and *pentarchai* leaned forward, covered their heads and a part of the horses' necks with their shield while holding the spear as high as their shoulders, and charged at the canter so that the array retained its cohesion. When this happened, the Persian ambushers moved to the flanks to outflank the Roman array. The second line under *stratelates* Procopius waited as long as was necessary to make certain that the Persians could not cancel their order before launching their attack against the outflanking Persians. Since the Roman ambushing force was large it is clear that it was regularly arrayed as four *mere*. The second line would have followed a similar procedure to the first for the assumption of the close order *pyknosis* for the battle, so that when the second line reached the range of bowshot to the enemy it was fully deployed in close order so that Procopius would then have given the command '*Percute!*' for the flanking *mere* and these would then have attacked the *Athanatoi* by using the canter and close order.

The Roman wings then apparently used their *kontaria*-spears as if these were *akontia*-javelins (sing. *akontion*) and threw these at the enemy. Socrates suggests that the thrown *akontia* killed all the Persians, but it is clear that this represented only one stage of the battle. It would have been the front rank that threw their *kontaria*-spears, after which they engaged the enemy with swords. The close-quarters fight with the *Athanatoi* wings must have been short, because they were surprised. The Roman pursuers of the fleeing *Athanatoi* would have consisted of the third line and the outer edge *koursores*. The command for this was the '*Cursu mina!*', while the *defensores* in the middle would have followed after them with the command '*Cum ordine seque!*'. It is probable that at the same time as this happened the *koursores* on the inner sides of the outflanking Roman *mere*, together with the fill-up *banda*, attacked the Persian first line from the flank and rear with the orders '*Depone senestra!*' and '*Depone dextra!*' depending on the side. This resulted in the encirclement and killing of all those Persians who failed to flee in a timely manner. The outflanking of the Persian array resulted in the panicky flight of the entire Persian army.

The following diagrams present the main stages of the battle.

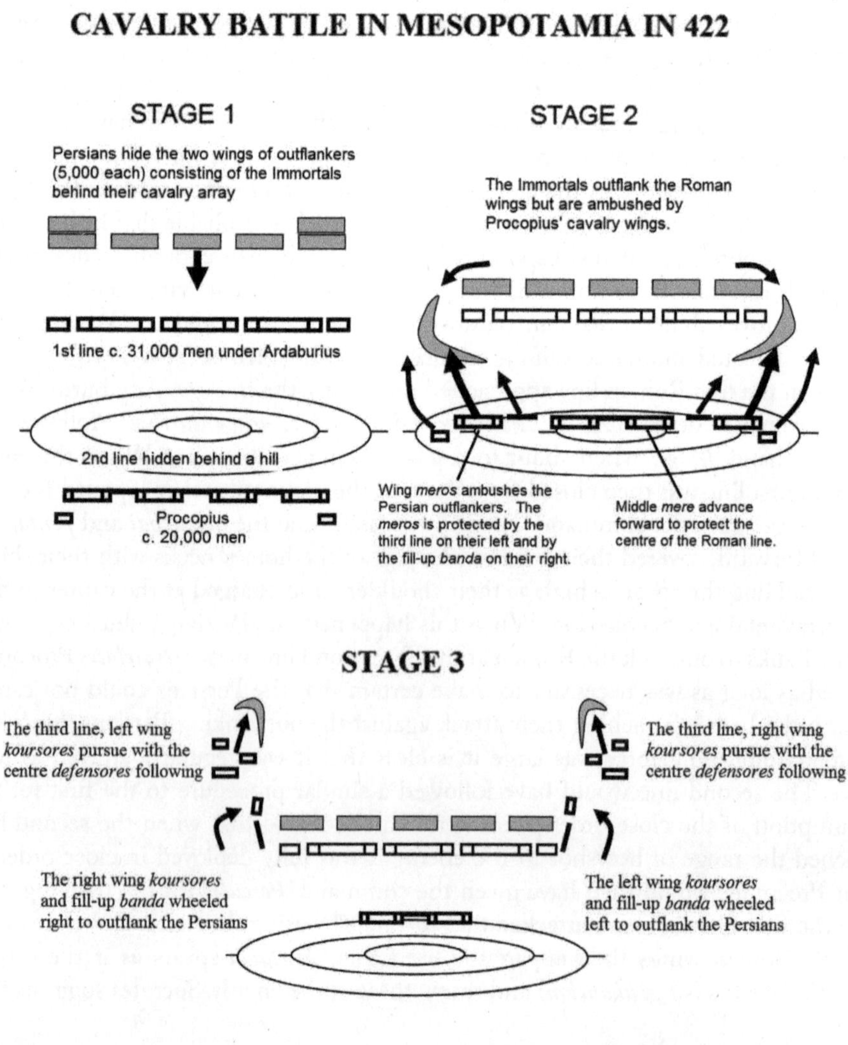

Appendix 2: The Battle of Mursa, 28 September 351[1]

Background

We find the very same battle formation as depicted by Syrianus Magister in the *Peri strategikes* in use already during the fourth century, and one may make the educated guess that this system had been in use ever since Alexander Severus had introduced the *clibanarii* cavalry into the Roman armed forces. The following analysis supersedes my previous analysis of the usurpation of Magnentius and the Battle of Mursa in *MHLR Volume 1* (311–2, 319–29) for those portions which are depicted differently here. The analysis of the Battle of Mursa between the forces of Constantius II and the usurper Magnentius on 28 September 351 stands as the best evidence for the use of the array depicted by Syrianus (and also by Vegetius) and I will analyze it in detail as a case study and example of how the Late-Roman phalanx with its different types of forces were used in actual combat. The battle took place on an unobstructed plain close to the city of Mursa, and the army of Constantius II had river Dravus (Drava)

Victorious Constantius II on the Kertch Missioriorum
Drawing by author. Note the equipment worn by Constantius and his bodyguard.

on his right. Constantius II had numerical superiority (Constantius's 80,000 men vs. Magnentius's 36,000 men) and superior numbers of horsemen, so the site of the battle was very favourable to him. Magnentius possessed high-grade infantry (elite Gallic legions, *auxilia palatinae*, and of Franks, Saxons, and other tribal units), but in the end this was not enough to tip the balance. For the different stages of the battle, see the diagrams on pp.442, 444–5.

The background to the battle was the usurpation of the *comes* Magnentius (commander of the *Ioviani seniores* and *Herculiani seniores*, both of which belonged to the imperial bodyguards, at this time the *protectores domestici* serving under the nominal command of the *comes domesticorum* and the *magister officiorum*; this means that Magnentius was probably the *comes domesticorum peditum*) with the help of Marcellinus, the *comes rei privatae* of Constans II. The principal reason for the usurpation was the homosexual paedophile behaviour of Constans II, together with the personal grievances of many important key persons in the territories controlled by Constans II. Marcellinus was duly rewarded with the position of *magister officiorum* by Magnentius. Magnentius and Marcellinus both hoped that Constantius II would accept the fait accompli and would divide the Empire between the two. This was a vain hope because they had murdered Constantius's brother.

After a series of moves and counter-moves not discussed here, the two sides prepared for combat, with Magnentius posting forces at the passes of the Julian Alps to protect the approaches to Italy. This posed serious problems for Constantius, who then decided to lure the enemy into leaving their advantageous positions by sending his scouts to the Pass of Adrana, where they then conducted a feigned flight back to Constantius's main force, after which both conducted a strategic feigned flight which convinced Magnentius to begin a pursuit. Constantius's plan was to lure the enemy into terrain favourable for his cavalry. In the course of the following war of manoeuvres, Constantius kept up the pretence of being afraid of the enemy by refusing to fight, which convinced Magnentius to seek battle himself.

Magnentius attempted to achieve this by threatening and then by capturing Siscia with an assault, but with no result. Constantius intended to engage his enemy on his own chosen battlefield at Cibalae. The location had symbolic importance because Constantine the Great, the father of Constantius, had defeated Licinius there. Magnentius attempted to lure Constantius away from Cibalae by threatening Sirmium, but with no result, after which he decided to advance against Mursa in an effort to convince Constantius to follow. It is clear that Magnentius did this by bypassing Cibalae via Malata and Teutoburgium. Magnentius did not have any siege equipment so it is clear that his principal aim was to convince Constantius II to leave his advantageous position at Cibalae, either by capturing the cities through surrender or assault or just by threatening them.[2]

This time Magnentius succeeded in his aim, because now Constantius abandoned his advantageous position and marched against Magnentius. Magnentius and Marcellinus were both convinced that it was in their advantage to engage Constantius's numerically superior army because they possessed elite infantry used to winning their battles. Magnentius's plan also included the use of ambush against the enemy. He placed four phalanxes of *Celtae* (four legions of Gauls?) inside the old

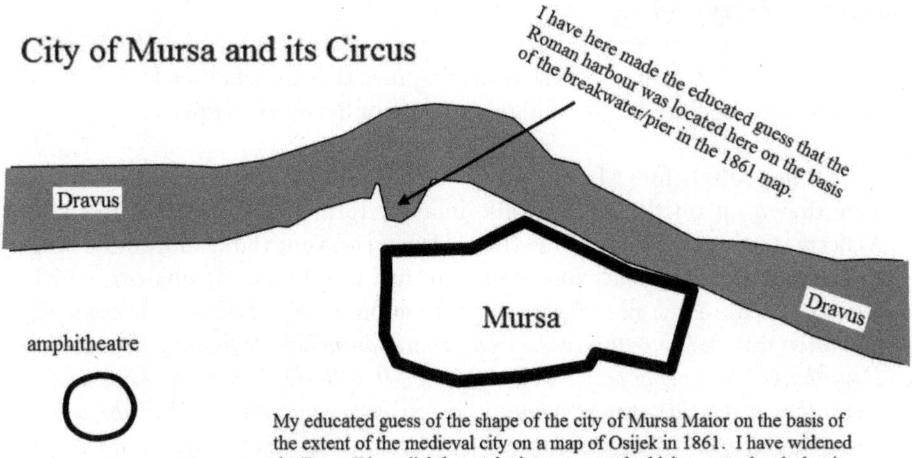

My educated guess of the shape of the city of Mursa Maior on the basis of the extent of the medieval city on a map of Osijek in 1861. I have widened the Drava River slightly to take into account the higher water levels, but it is still possible that the ancient river was still wider than depicted here.

Above: Coin of Constantius II in military gear struck in the city of Siscia. Source: Cohen. Note the gem-studded equipment worn by the emperor.

circus, which was located just outside the city and which was overgrown with woods. See the attached map. The idea was that these would attack the rear of Constantius's army as it bypassed the city of Mursa to engage the army of Magnentius, which had encamped to the west of the city. This failed because the defenders of Mursa saw this and sent a warning to Constantius – which gave him a chance to ambush the ambushers. Constantius sent the tribunes (*taxiarchoi*) Scudilo (a *tribunus scutariorum* of Alaman origins; the *scutarii* belonged to the *auxilia palatina*) and Manadus with the best *hoplitai* (*scutarii*) and *toxotai* (archers) from their forces to the scene. They closed all the doors and mounted the steps above, presumably by using ladders or ropes, and shot and killed all inside. When Magnentius's plan had failed he marshalled his army for combat.

Quotes from the key sources

The following two quotes from Julian's *Orations* and a quote from Zosimus with my explanations and analysis of the contents prove that Constantius II deployed his army exactly as we find it in the military treatise of Syrianus Magister:

> When the plains before Myrsa [*Mursa*] were in sight, the cavalry of both armies were drawn up on the wings, while infantry formed the centre. Then your Majesty [*Constantius II*] kept the river [*Dravus*] on your right, and, outflanking the enemy with your left, you at once turned and broke his phalanx, which indeed had from the first the wrong formation. [*Julian fails to state what the formation was, but since this was a frontal battle in which the forces of Constantius II outflanked the enemy only on the left flank, it is likely that Magnentius used a single thin elongated lateral phalanx in an effort to equal the width of the enemy phalanx, and that he also used his reserves for the extension of the width of the array. It is probable that Magnentius therefore deployed his phalanx eight deep with light infantry behind, because his forces were still able to form a double-front while Constantius deployed his phalanx sixteen deep with light infantry behind. This was indeed the wrong formation for Magnentius to use in a situation in which the other side was certain to outflank them. Magnentius should have chosen his battlefield better, for example by resting his left flank against the Dravus, and he should have used reserves and the double phalanx or hollow oblong from the start*]... he [*Magnentius*] could not even listen without trembling when the legions shouted their battle song [*probably the barritus, the rising shout behind a shield so that the shout was loudest with the long step forward*]. His ranks had been thrown into disorder [*most of Magnentius's cavalry on his right flank appear to have fled with Constantius's cavalry in pursuit, but with the katafraktoi remaining with the infantry phalanx*], but the soldiers formed into *lochoi* and renewed the battle [*the kata lochous renewal of the battle means that the rear half of each lochos-file turned about to face the rear as amfistomos falagx. Aelian (Matthew ed. 38) and its Byzantine Interpolations (Devine 38.1–39.5, Dain C1-D5) note that the amfistomos falagx was particularly useful when there was a need to face cavalry both in front and behind. The Interpolation notes that the cavalry countermeasure against this was to use the square formation from both the front and rear against the infantry*]. For they disdained to be seen in flight, and to give an example in their own persons of what had hitherto been inconceivable to all men, I mean a Celtic or Galatian soldier [*Gallic elite units*] turning his back to the enemy. [*The Gallic legions and auxilia were clearly very disciplined forces.*] The barbarians [*Saxons, Franks and other Germans, inhabitants of the British Isles*] too, who, if defeated, could not hope to make good their retreat, were resolved either to conquer, or not to perish till they had severely punished their opponents [*Maurice considered these traits to be characteristic of the Fair-Haired peoples*]. Just see the extraordinary daring of the usurper's troops in the face of dangers and their great eagerness to come to the close quarters.
>
> Our men [*Constantius's men*], on the other hand, had so far carried all before them, and were anxious to retain the good opinion of their comrades and

of the Emperor… So they charged again as though the battle had only just begun, and gave a wonderful display of daring and heroism. For some hurled themselves full on the enemy's swords, or seized the enemy's shields [*both of these groups were clearly footmen who charged forward to close quarters fighting with swords*], others, when their horses were wounded and the riders thrown, at once transformed themselves into hoplites [*these were clearly horsemen who fought on foot once they lost their horses*]. The usurper's army, meanwhile did the same and pressed our infantry hard. Neither side gained the advantage, till the *thorakoforoi* [*the* clibanarii, catafractarii *of Constantius*] by their archery [*the* thorakoforoi *cavalry were clearly modelled on the Persian model as their origin during the reign of Alexander Severus indicates. See Syvänne,* Gordian III and Philip the Arab, *48ff.*], aided by the remaining force of cavalry [*these would be the* hippotoxotai *and* thureoforoi *posted on Constantius's right flank, see below*], who spurred on their horses to the charge [*note that the* clibanarii *were used to break up the infantry formation as required by the military theory*], had begun to inflict great loss on the enemy, and by main force to drive the whole army before them. Some directed their flight to the plain, and of these a few were saved just in time by the approach of night. The rest were flung into the river, crowded together like a herd of oxen or brute beasts. [*This account condenses the events, because according to the second Oration Magnentius's army fled first to their marching camp and only after Constantius's army captured it with an assault did they flee to the plain and river.*] Thus did the usurper's army reap the fruits of his cowardice while their valour availed him nothing.

…What emperor can cite in the past who first planned and then reproduced so admirable a type of cavalry, and such accoutrements? [*As already noted, Constantius II was not the first emperor to introduce this type of cavalry, because it had been in use from the reign of Alexander Severus onwards, but he deserves full credit for having reintroduced this type of cavalry into such a dominant position as it had under him. His father Constantine the Great had favoured more lightly-equipped cavalry against the more heavily equipped cataphracts that his enemies fielded.*] First you trained yourself to wear them [*this proves nicely that Constantius's cavalry included clibanarii even before he increased their numbers*], and then you taught others how to use such weapons so that none could withstand them… Your cavalry was almost unlimited in numbers and they all sat their horses like statues, while their limbs were fitted with armour that followed closely the outline of the human form. It covers the arms from wrist to elbow and thence to the shoulder, back and breast. The head and face are covered by a metal mask which makes its wearer look like a glittering statue, for not even the thighs and legs and the very ends of the feet lack this armour. It is attached to the cuirass [*thorax*] by fine chain-armour like a web, so that no part of the body is visible and uncovered, for this woven covering protects the hands as well, and is so flexible that the wearers can bend even their fingers.

Julian, *Or.* 1.36A-37D tr. by W.C. Wright (93–7) with some changes, corrections and comments added inside parentheses in italics.

For when he [*Constantius II*] reached the open country and the plains of Paeonia and it seemed advantageous to fight it out there, then and not before the Emperor drew up his cavalry separately on both wings.

Of these troops some carry lances [*aichmoforoi*] and are protected by cuirasses and helmets of wrought iron mail. They wear greaves that fit the legs closely, and knee-caps, and on their thighs the same sort of iron covering. They ride their horses exactly like statues, and need no shield [*these are the clibanarii*]. These are followed [*translation corrected here*] by a large number of the other cavalry, those who carried the shields [*the* thureoforoi *cavalry of Greek military theory, which would mean the* scutarii *cavalry in Roman terminology*] and those who fought on horseback with bows [*in Greek military theory these would be the* hippotoxotai, *in Roman terminology the* sagittarii]. [*We can use the* Peri strategikes *to pinpoint what Julian meant with this. Since Syrianus placed the* katafraktoi *next to the infantry and the light-cavalry to the outer flank, it is clear that the shield-bearing cavalry was posted behind as reserves. This solution makes sense because most of the shield-bearing cavalry, the* scutarii *of the Late Roman period, would have belonged to the elite units well-suited to serve as reserves.*] Of the infantry, the hoplites [*hoplitai*] occupied the centre and supported the cavalry on either wing. In their rear were the slingers [*sfendonêtai*] and archers [*toxotai*] and all troops that shoot their missiles from the hand and have neither shield nor cuirass [*i.e. the infantry was deployed in the standard manner, hoplitai in front and psiloi behind, and no reserves are mentioned even if it is likely that at least the flanks had reserves in the same manner as is implied by the text of Syrianus. Although, I would not entirely preclude the possibility that there were at least some small numbers of footmen posted in the centre with the commander of the centre to serve as an emergency force.*] This, then, was the disposition of our phalanx. The left wing slightly outflanked the enemy whose whole force was thereby thrown into confusion, and their line broke [*given the discrepancy in numbers it is clear that Magnentius had made his phalanx shallow in order to extend the width*]. When our cavalry made a charge and maintained it stubbornly, he who had so shamefully usurped the imperial power disgraced himself by flight, and left there his cavalry commander [*the magister equitum Romulus according to Zosimus; since Magnentius fled with his right wing cavalry, Romulus must have been in command of the left wing*] and his numerous *chiliarchoi* and *taxiarchoi*, who continued to fight bravely, and in command of all these the real author [*magister officiorum Marcellinus, presumably in the centre of the infantry phalanx*] of that monstrous and unholy drama, who had been the first to suggest to him that he should pretend to the imperial power and rob us of our royal privilege… But the Emperor proclaimed an amnesty for those who should renounce the conspiracy… [*it is actually probable that Constantius proclaimed this again after the usurper had fled, because it was typical for armies to surrender to the victor when their commander had fled. In this case, however, it did not work as expected.*]

The man who had trained and tutored the usurper was neither among the fallen nor the fugitives [*i.e. he removed the signs of his office and died fighting.*] … For till the battle actually began, and while the troops were forming the phalanx, he [*Marcellinus*] was full of confidence and went to and fro in the centre of their line [*this confirms that Marcellinus commanded the centre.*]. But when the battle was ended as was fitting, he vanished completely …

For though their generals showed such cowardice, the courage of the soldiers was by no means abated. When their line was broken, which as due not to their cowardice but to the ignorance and inexperience of their leader, they formed into *lochoi* and kept up the fight [*the* lochoi-*files formed up double-front to face both front and rear*]. And what happened then was beyond all expectation; for the enemy refused altogether to yield to those who were defeating them [*this may imply that Constantius halted the attack against the remaining enemy force after Magnentius had fled in the hopes of inducing the remaining forces into surrendering on terms. See also Zosimus, below. If this is the case, then Constantius made a mistake by giving the enemy time to regroup their forces.*] while our men did their utmost to achieve a signal victory, and so there arose the wildest confusion, loud shouts mingled with the din of weapons, as swords were shattered against helmets and shields (aspis-*type*) against spears (dory-*type*). It was a hand-to-hand fight [*this shows the desperation, determination and courage of both sides*], in which they discarded their shields and attacked with swords only [*It is doubtful if this should be interpreted as a sign of the troopers discarding shields voluntarily, but rather as a sign that their shields had become useless so they discarded those because of this. Regardless, one cannot entirely preclude the possibility that some fotmen of Germanic or other origins threw away their shields and rushed with a sword in hand against the emperor's men in order to cause maximum damage with their self-sacrifice. i.e. it is possible that some of the usurper's Germans fought as berserks.*] while, indifferent to their own fate, and devoting the utmost ardour to inflicting severe loss on the foe, they were ready to meet even death if only they could make our victory seem doubtful and dearly bought. It was not only infantry who behaved thus to their pursuers, but even the cavalry when *dorata*-spears were broken and were now entirely useless. Their *xystoi*-spears are long and polished, and when they had broken them they dismounted and transformed themselves into hoplites [*i.e. some of the cavalry on both sides dismounted to defend themselves better against the cavalry*]. So for some time they held their own against the greatest odds [*it is very likely that this refers only to the left wing cavalry of Magnentius and that his entire right wing fled together with him*]. But since our cavalry kept shooting their arrows [*this would be Constantius's right wing cavalry, which included the Armenians under Menelaus; see the quote of Zosimus*] from a distance as they rode after them [*the cavalry battle on the right flank of Constantius's army clearly resembled the standard prolonged cavalry battle in which units from both sides advanced and retreated in turn until one side managed to force also those units that had remained behind into flight.*], while the *thôrakoforoi* [thorax-*wearers, i.e. the* clibanarii] made frequent charges [*the* clibanarii *were needed for breaking up the dismounted enemy cavalry*], as was easy on that unobstructed and level plain, and moreover overtook them, the enemy were glad at last to take to flight, while our men kept up a vigorous pursuit as far as the camp [*i.e. the fugitives fled first to their camp and not to the plain or river*] and took it by assault, together with the baggage and slaves and baggage animals. Directly the rout of the enemy had begun, as I have described, and while we kept up a hot pursuit, they were driven towards the left, where the river was on the right side of the victors. And there the greatest slaughter took place, and the river was choked with the bodies of men and horses, indiscriminately [*this appears to refer to the stage in which the army of Constantius captured the enemy camp and not to the battle, because the flight of Magnentius' army started on their left flank and the fugitives were pursued up to the enemy camp*].

Julian, *Or.* 2.57B-59C tr. by W.C. Wright (153–61) with some changes, corrections and comments added inside parentheses in italics.

… the armies met and engaged in the plain before Mursa; … and great numbers fell on both sides… Constantius, considering that as this was a civil war victory itself would be scarcely an advantage to him, now the Romans being so much weakened, as to be totally unable to resist the Barbarians who attacked on every side, began to think that it would be better to end the war by offering proposals of peace. While he was thus deliberating, the armies were still engaged [*this probably means that Constantius sent promises of pardon to the enemy after Magnentius and his cavalry had fled, but the enemy did not accept these proposals and kept on fighting*]; and that of Magnentius became more furious, nor would they cease fighting though night came on, but even their officers continued performing what belonged to common soldiers, and encouraging their men to oppose the enemy with vigour [*the situation was desperate so that officers of Magnentius had to show a personal example to the troops*]. On the other side likewise, the officers of Constantius called to mind the ancient bravery and renown of the Romans [*the massive casualty numbers, see below, forced them also into the desperate measure of leading the men from the front*]. Thus the battle continued until it was completely dark; nor did even darkness cause them to relax; but they wounded each other with spears, swords or whatever was in their reach; so that neither night nor any other obstacle which usually causes some respite in war, could put an end to the slaughter [*the length and physical limitations mean that this is an exaggeration. It is clear that there were short pauses in fighting during which one or the other side retreated a few metres and the other side did not follow, during which the men on both sides rested before resuming combat*], as if they thought it the greatest felicity that could happen to them to perish beside each other. Amongst the officers that showed great bravery in this battle and fell in it were Arcadius, commander of the legion called *Abulci*, and Menelaus, who commanded the Armenian horse archers [*it is probable that these were deployed as rank-and-file rhomboids in combat as was their native custom*]. What is said of Menelaus is worthy of being related. He could take three darts at once, and with one shot hit three men [*Menelaus was clearly an expert of Persian-style shower archery*], by which manner of shooting he killed a great number of the enemy, and was himself almost the cause of their flight. He was killed by Romulus, who was the first in command in the army of Magnentius [*i.e. Menelaus and his Armenians were on Constantius's right flank*], and Romulus himself fell at the same time [*the death of Magnentius's* magister equitum *undoubtedly contributed to the collapse of the left cavalry wing*]. He was wounded by a dart which Menelaus had shot at him [*the wound was clearly deadly*], yet continued fighting after he had received the wound, until he had killed the person who had given it to him.

Constantius now gaining the victory, by the army of Magnentius taking to flight, a terrible slaughter ensued [*this account suggests that even after the flight of Magnentius and his cavalry, Magnentius's right-flank infantry was able to hold its own against the* clibanarii *and infantry of Constantius and that the resistance of the Magnentian army ended only when their left wing cavalry were defeated at the same time as its commander Romulus fell*].

Zosimus 2.50.4ff., anon English tr. 1814, p.63–4
with my comments in italics inside parentheses

Reconstruction of the battle and tactics

In sum, it is clear that neither side rested their flank against the Dravus River but deployed their cavalry on the wings and infantry on the centre. We do not know if Magnentius had any *clibanarii/catafractarii* in his army but it is probable that he had at least some so that both deployed their cataphracts next to their infantry phalanx and light-cavalry as the outer wings, while the *scutarii* cavalry served as reserves for both flanks. In Constantius's army the entire cavalry force that was posted in the first line on the flanks was able to use bows while mounted because his *clibanarii* employed bows during this battle, while it is probable that the *scutarii* cavalry did not use bows in this battle. Both armies would have deployed the 'hoplites' in front and the archers, javelineers and slingers behind the 'hoplite' phalanx. Magnentius appears to have attempted to equal the width of the enemy line by making his formation shallower, without succeeding in this because the Constantius's left flank still extended beyond Magnentius's right flank. The same was not true on Constantius's right flank because the presence of the River Dravus prevented outflanking on that side.

It is probable that Constantius's Christian army followed the same set of commands as we find in the *Strategikon*, so when the cavalry approached the enemy it was first ordered to tighten the formation for the charge with the commands: '*Ad latus stringe, ad decarchas, ad pentarchas!; Junge!*'. Magnentius's cavalry would have followed the same set of commands. It is possible that when the cavalry was deployed alongside the infantry, and had reached the range of one bowshot where the charge was to start, that they followed the same set of commands '*Parati! Adiuta Deus!*' ['Ready! O God!'] as we find in the infantry sections of the *Strategikon* and did not start the charge at the canter or gallop immediately with the command '*Percute!*' – it is probable that the mounted archers charged mainly by using the gallop during this period, while the *clibanarii* would have used the canter in order to retain the cohesion of their formation. It is also possible that period cavalry sometimes used the same system as the infantry, even when employed separately from the infantry, because we find the cavalry using these in the *De militari scientia* (9–12).

The infantry was prepared for combat either by having the units assume the close order *pyknosis* with the command '*Iunge!*' (an option used mainly by the fully-armoured units) or by having them assume the *testudo/synaspismos/foulkon* order with

Battle of Mursa Phase 1: The Battle Formations

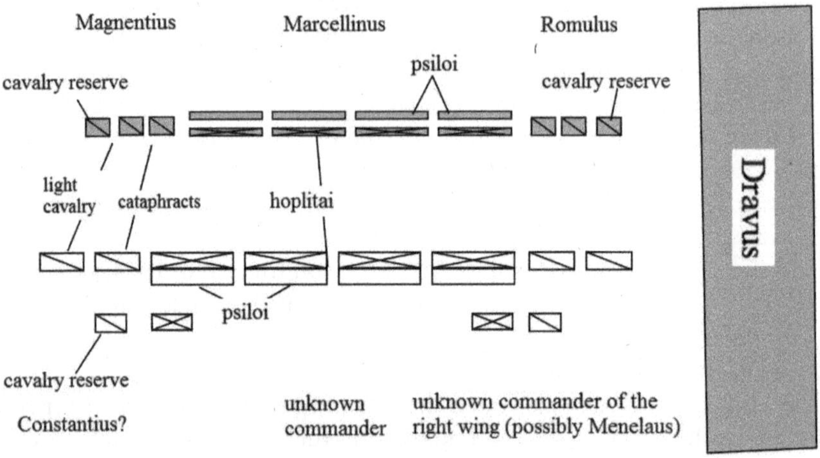

The battle formations

I have here made the educated guess that Magnentius posted his reserves on the flanks to extend his battle line to equal the enemy, which is likely to have been one of his major mistakes. He also adopted the shallower depth of eight ranks plus four ranks of light-armed for his phalanx for the same reason while Constantius used the standard depth of sixteen ranks plus eight ranks of light infantry. The reason for both of these conclusions is that Julian referred to the adoption of the wrong battle formation by Magnentius while Constantius's army still outflanked the enemy on the left wing. It is probable that all of the commanders still retained their personal bodyguards when they stayed behind their phalanxes but these were too small in number to be considered actual reserves.

I have arrayed Constantius's army in the same manner as we find in the *Peri strategikes* and in some of the sixth century battles, but the reader should keep in mind that it is possible that Constantius used infantry reserves for his centre.

Both armies appear to have had their marching camps behind them to serve as places of refuge.

the command '*Ad fulcon!*', an option mainly used by the unarmoured units, but which could also be used by the armoured units. In the case of Magnentius's army it is likely that he did not use the *testudo/foulkon* even for his unarmoured units because his purpose was to widen his frontage. It was possible to protect oneself against the volleys of arrows even in the close order by placing the shield above the head, even if this did not provide the same protection as the tighter formation. Once the infantry units had assumed the correct unit formations (*pyknosis* or *fulcum/foulkon*) and were at about the distance of one bowshot from the enemy, the general gave the command '*Parati!*' ['Ready!'], which the herald answered with '*Adiuta!*' ('Ready!'), after which the soldiers shouted in unison '*Deus!*' ['O God!']. As noted, it is possible that both infantry and cavalry did this when they reached the distance of one bowshot from the enemy. It was then that the cavalry was given the order to charge with the order '*Percute!*', and the mounted archers started shooting their arrows and it was at this point that the infantry archers also began their archery while the heavy infantry started its advance against the enemy.

It was thanks to the outflanking position that Constantius's left wing cavalry routed Magnentius's right wing cavalry immediately and easily with the result that the usurper fled together with it. Constantius's cavalry forces (mounted archers, *scutarii* and cataphracts) appear to have made the mistake of not attacking the flank and rear of the enemy immediately and concentrated solely on the pursuit of the enemy (these would be the mounted archers, with the *scutarii* acting as their reserves), which appear to have fled past their marching camp into the wilderness of modern-day Croatia. The set of commands used by the pursuers were as follows. The mounted archers were given the order '*Cursu mina!*' to begin the pursuit of the fleeing enemy by using the gallop, while the *scutarii* cavalry were given the order '*Cum ordine seque!*' so that they followed the mounted archers while maintaining their order by using the canter. However, it is possible that in this case the *scutarii* also galloped after the fleeing enemy, because these appear not to have contributed to the defeat of the enemy's right wing in any form or manner after this. When this happened it is probable that Constantius's cataphracts still retained their position on the left flank as expected and did not join the pursuit. The super heavy cavalry would have been useless in pursuit of the fleeing cavalry. The cataphracts would have been wheeled to the right with the command '*Depone dextra!*'. It is also probable that Constantius then attempted to lure the enemy into surrender with promises of pardon, but with no result. It only gave Marcellinus more time to reorganize his infantry phalanx and right wing against the eventual attack from both flank and rear. The infantry phalanx of Magnentius responded to the flight of their cavalry and to threat of being attacked in flank and rear by forming up a double-front (*amfistomos falagx, orbis*), which was undoubtedly done with the command '*Undique servate!*' ['All directions face!'], because the enemy was already so close that it was impossible to divide the phalanx into the double phalanx formation. The double-front meant that the rear half of each file turned about to face the enemy. If the light infantry was posted behind, as is probable, then the hoplites first formed up the double-front after which the rear portion advanced towards their new front and admitted the light infantry between the two phalanxes. It is less likely that the light infantry would have remained in front of the rear portion of the double-front. It is also clear that the rear portion of the Magnentian phalanx did not advance in this case a bowshot behind the front phalanx to form up an actual double phalanx, because this would have given Constantius's cavalry and infantry the chance of penetrating the interval between the Magnentian phalanxes when it was no longer protected by its cavalry. In other words, the Magnentian infantry phalanx retained the *amfistomos falagx* (two-fronted) structure but probably so that the light-armed formed the middle ranks as depicted in the Burnley diagram, and also protected the small interval needed for the light-infantry between the two heavy infantry phalanxes. This proved an effective countermeasure against the attacks by Constantius's left flank cataphracts and infantry for the entire duration of the battle. The *Byzantine Interpolation of Aelian* (Devine ed. 38.1–39.5; Dain ed. C1-D5) states that when cavalry faced the *amfistomos falagx* it was to use the square formation both against the front and rear

Battle of Mursa Phase 2:
1) Outflanked Magnentius flees together with his cavalry while Constantius's mounted archers pursue them with the support of the *scutarii* cavalry.
2) Constantius offers terms of surrender which Marcellinus esploits by reordering his infantry phalanx into two-fronted formation.

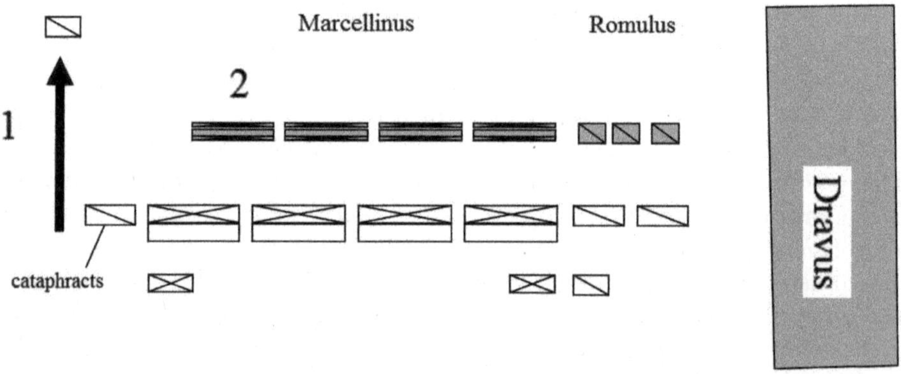

Battle of Mursa Phase 3:
1) It is probable that Constantius attempted to use his *clibanarii* on the left flank for the breaking up of the double-fronted phalanx in the same manner as advised in the Byzantine Interpolation of Aelian, namely so that one half attacked it from the front and another from behind. I have here made the educated guess that Constantius could also have used his reserve infantry against the flank while keeping his left wing infantry phalanx behind so that the *clibanarii* could concentrate on their task. However, it is probable that the light-armed archers and slingers still supported the *clibananarii* with their missiles. The Magnentian right wing withstood all of the attacks against its front, flank and rear.
2) The phalanxes of Constantius in the centre supported the wings by tying up the Magnentian phalanx in place.
3) The prolonged battle was finally decided after the darkness fell when the Armenian mounted archers of Menelaus inflicted massive damage on the opposing cavalry with arrows while also killing the opposing commander Romulus. Romulus killed Menelaus before he succumbed to the wounds.
4) The collapse of the left wing demoralized the Magnentian army which fled to its marching camp. Marcellinus died while protecting the flight of his army.

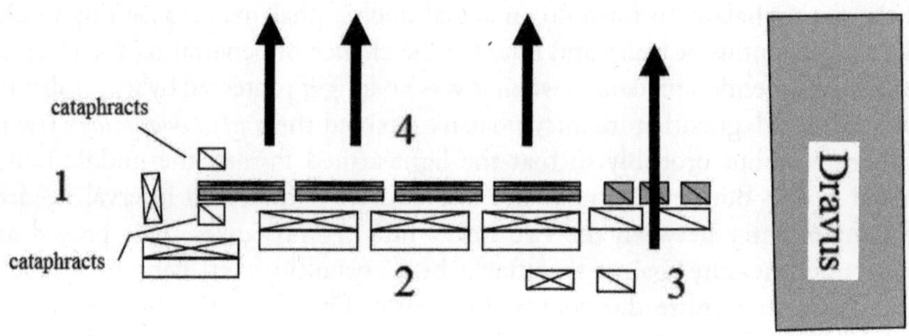

of the enemy formation and one may make the assumption that the cataphracts of Constantius indeed did this, but without success because the right wing infantry forces of Magnentius were clearly the crème de la crème of the Roman army. The use of the square cavalry formation means in practice the posting of the separate cavalry units in ranks and files formations side-by-side so that the separate square units formed a single long oblong, so that the entire cavalry formation attacked the enemy phalanx simultaneously both from front and rear along the entire length of the targeted portion of the enemy line.

The end result was that the battle became a stalemate in a situation on which the rest of the two armies were locked in a desperate frontal hand-to-hand combat in which neither side had any real advantage. This proves that Magnentius and his advisors were not entirely mistaken in their effort to seek battle with their numerically inferior forces. The very high fighting quality of both armies is not only demonstrated by the length and desperation of the battle, but also by the casualty figures. According to Zonaras (13.8), Constantius lost about 30,000 dead while Magnentius's army lost about 24,000 men – only armies which are highly disciplined, determined and well-motivated can suffer this many casualties. The length of the battle means that there were also lulls in combat, because even professional athletes are able to fight intensely only for a couple of minutes without having to rest. When this happened, the men probably withdrew a few metres from each other, or one side retreated while the other did not follow or followed only a little, after which both sides rested and then resumed combat again. This lasted until night fell. It was presumably thanks to the darkness that the battle was decided by the Armenian mounted archers under Menelaus on Constantius's right wing. When the darkness fell, the arrows could no longer be seen

Battle of Mursa Phase 4:
1) Constantius's army captures the enemy camp with an assault.
2) Some of the Magnentians who flee to the plains survive.
3) Almost all of the Magnentians who fled to the River Dravus die.

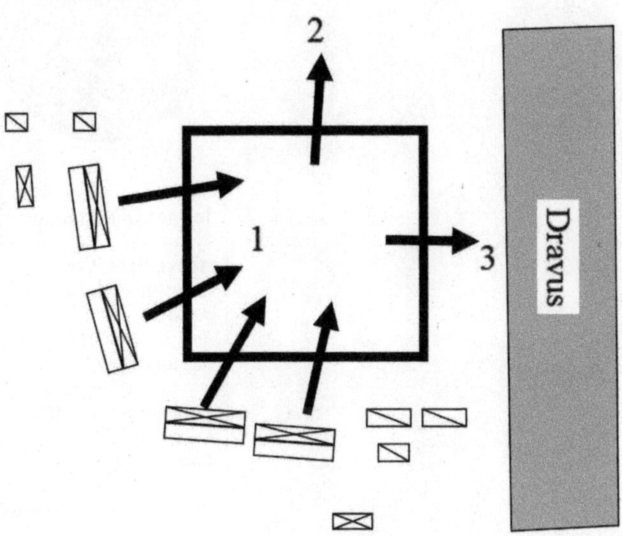

Above: Coin of Constantius II, Victoria Augustorum.
Source of image: Cohen.

as well as in daylight, with the result that the arrows caused so many casualties that the left wing cavalry under Romulus was overwhelmed by Constantius's *clibanarii* with the support of the *scutarii*. The decisive charge was apparently conducted by the *clibanarii*, but since the *scutarii* and mounted archers are credited to have assisted this, it is probable that the *scutarii* conducted some of the repeated charges under the protective fire of the mounted archers (their units must also have rotated those in front and those left behind to rest) so that the *clibanarii* could rest themselves and their horses. Nightfall also contributed to the defeat of the Magnentian forces in another manner. Thanks to the darkness it was possible to flee unnoticed.

Constantius' army pursued the fugitives up to their marching camp and then took it with a frontal assault. The very high quality of Magnentius's western forces is evident also from this. Magnentius's army was able to retreat back to its marching camp even when the enemy had outflanked both of its flanks. This means that some of the Magnentian units forced their way through the Constantian cavalry that had encircled them. After Constantius had captured the enemy's marching camp, the remnants of Magnentius's army then fled either to the plain or to the river. This suggests that the army of Constantius was deployed for the assault of the enemy

Image of Constantius

Above right: Coin of Constantius II, *Perpetua Aug; Bellator hostium*. Source of image: Cohen

Left: Constantius II in Calendar, Biblioteca Barberina, Rome.

camp in such a manner that the open side where the enemy could flee was located on the side where the River Dravus was. This is logical, because it would have been dangerous to place the pursuing army between the river and enemy marching camp. Magnentius's army appears to have suffered most of its casualties among those who fled to the river. This means that the Magnentian army had put up a really good and effective fight against Constantius and that Constantius's army had won the battle only through its unbelievable ability to keep on fighting despite suffering horrendous casualties. Excluding the Pyrrhic victories of Pyrrhus of Epirus, it was a very rare event for the victorious army to suffer this many casualties, because in ancient battles most of the casualties were usually suffered by the losing side only during the rout, and not by the side which won the battle.

This battle stands as a good indicator of the very high quality of the Late Roman forces during this period and also of the tactics used by these armies. The field army of Constantius II showed particular devotion to their emperor in several battles. Another very good example of the readiness and steadiness of his men in combat is the Battle of Singara in 344, for which see Syvänne, *MHLR Volume 1*, 314–6. The phalanx tactics remained largely the same during the entire period, even if the quality of the armies using these varied greatly from one time and place to another. The general fighting quality of the third- and fourth-century armies until the 360s remained very high, but the ever-increasing and rampant corruption of the upper echelons of the military that began during the reigns of Valentinian and Valens eventually diminished the overall effectiveness of the Roman army, so that the army that Valens led to the plains of Adrianople in 378 was a mere shadow of what it had been when the Romans fought at the Battle of Mursa. For this, see Syvänne, *MHLR Volume 2*, especially pages 182–203.

Appendix 3: The Battle of Casulinus River, AD 554[3]

In the winter of 554, 75,000 Franks invaded Roman-held Italy. The commanders of the Frankish army were two Alamannic *duces*, Lothar/Leutharis and Butilin/Butilinus. The Romans were commanded by Narses the Eunuch. Since the invaders outnumbered him and it was also generally speaking wiser to prolong the conflict when fighting against the 'Light-Haired' peoples, Narses resorted to the use of a guerrilla campaign to destroy the invaders. Consequently, Narses waited until the onset of spring before he started operations in earnest. The Franks had divided their forces into two parts to facilitate their provisioning, which enabled the Romans to ambush the vanguard of Leutharis and then defeat his force through a combination of hunger and disease. Butilinus did not want to suffer the same kind of ignominy so he sought to engage the Romans in pitched battle. This time Narses was ready to do so, even though the Franks outnumbered his army. The Romans had about 18,000 men and the Franks about 30,000 men. The situation was still opportune. The Romans were in high spirits, while the Franks were reduced in strength and the weather at the beginning of autumn was already so warm that the Franks suffered from the August heat wave. The Franks encamped by the Casulinus River, and waited for the approach of Narses. Narses, however, posted his forces in such a manner that the Franks were forced to advance into his chosen site of battle if they wanted to retreat from Italy.

The battle plan of the Franks was simple. They intended to fight their way through the Romans by using the infantry wedge. The Frankish infantry tactics were based on the use of thrown missiles (javelin and/or axe) with the intention of disrupting the enemy ranks followed up with a charge into contact. Narses countered the wedge by using the *epikampios emprosthia* (forward-angled formation), but the details of the battle array prove that the original form of the array was actually the so-called *taxis allê*, which became an *epikampios emprosthia* in the course of the battle. Narses also used a stratagem. He sent two Herul deserters to tell Butilinus that the Heruls would desert when the Franks attacked. It was because of this that there was a hole in the middle of the Roman infantry phalanx to lure the enemy. Since the Romans with their 18,000 men was outnumbered by the 30,000 Franks/Alamanni, Narses augmented the numbers of his army with camp followers.[4]

The battle began as a result of the Romans intercepting a foraging expedition by the Franks, which resulted in the Romans taking the bridge which separated the armies. This affront roused the angered Franks into action so that they arrayed their army as a wedge and moved towards the Romans. Narses responded by stationing his men in battle formation. Narses allowed the Franks to cross the river so that they would enter the trap he had planned for them. Narses posted the cavalry on the wings. The location of the cavalry means that the initial array was not yet the *epikampios emprosthia* mentioned by Agathias: this was clearly the final shape of the Roman formation. Some of the horsemen were armed with short spears (*doratia*), shields, bows and arrows, and swords, while others carried pikes (*sarisae*). The last-mentioned detail suggests that Narses had *clibanarii*-type cavalry using the *sarisae* (*contus*) with the two-handed grip which was designed for use against enemy infantry. These would have been posted next to the infantry, with the multipurpose cavalry further out.[5] Narses posted his *bucellarii*, the camp followers, and the servants on the right wing. The infantry phalanx occupied the centre. The front of the phalanx consisted of the heavy-armed infantry, clad in mail reaching their feet. They formed a solid wall of shields, which implies the use of the extra-large shields which we find in the *Peri strategikes*. The light infantry consisting of the archers, slingers and javelin throwers was posted behind them. According to Agathias (2.8.5), there was an empty place left for the Heruls in the middle of the phalanx because they supposedly hesitated to fight. It is clear that this was actually a stratagem of Narses meant to induce the Franks to attack the Roman centre. It is probable that the regular Romans were actually formed so that they were ready to engage the Franks from both sides (*antistomos difalaggia*) when they entered the 'hole'. The Heruls were posted behind the centre in reserve position so that they could advance forward to form the plug to the enemy advance. Narses had also posted on his left an ambushing force hidden inside the woods. It consisted of the cavalry forces under Valerian and Artabanes. They were directed to attack the flank and rear of the Franks. Narses and Zabdalas commanded the right flank. The Heruls were commanded by Sindual. Agathias fails to mention any commanders for the infantry, but it is obvious that they had their own commanders.

It is clear that Narses had also some reserves. Since Narses probably used the *taxis allê* formation (it explains the details best), the wing reserves consisted of the regular infantry and camp followers used to bolster their numbers. Since Narses took his place on the right wing with his *bucellarii* cavalry, it is probable that these formed the cavalry reserve for that flank. It is probable that the left wing had a similar small cavalry reserve behind that wing. As already noted, the reserves of the centre consisted of the dismounted Heruls and possibly of some mounted Heruls.

The Frankish wedge attacked the Romans impetuously, as was their habit. According to Agathias, the Franks pushed through the Roman phalanx without causing many casualties – this clearly resulted from the use of the *antistomos difallagia* and the stratagem of having the Heruls posted behind the hole. This was the Cannae tactic. In the *Byzantine Interpolation of Aelian*, the *antistomos difalaggia* was used against a cavalry wedge, but it could equally well be used against an infantry wedge, as we see here. The easy breakthrough of the Roman formation disordered the Franks because they continued their impetuous attack with a run towards the Roman camp, with the result that they came face-to-face with the dismounted Heruls who formed the bottom-half of the hollow wedge, which consisted of the *antistomos difalaggia* and Heruls.

At that precise moment, Narses launched his trap. The wings were lengthened and turned into the *epikampios emprosthia* formation. The Roman cavalry wings and ambushers encircled the Franks so that the mounted archers poured volley after volley of arrows crosswise from either side to the backs of the enemy while the infantry closed its trap, the *epikampios emprosthia*. It is probable that the *clibanarii* (*sarisoforoi*) cavalry supported this by charging against any parts of the Franks' formation that tried counter attacks. The Franks and the Alamanni panicked and became a helpless mass of men. Butchery followed, but it is still clear that even the one-sided butchery was punctuated by periods of rest, because it was physically impossible to continue

The Battle of Casulinus River in AD 554: Phases 1-2

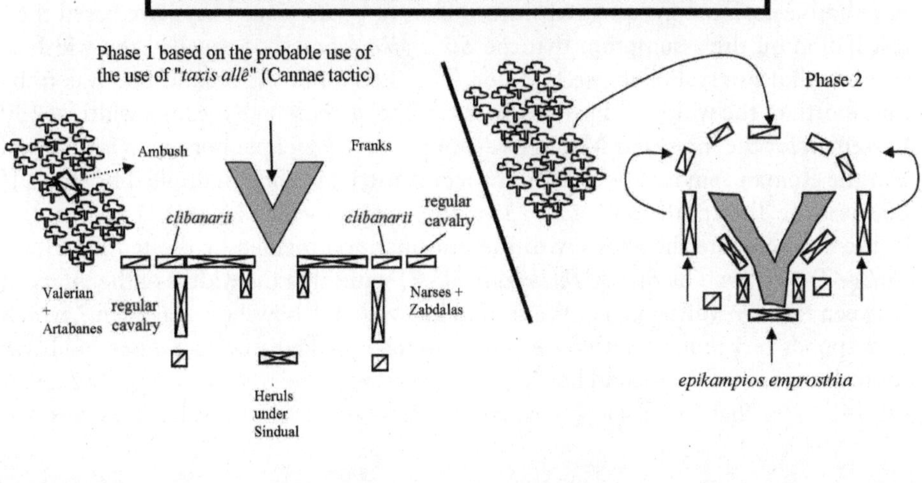

to kill for hours. It is clear that the frightened Franks were massed together so close as to have difficulty using weapons. It is probable that during the lulls in the killing some of the Franks attempted to fight their way through, only to be met with volleys of missiles or by cavalry charges. The Romans would have rotated the units so that while some killed others rested.

There were some Franks who managed to fight their way through to the river behind them, only to meet their maker there. According to Agathias, only five Franks were able to escape. The Romans lost only 80 men killed during the initial contact between the armies. Agathias noted the bravery of Aligern the Goth and Sindual, the commander of the Heruls, in this battle. The Romans held the Germans in high esteem for their fighting skills and bravery in hand-to-hand combat, which is the reason why Narses had lowered the morale of the Franks before engaging them and why he also employed Germanic warriors in his army. Regardless, it is clear that the battle demonstrates the versatility and superiority of the Roman armed forces when under a skilled commander against Germanic peoples who specialized only in the use of the close quarters combat.

Appendix 4: The Bowshot in the *Strategikon*[6]

As I noted in my doctoral dissertation *The Age of Hippotoxotai*, the length of the bowshot in the *Strategikon* is highly problematic, because there is conflicting evidence in other sources to support two different alternatives. Schilbach measures the Byzantine bowshot in the tenth-century *Sylloge Tacticorum* at 328.84 metres, which I have here rounded up to the distance of ca. 330 metres. This would mean that the distance of 3–4 bowshots would be about 990–1,320 metres. The 330m corresponds roughly with the actual flight range of the military bowshot, while modern research has established that the actual maximum effective range was about 75–120 metres.

Modern-day historians have usually calculated the length of the bowshot in the *Strategikon* to be significantly less than the ca. 330 metres it was said to be in the tenth century. Agostino Pertusi (table III between pages 672 and 673) calculates the bowshot to be about 150m, while Aussaresses (p.75, Tableau 6) and Bivar (p.283) have both calculated the range of the bowshot to be 133.2m. They have based their calculation on the assumption that the *Strategikon's* bowshot equalled the width of one bowshot interval of the second line. The interval of the second line was to be one-fourth of the width of the 600-man first-line *meros*, which means a width of 150 horsemen for the interval if Maurice's assumption of 3 feet per horseman is followed and the Roman standard of 29.6cm for feet is used and then multiplied by three. If one uses the 'Byzantine foot' of 31.23cm, the bowshot would be ca. 141 metres.

We can compare the above with the information provided by the tenth-century *Sylloge Tacticorum*. The *Sylloge Tacticorum* (43.8) states that the widths of the intervals between the *mere* (divisions) of the second line were 0.5 bowshots (ca. 165m), which corresponds very roughly with the estimations made by Pertusi, Aussaresses and Bivar for the intervals of the second line in the *Strategikon*. Similarly, the *Sylloge Tacticorum* (46.14) states that the distance between the first and second cavalry lines was two

Infantry and cavalry archers in Ilias Ambrosiana (5th century). Drawings by Mai (1819).

bowshots (ca.660m), but we have to remember that the tenth-century cavalry array also had a third line behind the second, so the tactical usages of the different battle lines had changed radically. Regardless, if we follow this line of thinking it would mean that the distance between the first and second lines was approximately equal to the distance of four times 165m (= 1/2 bowshot), which lends support for the shorter bowshot for the sixth century. If we assume that the distances between the lines were roughly the same in the *Strategikon* and *Sylloge Tacticorum*, this would indicate that the sixth-century bowshot was about half of the tenth-century bowshot. This would mean that the distance of three-to-four bowshots from the enemy meant the distance in which the Roman cavalry was expected to be able to shoot three to four times before they would come into contact with the enemy.

On the other hand, there is very convincing evidence that can be used against the shorter bowshot for the sixth century. In the *Strategikon*, if the attack by the first line had not succeeded, it was expected to withdraw back one or two bowshots and wheel to face the enemy, and if even that was unsuccessful to again withdraw and renew the fight or find security behind the second line. The two retreats and counter attacks definitely demanded a significant amount of room to complete. Furthermore, Procopius's description of the Roman battle formation at Ad Decimum supports the interpretation of the bowshot being approximately 330 metres. In this instance, the support line was at a distance of about 1.3km (7 stadia = 7 times 185m)[7] from the first battle line, which corresponds with the distances between the first and second line in the *Strategikon* if we accept the tenth-century 330m as the distance of bowshot in the *Strategikon*. The *Strategikon* (2.13) instructed the second support line to be at a distance of a mile (Roman mile ca. 1.5km) or more from the first line during the approach march, after which it advanced to the distance of four bowshots (1,320m) from the first line for the battle. The latter figure is practically the same as we find in Procopius.

My final conclusion is therefore that the tenth-century and sixth-century bowshots were the same.

Notes

Introduction

1. This is based very loosely on the method of analysis that I used in my doctoral dissertation, which in its turn was loosely based upon the concepts put forth by Allan R. Millett, Williamson Murray, and Kenneth Watman for the modern period. I do recognize the potential danger of bringing modern theoretical concepts into the past. However, in this case the subject matter merits this approach and the list given here should help the reader to assess the effectiveness of the Roman armed forces during the different periods of time: i.e. the readers should use these tools when they assess the effectiveness of the Roman armed forces during the Late Roman period in the eight volume book series *Military History of Late Rome*. I give in this monograph only some examples drawn from the series. The effectiveness of the Roman armed forces obviously varied throughout the Late Roman period, so sometimes everything was done by the book (i.e. as the Roman military doctrine required), while at other times there were serious problems with this.
2. For an overview of the Late Roman and 'Byzantine Roman' military treatises see: Syvänne (2013b) with Dain, 1967.

Chapter 1

1. For an overview of the Late Roman and 'Byzantine Roman' military treatises see: Syvänne (2013b) with Dain, 1967. For analyses of Julius Africanus, see Syvänne, *Caracalla with Gordian III and Philip the Arab*.
2. For additional information, see Syvänne, 2013a with Syvänne, *Severus*, Appendices 1–2.
3. I owe the referral to the *pyrgos* to D'Amato (personal correspondence).
4. For an analysis of these developments, see Syvänne ((2011a-b, 2011c, 2011–2, 2017a, 2020a, 2021a) and forthcoming *Septimius Severus*).
5. For a fuller analysis, see Syvänne, 2020a, Appendix 1 with Syvänne, *Severus*.
6. For the events that took place during the reign of Aurelian, see Syvänne (2020a, *Aurelian and Probus*; for the Battle of Placentia, see 79–82).
7. Vegetius (3.14) places in the fifth rank the *carroballistae* (horse-drawn carts with *ballistae*), *manuballistarii*, *fundibalatores*, and *funditores*. i.e. Vegetius misunderstood the *ballistarii* to be artillerymen using the *carroballistae*. Similarly, Vegetius understood the *tragularii* to be the *manuballistarii* of his own day. This is one of the pieces of evidence which suggest that Modestus has preserved the original meaning of the source he has used, while Vegetius has misunderstood it to mean the weapons of his own day: 390–450.
8. *The clibanarii* were fully-armoured (both men and horses armoured so that usually only the eyes were visible), super heavy cavalry armed with the long *contus* that had to be used with two hands. In addition to this, they were equipped with swords and daggers and sometimes also with bows. The third-century novelist Heliodorus (9.14) describes the Sasanians using the mixed formation with *clibanarii*, which suggests that the Romans may have copied the tactic and formation from the Persians. However, it should still be kept in mind that the Romans could also have found an inspiration to this array from any of the historical texts describing the Achaemenid or Seleucid battle formations which had cataphracts. The first units of cataphracted super-heavy cavalry had been introduced into the Roman cavalry either

by Trajan or Hadrian, but this type of super-heavy cavalry became more fashionable only in the course of the third century after the emperor Alexander Severus had equipped his cavalry with the captured Persian *clibanarii* equipment. For the fourth-century use of the mixed formation and super-heavy cavalry *clibanarii* deployed as a massive wedge, see Syvänne (*MHLR 1*, 246–8).

9. See Syvänne, *MHLR Vol.1 (284–361)* with *MHLR Vol. 2 (361–395)* pp. 81–118.
10. This analysis reflects my opinions as presented in *MHLR 2* (133–8), except when stated otherwise.
11. Sánchez-Ostiz, 18; Lenski, 6, 127, 287ff.
12. Sánchez-Ostiz (16ff.) has collected the different views including these.
13. Almost a quote from *MHLR 2*.
14. Lenski, 286ff. with above.
15. The emperors also introduced imperial legislation that restricted trading with the barbarians. E.g. in 369 trade with the Goths was restricted to two locales; export of wine, oil and *garum* to the barbarians was outlawed in 370/5; marriage with barbarians was made punishable with death in 370/5; export of gold to barbarians was forbidden in 374. For this, see Lenski, 135ff.
16. Julius Africanus's *Kestoi*, the *DRB* and Vegetius in Syvänne, *Caracalla*, 323. The wings noted by Hassall, 78–80.
17. Hassall, 1979, 84–89.
18. Syvänne, *MHLR 2*, 245–50. Somewhat similar deployment patterns for the infantry can be conjectured also for the the rest of the late-fourth to early-sixth century battles in which the different divisions were all arrayed as hollow oblongs or in which the entire army consisted of a single hollow square/oblong (e.g. Syvänne, *MHLR 2*, 143–7, 171–3, 199.; Syvänne *MHLR 3*, 86, 121–8, 265; Syvänne, *Britain in the Age of Arthur*, 151–63; Syvänne, *MHLR 4*, 91, 112; Syvänne, *MHLR 5*, 66–7, 69, 256–7).
19. See Greatrex, Elton and Burgess.
20. See Syvänne, *MHLR Vol.2* (121ff.), *MHLR Vols. 3–5*, and *MHLR Vol. 6* (1–105).
21. Based on Syvänne (2004) which was based on personal analysis plus: Bury 2.75; W.A. Oldfather and J.B. Titchener, 1921; Dain, 1946, 29–40; Dain 1967, 332, 338. According to Oldfather and Titchener, the unknown author had used Aelian, Arrian, Asclepiodotus, and someone unknown (6 end, 12 beg., 14 end, 15 beg., 17 mid., 20 beg., 22 end, 23 end, 38 end, 57 end) as his sources. I have here (in 2022) corrected Dain's claims concerning the entries 29 and 63.
22. Based on Syvänne, *MHLR Vol.5*, Appendix. The latest edition, translation and analysis of the *Epitedeuma* is by Geoffrey Greatrex and Hugh Elton (hereafter *Epiteudeuma* when referring to the source and Greatrex & Elton when referring to comments). The latest edition of the *Taktikon* and two of the epigrams (p. 150, 162) is by Förster.
23. I therefore agree with Greatrex, Elton and Burgess that the *Epigrams*, *Taktikon* and *Epitedeuma* made up one book originally. Urbicius has also sometimes been claimed to have composed the *Strategikon* of Maurice or its infantry section (Book 12) or its hunting section (12.D, the *Cyneticus*) and a fragment on river crossings. However, I agree with Greatrex, Elton, and Burgess that this is very unlikely.
24. For further infotmation, see Syvänne (2004, 26; 2021, 367–70).
25. See Syvänne (2004, 14–6), who favours a sixth-century date for the treatise, however Theotokis and Sidiropoulos (1–54) like most modern historians favour a Middle-Byzantine date. The other modern historians who prefer the Middle-Byzantine period include Douglas Lee with Jonathan Shepard, Cosentino, John Haldon (2014, 39), and Philip Rance (2007). Most of these prefer a ninth-century date. However, I do not agree with this 'consensus view'. I agree with the dating scheme suggested by Köchly & Rûstow, and Dain and Zuckerman. Köchly & Rûstow and Dain place the *Peri strategikes* in the sixth century, while Zuckerman (e.g. 1990,

216) places Syrianus's military treatise (*Peri strategikes, Rhetorica militaris and Naumachica*) to the period of the late-sixth/early-seventh century.

26. For the use of Syrianus as a source for the *Taktika* of Leo the Wise, see John Haldon (2014, Index Syrianos *magistros*, as a source for the *Taktika*).
27. Referrals to other modern studies can be found above and in Syvänne, Eramo, and Theotokis and Sidiropoulos.
28. See *MHLR 6*, esp. pp. 256–60, 328–37.
29. Rance (2007, p.723 n.67) after Zuckerman (1999,373-
30. The Dura-Europos shields date from the period 253–6.
31. In this case Corippus calls the shields with two terms, *clipeus* and *scutum*, but he did this to avoid repetition. The details (the men's bodies hidden behind the shield wall so that only battle-axes, tops of conical helmets and spears could be seen) make it clear that he meant huge shields. See also Syvänne, *MHLR 1*, 379 (Romans with conical helmets in 360), with *MHLR 6*, 273–4.
32. *STR* 12.B.20.7–9: '... *alla soutariois tous men skoutatous meizosi, ... tous de psilous skoutariois mikroterois...*': my free translation: '... but on the other hand the *skoutatoi* should carry larger shields... the light-armed should carry smaller shields...'. G.T. Dennis translates the size of the shield of the *skoutatoi* somewhat misleadingly as moderate-sized shields, but in a sense he is correct to do so because the *meizosi* is a comparison (larger) which means that there also existed the largest shields.
33. Notably the same treatise (39.30–5) also suggests that the leading footmen were also to use iron armour on their lower legs and feet, and iron soles under their feet to protect these against caltrops and stakes when approaching enemy encampments during a night attack: i.e. the use of metal protection was adopted by everyone when there existed the danger of caltrops. The *PST* (38) also suggests that in retreat the men who guarded the rear were to throw caltrops behind them to slow down the enemy. This was not a new tactic and certainly not as late as Middle-Byzantine, for we find the Roman cavalry using this same method at the Battle of Nisibis in 217. For this, see Syvänne (*Caracalla*, hardback 2017, 283ff.; *Caracalla*, paperback 2022, 238ff.).
34. The different variants of the rhomboids were an ancient cavalry formation, which Syrianus (*PST* 17) includes in only very muddled form as one of the alternatives for deploying cavalry – if one decides to understand his text in this manner (the heads of the horses of the second rank aligned with the shoulders or flanks of the horses of the first rank). The *kyklos*-circle meant several different formations that had a rounded shape. This could be the two-fronted (*amfistomos falagx*) array in which the flanks also pointed their spears outwards, or it could be a hollow square/oblong in which the corners became rounded, or it could be an actual circle (e.g. one of the traditional forms of the *testudo* was formed in that manner. In naval warfare it was also an actual circle in which the bows were pointed outwards towards the enemy.
35. *PST* 15.11–20, 25.1–7, 29.1–17, 31.7, 31.39–48, 32.55–78, 34.18–23: *tetragonos falagx* (*plinthion*), *heteromêkes synistatai falagx* (*plaision* and *difalaggia* with *psiloi* on the flanks). The hollow square emerges as the principal battle formation again in the tenth century, but the fact that the *Strategikon* remained the Bible of the East Roman armed forces until the writing of Leo's *Taktika* means that this piece of information is not particularly suitable for the period before it and we know that Leo borrowed directly from Syrianus. This, however, is not conclusive because the hollow square was obviously used in combat also during the period between the *Strategikon* and Leo's *Taktika*. However, the referral to the hollow square is still more relevant to the period before the writing of the *Strategikon* than after it.
36. *Str.* 7.B.11.45–52, infantry *tetragonos/plinthion* or *difalaggia* during retreat; *Str.* 12.A.7.21, 12.A.7.66 infantry rear guards in the *epikampios opisthia* formation, deployed as a hollow square *plinthion*; *Str.*12.B.22, esp. 12.B.22.52–8, four-sided formation (*tetragonon*) for the marching camp: *Str.* 12.D.138–9 infantry used in hunting to surround the game inside a

four-sided hollow square (*plinthion tetrapleuron*). The standard marching formations in the *Strategikon* (12.B.19–20) were the *plagia* (lateral) phalanx in open terrain and the phalanx in column formation so that it could consist of a single column (narrow passes), double column (wooded terrain) and four columns (a very large force in favourable terrain). In practice, however, the *plagia* phalanx and the column marching formations of the *Strategikon* can be equated with the hollow square array because these were surrounded by light-infantry, or dismounted cavalry, or mounted cavalry on all four sides, and the wagons of the baggage train could also be used as protection when so decided (*Str.* 9.3.87–9.4.65 with *Str.* 9.3.87–9.4). The other uses that the *Strategikon* has for the hollow square were: 1) to act as rear guards for the flanks of the *epikampios opisthia*; and 2) to surround the wild game inside it when using the hunt as a military exercise.

37. Note the existence of the Alan, African and Illyrian Drills, and see e.g. *MHLR 6*, 67–70, 98–101.
38. This bears some resemblance to the nomadic tactic as described by the *Strategikon*, but with the difference that the nomads used the reserve posted two-to-three miles behind as ambushers and reserves, i.e. they posted sizable forces behind to act as real reserves while Syrianus recommended these merely as a safety precaution.
39. The cavalry formation in question will be described in greater detail in the chapter dealing with cavalry tactics.
40. *PST* 16, 27.11–22, 44. The *PST* (27.11–22) names the following pieces of equipment for the soldiers: *doru*-spear, *aspis*-shield, *thorax*-breatsplate, *perikefalaios* (helmet), *rhomfaia* (sword), *toxon*-bow, *brachiôn* (shoulder-piece), and *faretra* (quiver for arrows). The *perikefalaios* ('tied around the head') can mean either a covering for the head or a helmet. According to Liddell-Scott (p.1159), the word was already used to mean the helmet by Polybios, so the use of this word does not date the text to the Middle-Byzantine era when the Romans were certainly using turbans – here the meaning is clearly the helmet because the men are armoured. The key word pointing towards the sixth century here is *rhomfaia*, which during the ancient period meant either the large sword used by the Thracians or a sword in general (Liddell-Scott, 1323). During this era (Isidorus, *Etymologies* 18.6.3), the word no longer referred to the extra long sword used by the Thracians, but to the double-edged broadsword *spatha/spathion*. However, when one remembers the older usage of the *rhomfaia* to mean any sword and the fact that Sophocles's *Dictionary of Roman and Byzantine Greek* (972) also identifies the word *rhomfaia* with a sword, and in particular with the rapier, the dating of the text to the sixth century merely on the basis of this word is not conclusive, but when taken together with the multipurpose footmen and other details it does point to a sixth-century dating. The tactics of using the *sarissa* type of pike in the phalanx, or the use of spears and pikes of different lengths to present a wall of spears in which the spearheads of all ranks reached equally far away from the front rank, do not really date the work (*PST* 16). These were clearly copied from ancient Hellenistic texts, but with the significant difference that the *PST* has only the spears of the first four ranks protrude outside the formation.
41. Based on Syvänne, 2004, 16–9.
42. Belisarius-Procopius incident in *MHLR 6*, 170. Heraclius consulting military treatises in *MHLR 8*, 151. The reason why the *Strategikon* was not studied during the reign of Phocas/Phokas I (602–10) is simple: he was a usurper who had taken the throne from Maurice. It is quite easy to see why the study of Maurice's treatise was neglected under his successor.
43. For earlier use, see e.g. Syvänne (*MHLR 1*, 21, 91; *Gallienus*, 35; *Gordian and Philip*, 17, 131, 172) with the forthcoming Syvänne (*Septimius Severus*).
44. Syvänne (2004, 19–20 and further references therein) with Eramo (2018).
45. There are three clues for the date of the text. The text (16, p.95) mentions all the enemies recorded by Maurice (the 'fair-haired races', the Antes and Slavs, the Persians, and the 'Scythians'), but adds the Saracens to the list. This has been seen as an indication of Islamic

Arabs, but is obviously not conclusive because the Romans had faced Saracens before. The second clue is that the number of Latin commands has been reduced and these are sometimes used incorrectly. The third clue is the position of the text after the *Strategikon*. Taken together, these instances indicate a period after the Muslim conquests had begun.

46. Edition of the text in Dain, 1944, 92–100; translation of Devine based on Dain and *Codex Burnley*, including the tenth century *Syntaxis armatorum quadrata* (=Dain, 1944, 155–57) in Devine, 59–64; For dating the text see: Dain, 1944, 61–115, esp. 107–115; Dain, 1967, 339–340, 346–347; Devine, 35–39, 59–60. In his latest edition of *Aelian* Christopher Matthew considers many of the parts of the so-called *Interpolated Byzantine Recension* to belong to the original text. I am inclined to agree with this view. The term *Corpus Élianique* comes from Dain. *Aelian* was later, for example, copied and adapted by the Arabs and was found very useful also by Alexios I Komnenos, not to mention the military theorists of the early modern age. In addition, the Byzantine treatises the *Taktika* of Leo and the *Sylloge Tacticorum* also include parts of the treatise.
47. The following is a summary taken from my doctoral dissertation. Edition of the *Excerpts* in Melber, J., 1887, *Excerpta Polyaeni*, 427–500; Edition and translation in Polyaenus, *Stratagems of War*, ed. and tr. by Peter Krenz and Everett L. Wheeler, 1994, Vol. II *Excerpts of Polyaenus*, 851–1003. For the opinions of others on the problem of dating see: Dain, 1931, 321–45; Dain, 1967, 337, 364/5; Krenz/Wheeler, vi-xxiv, esp. xx-xxi; Schindler, 1973.
48. Syvänne (2021, 33ff.; 2017, 311ff.,), Dain (1967, 359–60), Zuckerman (1994), and Lammert (1951).

Chapter 2

1. Based on: Syvänne (2004) and the book series *MHLR* vols. 1–8.
2. Originally the *magister equitum* (Master of the Horses) was the senior *magister*, but as a result of the predominance of such persons as Stilicho and Aetius the precedence changed in the western half by the turn of the fifth century
3. I have speculated in my books (e.g. *MHLR 2*, 79) that these included originally many of the units that were later transferred into the praesental forces and also that some of the *domestici* served as officers of the imperial bodyguard units that consisted of footsoldiers (e.g. *urbaniciani, teiche*/Walls and *vigiles*). The *protectores* were the members of the unit of *protectores domestici/domestici* that had been dispatched to serve as detached officers in the regular units outside the imperial household, while the *protectores domestici* or just *domestici* meant that these *protectires* served in the household of the emperor, presumably mainly as officers of the *scholae* and other imperial bodyguard units.
4. Based on Syvänne, *MHLR Vol.1* and the edition and translation of Bundy pp.69–75. This version, however, contains some changes and additions.
5. For the shields in general, see Kolias 88–131.
6. See e.g. Syvänne, *Emperor Septimius Severus. The Roman Hannibal*, with *MHLR 6*, 272–95.
7. See e.g. Syvänne, 2020a, 9–13 with *MHLR 1*, 84–6 (mostly after Procopius).
8. Haldon, 2014b, 148–9. Haldon suggests that there existed naval commanders with the title of *drouggarios* already in the seventh century. This is indeed quite probable. It is possible that these were initially commanders of temporary fleets that consisted of detachments drawn from the permanent fleets.
9. The *nauarchus* was originally captain of a ship and the *trierarchus* a captain of a trireme, but the former came to mean admirals and the latter already meant captains during the Hellenistic epoch.
10. For a fuller analysis, see *The Age of Hippotoxotai*, the *MHLR* series, together with the discussion on tactics in this book.
11. Based on Syvänne (2004, 42–55) and *MHLR 6* (Appendix), with Syvänne (*Emperor Septimius Severus. The Roman Hannibal*, Appendix 1).

12. The period sources and archaeology are unhelpful in this case. However, the tenth-century military treatises contain specific measurements for cavalry shields. In those treatises, the length of the cavalry shield usually varied from 93.6 to 117cm, while the round shields varied from 70.3 to 93.6cm, the small *pelte* being 70.3cm in diameter.
13. E.g. in 363 some of the Roman cavalry units could cross the Tigris while others did not. In contrast, in 388 the entire cavalry force of Theodosius I was able to cross a river. The Batavian, Gallic and 'Arctic German' units were particularly famous for their ability to cross rivers by swimming. For these instances, see *MHLR 2*, 114–6 and 244–5.
14. The demand to fight as infantry e.g.: *Str.* 11.1.64–7, 12.B.23.14–8 (implied); Procopius, *Wars* 4.11.14–56, esp.50ff.
15. See e.g. Syvänne, *Caracalla*, 324–5.
16. Based on Syvänne (esp. 2004, 52–4; 2013b, 332; 2021b, 170–1; 2022b, 151) and *MHLR* series in general.

Chapter 3

1. The following is based on Syvänne (2004, 56–70) and sources therein, plus sources mentioned separately here.
2. See for example Procopius, *Wars* 2.19.6–14, 2.6.3–8; *MHLR 7*, 244–9.
3. For example: Vegetius (3.9); *PST* (33); *STR* 8.A.44, 10.2–3, 11; *De velitatione* (a 10th-century treatise with similar instructions); Procopius, *Wars* 2.19.6–14, 2.6.3–8, 6.10.1–11; Sebeos 34; Theophanes AM 6114–6118; Syvänne, (2004, 60ff.; *MHLR 7*, 244–9).
4. See e.g. *PST* 6; Procop. *Wars* 1.7.3–4, 1.13.9–15.17, 1.15.1–9, 1.21.4–28, 2.14.13, 2.16.1–4, 2.24.12–27.46, 2.26.1–27.46, 8.18.13–19.22, 8.25.7–15; Procopius, *Secret* 8.5–6, 11.5ff, 21.26ff.; Agathias 1.11.1–2.10.7, 5.23.7–25.6; Menander fr. 5.4.23–6, 15.1.21–23, 18.6.13–20.2, 21–3, 26.1.72–138; John of Ephesus 6.2–18, 6.28; Theophylact 1.15.14–2.10.7, 2.10.8–17.13, 3.10.4–11.3, 3.12.11–15.9, 3.17.5–18.3, 6.3.9–5.12, 7.2.1–5.10, 7.13.1–15.8, 8.2.1–3.15, 8.5.5–6.1; Malalas 17.10; *Chronicon Paschale* a.522; Theophanes AM 6013, 6020; Constantine Porphyrogenitus, *De administrando imperio*, 29–36; Syvänne, 2004, 60ff.
5. *STR* 8.A.44, 8.B.4; 9.1, 9.3–5, 10.1, 10.4; John of Ephesus 6.2–8; Theophylact 1.15.1–2.10.7, 3.5.11–6.5, 3.10.4–7, 3.11.1–2, 4.2.1; Procopius, *Wars*, 2.14.13–19.49, 3.6.5–27, 3.9.10–25.26, 5.2.1–16.21; Agathias 1.11.1–18.2; Syvänne, 2004, 60ff.
6. For the modern theory of limited and unlimited wars see: Clausewitz, *On War* 8.7–9 (Howard and Paret ed., pp.611ff.; Graham and Maude ed. vol. 3, pp 131ff); Corbett, Chapters 3–6 (original publication pp. 38–83; *Roots of Strategy 4* ed. pp.190–228). For the various different goals and methods of warfare adopted by the Romans and their enemies, see the *MHLR* series.
7. Syvänne, *MHLR 7*, 244–9.
8. The following is based on Syvänne (2004, 70–92), sources therein, and sources mentioned separately.
9. *PST* 33; Vegetius 2.4, 3.1; and the *MHLR* series for individual examples.
10. The feasibility of these numbers is confirmed by the list of units in the early-fifth-century *Notitia Dignitatum*. For an estimate of the troop sizes in that document see the relevant entry in the index of the *Later Roman Empire* by Jones.
11. For another good example of this, see my biography of Septimius Severus.
12. Zosimus 2.22 with Syvänne (*MHLR 1*, 269–73).
13. See Syvänne, *MHLR 4*, 97–117.
14. E.g. Napoleon, 5–7, 10, 15, 17, 19, 45–8, 69. Napoleon's text is also very important for his analysis of the difference between his own era with firearms and Caesar's era. He saw no real difference in the ability to field large armies, but he saw a major difference in the influence of firearms on combat and sieges. For a modern analysis of Napoleon's comments about Julius

Caesar, see Eramo (2022) and for Napoleon's obsession with numbers in particular Eramo (2022, 263–79).

15. With only a minimum number allowed for the flanking units and rear guards and 310 per *tagma*. However, in the actual diagram the flanking units and rearguards contain more files.
16. The text of the *Strategikon* (2.4.11–12) would allow only 2 *banda* as rear guards making 108 *banda* (33,480 men) altogether, but the diagram contains three files for each of the rear guards which would indicate that there were three rear guard *banda* on both sides instead of the two allowed by the text. This would indicate that the first and second lines had equal strengths when one takes into account that the rear guards were taken from the flanking *mere* of the second line and the flank guards and outflankers from the first line. In other words, the two lines had 56 *banda* each.
17. The commanding officer of each *tagma* was required to keep his unit at full strength: *STR* 7.A.2.
18. *STR* 1.4.26ff.
19. *STR*.1.2.70ff.
20. The following is based on Syvänne (2004, 93–113) and sources therein.
21. This estimate is based on the Roman mile. With the Byzantine mile the figures are c.24.1–32km.
22. *STR* 1.9.18–23, 9.3.132ff, 12.B.20; *PST* 6–9, 29, 41–3. See also Austin&Rankov, 9–86. The fact that the surveyors and quartering parties rode one day ahead of the main army shows the minimum distance covered by the scouts. In hostile country the quartering parties stayed closer to the army (*PST* 26), but the scouts and spies had to be deployed further out. Note also *PE* 7; *DV* 2, 7, 15; *DRM* 3–4, 18.
23. *PST* 7; Vegetius 3.6; Procopius, *Wars* 3.23.5–18, 4.13.2–3; Theophylact 2.8.2, 7.4.8–11, 7.12.4–5. The *MHLR* series also contains several examples of the use of scouting patrols.
24. *De velitatione* (9.1–29, Dennis ed.; 9.3–6 Dagron/Mihăescu ed.); The diagram of the scouts is based on the reconstruction of the scouting pattern in *De velitatione*/Dagron/Mihăescu, Scéma no III, p.201.
25. *STR* 2.6.40ff., 2.13, 2.20, 3.5.77ff, 6, 7.A.6, 7.A.15, 7.B.4/8/14–15/16.38–41, 8.B.72/77/80.
26. Syvänne *MHLR 1*, 355. Compare with Syvänne (2011a-b; *Britain in the Age of Arthur*, 152–63).
27. For the Roman use of wagon laager during the early-sixth century, see e.g. Syvänne (*MHLR 5*, 175, 224, 239–41, 257; *MHLR 6*, 155–7). The *Strategikon* (5.5, 12.B.8, 12.B.13, 12.B.18, 12.B.22, 12.C) included the regular wagons as part of the fortified camp but posted these either in the middle of the marching formation or behind the infantry phalanx together with the *ballistae* wagons and carts.
28. See e.g. *MHLR 2*, 114–6, 244–5.
29. *STR* 5.3–4, 7, 9.3–4, 12.B.19–22, 12.C; *PST* 19, 37–8; Vegetius 3.6–7. The fact that we find the same instructions for using human shields during marching in Frontinus (Aemilius Paulus at 1.4.1. and by Agesilaus at 1.4.2.) and Polyaenus (2.1.30) makes it clear that this stratagem was always employed when the commander was aware of it.
30. The following discussion is based on: *STR* 5.1–5, 12.B.6, 12.B.20; Vegetius 3.1–8; Zachariah 7.6. According to the *Strategikon* (5.4), the cavalrymen were to carry 20–30 pounds of hardtack etc. worth 7–10 days' rations (3lbs/day) and infantry (12.B.6.12–5), when traveling light, was required to carry 8–10 days' worth of rations.
31. The reason for this is that the *vexillarii* mean the legionary cavalry (Syvänne, *Caracalla*, 9) so it is preferable to place the date later. In fact, the Severan era is probably the likeliest date.
32. Pseudo-Hyginus; *STR* 5.1–5, 7.A-B.10, 9.3.96ff, 12.B.18–22, 12.C; *PST* 27–9; Vegetius 1.21–5, 2.25, 3.8.
33. Note that the 1,600 *vexillarii* are actually legionary cavalry and not detachments of legionary footsoldiers as usually assumed. See Syvänne, *Caracalla*, 9.

34. See e.g. Syvänne (*Caracalla*, Appendix 2; *Gordian III and Philip the Arab*, 15, 35–7, 51–2, 94–5).
35. It was an old Roman practice to post the army in combat readiness while others built the fortified camp if the enemy was close by. See e.g. Frontinus, *Stratagems* 1.5.9.
36. The instruction to place the tent of the general in one of the four quarters and not in the crossroads can be found in the texts that followed Maurice or which were based on the longer treatise on encampments. However, in the late-tenth century *De re militari* the general's quarters were located in the crossroads but in such a manner that these were separated from the rest of the camp with a road that ran along its rectangular perimeter so that it no longer interfered with traffic and probably also gave the general even greater personal freedom from distractions because the general's tent was in the middle of this quarter. For the *De re militari* see e.g. *Campaign Organization and Tactics in the Three Byzantine Treatises* (p.246ff.).
37. Syvänne, *MHLR 7*, 254.
38. For the exploits of John Troglita in North Africa, see Syvänne, *MHLR 6*, 272–96. John's other exploits can be found by using the Index.
39. For Maurice's exploits as a general, see Syvänne, *MHLR 7*, 117ff.
40. For examples, see Syvänne, *MHLR 8*, Index.
41. *STR* 9.3.96ff; Vegetius 3.2, 3.26; al-Tabari i.2126, 2260, 2262–3, 2291; Corippus *Ioh.*,7.375–497; 8.49–163; Procopius, *Wars*, 4.28.1ff, 6.8.1–18, *Secret* 1.24–30.; Theophylact 3.1.1–15, esp. 3.1.10–13; see *REF1*, 15, 311–3. The Romans also sent spies into enemy camps. For example, in 530, two bodyguards (*doryforoi*) entered the Persian marching camp. After their departure, one was caught (and later ransomed) while the other delivered his intelligence report and the Romans were able to surprise the Persians in their encampment. See Procopius, *Wars* 1.3–8. Similarly, we find Hannibal using wigs and varied styles of beards to fool possible spies and conspirators, which has found its way also into the maxims of the *Strategikon* (8.B.87), undoubtedly because Maurice was worried for his personal safety and for the safety of his generals. Note also Frontinus 3.16.4. There exists one instance in which a general was assassinated at a banquet, even when he was in the midst of his bodyguard. This took place when the bodyguards of Artabanes assassinated the rebel Gontharis in a banquet in the imperial palace of Carthage in 546. Strictly speaking, this was therefore not an instance in which the general would have been killed in his tent. For the incident in question, see Syvänne. *MHLR 6*, 197–8.
42. See Syvänne, *MHLR 7*, 113–4.
43. The horses were to be watered the night before. The men were to carry provisions (food and water) in their saddlebags in case of a prolonged battle. See *STR* 7.A.9–10.
44. Based on Syvänne (2004, 113–7) with further references therein; and in: *STR* 2–4, 7, 8.A.9, 8.B, 9–11; Theophanes AM 6115/8; Procopius, *Wars* 2.29.27–30.14, 2.30.21ff, 8.16.6, 8.16.15ff, 8.17.11–19; Theophylact 6.4.7–12. Menander frg.5.4.23–6, 15.1.21–3; Theophanes AM 6116; *PST* 18, 20, 26–29, 33, 41–43; Vegetius 3.9–10; *MHLR* vols.1–8. Vegetius lists the following set of considerations that had to be taken into account before committing the forces to a battle: 1) Which side had more men? 2) Which side had better-armed men? 3) Which side had better-trained men? 4) Which were braver in dangerous situations? 5) Which side had better infantry and which better cavalry? 6) Whose cavalry excelled in archery and whose in lancing, and which side had better horses and who wore more armour? 7) Furthermore, one had to determine whether the location favoured the Romans or the enemy. If the Romans put their trust in cavalry, then they were to choose an open field. If one was to use cavalry against infantry, one was to obtain a higher position but with no obstacles to cavalry. If on the other hand the Romans preferred to use infantry, they were to choose confined or rough areas obstructed by ditches, swamps, woods or even mountains. The sun, the dust and the wind were to be behind one's back, which was, according to Vegetius, known even to the unskilled. One can find these or similar instructions about the sun and the wind for example from

the *Strategikon*, Ayin Nameh, and Tafrij, and undoubtedly from a number of other military treatises. 8) One had to investigate whether it was advantageous to delay the confrontation or whether it was better to handle it quickly. It was important to investigate which side could stay in the field longer. Especially relevant to this was to determine which side had the better supply situation. The larger the army the slower it moved and the more difficult its supplying would be. One could perhaps simply prevent the gathering of fodder and provisions and win an easy victory. 9) What were the qualities and capabilities of the officers in the enemy camp? Were they bold or cautious, experienced or inexperienced? 10) Which of the tribes were cowardly and which brave? 11) How loyal were the Roman auxiliaries? and 12) What was the fighting spirit in the Roman and in the enemy camp?

45. For these and other instances, see Syvänne, *MHLR* vols.1–8.
46. See also the *Byzantine Excerpts of Polyaenus* (6th cent-c.850?) 13.1, Lycurgus recommended not to campaign many times against the same foes so that they wouldn't learn how to fight back, similarly in the original work of Polyaenus (161–164 AD) 1.16.2, and also in the later Leo the Emperor, *Stratagems* 9.2 (*Sylloge tacticorum* 83, Dain).
47. For the exploits of Philippicus, see Syvänne, *MHLR* vols.7–8 with Index Philippicus.
48. See Syvänne, *MHLR* vol.2.
49. See Syvänne, *MHLR 1* with *MHLR* vols.7–8.
50. *STR* 9.1–3, x.2; *PST* 33; with Syvänne *MHLR* vols.1–8.
51. It is impossible to date this maxim, because Roman history knows many such instances. For the Late Roman period the best known examples are the granting of federate status to the Visigoths, first by Valens and then by Theodosius I. The Visigoths were then given similar status by successive emperors until they took the land given to them. The same was true of the Burgundians, Suevi and Alans. Similarly, Tiberius II had also resorted to the same emergency measure when he brought to the Roman territory the Avars when there were inadequate numbers of native forces in the area.
52. *STR* 8.B.80 (note the similarity with Aineias 12.2, 4); Syvänne, 2000, 34ff; Sebeos 12, p.24.
53. Based on *STR* 11.1 and Syvänne (2004, 328–50) and *MHLR* series.
54. *STR* 11.1. The *Strategikon* (2.1) notes the continued use of the single line by the Romans and Persians and argues for the use of reserves, but at 3.16.9–12 the *Strategikon* admits that the enemies usually used two lines.
55. The evidence supports these conclusions. Partial list in Wiita, 98–106.
56. See Syvänne, *MHLR 8,* 206–7.
57. Based on *STR* 11.2; Syvänne (2004, 351–65: *MHLR* vols. 6–8).
58. See Syvänne, *MHLR* vols. 4 and 7.
59. See in particular Syvänne, *MHLR 1*, Chapter 'Enemies'.
60. Based on *STR* 11.3 and Syvänne (2004, 366–88) with Syvänne, *MHLR* vols.1–8 (see the analyses of the enemies).
61. Based on *STR* 11.4 and Syvänne (2004, 389–96; *MHLR* vols. 6–8).
62. Based on Syvänne (2004, 397–418) and Syvänne *MHLR* vols.1–8 (more detailed analyses of the types of enemies offered).
63. The first extant treatise to give instruction on how to deal with the Hagarenes (i.e. Arabs who were at the time Muslims) is the *Taktika* of Leo, but these instructions refer to entirely different circumstances than were prevailing in the early-seventh century or before.
64. The text is partially a quote from Syvänne (2004, 419–22).

Chapter 4

1. A quote from Syvänne (*Hippotoxotai*) footnote: The Roman cavalry formations were to have their counterparts in modern Europe, which only goes to show that ancient military men had already discovered the basic cavalry formations. The principles of cavalry warfare remained practically the same through the centuries – only some of the details like the equipment

changed, but the use of the different unit formations retained their usefulness. For example, during the eighteenth century the close-order was used for the main attack. The troopers in the eighteenth-century close-order cavalry formation were deployed knee to knee, with a file depth of two or three ranks, or as columns with depths of 15 men. They charged either at a trot or at a gallop and were either deployed without intervals as *en muraille* or with small intervals. The equivalent of the open formation was to fight *en fourrageur*, which was used whenever speed was of the essence in attacking. In this formation, the soldiers used the gallop and allowed the files to open up so that there were one or two paces between individual horses. The equivalent of the irregular order was the skirmishing order, which was used for skirmishing and for pursuing the defeated opponent. And just like in Roman practice, the pursuing troops were to be protected by the close-order troops. For detailed analyses of seventeenth to nineteenth century cavalry tactics see: Nosworthy, 1990/1992, ibid. 1997, 263–285; Rothenberg; Chandler, 1976, esp. 27–61; Chandler,1973/1979; Duffy, 221–230. See also: Picq; Jomini.

2. See Syvänne, *Septimius Severus*.
3. For the Macedonian and Muslim uses of the rhomboid, see Syvänne, 2010, 2014, 2017b, 2019a, 2022b (267).
4. The *Equites Illyriciani* formed part of the local frontier defence forces in the provinces of Phoenicia, Syria, Palestine, Osrhoene, Mesopotamia and Arabia. The *Strategikon* expected these to be included in the main field army. See: G.T. Dennis, 1984, 28, note 1 with the *Notitia Dignitatum, Or.* 32–37.
5. See Syvänne, *Septimius Severus*, Appendix 1.
6. The use of the *optimates* as an elite group of soldiers posted in the second line right next to the *strategos* appears to have been an old practice, because we find the elite Moorish cavalry posted in this manner already during the reigns of Philip the Arab (Syvänne, *Gordian III and Philip the Arab*, 171–3) and Gallienus (Syvänne, *Gallienus*, 73).
7. *STR* 2.6–8, 3.6–8.
8. *STR* 1.1–2, 2.6–8, 3.1–6; *PST* 17.1–19; *DMS* 1 (esp. 1.2); Procopius, *Wars* 1.1.12–3. Dennis (1981, 148–53; 1984, 36–7) has amended the diagrams of the *tagmata* to contain flanking sections with intervals and not only the central section. The opposite view is held by Mazzuchi (112–5), who has reconstructed the *banda* as single units without intervals. In light of the text, which assumes compacting towards the centre, his emendation is far more likely. Since the *bandon* served as the smallest tactical operational unit, it is clear that it had to operate as a single entity. This was facilitated by the use of *hekatontarchai* and *ilarchs* on the edges to direct the single entity in manoeuvres. The placing of the better armoured horsemen in front is just common sense, so it is not surprising to learn that the method was age-old by the time the *Strategikon* was written. The importance of equipping the horses with protective armour was already recognized by Xenophon (*Hipparchos* 1.6–7; *Peri hippikes* 12.1–10) and as is well-known the same system was also used by the Huns and Mongols. It is likely that the men assigned as archers carried the shields with them because this would have been needed in melee situations, just as we find in the later *Taktika* of Leo the Wise (12.36) in the section which refers to the internal composition of the cavalry *tagma*.
9. As I note in my doctoral dissertation (2004, 135) a good example of this tendency can be found from the sixteenth and seventeenth century *reiter* tactics, which in the end degenerated into a caracole style of fighting. Gustavus II Adolphus of Sweden went on to copy the charging tactics of the Polish Hussars and reintroduced the charge into the Central- and West-European theatres of war. As a result of his changes in tactics, the Swedish cavalry were at times altogether forbidden to use their pistols and were ordered to charge with sword in hand.
10. *STR* 3.4–5. According to McGeer (1995, 280ff.), the practice in the tenth century was to wait for the enemy arrows to fall so that they could calculate the distance to begin their archery.

The *Strategikon* does not contain such instructions, which means that it is irrelevant for this period. The Late Roman cavalry could not wait to be hit by enemy arrows, because they faced enemies who did not necessarily employ archers or mounted archers. Therefore, it is clear that the Late Roman officers simply estimated the range, because they couldn't rely on the enemy to do it for them.

11. The term was used e.g. by Ammianus Marcellinus (see Syvänne, *MHLR 2*, 95–6). In the example given by Ammianus, it referred to the posting of the weaker men in the middle in a battle formation of three lines, but the same principle was also used in the arrangement of the units and soldiers in width. In other words, the centre and edges consisted of better men and units so that the weaker units were posted between these.
12. *STR* 2.3, 2.6, 3.4–16, 4.5; *PST* 17; *DMS* 2. It should be noted that the late George T. Dennis's translation (1984, 42, *STR* 3.6) and diagrams give misleading information concerning the *koursores*. Dennis continually referred to the *koursores* as being troops in open order and amended the diagram of the Federates to reflect this. This is contradicted by the text. The whole *meros* was to be compacted (e.g. 3.5.20–1) and the *koursores* were the troops used for the pursuit. Note also the demand made for giving support to neighbouring *mere* in the *DMS* 9–11.
13. For example, it is clear that the Germanic commanders in Roman service used the single line version.
14. Based on Syvänne (2004, 136–7) but with some changes and additions.
15. *STR* 1.3, 3.2–5. The late Dennis (1984) interpreted the *koursores* to have been always deployed in open order, but this is unlikely in light of the instructions (*STR* 3.5.15–25) to compact the whole *meros* before the attack. See also above. The federate Heruls used the loose, scattered cavalry formation at the gallop in their attack (see Syvänne, *MHLR 1*, 84–6; *MHLR 6*, 183–5)
16. The most 'advanced' state of this formation was, as already stated, the *en mureille* of the European eighteenth-century cavalry. The continuous line without any intervals could only be used on an open unhindered field, because any minor obstacle could cause a ripple effect throughout the line when there were no intervals to soften up the wave. This continuous line usually demanded the use of the trot instead of the gallop, even though the Prussian cavalry was trained to use it with the gallop. There always existed the danger of disorder in too long a continuous line.
17. In Roman feet (29.6cm) the width of three feet is 0.888m and in 'Byzantine' feet (31.23cm) (Schilbach, 13–16) it makes 0.9369m. In Roman feet the depth of eight feet is 2.368m and in 'Byzantine' feet it makes 2.4984m. I have here used 'Byzantine' feet because it clearly fits better the actual distances depicted.
18. As I noted in my doctoral dissertation, the actual width and depth taken by each horse/rider combination may have varied. The following list shows the great differences in the estimates made by modern theorists and historians. For example, in the sixteenth century the Duke of Alba estimated that each horseman/horse occupied 1.5m by 4.5m in close order while in 1658 Melder estimated that each horse/horseman occupied, 0.9m by 3m (this is quite close to the figure given by Maurice) and in 1909 one theorist estimated that the cavalry file had a width of 0.75–1.5m in close order. For additional details of the space taken by horsemen in the sixteenth to twentieth centuries, see Delbrück 4.117–45, 281–4. Delbrück correctly notes the great variance in the figures given by experienced cavalry officers for the same close order. In the seventeenth century John Cruso (*The Militarie Instructions for the Cavall'rie*, 45–6) specified the same distances as above for the close formation but made these to be the spaces between the horses, probably intending to include the horseman in the figures. In 2001 (128–9) Bachrach estimated that each horse/rider combination using the gallop in tight formation during the early Carolingian period occupied about 2m in width and had about 4–5m between each rank. This may have been true for the Carolingian cavalry when it used

the gallop, but this was certainly not true for our period Roman cavalry. The Romans used the trot/canter while the Carolongians used the gallop. The need in the sixth-century Roman formation was not as pronounced since they used the trot/canter. Regardless, I would still suggest that Bachrach's figures are too deep for the space between the ranks. The use of the Roman open order intervals would also have been sufficient for the Carolingian cavalry.

19. *STR* 1.3, 2.2–6, 3.2–5, 6.3, 9.5. For the use of irregular (*droungos*) order by the *koursores*, see: *STR* 3.5.37–50, 4.5.15–26, esp.20–22.
20. Based on Syvänne (2004, 137), but with a new suggestions regarding the use of rhomboids and wedges. See also Appendix 2.
21. For the origin of the word *drouggos*, see Rance (2004a, 97–105) with Haldon (2014, 145–6).
22. Rance (2004a) with Syvänne (2004, 137), Haldon (2014, 145–6) and Eramo (2018, 129–31).
23. The *STR* bears Leo's name, but his authorship has not been accepted by most historians. I do not see reasons to doubt it. For a fuller argumentation, see Syvänne with my additional comments available online at academia.edu.
24. The referrals to *drouggos/drouggisti* are conveniently collected by Eramo (2018, 129–31). Note, however, that the *Taktika* 7.34 includes also another referral to *drouggos* not included in the list, but which is essentially the same as the ones included.
25. *STR* 3.4–4.5. This is partially a quote from Syvänne (2004, 137).
26. Based on: Syvänne (2004, 137); *STR* 3.10, 3.13–14, esp. 3.13.5–8, 3.14.17–18, 4.5 (the units guarding against outflanking and units used for outflanking were required to be adept at using the *drouggos*), 12.D, esp.68–71.
27. Based on Syvänne (2004, 137–9) but I interpret the width of the cavalry formations differently. My previous interpretation took the loose formation literally, but on closer look it is likelier that it meant the close order of Maurice.
28. In my *Desperta Ferro* article (2014) I dated the battle to the traditional year 636, but it should be redated to the year 634. For the reasoning, see Syvänne, *MHLR 8*.
29. Based on Syvänne (2004, 139–41) with new information added in particular concerning the era before the sixth century.
30. For a fuller analysis of Arrian, see Syvänne (*Septimius Severus*, Appendix 1).
31. For a summary of these observations, see: Nosworthy 1990/92, index cavalry; Nosworthy, 1995/97, 263ff.; Delbrück 4.117–45, 281–4.
32. See Chauviré (30–53) for a fuller discussion of early modern era concepts.
33. See: *HA Maximini Duo* 2.1.-3.6 with Syvänne (*Septimius Severus*, 45); Syvänne (*Aurelian and Probus*, 12, 77, 100–3); *MHLR* series (see the index referrals to the existence of cataphract/ *clibanarii* cavalry). For the special qualities of the Arab horses, see Syvänne (*MHLR* 7, 71). Note, however, that the Romans also possessed extraordinarily fast horses in their imperial stables too (Syvänne, *MHLR* 7, 334) even if their numbers appear to have been too small to be of strategic importance.
34. *STR* 11. In *The Age of Hippotoxotai* I also noted the following: 'The cavalry formations of the age exploited the herd instincts of the horses in massed charges. In this the stallion aggression could be useful and the Romans and their enemies used primarily stallions. The only exception to this rule was the steppe nomads who primarily used geldings, but at least the Avars also used stallions in their elite cavalry. Stallions saw other stallions as their enemy. This was a useful trait in a wild charge or after the troopers had mingled with each other to fight in melee. The horses would not necessarily register until later that they had received a wound from a missile or other weapon, unless crippled instantly. As a downside, this trait made spirited horses difficult to control in formation. It was difficult to maintain ranks in the charge, at the trot in close order and even more so at the gallop in the open or irregular order. The nature of horses varied just as the nature of humans. Some horses were courageous while others were not. Some of the routs were not the result of human actions or of the unit orders. The horse knew instinctively if its rider was nervous and could join any rout without

the rider's will. In consequence, it was important to have well-trained warhorses that trusted their riders. In the course of a long campaign their numbers would diminish. In sum, the use of the close order diminished the probability of the horses or riders acting on their own and therefore gave better chances of success in cavalry battle. The Romans were clearly aware of this fact. In general, for the behavioural characteristics of the horses and their use by Romans/Byzantines and their enemies see: Hyland, 1994, xi-xii, 18–53. It should be noted that the Roman battle array could also contain geldings and mares since the allied ethnic contingents such as the nomads and Arabs used their own horses'.

Chapter 5

1. This chapter is based on Syvänne (2004, 118–31), but it contains some changes and important additions.
2. See the chapter on military treatises together with Syvänne (2004, 2011a-b, 2019c, 2021a).
3. The following is based primarily on Syvänne (2004, 118ff.) and *Strategikon*, books 2–7.
4. Note that there is a superficial discrepancy in the maximum numbers for the *moirai* and *meros* because the maximum size for the *moirai* would allow *mere* of 9,000 men whereas the maximum for the *meros* was 7,000. However, this discrepancy is superficial because it is clear that when the centre *moira* was made 3,000 strong, the flanking ones were to consist of at most 2,000 men. However, this system was later changed so that in the *Sylloge Tacticorum* (35.1–2) the *tourmaches* (also called *merarches*) had 3,000 to 9,000 men under him so that his subordinates, the *drouggarioi* (formerly *moirarchai*) had 1,000 to 3,000 men each. In other words, the later Roman practitioners of the military art accepted larger divisions (sing. *meros, tourma*) than Maurice in the sixth century.
5. See also Eramo's comments in the *DRM* (p.141 n.80). For the use of the *comes foederatorum* title, see Syvänne, *MHLR* vols. 3–8. It is possible that the requirement of the *DRM* to post 1,000 extra men in the centre *meros* resulted from the inability of the Romans to recruit enough men of Germanic origins into the *foideratoi* after the Muslim conquests began, so the centre required more men to be as effective as it had been previously. As noted, the *DRM* is likely to have been produced during the latter-half of the seventh century.
6. The medics used stirrups on the same side of the horse to carry the wounded and the unhorsed.
7. *STR* 1.3–5, 2.2–4.5. The centre *meros* composed of stronger elite troops: *STR* 3.8. For the strengthening of the centre *meros* see also the later military treatises: *DMS* 4; *ST* 46 (wedge); *Praecepta* 3–4 (wedge); Ouranos, *Taktika*, 60–61(wedge); McGeer, 1995, 280ff. The *DMS* appears to use different terminology for the divisions of the cavalry battle array. The front consisted of three *mere* a bowshot apart from each other (4). Behind them were posted the *plagiofylakes* (= 2nd line wing divisions to protect the flank *mere* of the first line 4) and *notofylakes* (protected the rear of the centre *meros*/11) as the 2nd support line at a distance of 4–5 bowshots or one mile. These terms appear to reflect the actual usages of the 2nd line divisions.
8. *STR* 2.13, 3.1–10. In the tenth-century Byzantine military manual *Sylloge Tacticorum* (ca. 904) the intervals between the first line *mere* were 13 *orgyia*. Schilbach (22–6) equates 1 *orgya* = 1.87m, therefore 13 *orgyia* = c. 25m. One may make the guess that given the use of continuous lines the widths of the sixth-century intervals were likely to be about the same. The *DMS* (4) separates the three divisions by single bowshot intervals, but this is clearly too wide an interval to fit the description in the *Strategikon* which separates the *mere* of the second line in this manner, but it is possible to think that in the *DMS* the division of the *mere* into three *parataxeis* each separated by a bowshow actually means the three combat lines, the first line, the second support line (in the *DMS* called rear guards) and the the third line (in the *Strategikon* the rear guards). The key problem with this is that the distances between the three lines would then be roughly the same as in the tenth-century *Sylloge Tacticorum*

(43.9, 46) which required the lines to be separated by about one or two bowshots, with the implication that the tactics of the *DMS* would not be the same as in the *Strategikon* which expected the first line to attempt to regroup in the interval (three or four bowshots) between the lines. If the lines would have been separated by only one or two bowshots, the first line would have not attempted to regroup in the interval but would have fled directly to the second line where it would have attempted regrouping. If we interpret the *DMS* in this manner, there still exists the problem that the *DMS* does not give the third line (if the text meant this) any role in combat. In addition to this, if the three divisions of the frontline would have been separated by bowshot intervals, the line would have given the enemy chances of penetration and outflanking, and would not have matched the expectation that the second line had intervals of one bowshot. The intervals for the first line in the Alan Drill were 200- to 400-feet (ca. 62–125m) and it has the widest intervals between units mentioned by the *Strategikon*, so it is clear that the 330m interval is already extremely wide. I would therefore suggest that it is probable that the text of the *DMS* is defective in its current form.

9. *STR* 3.5.15–25; Mazzucchi, 111–138, esp. 111–115. I accept Mazzuchi's (113–115) interpretation of the formation of the *tagma* without any intervals between its different sections, but I do not accept Mazzuchi's diagram of the *meros*, which has no intervals between the *koursores* and *defensores*. The reason for my different interpretation is that the manusripts have these intervals and their separate usage demanded their existence. Furthermore, the smallest operational battlefield unit, the *tagma*, also had to have a small interval to separate it from the other *tagmata* so that it could operate independently of the others. The *Strategikon* clearly has instructions indicating manoeuvres done by the *bandon*. For this see the discussion about the manoeuvres of the flank guards below.
10. E.g. *STR* 2.5, 2.11–2, 7.3, 7.4, 3.16; *DRM* 5–6.
11. See e.g. the use of a single line by the Ostrogothic king Theoderic the Great (he was Roman *magister militum*) and the regular Roman army at the Battle of Nymphius River in 582 in Syvänne (*MHLR 5*, *MHLR 7*, 205–6). The Romans later abandoned this system and started to use the flank divisions of the second line (and also of the new type of third line) for the outflanking. For this development, see the forthcoming biographies of Nikephoros II Phokas and John I Tzimiskes by Syvänne.
12. See Syvänne (*MHLR 3*, 274–6).
13. Based on Syvänne (2004, 128; *MHLR 6*, 51–3, *MHLR 7*, 154–8); *STR*. 4.3.35ff; *Excerpts of Polyaenus* 37. Procopius, *Wars*, 1.4.1–14, 1.4.32–35; Gregory of Tours (Thuringians against the Frankish cavalry in 531).
14. *STR* 2.6.33–5.
15. *STR* 1.4.26–31.
16. Based on Syvänne (2004, 124–5); *STR* 2.1–20, 3.8–10, 3.8.36–9.25. In *The Age of Hippotoxotai* I noted the following: 'In case the second line consisted of only a third of the army, while the first line made up 70 per cent of the army just like in the accompanying diagram, the support line would have been considerably shorter than the first if the intervals were only one-fourth of the width of the 2nd line. If we accept the width of one-fourth for the interval, which by the way is not its width in the diagram, we would have to expect that the left *meros* of the front line would have retreated towards the left flank of the left *meros* of the second line, and that the centre *meros* would have retreated towards the interval between the *mere* of the second line and that the right *meros* of the first line would have retreated towards the right flank. If this were the case, then the flanks of the first line would have been vulnerable because the second lacked the length for its support. In the defective diagram the second line doesn't cover the whole length of the first line and is only about 30 per cent of the whole strength and only about 42 per cent of the strength of the first line. Furthermore, the interval of the second line is about 43 per cent of the first line. In fact, one cannot find the ratio of one-to-four for the intervals of the second line in any of the existing diagrams as was noted by Mazzucchi

(p.117) in his article'. The weakness of having only two divisions in the support line was that it did not provide the same measured security for each section of the first line as the version with four divisions. The fact that the version with four reserve divisions was later adopted by armies as small as 4,000 men proves that the Romans came to the same conclusion later. For a reconstruction of Leo's formation for 4,000 men, see Oman (*A History of the Art of War in the Middle Ages*, 1.197–99).

17. Similarly in Tacitus (*Histories* 3.15–17).
18. Based on Syvänne (2004, 125–6) with some changes; *STR* 2.1–20, 3.10.1–20.
19. Four reserve divisions in a cavalry army of 4,000 men depicted by Leo VI in the early-tenth century (reconstructed by Oman, 1.197–99). Several reserve divisions in cavalry armies as small as 3,000 or 1,000 horsemen in the *Sylloge Tacticorum* (reconstruction by Syvänne, 2008).
20. Based on Syvänne (2004, 126) with the *MHLR* series (note e.g. *MHLR 5*, 112ff.; *MHLR 6*, 132ff.; *MHLR 8*, 253ff.).
21. Based on Syvänne (2004, 126–7) with the *MHLR* series (note e.g. *MHLR 6*, 63–5, 91–5, 112ff., 183–5, 202–4); Procopius, *Wars* 2.3.8ff., 5.18.1–29, 27.1–29; *STR* 6.1, 11.1–3, 12.D; Procopius, *Wars* 5.18.1–29, 27.1–29. The military lexicon *Hermeneia* (59) that has sometimes been dated to this period also contains a word *esparmene* [phalanx?] for the scattered/spread/loose formation. However, this appears to have been an infantry formation because the *esparmene* can also be found from Asclepiodotus' treatise (11.7). According to Asclepiodotus the *esparmene* formation was a chequered (infantry) marching formation. The scattered formation used by Sittas was improvised on the spot while the loose formation adopted by the Heruls was their native fighting formation.
22. Based on Syvänne (2004, 127–8) with updating (e.g. *MHLR 6*, 304–5, *MHLR 8*, 253–72). Cavalry with two lines in front of infantry: *STR* 12.A.3. Cavalry with a single line (3 *mere*) in front of infantry: *STR* 12.B.23.19–27. See also the Chapter dealing with Infantry Tactics.
23. For examples of the earlier use of cavalry together with infantry, see e.g. Syvänne (*Gallienus*,141ff.; *Aurelian and Probus*, 105–15). For the Late Roman period, see Syvänne (*MHLR* series, e.g. *MHLR 2*, 243–5 and *MHLR 6*, 304–5). Note, however, that it is possible that infantry was actually present also in cavalry battles in which their presence is not mentioned by the narrative sources because the infantry was left behind in front of the marching camp while the fighting was done by the cavalry. The problem here is that if the infantry contingent was small, it was left inside the marching camp while if it was strong it was posted in front of the infantry camp. This means that when the sources do not give details of the way in which the accompanying infantry was used in each particular battle it is impossible to know for certain what version of the cavalry battle formation the Romans used. However, on balance, in most cases the infantry contingent appears to have been considered so small that it was left inside the camp so that the actual battle was a pure cavalry battle. This is clearly the expectation in the cavalry portion of the *Strategikon* (books 1–8).
24. See Syvänne (*MHLR 6*, 304–5; *MHLR 7*, 310–5).
25. Based on Syvänne (2004, 128–31; *MHLR* series); *Strategikon* 2.1.1–2.10, 6.3.1–2. For examples of the Roman use of two lines see e.g.: Caesar, *Gallic War*, 7.13, 8.28; Tacitus, *Annales* 1.68; Syvänne (*Gordian III and Philip the Arab*, *Gallienus*).
26. See Syvänne, *MHLR* series in general for examples of cavalry battles. Good examples of the Romans hiding the presence of their second cavalry line from the Persians are the Battles of Mesopotamia in 422 (Syvänne, *MHLR 3*, 276–8) and Melitene in 576 (Syvänne, *MHLR 7*, 103–12). It is possible that Heraclius also hid his second line in wooded terrain in 625 (Syvänne *MHLR 8*, 168–9), but this is uncertain because the sources do not give us enough details. It is equally possible that Heraclius employed infantry and an ambush inside the woods. Therefore, the Battle of Melitene in 576 remains the last certain example of the Romans successfully hiding their second cavalry line.

Chapter 6

1. Based on Syvänne (2001, 34ff. *Master's Thesis*; 2004, 142–3; with the *MHLR* series); *STR* 2.11, 5.1–6, 7–8, 12.B.23.33ff. A good example of the pre-battle procedures and stress can be found in the context of John Troglita's battles in Africa (*MHLR 6*, 272–96). Instructions for the day of the battle: *STR* 2.18–20, 3.5, 4.3.85ff., 7.
2. See e.g. Syvänne, *Septimius Severus* for the Battle of Lugdunum/Lyon in 197.
3. Translation is based on Dennis (*Strategikon*, p. 37), but includes some changes; '*Silentium, nemo demittat, nemo antecedat bandum, sic venias vero aequalis facies, bandum capta, ipse seque cum banda milix, talis est comodum miles barbate, si vero bandum demittes eo modo non vero vices, serva milix ordinem positum. Ipsum serve et tu bandifer, sive pugnas sive seques inimicum sive aequalis facies, non forte minare ut ne sparges tu suum ordinem*'. In the translation given, *bandum* means the flag/standard and *miles* the soldier. The *DMS* (7) contains quite similar instructions in Greek for the Herald to shout. On the basis of the fact that Maurice considered it beneficial for the herald (*mandatōr*) to be able to speak Persian (*STR* 12.B.7.3–4), one may surmise that the herald could also be used to interrogate prisoners, to insult the enemy before the battle, to conduct negotiations, to give false commands to the enemy (the collections of stratagems included this), and to inform the Roman officers of the commands given by enemy officers. Unfortunately, one cannot confirm these speculations because Maurice fails to explain why he considered the knowledge of Persian to be useful. Similarly, the narrative sources offer no help in confirming these hypotheses. However, I would still suggest that the knowledge of the foreign language was useful for the reasons stated.
4. *STR* 2.17–19; al-Tabari i.2190; Syvänne (2004, Appendix 1/Tricamarum/Mugello). However, as I noted in *The Age of Hippotoxotai*: '…in practice, one can imagine that the silence wasn't always followed. After all they were only humans. There could have been the neighing of the horses and sometimes some loud jokes made to alleviate the nerves. It is more than likely that the veterans would have whispered some encouraging words to young green soldiers amidst the veterans (which in turn would have soothed their own nerves), and the officers would have whispered or uttered encouragements and orders to be silent. However, in general, the silence would have been kept.' The silent approach to fighting was Maurice's own personal opinion, which was not shared by everyone. The shouting etc., returned back to the Roman combat later. See for example the forthcoming biographies of Nikephoros Phokas and John Tzimiskes by Syvänne.
5. The translation of the command is by Dennis. I could not come up with a better myself.
6. *STR* 1.8 (capital punishment for charging out of ranks), 3.1–5. The duals were not the only thing that broke combat doctrine and military laws. For example, in 360 AD, during the Siege of Bezabde, the Roman soldiers fought without their helmets (in true Hollywood style) in the hope that the emperor Constantius II would recognize and reward them and as a result died under the hail of arrows (Ammianus 20.11.12). During the third century even the emperors fought several duels. For this, see Syvänne (the indexes in *Caracalla, Gordian III and Philip the Arab*, and *Aurelian and Probus*). Note also that the Late Roman commanders themselves could also fight duels despite the laws and instructions against this. Good examples can be found e.g. in Syvänne (*MHLR 3*, 271; *MHLR 4*, 31).
7. Based on Syvänne (2004, 143–6); *STR* 1.Pr.36ff, 2–4; 8.A.1–44, 8.B.2–3, 7–9, 15–21, 24–28, 37, 39–45, 48–101; *DMS* 15; Chapter 3.4. When and Where to Fight; Onasander 13.3.
8. Impact of fame and generosity of the commander is in evidence e.g. in Procopius, *Wars* 7.1.8–24 (esp. 8), 7.39.9–20, 40.9, 7.26.14–17; Whitby, 1995, 84–5.
9. For examples, see Syvänne (*MHLR* series) and for example Procopius, *Wars* 5.18.1–29, 5.22.1–9, 7.1.8–24 (esp. 8), 7.39.9–20, 7.40.9, 7.26.14–17; 8.11.39–47. For example, Constantine the Great, Areobindus, Bonifatius, Aetius, Belisarius, Bessas (even at the age of 70) and John Troglita all demonstrated their skills with weapons. However, there were

obviously also commanders who could not lead by personal example thanks to infirmity, old age or otherwise (eunuchs had less testosterone and consequently less muscle mass).

10. See e.g. Syvänne (2013c; *MHLR 6*, 63–4).
11. *STR* 2.14–20, 3, 8.B.100; *PST* 30; *DMS* 3. For an analysis of the 'Byzantine' flags see: G.T. Dennis, 1982; Ravegnani, 38–9.
12. For examples of all of these, see Syvänne (*MHLR* series). For an example of the imperial investigation of the conduct of different officers after a defeat or victory, see in particular *MHLR 6*, 67–71, 171–2. For the tendency of the historian to praise a subordinate, see in particular *MHLR 7* (Theophylact trying to praise Heraclius Sr. and Priscus at the expense of others). Belisarius's prudence with wine: Procopius, *Wars* 7.1.13–15. During the fifth century there was an abundance of generals who misused alcohol, see for example: Eunapius fr.65.3, 67.5, 67.8, 71.1. In the infantry section Maurice (*STR* 12.B.23.33ff.) forbad the distribution of wine to soldiers. It is probable that the same was true for the cavalry as well. The soldiers drinking wine and neglecting guard duty in Syvänne, *MHLR 7*, 297.
13. Based on Syvänne (2004, 146–9) with some changes; *STR* 2.1–3.14; *DMS* 3–11.
14. Translations in the chapter are usually based on Dennis's translation, but I have tried to improve those when possible. In other words, the translations are usually modified versions.
15. The *Strategikon* 3.5 states that the different manoeuvres (i.e. assumption of close order, wheeling, about turn, countermarch etc.) practised by the individual *tagmata* were also to be trained by *moirai, mere, protos parataxis* (first line) and *deuteros parataxis* (second line). This means that the entire battle line trained contracting towards the centre in such manner that all soldiers were in close order at the end of it while distances between *mere* and *moirai* were retained.
16. For the use of the gallop by Germanic and nomadic cavalries, see Syvänne, the *MHLR* series. For the use of the gallop by the cavalries of Belisarius and John Troglita, see *MHLR 6*. Maurice (*Strategikon* 4.5) appears to have belonged to the school which sought to obtain the maximum benefit from both approaches, because he felt in necessary to defend his instructions for the cavalry to train in the use of the *drouggos*-order and gallop in ambushing. Maurice quite rightly considered it overcautious to use only the rectangular rank-and-file formations and not the *drouggos*-order at all. Since the gallop was used with the *drouggos*-order (the formation broke up when the *koursores* or ambushers galloped), the implication is that there were also such military thinkers who did not approve even the use of the gallop by the *koursores* because this resulted in the loss of unit cohesion (i.e. the unit became *drouggisti*).
17. For example the *Strategikon* (1.5.8–9) and Vegetius (2.14) both demanded that the front line cavalry officers were able to use the bow. This enabled the Romans to derive the maximum benefit from their archery in situations in which the enemy lacked adequate numbers of archers. Good examples of the use of this tactic are the battles between the Ostrogoths and Romans in the sixth century. This tactic was made famous by Belisarius. For this, see Syvänne *MHLR 6*. It is also clear that the Roman federates and possibly also some of the regulars continued to use different types of specialist equipment so that there were units and individuals who did not use shields with their spears, but continued to use the Sarmatian style *contus* with two hands. See for example: Agathias 2.8.1; Corippus 5.295, 6.638–42, 8.389–427.
18. For the Heruls see Syvänne *MHLR 1*, 84–6. Regulars using swords: Procopius, *Wars* 4.17.18. The *elafroi* is also included in the close-to-period dictionary of military terminology the *Hermeneia* (25), just like in Aelian (Devine, 2.9–11; Köchly, 2.11–13), Arrian (*Tactics* 4), and Asclepiodotus (1.3).
19. *STR* 3.5.37–40 with Syvänne *MHLR 6* (see the index for the battles of Belisarius). Translations of the commands by author.
20. Both the *Strategikon* (3.5.41ff.) and *Taktika* of Leo (12.61) call the retreating unit(s) of the first line *koursores*, which should simply be understood as the units which retreated at the

gallop towards the rear could all be called 'runners' (*koursores*) regardless of their original assignment in the first line.

21. *STR* 3.5.41–50; *DMS* 10–11; Syvänne (*MHLR 6*, 98–101; *MHLR 7*, 103–12, 212–6; *Gordian III and Philip the Arab*, 170–3). For examples of different cavalry battles, see in particular *MHLR vols. 2–8* (Index) and *Gallienus* (Index).
22. Based on Syvänne (2004, 149–50) with some changes; *STR* 2.13, 3.5–13, 6.5, 11.2.90ff.; *DMS* 3 esp. 3.9, 3.11.
23. The *DMS* 3.11 includes the infantry commands *Ad scuto clina* (Shieldward face) followed by the commands *Move* (Move, Advance) and *Redi* (Return). This would mean that each trooper in the *bandon* turned individually towards the left, after which the *bandon* moved to the left and then when the *bandon* reached the desired place each soldier turned to the right with the order *Redi* (Return). The command to turn towards right on its own was *Ad conto clina* (Spearward face). See the discussion in *DMS*/Æramo (62–3, 135–6). The positioning of the flank guards about a bowshot to the left makes it possible to speculate that all or some of the flank guards would have been left-handed just like their Persian counterparts in the Persian formation. The reason for this speculation is that the left-handed trooper could shoot to his right by holding the bow in the right hand while drawing with his left. In addition, he could shoot more easily to his right than a right-handed person because he could easily hold the bow in his right hand while drawing the cord with the left. This, however, is only speculation because none of the period sources refer to such specialization. For the Persian practices see: *Ayin-nameh*/Inostrancev, 7–52, esp. 13–14, 24–28; Syvänne (2004, 334).
24. Based on Syvänne (2004, 150–1) with some changes; *STR* 3.5.51–55, 110ff, 3.10, 3.14; *DMS* 3. There is one instance of the cavalry attack when the enemy tries to adjust their line during the sixth century. See Syvänne (2004 Appendix 1/Bagradas; *MHLR 6*, 180–1).
25. The *DMS* 3.11.
26. The principle of hiding cavalry behind the cavalry line to encircle the enemy was known at least since Xenophon's day (*Hipparchos*, 8.17–8).
27. One may suspect that this had happened at some point in the past so that the Romans had extended the width of their line by reducing the number of ranks in the battle formation rather than by spreading the intervals between the units if the Romans intended to maintain a continuous front. Similarly, one may suspect that some commanders had overextended their battle line by having units spread out so that the formation had too wide intervals between the units to be effective. It is also clear that in the past there had been *strategoi* who dispatched some of their reserve forces to the flanks to widen these because the *Strategikon* warned against this practice. The internal structure of the cavalry formation and the use of two ranks for the *banda* posted in the intervals of the second line show that the cavalry was used to dividing and doubling of their formations.
28. Based on Syvänne (2004, 151–2) with some changes; *STR* 2.5, 3.16, 4; *DMS* 16.
29. Based on Syvänne (2004, 151–2) with some changes; *STR* 3.5.56–62, 3.8, 3.15, 7.B.5; *DMS* 3, 12.
30. The *DMS* 3.3 uses the infantry command '*Muta locum*' ('countermarch, change place'; *STR* 12.B.16.109) instead.
31. Based on Syvänne (2004, 152–3) with some changes; *STR* 2.4.25–8, 2.13.24–26, 3.5.56–62, 3.15.19ff.
32. There are obviously some examples in which the Roman spies and scouts performed their duties inadequately so that the Romans were surprised, even without being fooled by an enemy's feigned retreat, into an ambush. The unfortunate thing about these instances is that in most cases the sources do not provide enough details for us to make a judgement about the effectiveness of the rear guards when the enemy ambushed the Romans. The best example of the problem is the battle between the Persians and Romans at Erginay in 605. In that instance, the Romans had advanced towards the Persians while the Persians bypassed the Romans and

approached them from the rear. When the Romans learnt of this, they turned around and attacked the Persians. In this case it is actually probable that the Romans turned their entire battle formation around so that they attacked with their regular formation, rather than by using the second line and rear guard as their new frontline. However, if the Romans for some reason failed to do this (despite the distances allowing this), then the battle demonstrates nicely that the safeguards did not work when the Roman army was demoralized due to political reasons (Phocas's usurpation resulted in demoralization). See Syvänne (*MHLR 8*, 53–4). Other good examples of this problem are the battles between the Armenians in Roman service and the Avars in 595, and between the Romans and Muslims at Heliopolis in 640. In the first instance the Avars ambushed the Armenians and in the latter instance the Muslims ambushed the Romans. The problem with these instances is that we do not know all of the details and the composition of the Roman forces, so that we could make some judgements about the efficiency of the rear guard when the enemy managed to ambush the Romans. In the former case it is very likely that the Armenian force consisted solely of cavalry but, in the latter instance at Heliopolis, we know that the Romans had infantry present but we do not know how it and the cavalry were used in order to assess the possible role of the rear guards – if these were even used when there was an infantry force present. See Syvänne (*MHLR 7*, 302–3; *MHLR 8*, 306–7). However, we have one certain instance in which the second line and possible rear guards failed in their task. This took place at the Battle of Faventia in 542 (Syvänne, *MHLR 6*, 246–7). In that case Totila, the king of the Ostrogoths, sent ambushers behind the Roman cavalry formation with the result that the Romans panicked and fled.

33. Based on Syvänne (2004, 152–3) with some changes; *STR* 2.9, 3.8. The tenth-century military manuals add that if someone's horse was injured in battle, then the rider was to take it back to the pennant designated for the spare horses of his unit, exchange the horse and ride back to take his place in the line. Each unit also kept its horses separate to the rear with a pennant to ease recognition. This is all common sense and it is clear that the procedure was the same also during the Late Roman period and before. See: Nikephoros II Phokas, *Praecepta militaria* 4.7 or 4.76–84; Nikephoros Ouranos, *Taktika* 61.7 or 61.123–135.
34. Based on Syvänne (2004, 153–5) with some changes; Syvänne *MHLR* vols. 1–8.
35. It should be noted, though, that the Arabs used a mixture of mares and stallions, which meant that if the Arabs were part of any ambush force including regular Roman cavalry they would have been required to use only stallions, since the mixing of stallions and mares was bound to cause neighing and snorting. See Tafrij 15.1, p.97. The mixing of mares in the cavalry battle array could also cause problems, which meant that the Arabs were better employed separately in their own units. The Romans recognized this problem by deploying the Arabs in their own separate units. The Persians could also use a mixture of stallions and mares (Chosroes II riding a mare: Theophylact 4.10.3–4. Pictorial evidence of stallions: Gall, Abb. 12.c; Nicolle, 1996, p.11C, p.22C). This means that the Persians needed to be posted in their own units too if the intention was to post an ambush.
36. *STR* 1.3.27–9, 12.B.23.33ff; Synesius, *Constitutio* and *Letter* 78; Zosimus 5.45.5–6; Procopius, *Wars* 3.12.8, 4.4.15–25, 4.4.30, 6.1.21–34, 7.11.22–25, 7.12.3–10, 7.27.3–6; Ammianus 15.12.4; Zosimus 4.25.2. The practice of using alcohol and other intoxicants to create willingness in the men to fight is known throughout history. For example, the Turks used hashish-eaters, the Soviets and Russians have used alcohol throughout history, and the Russians in Chechnya and Ukraine have used tranquillizers and other substances. The Germanic peoples, and in particular the Vikings, are famous for their berserkers which may also have had its earlier precedents. However, when used in moderation alcohol was and is beneficial for archery and fencing, and it is because of this that it is considered as a form of doping in modern competitive sport.
37. See e.g. Syvänne (*MHLR 6*, 304–5, 346; *MHLR 7*, 212–6); *STR* 12.A.7 (esp. 12.A.7.81–3). For the techniques of dismounting individual horsemen, see *Münyetü'l Guzat*, 66–7.

38. Based on Syvänne (2004, 155–69) with changes based on the *MHLR* vols. 1–8.
39. For examples, see Syvänne, *MHLR* vols. 1–8. For dismounting, see e.g. Syvänne (*MHLR 6*, 304–5, 346; *MHLR 7*, 212–6); *STR* 12.A.7 (esp. 12.A.7.81–3). For the techniques of dismounting individual horsemen, see *Münyetü' l Guzat*, 66–7.
40. Based on Syvänne (2004, Appendix 1/Tricamarum/Suenia 589); *MHLR 6*, 98–101; *STR* 3.5, esp.3.5.86–109, 6.
41. The tenth-century treatise of Phocas (*Praecepta* 2.3) contains this distinctive possibility for the use of the *prokoursatores* who were used for skirmishing 2–3 miles (*ST* 46.23) in front of the main battle line. In this instance the *prokoursatores* as vanguard were to ambush the vanguard of the enemy if possible and cause by their flight the flight of the main enemy army. This resembles closely the instructions of Maurice to the vanguard.
42. *STR* 3.5, esp.3.5.86–109, 6.
43. Based on Syvänne (2004, 162) and text above with some changes and additions made on the basis of the different versions of the *Strategikon*. The use of the hidden cavalry unit was age-old, as Xenophon testifies, but he adds to this list hidden infantry units behind cavalry. Xenophon also noted the importance of reserves, or fake ambushers, or actual ambushers behind the first cavalry line (Xenophon, *Hipparchos* 3.8–14, 4.13–5, 5.7–8, 8.12–5, 8.17–20, 8.23–5). In fact, Xenophon's instructions (8.23–5) to leave some men behind when others attacked and retreated so that those left behind could charge against the enemy at the same time as those who were retreating reached them resembles the *koursores/defensores* system too.
44. Based on Syvänne (2004, 161 with Appendix 1/Tricamarum/Suenia 589; *MHLR 6*, 98–101; *MHLR 7*, 226–8; *STR* 3.10.40–46. The centre *meros* composed of stronger elite troops: *STR* 3.8; *MHLR* 7) with some changes; Later examples: *DMS* 4; *STR 46; Praecepta 3–4; Ouranos, Taktika*, 60–61; McGeer, 1995, 280ff.
45. Based on Syvänne (2004, 162) with some changes; *STR* 310.47–55, 3.12–14. In one of the formations used in hunting (*STR* 12.D.112ff) the flank divisions advanced past the centre so that it became '*epikampion taxi*' (forward-angled formation) during outflanking.
46. Based on Syvänne (2004,) with some changes; *STR* 2.13.10ff, 3.10.55–8, 3.15, 7.B.3–5, 7.B.14–15. The diagrams in Dennis, 1981, 506–7.
47. Based on Syvänne (2004,164–5, 282ff.) with some changes; *STR* 2.5, 4; *MHLR* vols.1–8 (Indexes on ambushes, surprise attacks); *DMS* 16. The feigned flight was certainly not a new tactic. Sensibly, on scouting duty (and therefore in ambuscades) it was recommended to use only geldings (*PST* 7.20–2).
48. *STR* 4.3.35ff. Notably, the Persian cavalry employed caltrops in similar manner at the Battle of Nihawand in 641, but in that case the Muslims managed to force the retreating Persians into their own caltrops (see al-Tabari i.2603–4).
49. Based on Syvänne (2004, 165–6 with Appendix 1/Solanchon; Appendix 2/Mamma/Taginae/Mons Lactarius/Hippis River/Casulinus River; Appendix 3/Peter 594); *STR* 11.1.64–7; Procopius, *Wars* 4.11.14–56, esp.50ff., 6.1.4–10 ; Theophylact 2.18.7ff (esp. 2.18.11), 7.2.1–9; Syvänne *MHLR* 6–7 (indexes for the battles mentioned in Syvänne 2004). The dismounting of cavalry was obviously an old method. We find this e.g. in Frontinus (*Stratagems* 2.3.23). The requirement remained the same, so we find the 10th-century treatise *De velitatione* (10.49–65) requiring the horsemen to be able to dismount when the enemy had decided to fight where they stood and had formed a rampart out of baggage and animals.
50. Based on Syvänne (2004, 157–9) with some changes.
51. Calculation is based on J. M. Smith, Jr. Ayn, p. 314 note 19. The polo game is divided into 6 '*chukka*' which reflect the demands of all-out effort by the horses. When the modern version of the game was developed, the chukka often resulted in 12–13 minutes of play which was found to be excessive. On the basis of this the chukka was at first reduced to 10 minutes and then later to 7½ minutes. During this time a horse may cover as much as 3 miles at a gallop.

Consequently, the minimum requirement for the entire length of the polo game is at least three horses, but most of the players use more.

52. For example, the cavalry parade in Rai Uno 6.6.2001.
53. Based on Syvänne (2004, 166–8, Appendix 1) with some changes; *STR* 1.3.26ff, 3.5.37–50, 3.5.86ff, 3.12, 6, 7.B.12, 8.B.44–5; *MHLR* vols.1–8.
54. See the indexes in Syvänne (*MHLR* 6–7).
55. See Syvänne (*Aurelian and Probus*, 78–80; *MHLR 6*, 51–3, 237–9).
56. *STR* 7.B.12. See for example: Procopius, *Wars* 4.11.50ff; Theophylact 7.2.1–9; indexes in *MHLR* vols.1–8, esp. vols. 6–8.
57. Based on Syvänne (2004, 168–9) with some changes.
58. *STR* 7.B.11; Procopius, *Wars* 5.29.16–6.1.1.
59. Based on Syvänne (2004, 185–9) with some changes. See also the chapter on Infantry Tactics.
60. Quote from Syvänne (2004, 185): 'In modern day context, the riot police with their well-trained horses exploit this particular psychological tendency among the masses. In general, modern day rioters are not trained to face charging or even slowly-moving horses. On the other hand, the horses are well-trained to face thrown stones, noise and unsteady surfaces as I was privileged to observe during my compulsory stay as a conscript in the Finnish army as the conscripts were used in the training of the police riot squads.'
61. Most of these variants can be found in Agathias, 1.21.4–6.
62. A quote in Syvänne (2004, 185), based on Archer (144–5), provides a summary of the different qualities of different arms of service in Europe in about 1200 AD: 'His conclusions are: Heavy cavalry had attack superiority against light infantry; Heavy infantry defended successfully against heavy cavalry; Light cavalry (bows) had attack superiority against heavy cavalry and heavy infantry; Light infantry (bows/javelins) defended successfully against light cavalry and had attack superiority against heavy infantry. These conclusions are roughly accurate in the open terrain, but the combination of the arms in a single unit like the East Roman multipurpose cavalry with ability to dismount changes the equation as did changes in terrain. It is clear that the best effectiveness of the weapons systems could only be achieved through a combination of the different arms in a single army or alternatively in multipurpose troops. The Romans used both systems and therefore achieved the greatest effectiveness from the weapons systems'.
63. See in particular Syvänne, *MHLR* volumes 4–8.
64. Syvänne (*Aurelian and Probus*, 99–103).
65. Nazarius 22.4, 23.4 with Syvänne (*MHLR 1*, 246–8, 324–5).
66. Syvänne (*Septimius Severus*, 40–2; *Gallienus*, 70–5, 142–6; *MHLR 8*, 204ff.)
67. Syvänne, *MHLR* vols.1–8. Good examples of the Late-Roman use of cavalry for frontal charges against infantry: Syvänne (*MHLR 1*, 246–8; *MHLR 2*, 244–8; *MHLR 5*, 140–2, 148–61, 168–72, 232–46, Appendix; *MHLR 6*, 351; *MHLR 8*, 140ff.).
68. Based on Syvänne (2004, 169–85) with changes and additions.
69. During the tenth century the Roman cavalry waited until the enemy arrows began to drop on the Byzantine formation to give them the distance at which to begin firing. See: *Praecepta militaria* 61.13, p.128.192–6. This system was not used during this era because the Romans also faced enemies that did not employ horse archers. It is because of this that the *Strategikon* simply instructed the horsemen to start firing when they reached the right distance.
70. It should be stressed that the Persians also had stiffer bows and archers who specialized in powerful shots with various kinds of thumb-locks, but this refers to the standard differences in the use of archery in battles. The Persians concentrated on the volume of bowshots instead of the power of shots.
71. Training: *STR* 1.1; *PST* 44–47; Enemies/Persians. In *The Age of Hippotoxotai* I assumed that the length of the bowshot in the *Strategikon* was about half of the tenth-century distance, so that the range in which the Roman cavalry started its charge by first tightening the files

followed up with tightening of the depth – in other words three to four bowshots – would have meant the distance in which the Romans were able to shoot three or four arrows per man. As noted, I consider this now to be unlikely, because other evidence suggests that the bowshot meant the maximum distance, which was the same as in the tenth century. The tightening of the formation was started at the distance of three- to four-bowshots, not because this was the distance in which the Roman cavalry could shoot that many arrows, but because tightening the array safely required time to perform. This obviously changes the calculation regarding the number of arrows shot before contact. The Persian shower archery: assuming that 5 arrows were shot for each Roman style shot = 10 arrows. The Persians supposedly shot in fives (5 arrows = 5 per each Roman style shot).

72. *STR* 2.6.9–16.
73. John Lydus, *Magistrates*, pp.20–22,
74. *STR* 3.5; *PST* 36; Nazarius (*Pan.* 4.29.5); Julian (*Or.* 2.57B-D); Ammianus 25.3.3; Claudian (*The Gothic War* 356); Corippus (e.g. *Ioh.* 2.224–34, note how he describes the cavalry equipped also with *tela, spicula*; 4.487–8); Procopius, *Wars* 1.1.12–3; Agathias 2.8.1; Syvänne, *MHLR* series. For the shields in general, see Kolias 88–131.
75. The word used to describe the shape of the shield is *epimêkês*, which usually means long/length. However, Chatzelis and Harris translate (p.54) it as oblong, as do Kolias and Haldon. Even if it is clear that this says four and a half *spithamai*, I would still suggest that four *spithamai* refers to the width of the shield, in this case because it's the width of the file in the *pyknosis* formation, and the description of the use of the shield in combat suggests a large infantry type of shield (*scutum/thureos*). We know that Leo the Wise (the author of the *ST*, see Syvänne, *academia.edu*) meant the royal *spithame* (span), because in the context of infantry he stated that the infantry shield of six *spithamai* was almost as high as man. The *spithame vasilike* was 23.4cm and the *spithame koine* 19.5cm, so it is clear that it was the royal span (140.4cm vs. ca. 117cm).
76. The shield finds from Dura Europus (mid-third cent.) certainly confirm the existence of shields that fulfilled these dimensions. The length of the oval shields varied between 1.07–1.18m and the width between 0.92–0.97m, while the rectangular *scutum* had a width of 0.83m and length 1.02m. See Bishop and Coulston (179–80) with James. However, since it is known that the cavalry (just like infantry) was deployed with the width of ca. 94cm per file, it is quite probable that this was the approximate width of the large cavalry *scutum/clipeus*. Therefore, the horsemen brought their horses about rim-to-rim when the formation was tightened in width and the troopers were using the large *scutum* and not the smaller round shields. However, it should be remembered that on the basis of the narrative sources and works of art most of the cavalry was equipped with round (or during the early period also with sexagonal, hexagonal, oval) shields. This means that the tightening of the array in width was not usually done on the basis of the shields, but on the basis of the training and experience.
77. The shield was clearly important during cavalry combat as a form of protection against arrows. It was because of this that I speculated in *The Age of Hippotoxotai* that one of the reasons for the failure of the French knights to carry home their attack in the Battle of Agincourt in 1415 was the fact that the French knights did not use shields. Despite their protective covered helmets, their eyes were still exposed enough to make them psychologically less willing to continue. The medieval Mamluks also recognized the usefulness of the shield during the approach, but did not consider the shield to be useful in the actual melee when used in conjunction with the lance which in their opinion required the use of two hands to be effective. See *Munyatu'l-Ghuzat,* 64. The *pila* of the Caesarian infantry against the Pompeian cavalry at the Battle of Pharsalus in 48 BC shows the same tendency. The sight of the incoming missile was frightening. The tenth-century Byzantine military treatises *Praecepta militaria* (3.5, p. 36.45–6) and Ouranos's *Taktika* (60.5, p.114.57–8) both noted the ability of the cavalry shields (*skoutarion*) to block enemy arrows. The tenth-century *skoutarion* measured

93.6–1.27m in height and was either round, oval or kite-shaped. We can find support for the usefulness of the shield in cavalry combat also in the *Gododdin of Aneirin* (ca. 600 AD, north Britain). Also of note is the fact that the British cavalry of the period used their shield bosses for shield bashing while mounted. See for example (note that these also usually show the use of the spear for both thrusts and throws, as well as the use of the sword): Aneirin B2.24, 26, 28, 31, 34; Aneirin B1.4, 16–7, 21–2; Aneirin A.3, 14, 19, 23, 29, 33, 36, 39, 50, 64, 71–2, 77, 82, 84. Corippus also confirms the importance of the shield with the statement that when the shield was broken, the trooper had to flee (Corippus, *Ioh.*, 5.321–5).

78. For the development of Roman mailed cavalry in general see: Eadie; Bivar; Coulston; Mielczarek (48–50, 84–5). See also Syvänne (*Gordian III and Philip the Arab, Gallienus*) for the Roman adoption of the Persian *clibanarii* type of cavalry and the increased importance of cavalry after that. The question of how and when the Romans adopted the *cataphractarii* and *clibanarii* types of cavalry has divided the academic community, as has the question of differences between these two types of troops. There is no easy answer to this because some of the ancient authors equated the two types of cavalry while others did not.
79. Cavalry charge vs. the Persians: *STR* 11.1.54ff. The Goth Aligern shot an arrow through the shield, breastplate, and body of Palladius: Agathias 1.9.4. See also: Procopius, *Wars* 6.17.14–8, 7.26.7–8; Syvänne (2004, Appendix 1).
80. Vibrating/brandishing/balancing of spears, lances and javelins during approach: Corippus, *Ioh.*, 4.487, 5.335, 6.722, 7.430, 7.438, 8.382–3, 8.390, 8.544.
81. On the basis of one old Lombard poem, *the Hildebrandslied* (ca. AD 650), one of the versions of the cavalry charge used by the Carolingian Franks, Lombards and Avars was to advance into the range of thrown spears, to halt, to place the shield on the back, to throw the spear, to draw the sword and place the shield in front and then charge the enemy in *cuneus* (wedge) formation (i.e. point formed against the spot where an opening appeared). See: Bachrach (2001, 196–199) and Hilda Davidson (1998, p.195). I would suggest one change to this, which is that the shield was never placed on the back, but kept on the left hand or arm as depicted in period illustrated manuscripts. This is how the Romans would also have charged into the gap appearing in front of them.
82. Syvänne, *The Age of Hippotoxotai*: 'The wielding of the lance with two hands allowed a great variety of shorter and wider grips, the positioning of the hand near or on the butt of the lance in various different ways (behind the belt, next to the belt, in front of the belt and so forth) and similarly in various different positions by the other hand, the use of parries from high or below, the transferral of the lance from left to right for feinting purposes during the approach and so forth'. For the use of the lance with two-hands for feints, parries and thrusts, see the 14th-century Mamluk-Kipchak military treatise *Munyat'l-Ghuzat*, 56–71.
83. Targeting of the horses in Corippus, *Iohannidos*: 5.114–5, 5.311–14, 6.672–5, 8.551–9, 611–3; The instructions in *Peri strategikes* (36) for the infantry to target the horses also points to the same direction. We find similar instructions in modern-era cavalry treatises, for which see e.g. the works of Wallhausen.
84. As I noted in my *The Age of Hippotoxotai* (2004, 175–7), in order to fully understand how the cavalry troopers fought in hand-to-hand combat it is necessary to compare the period sources with modern observations. The reason for this is that, excepting very few cases, the period sources do not contain empirical observations of the fighting systems in action, whereas the later manuals give detailed accounts of the actions and give empirical observations of the differences. The eighteenth- and nineteenth-century observations by European observers regarding the use of the lancers in cavalry combat shows that the lance was basically a weapon of shock, which was most useful against fresh inexperienced cavalry units, whereas against seasoned veterans armed with only swords, the lance could be a hindrance in close-combat where the easier manoeuvrability of the sword was found more useful. The lance was at its best against infantry. Regardless, the lance was still considered very valuable, the best proof

of which is the reappearance of lancers in Western Europe during the Napoleonic Wars and the subsequent proliferation of lancers. For this, see Rothenberg, 72–73; Duffy, 227–228. Regardless, it is clear that the sword was better in crowded conditions. See for example *Munyat'l-Ghuzat*, 67. Note also how in 946, the young Svjatoslav, prince of the Rus, threw his spear between the ears of his horse (*Nestor's Chronicle* AM 6454, AD 946). According to Birkenmeier's estimate (2002, 220ff. and after him G.T. Dennis, 2001, 6) which he based on period sources, in the battles of the Komnenian era c.10 per cent of the wounds (removed soldier from battle) were caused by archery and c.20 per cent of the casualties resulted from lance attacks in the first few moments of the armed clash, while the subsequent melee with swords produced 40 per cent of all wounds. In other words, the sword is claimed to have caused most of the fatalities. Horses suffered 6 per cent of the wounds while other weapons caused the rest of the wounds. Lances caused most of the serious wounds to the torso, while the swords caused most of the head injuries. It is possible that the numbers would be similar for this era, but as I stated in *The Age of Hippotoxotai* I am not inclined to give exact numbers for the amount of casualties caused by different types of weapons, because it is clear that the numbers would have varied greatly according to the situation, the armament, and the amount and quality of armour carried. The narrative sources do not give us an exact rundown of each different cause of death, but give us only such details that they considered worth telling in their context.

85. See Syvänne (*MHLR 6*, 133).
86. Arrian (*Tactics* 17.3–5), Onasander (16), Aelian (Matthew ed. 44); *Byzantine Interpolation of Aelian* (Devine ed. 44.1–2 and Dain ed. I1–2). The tendency of deeper formation to puncture a single cavalry rank was also noted by the tenth-century treatise of Ouranus (68, Foucault TM 1973).
87. Passwords used in battle: Theophylact 6.10.4–5. Officer alone during pursuit: Appendix 1/ Rome 3.
88. For third century cavalry and the early adoption of stirrups, see Syvänne (*Gordian III and Philip the Arab, Gallienus, Aurelian and Probus)*. For cavalry combat during the Late Roman era, see *MHLR* vols.1–2.
89. Procopius, *Wars* 5.18.6–11; Agathias 1.21.5 (trained horse of Narses); Theophanes AM 6118 also names the horse (Dorkon) of the emperor Heraclius. It was not only the horse of Alexander the Great, Boukefalos, who was famous in antiquity, even if he is the best known of these. The hippodromes undoubtedly had their own celebrity horses too.
90. The proto-Sabres used by some of the nomadic peoples and the scramasaxes obviously had a single edge, but these were rare on the Roman cavalry battlefields.
91. For example, in the short section provided by Corippus (*Ioh.* 5.104–125) there are four cuts and only one thrust or stabbing technique. This was the result of the sword's structure. It dictated the way in which it was used. Kolias G. Taxiarchis has collected the evidence concerning swords and their use and has reached a similar conclusion, pp.133–161, esp. 159–161. I disagree with Kolias only on one point. He claims that the sword was not handy due to its weight and size. This is not true. In the hands of a strong trained man, the sword was the perfect tool for the things described. And we should not forget the fact that the use of armour and shields required the use of heavier swords for cutting. The rapiers are not particularly good for that.
92. The observation on cavalry sword technique of the renaissance masters in Anglo, 253–270. Most of these are applicable to any age using similar weapons while mounted.
93. The observation on the cavalry sword technique of the renaissance masters in Anglo, 253–270. Most of these are applicable to any age using similar weapons while mounted. Wrestling in the widest possible sense on horseback: 1) Persian rock relief: Gall von, Hubertus, figures between pp. 35/36; 2) Renaissance figure: Wallhausen, 97 figs 2–3, 106 fig 3–4, 5–7; 3)

Xenophon, *Peri hippikes* (On the Art of Horsemanship) 8.11; 4) Corippus 7.463–8; Tabari i.2297, 2423.

94. Partial quote from Syvänne (2004, 178–9); Procopius, *Wars* 3.19.7; *STR* 7.A.11.
95. Enemy killed John as he was attempting to remount his horse: Procopius, *Wars* 4.24.13–4. Other examples mention the bodyguards giving a new mount to their commander (for example: Procopius, *Wars* 4.21.27–8; Syvänne (2004, Appendix 1/Sisarbanon; *MHLR 7*, 233; *MHLR 8*, 246). Halsall (211–2) has made the interesting observation that in one instance the undated Anglo-Saxon cemetery of Eccles in Kent has two types of skeletons, men killed by single blows to the front of the skull and others who were cut in the head from behind that had received multiple wounds. He suggests that the former had received their fatal head wounds in close combat whilst the latter group had received theirs in pursuit.
96. The *Georgian Chronicles* describe the heroic exploits of Vaxtang, who was the grandson of Mirian who lived during the reign of Constantine the Great. His reign has been misdated by the Georgian Chronicles by about 100 years. In truth he lived at the turn of the fifth century. See Syvänne (*MHLR* vols.2–3, the index).
97. *Khuzistan Chronicle/Anon. Guidi* (p.16): one Roman trooper tried to capture the king of Persia, Chosroes II with a lasso (also in Syvänne *MHLR 8*, 51). See also Malalas 18.21: the Huns captured two Roman generals with lassoes.
98. See Syvänne *MHLR* series with Agathias 5.19.4, 5.19.7.
99. John Masson Smith, 1984, 307–345. The ca. 300m being considered to be the maximum range, whereas the maximum of effective fire has been put variously at 150m or even as high as c. 160–201m. For this see the *Saracen Archery* (pp.71–84) and the discussion by John Masson Smith Jr., 1997, 249–266; and 1984 Ayn Jalut.
100. According to Smith Jr., the Mongols used a wave method that meant that each rank fired their bows at great intervals before turning back head to tail to give room for the next wave, rather than the method in which the troopers fired in formation. The latter method was in the opinion of Smith Jr. dangerous, wasteful and ineffectual. One can exclude the use of the wave method by the Romans and also by the Mongols. For this, see Syvänne 2012. The Mongols attacked as units in which the front ranks consisted of the armoured men, so when they retreated each soldier turned about so that the better armoured remained between the enemy and the not-so-well-armoured Mongols. As far as danger is concerned, according to Smith it resulted from the possibility of the accidental loose of an arrow if the archer behind drew his bow hands relatively close together at waist level and began simultaneously to push against the bow, pull on the string and raise his hands upwards to the shooting level, as a consequence of which the bow at some point in time would have been pointed at the man or horse in front. The other two known methods of drawing a bow avoid this danger. In the first, the bow-arm was almost fully extended in the intended direction before drawing the string and in the other the arms were raised and the draw happened as the archer lowered his hands to the correct elevation. It is clear that the Romans used the latter two, but when one takes into account the nomad influences on the Romans, the likeliest method is the lowering of the bow to the intended level. See: *Saracen Archery*, 57.

Chapter 7

1. Based on Syvänne (2004, 190–2) with updates and changes.

Chapter 8

1. The following discussion is based on Syvänne (2004, 201–7), but I have updated it with additional information.
2. *STR* 12.B.16 describes open and compact formations as well as the attack form of the *foulkon*. *STR* 12.A.7.49ff describes the defensive *foulkon* used against cavalry. *STR* 12.B.20 describes the irregular formation used when deployed in wooded or rough terrain.

3. *Ibid.* with *ST* (43.6–7). We can find roughly the same information already in Asclepiodotus (4) and Aelian (Matthew ed. 11). According to them, each footman occupied 4 *pêcheis* both in width and depth in open order (*araios*), two *pêcheis* in the *puknosis*-order, and one *pêchus* in the *sunaspismos*-order. The *pêchus* of their era has been interpreted differently by different scholars. The different Greek city states had different values for the *pêchus*: Attic 44.4cm; Alexandrian 45.12cm; Doric 48.72cm. Modern historians have also suggested their own versions for the *pêchus*, with the Loeb edition and translation of Asclepiodotus suggesting that it equalled 18 inches (45.72cm, p.267). In my opinion the likeliest is the Doric/Peloponnesian 48.72cm because it is closest to the later Romano-Byzantine figures. This means that the Macedonian phalangite when equipped with the *pelte*-shield occupied ca. 49cm in width when using the *sunaspismos*-order, 97.44cm (Roman version ca. 94cm) in the *pyknosis*-order and ca. 195cm (Roman version 192cm) in the open order. The *sunaspismos* version is different because the Macedonians used the smaller *pelte*-shield. However, in practice there was no difference because when the Romans used the smaller shield variants, for example the *parma/clipeus*, the width of their files also diminished.
4. *STR* 12.B.17.24–33.
5. Based on Syvänne, 2004, 202.
6. *STR* 12.B.16.17–29. The use of *boukoulon* in the *Strategikon* for the shield is somewhat problematic since the word could mean the whole shield as well as the boss of the shield (see Kolias, 1988, 100–2). Consequently, when *boukolon* was to be brought close to touching *boukoulon*, this could either mean that one shield was rested on top of another in width, or that the rims of the shields touched each other (rim-to-rim order). I have here chosen to interpret *boukoulon* as a shield because this corresponds with the known unit orders as practiced since Macedonian times and which continued to be used in the tenth century (see e.g. *Sylloge Tacticorum* 43.6–7). However, I translate *boukolon* differently in the context of the *foulkon* vs. infantry. In that case, I translate it as boss. The reason for this is that the *foulkon* was clearly a tighter formation than the order assumed with the *iunge/iougge* because it was to be assumed when the soldiers lacked armour and shin-guards. It is unlikely to be a coincidence that in the context of assuming the *foulkon*, Maurice uses both words *skoutarion* and *boukolon* simultaneously to make it clear that in this case the *boukolon* meant the shield boss and not the shield. *STR*. 12.B.16.22–25: '*Paraggellei: iougge. Kai puknoumenoi / sfiggontai pros ton meson topon kata bathos kai mêkos tosouton, ina men emprosthen tetagmenoi ek plagiou eis ta boukoula allêlois eggizousin, oi de opisthen kata nôtou allêlois scedon kekollêtai*'. Ernst Gamillscheg (Str. ed. Dennis 1981, 441, 443) translates *boukoulon* consistently as shield boss (schildbuckel/buckel) and therefore interprets this order as rim-to-boss order, while Philip Rance (2004b, 270–5) suspects that we should not take the text literally to mean rim-to-boss formation, even in this case. I disagree with Rance. Dennis (1984, 146) avoided the whole problem in his translation. There also exists pictorial evidence for the use of the rim-to-rim formation: Citadel of Faith, a Coptic wood carving, fifth to seventh centuries (Nicolle, 1996, 57).
7. Based on Syvänne, 2004, 202–4. For earlier instances of the Roman use of the *testudo*, see basically any of my publications. The Romans used all of the variants of the *testudo* or interlocking of the shields throughout their existence.
8. This analysis builds upon the information included in my doctoral dissertation (2004, 201–7) and Rance's (2004b) analysis of the meaning of *fulcum/foulkon*, so that I correct Rance's misunderstandings concerning the file depths of both the advancing *foulkon* and the *foulkon* used against cavalry and add three other close-order combat variants (two variants in *Peri strategikes*, one of which is the Macedonian *sunaspismos/suskouton*, and the other the all-men-kneeling *testudo*) not noted by Rance or misunderstood by Rance to represent the *foulkon* against cavalry. However, Rance is correct about the fact that all of the basic unit order variants used by the Late Romans were old and traditional and had not been introduced by Germanic soldiers during the Late Roman period. He has demonstrated this conclusively

with the earlier examples in his article, but with the caveat that he has misunderstood the actual organization and structure of the unit orders that he has used as evidence for this. It is because of this that the following corrects his reconstructions of the two *foulka* in the *Strategikon* (advancing *foulkon* vs. infantry and stationary *foulkon* against cavalry) and brings to light other types of shield interlocking orders that he has failed to recognize.

9. If this educated guess is correct, then the advance of Paulus and Ansilus in front of the array does not tell us anything about the tightness of the spear-phalanx, but if they advanced in front of the spear phalanx in order to be able to use their bows better it suggests that the width of the multipurpose infantry phalanx that Syrianus envisaged for the infantry when facing the enemy cavalry was not necessarily the shield-interlocking phalanx with file widths of ca. 62cm but rather the regular close-order *pyknosis* with file widths of ca. 94cm as in the standard *foulkon/testudo* against cavalry depicted by both Arrian and Maurice. However, the latter option is less likely because Syrianus gives only one unit order for his infantry phalanx and does not state that the infantry when employing bows and spears against the enemy was to use a wider formation than normally. Furthermore, it is possible that the two men just advanced in front of the array because they had broken their spears in the previous action with the Gothic cavalry and because they wanted to demonstrate their personal bravery in an effort to gain promotion – which they got.
10. Based on Syvänne (2004, 202–4); *MHLR* series.
11. The resting of shield-bosses on shield-bosses was obviously impossible, and the same was also true of the resting of the shield on top of the shield-boss in front because the shields could be brought only so far forward as the presence of the hand behind each shield-boss allowed.
12. *STR* 12.B.16.30–38; 12.B.20.6–10; an example of the men wearing the wrong type of equipment in Syvänne, *MHLR 7*, 306–7. In this case the use of different words by Maurice implies that *skoutarion* means the whole shield and *boukoulon* the boss of the shield. The order simply sacrificed personal freedom of movement for safety and group effort. However, the *foulkon* could still be opened up in preparation to melee by the use of the *largia ad ambas partes* command. *STR* 12.B.16.33–8: '*Paraggellei: ad foulkon. Kai tôn emprosthen kata to metôpon tetagmenôn puknountôn ta skoutaria autôn mechri tou eggizein tois boukolois kataskepontes prospeplasmenôn tas gasteras autôn mechri tês knêmes, oi parastestôtes autois opithen uperanechontes ta skoutaria autôn kai anapauontes ei ta boukoula tôn emprosthen skepousi ta stêthê kai tas opseis autôn kai outôs sumballousin*'. Rance (2004b, 270–5) interprets this to mean that only the two front ranks formed up the *foulkon*, but this is unlikely in a situation in which the soldiers assumed the protective unit order because the soldiers lacked armour. It is far likelier that the command *ad fulco/ad foulkon* was meant for all of the men in the same advancing unit. Note that, for example, Arrian (*Tactica*, 11) equated the Roman *testudo* with the Macedonian *synaspismos* formation, and it subsequently did at the turn of the eleventh-century Ouranos (67. 1–4, p. 304–305).
13. For an example of this, see Dio 49.29–30, which is also quoted here later. He describes the Romans kneeling under Mark Antony when surrounded by the Parthian mounted archers and then rising up when the enemy approached close enough. One may imagine that the Late Romans could also resort to this tactic when there was a need to rest before coming to grips with the enemy. See also: Claudian (*Hon VI*, 612ff.); Syvänne (*MHLR 3*, 131); Ammianus 16.36–7.
14. Ammianus 24.15.
15. *STR* 12.B.16.46–55.
16. The use of the shield roof by the rear ranks during the melee: *STR* 12.B.16.49–51.
17. For the Germanic shield castle, see e.g. Caesar, *Gallic War* 1.52.
18. Based on Syvänne, 2004, 205.
19. If one interprets the word *akia* as a file (as Dennis and many others do), then the width of the order was kept wider than in the regular *foulkon* to facilitate the use of spears. However, in this

case it is more likely that *akia* should be equated with *acies* (battle line). See Mihaescu (1969, 495: *akia* = *acies*). However, the change in meaning does not change the interpretation of the text, because it is clear that the width of the formation was indeed wider, roughly rim-to-rim to enable the soldiers to point their spears against the approacing horses.

20. E.g.: Corippus, *Ioh.*, 8.446–52; Menander fr. 12.3; Syvänne, *MHLR 6*, 70.; Arrian, *Acies*, (16–18, 21 25–29).
21. Syvänne, 2004, 205.
22. Based on Syvänne, 2004, 206.
23. Theophylact 2.18.7–25, esp. 2.18.11–12 and 2.18.18; Livy 44.9; Ammianus 26.8.9; Syvänne, *MHLR 7*, 220
24. Based on Syvänne, 2004, 206.
25. *STR* 12.B.20.
26. For a fuller analysis, see Syvänne, Septimius Severus based on analysis of the *Historia Augusta* (Pert. 11.1) and Dio (74.2), in conjunction with the other facts known about the Roman armed forces, e.g. the number of bodyguards and the number of light infantry in the legions (ca. quarter to a third of the force, i.e. here 100 men.)
27. Vegetius puts the strength of the cohort at 555 men, while Modestus puts it at 560 men (includes the centurions). See Syvänne (*Britain in the Age of Arthur; Aurelian and Probus; Septimius Severus*). Notably Dio (86.12.5) mentions the 550-man group (a cohort) for the reign of Septimius Severus, and it conforms to the texts of Modestus, Vegetius and the Viking *svinfylking* which consisted of two 550-man groups for a total of 1,110 men.
28. Based on Syvänne, 2004, 206–7.

Chapter 9

1. For an analysis of this and the following, see Syvänne (*Septimius Severus*, especially Appendix 1; *Caracalla, Gordian III and Philip the Arab, Gallienus*, and *Aurelian and Probus*). The analysis of the new type of legion in Syvänne (*Septimius Severus, Aurelian and Probus*) is based on Modestus 12–4 and Vegetius 3.14–7.
2. It is the inclusion of the *carroballistae* in this context by Vegetius (3.14) that suggests the likelihood that he has understood the *ballistarii* in the Late Roman sense as artillerymen and not as crossbowmen as these were originally.
3. Syvänne, *MHLR 1*, 346–9, 355; *MHLR 6*, 346–53; 2011a-b. We find the Romans using wedges in like manner in the text of Geoffrey of Monmouth, which may preserve a muddled account of the last stages of Roman Gaul. The reason for this conclusion is that the text includes information that explains the blank spots in the other defective sources, and it is known that Geoffrey based his account on an older chronicle. Furthermore, the tactics Geoffrey describe fit only the time period from the 360s until the end of the fifth century. For this, see Syvänne, *Britain in the Age of Arthur*.
4. Vegetius 3.14–20, 3.23, 3.26.
5. The following is based on Syvänne (2004, 193ff.); Chapter 1, Chapter 7.
6. For an analysis, see Syvänne (*MHLR* series) and for the Battle of Ctesiphon in 363, Syvänne, *MHLR 2*, 95–7.
7. For the worsening quality of the Roman infantry, see esp. Syvänne, *MHLR 5*, Appendix.
8. The following discussion is based on Syvänne (*Caracalla*, Appendix 2; *Gordian III and Philip the Arab; Aurelian and Probus*, Appendix 1).
9. This implies that these soldiers also wore armour on their arms (*manicae?*) and probably also on their legs so that they may have looked like the gladiator *crupellarius* or '*arbelas*' or a medieval knight with the addition of spikes.
10. See Syvänne, *MHLR 2*, 95–114; *MHLR 7*, 277–80; *MHLR 8*, 51–3; and the *MHLR* series in general.
11. See Syvänne, *MHLR 6* with Chapter 1.

12. See Syvänne, *MHLR* vols.6–8.
13. The following is based on Syvänne (2004, 193–218) and on some new interpretations of the same material and on Chapters 1 and 7.
14. The tenth-century *Sylloge Tacticorum* (33–4) includes the same, but adds information from earlier, no-longer extant Hellenistic treatises. According to the *STR*, when the Macedonian cavalry was deployed on its own, it was deployed as four divisions (left, left-centre, right-centre, right) so that the flanks consisted of the light cavalry (javelineers and archers) and the centre of heavy cavalry. The flanks were deployed slightly in advance of the centre. The depth of the cavalry formation was four ranks, with the front two ranks consisting of heavy armed and the last two of light-armed. This is in disagreement with the divisional pattern, so it is possible that it concerned only the two centre divisions, or alternatively that it meant all of the divisions, so that in each case the better-protected men were placed in the two front ranks and the archers behind – the last mentioned is probably the likeliest. The flank divisions used the open order (187.4cm per file), while the two centre divisions were deployed so that each file in close order occupied ca.94cm. This system is clearly the precursor of the Roman *koursores* and *defensores* system.
15. E.g. in Arrian (25) the single lateral phalanx formation was lengthened to outflank the enemy or to prevent the enemy doing the same, or alternatively the light-armed and cavalry were used for this. See also Aelian 29 and Asclepiodotus 6.
16. The analysis of this treatise is based on its analysis in Syvänne, *MHLR 6*, Appendix, 367–70.
17. See Syvänne, *MHLR* vols.1–8; and for Ctesiphon (*MHLR 2*, 95–7); Modares's attack in 379 (*MHLR 2*, 211); Dara in 530 (*MHLR 6*, 56–63); Taginae in 552 (*MHLR 6*, 340–4).
18. Asclepiodotus 8–10; Aelian 33.1–6 (Matthew ed. 33).
19. Aelian (Matthew ed. 44); *Byzantine Interpolation of Aelian* (Dain I1–2; Devine 44.1–2). Notably, Arrian (*Tactika*, 17.) saw only one use for the deep cavalry column formation, which was precisely the breaking through of the enemy array.
20. Syvänne, *MHLR* vols.1–8 with Chapter 1. For the wedge, see especially Aelian (Matthew ed. 47); *Byzantine Interpolation of Aelian* (Dain L1–5; Devine 47.1–5). For the use of the club-bearers and mace-bearers in open formation (sent in front of the phalanx so that they may have sometimes had behind the wedge sent in front of the phalanx), see esp. Syvänne, (2004, 190); *Caracalla* (286–7); *Aurelian and Probus* (vii, 16, 37, 106–9, 114); *MHLR 1* (16, 22, 27, 248, 292, 303, 314, 349, 405), *MHLR 2* (10, 194, 311); *MHLR 7* (317).
21. However, it should be remembered that Maurice was not strict in his instructions in the cavalry section either. The *banda* of the *Optimates* were allowed to possess more than 400 horsemen. Regardless, it is still clear that Maurice allowed greater flexibility for the infantry *tagmata* than was the case for the cavalry *tagmata*. Notably, we find somewhat similar differentation between the sizes of the infantry and cavalry *tagmata* in the tenth-century *Sylloge Tacticorum* (35.4), where the cavalry *bandon/allagion* had 50 to 350 or 400 horsemen whereas the infantry *tagma/bandon* had 200 to 400 men.
22. This and the following is based mainly on the *STR* 12.B.1–20, 23; Syvänne (2004, 194ff.), *MHLR* vols.1–8.
23. It is not surprising to find out that in the tenth century the *turma* (*meros*) could have 9,000 men (*ST* 35.2). In Maurice's *Strategikon* the infantry *meros* does not even have maximum size. This took into account the fact that the division could be larger than the earlier maximums based on the Hellenistic military theory of 16,384 men phalanx with its 8,192 men light-infantry addition.
24. Ancient Hellenistic military theory also included the version in which the baggage train was posted behind the lateral phalanx, but in this case the baggage train was posted behind to protect it from the enemy rather than to serve as a form of barricade against the enemy. In other words, the Romans used it differently. Good examples of the Roman use of the baggage train and mobile artillery-carts (*carroballistae* etc.) can be found from the campaigns of

Germanicus onwards, so that the artillery-carts are prominently displayed in the Column of Trajan and mentioned for example in the Battles of Issus in 194 (Syvänne, *Septimius Severus*, 110–6) and Cellas Vatari (Syvänne, 2004, 465–6; *MHLR 6*, 183–5).

25. It was because of this that the tightening of the formation in the *Strategikon* was always performed towards the centre where the standard was, so that there was no longer variation left, right and centre in the tightening of the array. It is unfortunate that we do not know how the Romans performed the tightening of the formation before Maurice, because, as already noted, Arrian and Syrianus do not state how exactly the formation was tightened – they only mention that the tightening of the array was one of the variant unit orders.
26. Syvänne (2004, 196–7), *MHLR* vols.1–8 (esp. vols.6–8); Procopius, *Wars* 1.18.6, 4.17.4; Theophylact 2.10.8–12.8, 6.6.3, 7.3.8. The use of the separate commanders for infantry and cavalry can also be seen in the titles used by the Late Roman generals: *magister peditum* and *magister equitum*; *comes domesticorum equitum* and *comes domesticorum peditum*. Their original purpose reflected their duties in the battle formation when the officers in question accompanied the emperor and his personal field army into the battle.
27. The selection process in the elementary schools of choosing members of a soccer team by shouting one name at a time as done by the chosen team captains shows how even a child can detect those with better abilities. Everyone knows those whom they want and do not want. In the armed forces the soldiers had even keener eyes for talent because they were all trained martial artists able to separate the good fighters from the mediocre ones.
28. Liddell and Scott p.257): one of the meanings for the *bathus* was extended, long and broad. Another possible interpretation would be that the *psiloi* were placed in the middle ranks of the *acies*.
29. This and following is based on Syvänne (2004, 194ff.).
30. *STR* 12.B.8–9, 12.B.16.56–73, 12.B.17–8, 12.B.22.95ff.; Syvänne (2004, 194ff.). The throwing of caltrops to keep the enemy cavalry at bay clearly demanded that there were men posted behind the phalanx in reserve position, therefore the reference to the use of caltrops against outflanking horsemen in the *Peri strategikes* 32.53–4 probably refers to a similar usage as in the *Strategikon*.
31. For early use of the field artillery for concentrated salvoes, see Syvänne, 'Campaigns of Germanicus 13–16 AD', available online at academia.edu.
32. *STR* 12.B.22.95–123.
33. For the various different variants, see Syvänne, *MHLR* vols.1–8. For the use of rough terrain against cavalry etc. during and after Maurice, see e.g. *MHLR 7*, 250; *MHLR 8*, 82–6, 168–9, 173, 177–9.
34. According to *Munyatu'l-Ghuzat* (a fourteenth-century Mamluk-Kitchak military treatise, page 76) the flight-range of a bolt/arrow shot from the arrow-guide was 1,000 arshins (c. 68cm) = 680m.
35. *STR* 12.B.16.
36. *STR* 12.B.16. 97–105 with 12.B.17.34ff.
37. The set of commands differed from the ones used by the cavalry because in cavalry the command '*Iunge*' meant the closing of the ranks from the rear towards the front and was therefore preceded by the command '*Ad latus stringe, ad dekarchas, ad pentarchas!*'. The infantry apparently used only one command, '*Iunge!*' for both.
38. For this, see Syvänne, *MHLR 6*, Appendix; with Chapter 1 *De politica dialogus*.
39. For the *PST* and *PE*, see Chapter 1.
40. See Syvänne, *Britain in the Age of Arthur*, 21 (length of the shafts in the Nydam bog finds after Pollington 131).
41. *STR* 12.B.16.8–16, 12.B.1735ff. Aelian, *Taktika* (Aelian, Devine, 40.1–4, 41.1–3, 42.1).
42. *STR* 12.B.16.74–81; *Hermeneia* 61–2.
43. *STR* 12.B.17.56ff, esp. 68–71.

44. *STR* 12.B.16.91–6; *PST* 32.42–54; *Hermeneia* 61; Aelian (Devine, 29.4–5, Köchly ed. 29.6–7); Arrian 25. The instructions in the *Strategikon* were the standard operating procedure because generals usually avoided making their lines too thin, but there were exceptions. At least Conon (in 538) formed a thin phalanx (probably a cavalry phalanx, but one cannot know for certain) 5 stades in front of Ancon and paid for his mistake when the Goths attacked. See: Procopius, *Wars* 6.13.5–15. The importance of having enough room to wield the swords is best exemplified by Julius Caesar's similar order to his legionaries at the Battle of Sabis (57 BC) when these had crowded upon each other. See: Caesar, *Gallic War* 2.25 with Syvänne, 2009a.
45. It is noteworthy that contrary to the military principle, this command does not have the direction first and then the command. It seems likely that the command was so short that there was no possibility of misunderstanding. It should also be noted that the first word, *depone*, and the words *dextra* and *senestra* for the direction in this command separated the wheeling manoeuvre from the other commands.
46. *STR* 12.B.16.87–90, 12.B.18.17–21; *PST* 23, 32.50–4, 35.31–34; Aelian (Devine ed. 25.5–9/ Köchly, 81–3); Arrian, 21; Urbicius, *Taktika* 6; *Hermeneia* 34–7.
47. *STR* 12.B.16.56–73; *PST* 32.55–65.
48. *STR* 12.B.16.81–86; *PST* 31.50–51. The *PST* (31.49–51) has a different meaning for *amfistomos falagx* (*amfistomos falagx* meant a column formation in which the leaders were on both flanks; *antistomos* meant a line in which the leaders were in front and behind). In the *PE* (18.8) the men behind the third rank held their spears upright so that they could form a two-faced phalanx if necessary. This is undoubtedly true for all unit manoeuvres requiring turning.
49. See Syvänne, *MHLR* vols.1–2.
50. *STR* 12.B.17.65ff. See *MHLR 1*, Battle of Argentorate/Strasbourg AD 357.
51. Vegetius 3.20, 26; *PST* 31–32, esp.32.72–8; *Hermeneia*, 43, 60–2.
52. Based on *DRB* 9, 15, 19; *STR* 9.4, 11.4, 12.A.6, 12.B.20–22; Vegetius 3.6, 3.22; Syvänne (2004, 96–113 with chapter 6.1.8; *MHLR 2*, 169–73; *MHLR 6*, 13–8; *MHLR 7*, 14–15, 21–3; *MHLR 8*, 12–3; *Septimius Severus*, 207; *Gordian III and Philip the Arab*, 94–5;2009a; 2011a-b. The use of lightly-equipped troops in difficult terrain was not new in Roman history. By simply leaving out the heavy equipment one could lighten the forces. See e.g.: John Lydus 1.46; Zosimus 4.25.3; Eunapius fr.45.1. Note also Caesar, *Civil War* 1.44, 8.8.
53. *STR* 12.B.20.17–8: '*Ean de suneisi kaballarioi ê touldos…*'. The statement 'if cavalry and baggage train are present' means that those were not always accompanying infantry armies. Notably we find the Romans using separate infantry and cavalry armies precisely during the reign of Maurice (e.g. Syvänne, *MHLR 7*, 244–8) but it should be kept in mind that the Romans had always used separate infantry armies in situations in which there were no cavalry units present (e.g. Syvänne, *MHLR 1*, 381). For the tenth-century version of a pure infantry army, see *Sylloge Tacticorum* 45.
54. Syvänne, *MHLR 2*, 211.
55. For instances of these, see the *MHLR* volumes.
56. For the sixth-century examples, see Syvänne, *MHLR 6*, (Index). For Mundus, see also *MHLR Vol.5*. The guerrilla tactics employed by Sebastianus against the Goths in 377 (*MHLR 2*, 197) stands as another good example of the Romans employing small forces with great effectiveness. See also *MHLR 6*, 140.
57. See all Syvänne, *MHLR* vols and esp. *MHLR* vols. 1 (e.g. 350–2) and 6–7. For an early example of the use of this tactic, see e.g. Syvänne (2011a-b).
58. Based on Syvänne (2004, 218ff.; Chapter 1 in this book and above discussion of formations and tactics; *MHLR* vols.1–8) with *PE* 17.3, 19.
59. *Byzantine Interpolation of Aelian*, Devine ed. 36.8, Dain A8; Onasander 21.5–8; Vegetius 3.20, 26, his second, third, fourth and fifth formations. In Aelian the triple phalanx was used against hollow wedge (*koilembolos*), but it is clear that the same instruction was usable against

the flanks of the *epikampios emprosthia*. The triple phalanx is basically just another *epikampios emprosthia* formation.

60. This is based on the *Strategikon* (12.1.7); *Byzantine Interpolation of Aelian* (Devine ed.36.2–3 = Dain ed. A2–3; Devone ed. 36.5–6; Devine ed. 37.2 = Dain ed. B3; Devine ed. 46.1–2 = Dain ed. K1–2). Aelian (Matthew ed. 30, 35, 46); *Peri strategias* (15, 25, 31–2), *Apparatus Bellicus* and Syvänne (2004, 218–21, 445–7, 469–70, 473–4, 478–80). Other names for the *epikampios* were *tripleuron tagma* and *trifalaggia*. Other sources are mentioned where relevant. See also Asclepiodotus 11.1. 5–8 and Arrian, *Techne Taktike* 26.7 (31K), 28.4–6 (36K). Arrian equates the *hypotaxis* in which the *psiloi* were placed behind the flanks with the *epikampios*. The early-sixth century Urbicius (Förster ed. p.470) has the same information.
61. For Priscus, see *MHLR Vol.7* together with Syvänne (2004, 480).
62. As I noted in the 2004 analysis of the *Strategikon*, it is possible that the text, as it has come down to us, may have confused the mixed (*symmiktos*) formation and the *epikampios opisthia* proper with each other (*STR* 12.A.2, A.7.2–3, A.19–23, A.7.89ff.). The reason for this suspicion is that the sizes given for the flank guards agree with the needs of the mixed formation, but not with the principle given in diagram 12.1.7 for the *epikampios opisthia*. Even the name means rearward-angled formation. However, I would still suggest that the extant version of the *Strategikon* has not confused the two. It only provides the principles together with the typically adequate numbers for the flank guards of the *epikampios opisthia* array. The simplest solution to this problem is that the sizes of the flank guard units were in truth related to the size of the army and situation. For example, both Julius Caesar and Priscus employed an *epikampios opisthia* in which each of the three sides were of equal length. For Julius Caesar's use of the *epikampios opisthia* array ('almost square'), see Syvänne (2009a), and for Priscus's use, see Syvänne (2004, 479–80 with *MHLR* 7, 314–5). For Urbicius, see Förster ed. p.470 (after Arrian) with *MHLR 5*, Appendix. See also Syvänne (2004 445–7, 469–70, 477–80; *MHLR 6*, 304–5, 346–53; *MHLR* 7, 213–6, 307, 313, 315).
63. As noted in my comment to this section in 2004, if taken literally the *Strategikon* states that the leading half of each cavalry line/file (*akia*) pursued the enemy while the second half of each cavalry file followed along in close order with even front. As noted in my comments to the text, if this system had been followed, the front rank of the *defensores* would have consisted of its worst fighters. This is contrary to all that we know of the period practices and therefore impossible. It is therefore clear that the cavalry *acies* (battle formation) was divided into two separate lines for the pursuit, the first of which conducted the pursuit and the other which followed. Each of these lines would have had its best men placed in the front ranks. One may assume that the *banda/tagmata/arithmoi* which were usually assigned as *koursores* or simply the regular first line formed the front half of each '*akia*' in this case. I therefore agree with Mihaescu (1969, 495) that Maurice's *akia* means the *acies* (battle line) in this case. In short, the best way to translate the word *akia* here is to equate it with a single cavalry array consisting of two lines, with the frontline consisting of both *koursores* and *defensores*. Arrian's *Ektaxis* (27–9) also contains similar instruction. According to this text, when the enemy cavalry had been repulsed, the infantry opened its ranks and half of the cavalry units (*lochoi*) pursued while the other half of the units followed in good order to protect the pursuers. The infantry then followed the cavalry. In other words, Arrian's *lochos* is to be equated with the *akia/acies* of Maurice. In contrast to the *Strategikon*, Arrian appears to have meant the use of the Alan Drill Formation for the pursuit of the enemy so that the pursuers actually consisted of the half of the *acies* (four out of the eight *lochoi* making up the *acies/promachos taxis*). For this, see Appendix 1 in my bio of Septimius Severus.
64. *STR* 12 2.7. The formation bears some resemblance to two other formations from different eras. Arrian's formation against the Alans is quite similar, as is the one described by the tenth-century *Apparatus bellicus* (72b/73b-73/74; Zuckerman, 1994, 387). An analysis of the

Apparatus Bellicus is provided immediately after the analysis of the *Strategikon's epikampios opisthia.*

65. For an analysis of Caesar's use, see Syvänne 2009a. For the tactics used by Marcus Aurelius during the Marcomannic Wars, see the forthcoming biography of him by Syvänne, due to be published by *Pen and Sword Books.*
66. *Apparatus Bellicus* (72b/73b-74/75; Zuckerman, 1994, 367–8). Other sources mentioned where relevant.
67. For this, see e.g. Syvänne *MLRH 6*, 341–2.
68. For the marching camp in *Apparatus Bellicus*, see Zuckerman, 1994, 373–81.
69. The unknown author confuses the different terms *hippilarchia*, *rhombus*, *embolon*, *epilarchia*, *hipparchia* and *ilê* with each other, which proves that he was borrowing from some earlier treatise or treatises. For the confusion of the terms, see Zuckerman (1994, 385–9) with Lammert (1951). In the following analysis I will present my interpretation of the evidence which explains (in my opinion) how this confusion has come about.
70. The ability of the enemy to pursue the Roman cavalry makes it clear that the enemy is always considered to be cavalry.
71. The resemblance of the description with the *koursores/defensores* system in the *Strategikon* has already been noted by Lammert (*ekdikêtês* = *defensôr*) in 1951, 366–7.
72. *Byzantine Interpolation of Aelian* (Devine ed. 46.1–2; Dain K1–2); Aelian, Matthew ed. 46 (pp.118–9); *Definitiones/Hermeneia* (Köchly&Rüstow ed. 60, pp.232–3; Montfaucon ed. p.513); *Èlien le Tacticien* (Dain, p.330); *Definitiones* (Dain p.332); *Hermeneia* (Dain, p.338); *Le Corpus Èlianique* (Dain, 339–40); *Recension interpolée d'Élien* (Dain, pp.346–7); *Syntaxis armatorum quadrata* (Dain, 367); Asclepiodotus 11.1; Julian's use of the *kyrte* in *MHLR 1*, 102–3. Aelian's treatise has been variously called *taktikê theôria* (*Tacica theoria, Theory of Tactics*), *Peri Stratêgikôn Taxêon Ellênikôn* (On the Military Arrangements of the Greeks), or *Tactica* (Tactics).
73. The *Definitiones/Hermeneia* also claims that the advantage of the array was that it hid the men posted behind from the enemy. This is obviously true.
74. Based on *STR* 12.A.4; Syvänne (2004, 223–4, 473–4; *MHLR 6, 350–2*); Frontinus, *Stratagems* 2.3.7; *PE* 19.4.
75. Excluding his estimation of army strengths, I am personally more inclined to follow Hans Delbrück's (1.315–35) interpretation of the place of the Carthagian cavalry in the battle array. Delbrûck suggests that Hannibal's cavalry wings covered the infantry wings of the African heavy infantry. This agrees with the diagram in the *Strategikon*. As noted in 2004, the inclusion of this array in the *Strategikon* suggests that it possible that Philippicus, who was particularly interested in the Second Punic War, contributed to the writing of the treatise even if the emperor himself was responsible for the final form.
76. Based on: *Strategikon* 12.2.2; *Peri Strategikes* 35.8–18; Syvänne (2004, 224–6, 462–4, 466–9; 2013c; *MHLR 1*, 246–8; *MHLR 4*, 106–17; *MHLR 6*, 68–70, 276; *MHLR 8*, Appendix 3).
77. Based on previous chapters together with Syvänne (2004, 458–60, 473–4; *MHLR 1*, 346–9, 355; *MHLR 6*, 346–53).
78. Geoffrey of Monmouth refers to the Roman usage of wedges as battle formations in the Battle of Siesia, but even if we accept his version to reflect actual events, his wedge is not the exact equivalent of the wedge as a battle formation. Rather, his wedges are divisional wedges used by one Roman army against the hollow oblongs employed by the Roman army under King Arthur. For an analysis of this, see Syvänne (*Britain in the Age of Arthur*, 147–63). It is of particular note that Geoffrey of Monmouth's account of the Battle of Siesia in 471 uses the right counter tactic against the hollow oblongs. He could not have read this from the Latin sources, because Vegetius does not include this. This information could only have been found in the *Byzantine Interpolation of Aelian*, or indeed from the sources claiming to depict the battles of Arthur. Another point which suggests authenticity of detail is the fact

that the Romans are depicted using several hollow oblongs side-by-side and in depth, which was indeed a unique tactical development that took place after 364 and a tactic which was no longer in use during the sixth century.

79. Based on Chapters 1 and 3.3 together with Syvänne: *MHLR 7* (307); *MHLR 6* (4–5, 52, 90, 283–5, 348); *MHLR 5* (2, 67, 69, 170. 224, 239–41, 256–7); *MHLR 4* (2, 91, 112); *MHLR 3* (17, 19, 86, 121–2, 125, 128, 203, 265); *MHLR 2* (17–21, 83–7, 101, 106–8, 113–4, 136–8, 143–7, 156, 171–2, 189–91, 199, 245–50); *MHLR 1* (18, 27, 314, 405); 2021a (*Gordian and Philip*), 94–5; 2020a (*Aurelian and Probus*), 232–3; 2017b (Parthian Cataphracts vs. Romans); (2017a *Caracalla*, Appendix 2); 2009a, (Battle of Sambre/Sabis) almost a hollow square ('*paene quadrato agmine*'); 2009c (Battle of Teutoburg Forest); 2004 (221–3, 460, 477–9). The analysis also uses: Aelian/*Byzantine Interpolation of Aelian* (Devine ed. 39.1–2 = 50.1–2 *Interpolated Recension*, Devine ed.36.2–3 = Dain A2–3, Devine ed. 36.5–6; Devine ed. 37.2 = Dain ed. B3); Aelian (Matthew ed. 35–6, 41, 48, 50); Arrian (Teubner ed. *plaision*/hollow oblong 29.7, *plinthion*/hollow square 29.8), Asclepiodotus 11.6; *Definitiones/Hermeneia* (Köchly & Rüstow ed. p.232, *plaision*/hollow oblong 55, *plinthia*/hollow square 56; Montfaucon ed. p.512); Urbicius (*Taktikon*, Förster ed. p.30; *Epitedeuma*, esp. 3, 8–9).
80. For a good discussion of the use of the various variants of the hollow square during the modern era, see esp. Calwell (1906 square 256–277, laager 277–285, columns with flank units 298–299, elastic square in the bush and jungles 354–373).
81. The double phalanx with light infantry flanks was indeed one of the versions of the hollow square/oblong. See e.g. *PE* 46.2 (Xenophon placing cavalry, javelineers and peltasts behind the *amfistomos difalaggia*). Most importantly, it was the battle formation used by Alexander the Great at the Battle of Gaugamela/Arbela in 331 BC (see Syvänne, 2010). Therefore, it is not at all surprising to see the Romans use it too.
82. See Syvänne, *MHLR* vols.1–8. For the exploits of Heraclius's cavalry, see *MHLR 8*.
83. See Syvänne, *MHLR 2*, 18–20, 143.
84. *STR* 7.B.11.45–52.
85. The saw formation differed slightly because this array had the wedges in contact with each other.
86. Essen, 114.
87. For the Roman use of wagon laager during the early-sixth century, see e.g. Syvänne (*MHLR 5*, 175, 224, 239–41, 257; *MHLR 6*, 155–7). The *Strategikon* (5.5, 12.B.8, 12.B.13, 12.B.18, 12.B.22, 12.C) included the regular wagons as part of the fortified camp but posted these either in the middle of the marching formation or behind the infantry phalanx together with the *ballistae* wagons and carts.
88. *STR* 12.A.3, 12.B.23.19–27.

Chapter 10

1. See in particular *MHLR 6*, 272ff.
2. The following is based on Syvänne (2004, 234–7) with changes, additions and corrections, and on the previous chapters.
3. *STR* 12.B.16–17. The *hypostrategos* is usually fairly invisible in the sources as a result of being simply in charge of leading the infantry phalanx forward.
4. In his *Wars* (7.1.8, 8.26.14–17, 4.16.4–6, 7.37.24ff.) Procopius reserved particular praise for such commanders as Belisarius, Narses the Eunuch and Germanus for their generosity. In contrast, the eunuch Solomon suffered a defeat in which he was killed-in-action as a result of his unfair division of booty (Procopius, *Wars* 4.21.23–28).
5. The following is based on Syvänne (2004, 237–40) with changes, additions and corrections, and on the previous chapters.
6. The Roman military thinkers never reconciled their views regarding the effectiveness of the 'super heavy cavalry' against infantry. There were always those who favoured the use of these

against infantry and those who considered this ineffective. In the tenth century the East Romans were once again using the old *clibanarii* tactics so that expectation was that the cataphract wedge/triangle would smash the infantry spears and *menavlia*: *Praecepta militaria*, 2.11; *Taktika* of Nikephoros Ouranos, 57.13. See also McGeer's analysis of 10th century warfare, (1995) 197ff.

7. *STR* 12.B.13.1–12, 12.B.23.14–18, 9.4, 12.A.2.
8. *STR* 12.B.13.13–28; Syvänne, *MHLR* vols.1–8.
9. Syvänne (*MHLR 2*, 245–50 with *MHLR 3*; Poetovio, *MHLR 2*, 245–50; Pollentia, *MHLR 3*, 120–4). It is possible or even probable that the Romans used the side-by-side hollow squares/oblongs also at other battles during the fifth century, meaning that the Roman *testudines* that the Huns despised just before the battle of the Catalaunian Fields in 451 referred to the use of regimental and divisional hollow squares/oblongs in the Balkans by the East Roman forces serving under Aspar, Areobindus and Arnegisclus in 447. It is in fact possible that the Romans deployed all of their infantry *mere* in all of their battle formations (including e.g. also the mixed formation, *epikampios* etc.) as hollow squares or oblongs during this era so that each of the infantry divisions would always have been ready to face threats from any direction. See Syvänne (*MHLR* vols.3–4).
10. The following is based on Syvänne (2004, 240–45) with changes, additions and corrections, and on the previous chapters.
11. Syvänne, *MHLR 1*, 260–4.
12. A good example of this was when the Persian cavalry dismounted and engaged the Romans besieging the Fort of Akbas in 583 and then defeated them with a barrage of arrows. See Syvänne (*MHLR 6*, 207) with Theophylact 1.12.4.
13. *PST* 32.17–39, 32.51–4, 35.24ff, 36; *STR* 12.B.9, 12.
14. The following is based on Syvänne (2004, 245–50) with changes, additions and corrections, and on the previous chapters.
15. *STR* 12.B.16–17, 23; Syvänne, *MHLR* vols.1–8. The exploit of Theodosius's army in *MHLR 2*, 290.
16. *STR* 12.B.14.1–4: *Silentium, mandata captate, non vos turbatis, ordinem servate, bando sequute, nemo demittat bandum et inimicos seque.*
17. The Romans charging wildly for example: 1) The inexperienced Roman populace (paramilitary forces) charged wildly against the Goths, but even that involved the presence of a nearby camp with possessions to loot (Procopius, *Wars* 5.29.22–34; Syvänne *MHLR 6*, 145–50); 2) At the Battles of Rhizaeum in 557 (*MHLR 6*, 320) and Toumar in 540 (*MHLR 6*, 187) the soldiers attacked wildly. The personal example of seven Armenians charging out from the formation in the text of Procopius (*Wars* 6.27.16–17) is a good example of the importance of personal example in situations in which the battle became prolonged. The heroism of a single individual could also decide the result of a siege (Theophylact 2.18.15ff.) when the others were encouraged by this. The same was true of the heroism of the general in combat (see e.g. *MHLR 6*, 272–96). The use of the *Labarum* and *Labarum* Guard to overcome the enemy in prolonged battles was a special case, for which see *MHLR 1*.
18. *STR* 12.B.16.39–55.
19. This happened for example at the Battle of Turin in 312. See Syvänne, *MHLR 1*, 246–7.
20. E.g. Joshua (51); Syvänne, *MHLR 6*, 146–9.
21. For the use of the tortoise formation in melee, see e.g. Ammianus 14.12.37, 44.
22. Syvänne, (*MHLR 2*, 211; *MHLR 6*, 78, 138–9); Procopius, *Wars* 1.24.39–54, 5.10.1, 5.23.18ff.
23. E.g. the *Strategikon* (7.B.11.45ff, 11.1.67–70); *PST* 36. The specialist units armed with clubs and iron maces were considered a particularly effective counter measure against the enemy *clibanarii* cavalry (see: Zosimus 1.53.2; Nazarius 22.4–24.4; Syvänne, *Aurelian and Probus* with *MHLR 1*; Speidel, 2004, 87–97).
24. The following is a corrected and changed version of Syvänne (2004, 250–8).

25. *STR* 12.B.1.
26. Procopius, *Wars* 5.29.42–44. It is quite possible that there existed special techniques for the use of two javelins or lances simultaneously. For additional examples of medieval usage of two spears simultaneously, see Syvänne (2004, 251–2).
27. It was because of this that the early-sixteenth century European infantry spear formations often had swordsmen and halberdiers inside their formations to fight against the pikemen. See e.g. Machiavelli, *The Art of War*, 47–51. The use of the shield and sword pair recognized the superiority of the shorter weapon and shield over the unwieldy pike or spear. Renaissance martial arts treatises also noted the impracticability of the combination of the shield with the pike, spear, lance etc. because these weapons could not be handled effectively with only one hand under or above the shield for thrusts. Their recommendation was to abandon the shield altogether and use the shafted weapons with two hands. For a discussion of this, see Anglo, 159ff, esp. 161.
28. The following two examples demonstrate well how the spear could be used as a thrusting weapon in the right circumstances. For example, in 502 during the Siege of Amida one Peter held the battlement alone by using spear attacks (Zachariah 7.4). The dismounted Roman cavalry also used spear thrust alongside throwing at the Battle of Mons Lactarius (Procopius, *Wars* 8.35.23).
29. *ST* 38.5, 39.2.
30. Bishop and Coulston, 54–6, 78–83, 130–4, 154–63, 202–5; Fischer, 152–6, 326; Kolias, 136–137. See also: Southern & Dixon, 1996, 103–112; Stephenson, 61–80.
31. Bodyguards were allowed to carry only swords when escorting officers in cities (ProB 4.28.8). Obviously, the lack of shields made the *bucellarii* also vulnerable to arrows and javelins. It should also be noted that when Artasires intended to murder Guntharis, he cut some arrows into two parts and placed them between his wrist and elbow as a form of shield. At the same instance the rest of his bodyguards were to grab the shields of Guntharis' bodyguards and come to his help with these shields (see: ProB 4.28.9, 34). These instances show well the importance of the shield in defence.
32. *STR* 12.B.4, 12.B.16.47–8; Vegetius 2.15–16; John of Nikiu 109.5–7; Procopius, *Wars*, 4.28.29; Agathias (2.9.10) mentions the use of *machaira* by the 'hoplites' in the context of infantry melee.
33. Some of the examples of the usage of both fighting stances, left or right foot in front, are collected in Syvänne (2004, 254). Here it should be stressed that there has never existed a martial arts or combat system that was or would be definitely superior to another. Each combat system was and is just as good as its exponent and the familiarity of the fighter with the techniques used by his opponent/enemy. As I noted in my dissertation: 'All those who have put too much emphasis on the appearance of the couched lance technique or the use of saber during the eighteenth century, or the reappearance of the lance during the same period should take this fact into account. The initial advantage of a certain weapon or method of fighting disappeared fast.'
34. It is possible to speculate that when the soldiers used the rim-to-boss order, that it is possible that the placing of the foot varied according to the placing of the shield. If the left edge was placed on top of the shield on the left, the left leg was probably placed in front, whereas in the opposite case the right leg could be placed in front. However, this is unlikely because the rim-to-boss order was meant to restrict the freedom of the individuals to act separately from their unit and the placing of the right foot in front was designed as aggressive stance in which the fighter charged or lunged forward from his initial stance.
35. Ammianus (16.37) describes it as follows (tr. by Rolfe, 283–5): '… and the infantry stoutly protected their flanks by making a front of their bucklers joined fast together, clouds of thick dust arose. Then there were various manoeuvres, as our men now stood fast and now gave ground, and some of the most skilful warriors among the savages by the pressure of their

knees tried to force their enemy back, but with extreme determination they came to hand-to-hand fighting, shield-boss pushed against the shield, and the sky re-echoed with the loud cries of the victors or of the falling'.

36. I.e. *feinte, pression, battement, coulé, invite* etc.
37. E.g. Vegetius 1.26.
38. Roman combat doctrine had always demanded that the men stay in their position in the battle formation: The oath given during the Republican period stated that the soldier was not to leave the ranks except to seek a weapon, strike a foe, or to save a comrade (Frontinus 4.1.4); Livy 22.38.2–5.
39. E.g: *PST* 16.27–30; Corippus, 5.361–3.
40. The phenomenon of nominating a Germanic king as commander of Roman units was not new in the fifth century, because we find Crocus, King of the Alamanni, in this position in the court of Constantius I. For this, see Syvänne, *MHLR 1*, 229.
41. See in general Syvänne (*MHLR* vols. 1–8, esp. vols. 3–5). Some examples of this phenomenon e.g. in: *MHLR 4*, 106–17; *MHLR 5*, 28–31, 148–61; *MHLR 6*, (304–5, 341–53); Procopius, *Wars*, 5.28.23–29, 5.29.37–44.
42. See e.g. Procopius, *Wars* 5.28.23–27, 5.29.37–44; Agathias 3.20.9, 4.16.2 with Syvänne (*MHLR* vols.1–8).
43. The face of battle approach to the study of combat was first coined by the late John Keegan in his superb study *The Face of Battle. A Study of Agincourt, Waterloo and Somme* (London 1976). John Keegan basically reintroduced the analytical method first pioneered by Ardant du Pick in the nineteenth century. Keegan's approach was later adopted for ancient Roman battles by Adrian Goldsworthy, Brueggeman, Gregory Daly, A. Zhmodikov, A. D. Lee, Peter Conolly and Philip Sabin, and for the Late-Roman period by Ilkka Syvänne (2004). This study is therefore based on Syvänne (2004, 260–74) with some changes.
44. For examples, see Syvänne *MHLR* vols.1–8 (indexes) and for the PPE see *MHLR 5*, Appendix.
45. For example, Goldsworthy (2000, 203) and Rance (1993, 213–4) have seen the use of a sixteen-deep infantry phalanx as a sign of the unreliability of the Roman foot soldier. Ardant du Picq (78–80) similarly criticized the placing of the reserves (the ranks in depth) too near to the combat in the phalanx formation, while MacDowall (1994/6, 12) claims that the infantry phalanx was good only for the holding of a position. Spaulding et al. (249, 271–2) considered the Late-Roman phalanx not flexible enough to permit any maneuvering.
46. I have here considered battles with less than 1,000 footmen to be skirmishes.
47. The retreat was probably performed as in the Mamluk military treatise *Tafrij* (English tr. by Scanlon, 104–5); 'If a troop of men attack the enemy and then are required to retire to their [original] station, they should avoid anything unseemly or hurried in their retreat… The unseemliness of their retreat could enhance the covetousness of the enemy… He will pursue it, and rout will ensue. …they should retreat to place behind their backs, swerving and looking sideways, some inclining shoulders and heads, and [with] chests in the direction of the chests of the enemy. … If one of the men of the centre attacks and takes advantage of an opportunity against the enemy [i.e., for individual combat], and then has to retire, he should incline leftwards and towards the left flank or towards what is between the wings of the centre and left flank.'
48. Syvänne (2004, 270): 'The ability of the infantry to retreat depended upon the amount of cavalry that the enemy had. If they had plenty, and the place of refuge was far away, the horsemen were generally able to crush the speedily-retreating infantry before it could reform itself'.
49. *STR* 10.3.42–44. According to Angilbert (8), he was the sole survivor from the front line (*prima frontis acie*) when his side lost the Battle of Fontenoy in 841.
50. See *PST* 15.94–7.

51. *PE 34*; Tabari i.2319.
52. For a fuller analysis of this battle, see Syvänne *MHLR 6*, 273ff.
53. In fact, in modern society not even the people of the countyside engage in the butchery of animals because these are these days taken to slaughterhouses for this. It is only if the countryfolk engage in hunting that they are familiar with the blood and gore.
54. Vegetius 1.7; Procopius, *Wars* 4.26.16 (Areobindus); *STR* 12.D; Urbicius, *Cynegeticus*; Chapter Training. Some of the urban dwellers were also familiar with the sight of killing because they had witnessed executions or games involving the killing of wild animals (Evagrius 5.18; Sebeos 20, pp.39–40).
55. See e.g.: Procopius, *Wars* 4.3.24–4.8, 5.29.22ff.
56. *STR* 1.3, 1.6–8, 3.5, 3.12, 8.B.44; 12.A.7; Cavalry Tactics; Syvänne, *MHLR* vols.1–8; Procopius, *Wars* 6.10.14ff; John of Ephesus 6.9; Pacatus' *Panegyric of Theodosius* (36).
57. *PST* 38; Agathias 3.7.8–10; Vegetius 3.25; Syvänne *MHLR 6*, 282–5, 314.
58. See: Appian, *The Civil Wars,* 3.66–70 esp. 3.68.
59. Procopius, *Wars,* 6.5.24ff, 6.2.14–18, 6.2.21–4, 30–33; Ravegnani, 37–8; Haldon, 1999, 246–7; Gabriel and Metz, 138–143; Salazar, 74–83; Campbell, 2002, 65–9; Birkenmeier, 206ff.

Chapter 12

1. Based on: *STR* 4; *PST* 33.25ff, 40; *DMS* 16; *PE* in general and in particular 15–7, 19, 31, 35–7; Vegetius 3.6, 3.9, 3.19, 3.22; Syvänne (2004, 282ff.; *MHLR* vols.1–8); Polyaenus in general; Frontinus, *Stratagems* in general.
2. *ibid.*
3. Caracalla had popularized the fighting of duels so that we can find commanders engaging in such thereafter. For this, see Syvänne (*Caracalla, Gordian III and Philip the Arab, Aurelian and Probus*), but it can still be stated with a fair amount of certainty that the fighting of duels was not the recommended way of fighting among Roman military thinkers, even if we find some commanders and individuals engaging in such. During the Late-Roman period, the three most famous instances are the duel between Aetius and Bonifatius in 432 (*MHLR 4*, 31), the duel between Areobindus and a Persian champion in 422 (*MHLR 3*, 270–2), and the possible duel between Bonifatius and Athaulf in 412 (*MHLR 3*, 208–9). In these instances, the duels were fought by commanders who could decide to overlook Roman traditions after the emperor Augustus (who did not want his commanders to obtain *spolia optima*), just like Caracalla and some of the other soldier emperors after him in the third century. Roman combat doctrine expected the commander to stay out of direct danger so that he could direct the reserves, and combat, and was only to endanger his life in such situations in which the demoralized soldiers needed encouragement. It is because of this that we find the generals in Procopius's *Wars* attempting to prevent the fighting of duels. The sole exception to this rule of thumb is the *PE* 11 which does include two stratagems concerning duels – it is therefore clear that its author took into account the fact that single combats did take place even if this was not the accepted form of fighting.
4. *MHLR 7*, 310–2. Compare this with the result of the Emperor Heraclius's use of this method at the Battle of Sarus River in 625 and at the Battle of Yarmuk in 634 (the Roman army was deployed according to the instructions sent by Heraclius. See Syvänne, *MHLR 8*, 173, 253–72. In the former case, Heraclius's main camp was on the friendly side so that when his forces were forced into flight it was only with difficulty that he could save the situation by making a counter attack across the bridge. In the latter case Heraclius had ordered the building of the camp on the hostile side of the wadi, but this did not help the Romans who failed to defend the camp effectively.
5. The following is based on Syvänne (2004, 293–4; *MHLR 6*, Index); Procopius, *Wars* 7.19.1–22, 8.35.7–14; Agathias 2.4.7, 6.4–6.

6. See the index in *MHLR 6* for the following battles: Satala, Ancon, Edessa, Dara, Archaeopolis, Phasis.
7. *STR* 4 with 7.B.11–2. For ambushes and surprise attacks by cavalry (some examples of enemy ambushes and uncertain cases included) see e.g. Syvänne (2004, 282ff.; *Gallienus,* 32–5, 95–7, 142–3; *Aurelian and Probus,* 70–3, 100–4; *MHLR 1,* 308(?); *MHLR 3,* 138–41, 180, 269–70, 275–8; *MHLR 4,* 38–40, 58–60; *MHLR 5,* 15–7, 121–2, 191, 195, 199, 201, 211, 215; *MHLR 6,* 52, 58,63–5, 99, 191, 193, 221, 238, 246–7, 254–5, 263, 257, 282–3, 314–5, 329; *MHLR 7,* x, 22, 66, 68–9, 99, 105–12, 146, 149, 153, 158, 167, 181, 205, 213, 220, 224, 227, 246, 251, 256, 295–7, 307, 315; *MHLR 8,* 154–9, 168–9, 173, 177–9, 241, 330–1). Ambushes directed at individuals (i.e. assassinations) not included. For other examples, see the articles and books in the bibliography under the name Syvänne.
8. Notably, we find similar instructions in the *Excerpts of Polyaenus* (37), which lists the following measures against enemy cavalry: 1) the use of water discharged into a plain (equivalent of a swamp); 2) the use of eight-rank deep hollow square (*plinthion*) with the addition that the men placed shields above their heads, dug a pit with their daggers after which the hoplites advanced in front of those, but then retreated behind those before the enemy cavalry could reach them so that the enemy cavalry fell into the pit; 3) Ifikrates gave his horsemen torches when superior numbers of enemy horsemen pursued them so that when Ifikrates's few cavalrymen charged against their pursuers, the flames frightened the enemy horses and caused them to flee; 4) and the Phocians destroyed the Thessalian cavalry by digging a trench and by covering it so that the Thessalian horsemen fell into the trench when they attacked. It is clear that the chosen examples were once again relevant for the Late-Roman and 'Byzantine Roman' periods.
9. See Syvänne (*MHLR 5,* 115–6; *MHLR 6,* 51–3, 2021b).
10. For an analysis of this battle, see Syvänne, *Gallienus* (32–5).
11. Note also *PE* 36. For ambushes and surprise attacks involving units of infantry or dismounted cavalry (some examples of enemy ambushes and uncertain cases included), see e.g. Syvänne, *Gallienus,* 115, 117–8; *Aurelian and Probus,* 143; *MHLR 1,* 189, 262–4, 256–7, 308(?), 324, 328; *MHLR 2,* 87–8, 95, 115, 160–1, 192–5, 197, 211, 236–8, 250; *MHLR 3,* 92–3, 97–8, 121–4, 203–6, 286; *MHLR 4,* 19–20, 40, 91, 132; *MHLR 5,* 74–5, 125–7; *MHLR 6,* 62–3, 69, 108–9, 315–7, 319, 329, 333, 355–6; *MHLR 7,* 107, 112, 146, 152, 242, 248, 296–7; *MHLR 8,* 168–9, 177–9. For other examples, see the articles and books in the bibliography under the name Syvänne. The greatest and most successful user of the stratagems in Roman history was the underappreciated master of warfare, Caracalla. For his career, see Syvänne, *Caracalla* together with *Septimius Severus.* Caracalla was the best pupil of his father, who was similarly a master of dissimulation and treachery.
12. Instructions for ambushes: *STR* 4; *PST* 33.25ff, 40; *DMS* 16; *PE* e.g. 19, 35–6.
13. Based on Syvänne (2004, 287ff.; *MHLR* vols.1–8); *STR* 9.1–2; *PST* 39; *PE* 31. See (includes some examples of enemy night attacks and Romans fleeing during night) e.g. *MHLR 1,* 318, 363, 382; *MHLR 2,* 87–8, 95, 115, 197, 211, 224; *MHLR 3,* 117, 263; *MHLR 5,* 87–8, 128, 236; *MHLR 6,* 90, 107–9, 239, 303–4, 319; *MHLR 7,* 107, 112, 146, 152, 197, 205, 217, 220, 236, 242, 248, 252–3, 272, 275–6, 296–7, 306–8; *MHLR 8,* 44–5, 86, 111, 155, 168, 171–2, 208, 275–6, 306–7. The Romans did not usually face similar trouble from night attacks against their marching camp because they fortified these. A good example of this is the failed enemy attack against the fortified Roman camp in *MHLR 2,* 173. A good example of negligence by the Romans that enabled the enemy to launch a surprise attack in *MHLR 8,* 44–5.
14. The list of stratagems is from the *STR* but similar and other stratagems can be found in the collections of stratagems and works of history. It should be stressed that the works of history were also school books for the aspiring generals.
15. *STR* 9.1–2; *PST* 39; *MHLR* vols.1–8.

Appendices

1. Based on Syvänne, *MHLR 1*, 324–6; Julian *Or.* 1.29A-49A, 2.55.C-60C; Zosimus 2.50–53; Zonaras 13.8. For the *clibanarii/thorakitai* cavalry, see also Julian *Or.*1.37C-D, *Or.*2.57.B-C; Ammianus 16.10.8. For an analysis of the political circumstances and background to the battle, see also Drinkwater, 2022. My own reconstruction of the events both in the *MHLR Vol.1* and here is slightly different, but Drinkwater makes several good points about the political background which are not found in my short account of the war between Magnentius and Constantius. On the individuals mentioned, see also *PLRE 1*.
2. For a discussion of the circumstances leading to the killing of Constans II, see Syvänne (*MHLR 1*, 310–1) together with Drinkwater (2022). Note, however, that my analysis of the campaign, battle and the logic behind the moves differs from Drinkwater's analysis. Drinkwater thinks that Magnentius was not seeking a battle in Mursa, while I suggest that it was indeed his purpose to lure Constantius away from Cibalae by threatening Mursa. This proved to be a mistake because the site Magnentius chose was ideal for Constantius's army.
3. Based on analysis of Agathias's text in Syvänne *The Age of Hippotoxotai* (2004, 473–4) and *MHLR 6*, 346–53 with the addition of the *clibanarii* from the Chapter 6.7. This reconstruction supersedes the previous ones.
4. Agathias 2.4.10. Boss (28) estimates the strength of the Franks at 22,000 men. His estimates of the divisions of the Roman army are incorrect. For example, he assumes that the size of bodyguards of Narses is 500 men while Agathias clearly stated otherwise. In addition, his reconstruction leaves out the battle-ready camp followers and servants who must have numbered in hundreds, if not in thousands. Furthermore, the battle formations of both armies are clearly incorrect.
5. The Macedonians used the cavalry *sarissa* or *xyston* with one hand which makes it possible that this only meant the regular long cavalry spear while the *doratia* meant the javelin type of spears.
6. Based on my dissertation *The Age of Hippotoxotai*, 122.
7. See Syvänne, *MHLR 6*, 92.

Select Sources

SELECT PRIMARY SOURCES

Aelian/Aelianos, Devine, A.M., 'Aelian's Manual of Hellenistic Military Tactics. A New Translation from the Greek with an Introduction', *Ancient World 19 (1989),* 31–64; includes 'The Byzantine Interpolation of Aelian. Edition with Arrian': Köchly H. und W Rüstow, *Griechische Kriegsschriftsteller II.1*, Leipzig (1855), 241–471. Latest edition and translation by C. Matthew, *The Tactics of Aelian.* Barnsley

Agathias, *Agathiae Myrinaei Historiarum libri quinque*, CFHB Vol II, ed. R. Keydell, Berlin (1967). *Agathias, The Histories*, CFHB Vol. II A, English tr. by Joseph D. Frendo, Berlin (1975).

Ammianus Marcellinus, *The History of Ammianus Marcellinus*, ed and tr. by John C. Rolfe, Loeb ed. 3 vols, Cambridge, Mass.-London.

Asclepiodotus, *Aeneas Tacticus, Asclepiodotus, Onasander*, tr. by members of the Illinois Greek Club, Loeb ed. (1923). *Asclepiodoté, Traité de Tactique*, French ed. and tr. by Lucien Poznanski, Paris (1992). Giuseppe Cascarino, *Tecnica della falange. In appendice: Il trattato tattico di Asclepiodoto.* Il Cerchio 2011.

Ayin Nameh, 'The Sasanian Military Theory from the Russian of Prof C.A. Inostrancev', translated by Mr. L. Bogdanov, *The K.R. Cama Oriental Institute*, 7–52. The treatise is in 13–16.

Byzantine Interpolation of Aelian,

a) in Dain, *L'Histoire du texte d'Elien le Tacticien des origins a la fin du moyen age*, Thèse pour le Doctorat èt lettres, Université de Paris, Paris (1946), ed. Dain, 92–100, 102–106;

b) English tr by Devine, A.M., 'Aelian's Manual of Hellenistic Military Tactics. A New Translation from the Greek with an Introduction', *Ancient World 19 (1989*), 31–64, the Byzantine Interpolation of Aelian 59ff.

c) Christopher Matthew, *The Tactics of Aelian or On the Military Arrangements of the Greeks. A New Translation of the Manual that Influenced Warfare for Fifteen Centuries.* Revised, translated and edited by Christopher Matthew. Barnsley 2012; This is now the definitive edition and translation of Aelian, but it assumes that the so-called *Byzantine Interpolation of Aelian* formed part of the original. The text includes modern diagrams. For diagrams closer to the original, see Syvänne, *Septimius Severus*, Appendix on Arrian which includes diagrams from the Barnsley manuscript of Aelian to illustrate Arrian's text.

Chronicon Paschale, ed. L. Dindorf (Bonn, 1832). English Translation by Michael and Mary Whitby [TTH 7] (Liverpool, 1989). References in the notes to this translation (contains the pagination of Dindorf).

Constantine Porphyrogennetos/Porphyrogenitus, Constantine Porphyrogenitus: *De administrando imperio*, Greek text edited by Gy. Moravcsik, English translation by R.J.H. Jenkins, Budapest 1949.

De Ceremoniis, Constantine Porphyrogennetos: The Book of Ceremonies. Translated by Ann Moffatt and Maxeme Tall. Leiden and Boston 2012/2017.

Corippus, *Ioannidos libri viii*, ed J. Diggle and F.R.D. Goodyear, Cambridge (1970). *The Iohannis or De Bellis Libycis of Flavius Cresconius Corippus.* Studies in Classics Volume 7, New York 1998, tr. and introd. by George W. Shea.

De militari scientia, ed., Italian tr. and commentary by Immacolata Eramo, *Appunti di tattica (De militari scientia)*, Presses universitaires de Franche-Comté, 2018. This is the definitive edition and study of the text. Those who cannot get access to this edition can consult online the older text of Müller, K.K., *Ein Griechisches Fragment über Kriegswesen, Festschrift für Ludwig Urlichs* (Würtzburg, 1880), 106–138.

De re militari (c. 990's, *Anonymou Biblion taktikon/ Peri katastaseōs aplēktou/ De castrametatione*), ed. and tr. by G.T Dennis, *Three Byzantine Military Treatises*, CFBH 25, Washington D.C. (1985), 241ff, under the title Campaign Organization and Tactics.

***De Rebus Bellicis* (DRB)**, *Anónimo Sobre Asuntos Militares*, ed. Álavaro Sánchez-Ostiz. Navarra; *Anonymi auctoris de Rebus Bellicis*, ed. R.Ireland. Leipzig; *De rebus bellicis. Part 2: the text.* ed. R. Ireland. Oxford; Part 1, *Aspects of the De Rebus Bellicis*, ed. M.W.C. Hassall, Part 2, *de rebus bellicis*: ed. R. Ireland, *BAR Int. Ser.* 63, 1979.Tompson E.A., *A Roman reformer and inventor*. Oxford 1952; *Anonymi De rebus bellicis liber*, ed. R. Schneider. Berlin 1908.

De scientia politica dialogus, *see PPE*

De velitatione bellica, (*Peri paradromês*), ed. and tr. by G.T Dennis, *Three Byzantine Military Treatises*, CFBH 25, Washington D.C. (1985), 137–239 under the title 'Skirmishing'; ed. and French tr. by G. Dagron and H. Mihăescu with a study of the text, *Le traité sur la guérilla (De velitatione) de l'empereur Nicéphore Phocas (963–969)*, Paris (1986), 29–135.

Evagrius, *Ecclesiastical History*, ed. J. Bidez and L. Parmentier (London, 1898). *The Ecclesiastical History of Evagrius Shcolasticus*, Translated with an introduction by Michael Whitby, [TTH 33] (Liverpool, 2000). French translation A.J. Festugière, Évagre, 'Histoire Ecclesiastique', Byzantion 45 (1975), 187–488.

Excerpts of Polyaenus, *see Polyaenus*

Frontinus (Frontin), *Strategematon*, in *Frontinus The Stratagems and the Aqueducts of Rome*, tr. by Charles E Bennett, Loeb ed. Cambridge Mass. London (1925/1980).

Georgian Chronicles, The, tr. By Robert W. Thomson, in *Rewriting Caucasian History. The Medieval Armenian Adaptation of the Georgian Chronicles. The Original Georgian Texts and the Armenian Adaptation*, Oxford, 1996. English tr.also by Robert Bedrosian (internet archive).

Gotha ms., 'Das Heereswesen der Muhammedaner nach dem Arabischen von F. Wüstenfeld 1879, Die Arabische Übersetzung der Taktik des Aelianus 1880'. *Abhandlungen der Hist. Phil. Classe der Königlischen Gesellschaft derWissenschaften zu Göttingen 26.1.*

Gregory of Tours, *Historia Francorum*, ed. W. Arndt and B. Kruch, *MGH Scriptores Rerum Merovingiciarum*, Hanover (1884). *Gregory of Tours, The History of the Franks*, tr and introduction by Lewis Thorpe, Penguin Books (1974).

Hyginus, *see Pseudo-Hyginus*

John of Ephesus, *The Third Part of the Ecclesiastical History of John of Ephesus*, tr. by R. Payne-Smith (Oxford, 1860). *Iohannis Ephesini Historiae Ecclesiasticae pars tertia*, ed. and Latin translation by E.W. Brooks, CSCO 106, Scr. Syri 55 (Louvain, 1936, 1952)

Julian the Apostate, *The Works of the Emperor Julian*, 3 vols. tr. by W.C. Wright, Loeb 1914.

Menander the Protector, *The History of Menander the Guardsman*, ed. and tr. by R.C. Blockley, Liverpool (1985).

Munyatu'l-Ghuzat, 'Munyatu'l-Ghuzat. A 14th-Century Mamluk-Kipchak Military Treatise', Text, Translation and Index by Kurtuluş Öztopçu, *Sources of Oriental Languages and Literatures (Doğu Dilleri ve edebiayatlarinin kaynaklari) 13*, Turkish Sources XI, Harvard (1989).

Naumachica of Syrianus Magister, in Pryor J.H. and Jeffreys E.M. (2006), *The Age of Dromon*. Leiden and Boston.; and also in '*Syriani Magistri Naumachiae*', in *Dain, Alphonsus, Naumachica partim adhuc inedita in unum nunc primum congressit et indice auxit*, Paris (1943), See also the *Peri strategikes/strategias* and *Rhetorica militaris*.

Nazarius, *Panegyricus*, see *Panegyrici Latini.*

Nikephoros Ouranos, '*Taktika*', in *Sowing the Dragon's Teeth: Byzantine Warfare in the Tenth Century, Eric McGeer, Dumbarton Oaks, Washington D.C. (1995), Dumbarton Oaks Studies; 33.*

'Douce chapitres inédits de la Tactique de Nicéphore Ouranos', in Foucault TM 5 (1973), 281–311; 'Naumachica', in Dain, Naumachica, Paris (1943).

Onasander, 'Onasander, Onasandros, Strategikos', ed. and tr. by the Illinois Greek Club, *Aeneas Tacticus, Asclepiodotus, Onasander*, Loeb Cambridge Mass. and London (1923).

Panegyrici Latini, *Panegyrici Latini*, Rome (1976); *In Praise of Later Roman Emperors, The Panegyrici Latini: Introduction, Translation and Historical Commentary with the Latin Text,* R.A.B. Mynors, C.E.V Nixon and Barbara Saylor Rodgers (Berkeley, Los Angeles, Oxford, 1994).

Peri strategikes/strategias, *Three Byzantine Military Treatises*, G.T. Dennis, CFHB XXV, Washington 1985. 'The introduction, text and tr. of the Anonymous Byzantine Treatise on Strategy', 1–135. H. Köchly and W. Rüstow, 'Byzantini anonymi Peri Strategikes' in *Griechische Kriegsschriftsteller II.2*, Leipzig 1855, 1–209, notes 311–355. See also *Naumachica* and *Rhetorica Militaris*.

Polyaenus, Edition and translation in *Polyaenus, Stratagems of War*, Vols. I-II, ed. and tr. by Peter Krenz and Everett, L. Wheeler, (Chicago, 1994), Vol. II *Excerpts of Polyaenus*, 851–1003.

***PPE* = *Peri politikês epistêmês* (*De scientia politica dialogus*)**, *Menae patricii cum Thoma refendario De scientia politica dialogus,* edited and translated into Italian by C.M. Mazzucchi, Milano 1982. English tr. as 'Dialogue on Political Science' in *Three Political Voices from the Age of Justinian.* Agapetus, *Advice to the Emperor. Dialogue on Political Science.* Paul the Silentiary. *Description of Hagia Sophia*. tr. by Peter N. Bell (Liverpool 2009).

Praecepta militaria, Nikephoros Phokas, in *Sowing the Dragon's Teeth: Byzantine Warfare in the Tenth Century*, Eric McGeer, Dumbarton Oaks Studies 33, Dumbarton Oaks, Washington D.C. (1995),

Procopius, Wars, *History of the Wars*, ed. and tr. by H.B. Dewing, Cambridge Mass., Loeb Edition reprints, 5 Vols.

Pseudo-Hyginus, *Pseudo-Hygin. des fortifications du camp*, ed. and French tr. M. Lenoir, Paris —— (2002); *Polybius and Pseudo-Hyginus: The Fortification of the Roman Camp*, ed. and tr. by M.C.J. Miller and J.G. DeVoto, Chicago 1994.

Rhetorica Militaris. Siriano, *Discorsi di Guerra a cura di Immacolata Eramo con una nota di Luciano Canfora, Ed. and Italian tr. Immacolata Eramo. Bari 2010; Byzantine Military Rhetoric in the Ninth Century. A Translation of the Anonymi Byzantini Rhetorica Militaris*, English tr. and commentary by Georgios Theotokis and Dimitrios Sidiropoulos. London and New York 2021.

REF1: The Roman Eastern Frontier and the Persian (AD 226–363). A Documentary History, Compiled and edited by Michael H. Dodgeon and Samuel N.C. Lieu, London and New York (1991). Contains partial translations of several primary sources.

Sebeos, *The Armenian History attributed to Sebeos*, Translated with notes by R.W Thomson, Historical commentary by James Howard-Johnston, Assistance from Tim Greenwood, 2 vols, TTH 31, Liverpool (1999). *Sebeos' History*, tr. by Robert Bedrosian, New York 1985 (availabe online from Internet Archive). This text is also known with the names *Pseudo-Sebeos* and *History of Khosrov*.

Strategikon, *Das Strategikon des Maurikios*, CFHB XVII, Vienna 1981, edited by G.T. Dennis and German tr. by Ernst Gamillscheg. *Maurice's Strategicon. Handbook of Byzantine Military Strategy*, English tr. by G.T.Dennis, Philadelphia (1984). *Arta Militara* Mihaescu, (with Romanian tr.), Bucharest (1970); *Strategikon, Il Manuale di Arte Militare dell'Impero Romano d'Oriente,* Giuseppe Cascarino, Seconda Edizione Il Cerchio 2016 (edition and Italian translation). Edition of the tenth century Ambrosian *Strategikon*: Barbara Leoni, *La Parafrasi Ambrosiana dello Strategicon di Maurizio. L'arte della guerra a Bisanzio*. Milan 2003.

Sylloge Tacticorum, *Sylloge tacticorum, quae olim Inedita Leonis Tactica dicebatur*, ed. A. Dain, Paris, 1938; *A Tenth-Century Byzantine Military Manual: The Sylloge Tacticorum*. English tr. by G. Chatzelis and J. Harris. London and New York 2017.

Syrianus Magister, *see Peri strategikes, Naumachica, Rhetorica militaris.*

Tabari, al-, *The History of al-Tabari*, ed. E. Yar-Shater, several translators, Albany New York 1985-.

Tafrij, *A Muslim Manual of War being Tafrij al-Kurub fi Tadbir al-Hurub by Umar ibn Ibrahim al-Awasi al-Ansari*, edited and translated by George T. Scanlon, Cairo (1961).

Theophanes, *Chronographia*, ed. C. de Boor, Leipxig (1883–5). *The Chronicle of Theophanes Confessor. Byzantine and Near Eastern History AD 284–813.* Tr. by Cyril Mango and Roger Scott with the assistance of G. Greatrex, Oxford (1997).

Theophylact, *Theophylacti Simocattae, Historiae*, ed Carolus de Boor, Lipsiae (1887) Teubner.

Theophylact Simocattes, *The History of Theophylact Simocatta*, tr. by Michael and Mary Whitby, Oxford (1986).

Urbicius, '*Taktikon/Tactica*', ed. R. Förster in *Hermes 12 (1877),* 467–71. 'Epitedeuma', in *Arta Militara, ed. H. Mihaescu, Bucharest (1970), 368–372. 'Cyneticus', Arta Militara, ed. H. Mihaescu, 372–380; and Burgess R. (2005), 'Urbicius Epitedeuma: an edition, translation and commentary', Greatrex G., Elton H., in BZ* 98, 35–74.

Vegetius, *Flavius Vegetius Renatus, Epitoma Rei Militaris*, ed. with an English tr. by L.F. Stelten, New York, Bern, Frankfurt, Paris (1990). *Vegetius: Epitome of Military Science. Translated with notes and introduction*, N.P. Milner. TTH 16, Liverpool, 1993, 2nd ed. (1996).

Zonaras, Greek (e.g. PG) editions available online. Partial English tr. by Banchich, Thomas M., and Lane, Eugene N., *The History of Zonaras From Alexander Severus to the Death of Theodosius the Great*, Routledge, London and New York, 2009.

Zosimus, *Nea historia, Zosimus, New History*, English tr. by Ronald T. Ridley, Australian Association for Byzantine Studies, Byzantina Australiensia 2, Canberra (1982/1990). Follows the edition of Mendelssohn, Leipzig (1887). Bekker, I., *CSHB 20*, Bonn (1837). Paschoud, F., Paris (1971)

SELECT SECONDARY SOURCES

Anglo, Sydney, (2000) *The Martial Art of Renaissance Europe*, Yale UP, New Haven and London, Singapore.

Baldwin, B. (1988) 'On the Date of the Anonymous PERI STRATHGIKHS, *BZ 81*.

Birkenmeier, John W. (2002) *The Development of the Komnenian Army: 1081–1180.* Leiden.

Bishop and Coulston (2006) *Roman Military Equipment From the Punic Wars to the Fall of Rome.* Oxford.

Bosworth, C. E., see *Tabari.*

Bury, J.B., (1958) *A History of the Later Roman Empire from the death of Theodosius to the death of Justinian, 2 vols.*, Dover Edition, New York.

Butler, Alfred J., (1902) *The Arab Conquest of Egypt and the Last Thirty Years of the Roman Dominion 1–2.* Oxford.

Campbell, Brian (2002) *War and Society in Imperial Rome 31 BC–AD 284*, London.

Cascarino, Giuseppe (2010) *Castra. Campi e fortezze dell'esercito romano,* Il Cerchio.

Chandler, David, G (1976) *The Art of Warfare in the Age of Marlborough.* London.

—— (1973/9) *Marlborough as Military Commander*, London.

Chauviré. Frédéric (2021) *The New Knights. The Development of Cavalry in Western Euope, 1562-1700.* Warwick.

Clausewitz, Carl von, (1966–8/1976) *On War*, tr. J.J. Graham, intr. and notes by Colonel F.N. Maude, 3 vols, London (1966–8). *On War*, ed. and tr. by M. Howard and P. Paret. Princeton (1976).

Corbett, Julian S., (2004) *Principles of Maritime Strategy.* London and New York 1911 (reprint New York 2004);

—— (1999) 'Some Principles of Maritime Strategy', partial edition in *Roots of Strategy 4,* Mechanicsburg, 148–261.

Coșentino, Salvatore, (2000) 'The Syrianos's "Strategicon": a 9th century source?', *Bizantinistica. Rivista di Studi Bizantini e Slavi 2nd ser.*, 2 (2000), 243–80.

Cruso, John, (1972) *Militarie Instructions for the Cavall'rie*, Cambridge (1632), facsimile ed. Kineton (1972).

Dain, Alphonse (1946), *L'Histoire du texte d'Elien le Tacticien*, Paris (1946).

—— (1967) 'Les Stratégistes Byzantins', *TM 2,* the text completed by J.-A. de Foucault, 317–392.

—— (1938) *Sylloge tacticorum, quae olim Inedita Leonis Tactica dicebatur*, Paris, 1938.

Dawson T. (2007), '"Fit for the task": equipment sizes and the transmission of military lore, sixth to tenth centuries', in *BMGS* 31.1, 1–12.

Delbrück, H. (1990) *History of the Art of War 4 vols.*, translated by Walter J. Renfroe, Jr., 1975, reprint 1990.

Dennis, George T. (1981) 'Flies, Mice, and the Byzantine Crossbow', *BMGS 7* (1981), 1–5.

—— (1982) ‚Byzantine Battle Flags', in *Byzantinsche Forschungen 8* (1982), 51–9.

—— (2001) 'Death in Byzantium', in *Dumbarton Oaks Papers 55* (2001), 1–7. Available at www.doaks.org in PDF Format DP55ch1.

Devine, A.M., (1989) 'Aelian's Manual of Hellenistic Military Tactics. A New Translation from the Greek with an Introduction', *Ancient World 19* (1989), 31–64; Includes also 'The Byzantine Interpolation of Aelian'.

Drinkwater, John F., (2022) 'The Battle of Muras, 351: Causes, Course, and Consequences', in *Journal of Late Antiquity* 15.1. pp.28–68.

Duffy, Christopher, (1998) *The Military Experience in the Age of Reason*, 1987/1998 Wordsworth Ed. (1998).

Eramo, Immacolata, see *De militari scientia and Rhetorica Militaris*

—— (2022) 'Leggere Cesare a Sant'Elena. Il Précis des guerres de César', in *FuturoClassico 8*, 151–78.

Essen, Micheal Fredhol von, (2018) *Muscovy's Soldiers. The Emergence of the Russian Army 1462–1689.* Warwick.

Fischer, Thomas (2019) *The Roman Army. With contributions by R. Bockius, D. Boschung and T. Schmidts*, English tr. by M.C. Bishop. Havertown and Oxford

Gabriel, Richard A. and Metz, Karen S. (1991), *From Sumer to Rome. The Military Capabilities of Ancient Armies*, Contributions in Military Studies, Number 108. Westport.

Goldsworthy, Adrian (2000) *Roman Warfare*. London.

Greatrex G., Elton H., and Burgess R. (2005) 'Urbicius' Epitedeuma: an edition, translation and commentary', in *BZ* 98, 35–74.

Haldon, John F. (2014) *A Critical Commentary on the Taktika of Leo VI*, Washington D.C.

—— (1999) *Warfare, State and Society in the Byzantine World 565–1204*, UCL Press. London.

Hassall M.W.C. (1979), 'The Inventions', in *DRB1*, 77–91.

James, Simon (2004), *Excavations at Dura-Europos 1928–1937. Final Report VII.* British Museum.

Jomini, Baron Antoine Henri de, (1992) *The Art of War*, English Translation 1862, London —— (1855/1862/1992).

Jones, A.H.M., (1964) *The Later Roman Empire, 284–602*, 3 vols. Oxford (1964).

Keegan, John (1976/96), *The Face of Battle. A Study of Agincourt, Waterloo and the Somme.* London.

Kolias, Taxiarchis G., (1998) *Byzantinische Waffen*, Byzantina Vindobonensia XVII, Vienna (1988).

Lee, D. and Shepard, J. (1991) 'A Double Life: Placing the Peri Presbeon'. *Byzantinoslavica 52*, 15–39.

Lenski N. (2002) *Failure of Empire. Valens and the Roman State in the Fourth Century A.D.* Berkeley, Los Angeles, London.

MacDowall, Simon (1994/6), text and Gerry Embleton, plates, *Late Roman Infantryman 236–565 AD.* Oxford.

Machiavelli, Niccolò, (1965/2003) *The Art of War, A Revised Edition of the Ellis Farneworth Translation With an Introduction by Neal Wood, New York (1965); Art of War*, tr. by Christopher Lynch, Chicago and London 2003.

—— (1981) *The Prince*, tr. George Bull, Penguin Books, London (1961/1981).

Napoleon, (2018) *Napoleon's Commentaries on the Wars of Julius Caesar*, tr. by R. A. Maguire. Barnsley, 2018.

Nicolle, David, (1996) *Sassanian Armies. The Iranian Empire early 3rd to mid-7th centuries AD.* Dewsbury.

Nixon, C.E.V. and B.S. Rodgers, see *The Panegyrici Latini.*

Nolan, Louis Edward, (2007) *Cavalry. Its History and Tactics*, 2nd ed. London 1854 (republished by Westholme in 2007) is the same as the 1st American ed.1864; 3rd ed. London 1860 is slightly longer than the previous two.

Nosworthy, Brent (1995/7) *Battle Tactics of Napoleon and His Enemies.* Bury St.Edmunds.

—— (1992) *The Anatomy of Victory. Battle Tactics 1689–1763.* New York.

Oldfather, W.A. and Titchener, J.B. (1921) 'A Note on the Lexicon militare', *Classical Philology 16* (1921), 74–6.

Ostrogorsky, George (1969/2005) *History of the Byzantine State.* Rutgers UP.

Pertusi, A. (1967) 'Ordinamenti militari, guerre in Occidente e teorie di guerra dei Bizantine, saec. VI-X', in *Ordinamenti militari*, 631–700.

Picq, Ardant du, (1987) 'Battle Studies. Ancient and Modern Battle, (Etudes sur le combat)', tr. by Colonel John N. Greely and Major Robert C. Cotton, in *Roots of Strategy 2*, Mechanicsburg (1987).

Pollington, Stephen (1996, 2001, 2006) *The English Warrior from Earliest Times till 1066.* Anglo-Saxon Books.

Rance, Philip, (1993) *Tactics and Tactica in the Sixth Century: Tradition and Originality*, Unpublished Doctoral Thesis, University of St. Andrews.

—— (2004b) 'The Fulcum, the Late Roman and Byzantine Testudo: the Germanization of Roman Infantry Tactics,' in *GRBS 44*, 265–326.

—— (2004a) 'Drungus, ΔΡΟΥΓΟΣ, and ΔΡΟΥΓΙΣΤΙ: A Gallicism and Continuity in Late Roman Cavalry Tactics', in *Phoenix 58*, 96–130.

—— (2007) 'The Date of the Military Compendium of Syrianus Magister (Formerly the Sixth-Century Anonymus Byzantinus', in *BZ 100/2*, 701–37.

Ravegnani, Giorgio (1988) *Soldati di bisanzio in età giustinianea*, Materiali e riceche Nuova Serie, 6, Collana della Facoltà di Lettere e Filosofia dell'Univerisità di Venezia in San Sebastiano, Sezieno di studi storici, 4, Roma (1988).

Rothenberg, Gunther E., (1980) *The Art of Warfare in the Age of Napoleon*, 1978/Bloomington (1980).

Sabin, Philip, (2000) 'The Face of Roman Battle', *JRS 90* (2000), 1–17.

Sabin, P.A.G. (1996) 'The mechanics of battle in the Second Punic War', in *The Second Punic War: A Reappraisal T.J. Cornell, N.B. Rankov and P.A.G. Sabin eds.*, BICS Supplement 67 (1996), 59–79.

Salazar, Christine F. (2000) *The Treatment of War Wounds in Graeco-Roman Antiquity*, Leiden-Boston-Köln (2000), Studies in Ancient Medicine 21, Leiden (2000).

Smith Jr., J.M., (1997) 'Mongol Society and Military in the Middle East: Antecedents and Adaptations', in *War and Society in the Eastern Mediterranean, 7th-15th centuries*, ed Yaacov Lev (The Medieval Mediterranean Vol 9), 249–266.

—— (1984) 'Ayn Jalut, Mamluk Success or Mongol Failure?', *Harvard Journal of Asiatic Studies, 44.2* (December 1984), 307–345.

Southern, Pat & Dixon, Karen Ramsey, (1996) *The Late Roman Army*, London (1996).

Spaulding Oliver Lyman Jr, Nickerson Hoffman, Wright John Womack, (1925) *Warfare. A Study of Military Methods from the Earliest Times*, New York (1925).

Stephenson, I. P. (1999/2001), *Roman Infantry Equipment. The Later Empire.* Stroud.

Speidel, Michael P., (2004) *Ancient Germanic Warriors, Warrior styles from Trajan's Column to Icelandic sagas*, London and New York, (2004).

Syvänne, Ilkka (2023a) *Emperor Septimius Severus. The Roman Hannibal.* Barnsley.

—— (2015–2022) *Military History of Late Rome, vols. 1–8*. Pen and Sword. Barnsley.
—— (2022b) MHLR Vol.8 (602–641)
—— (2022a) MHLR Vol.7 (565–602)
—— (2021b) MHLR Vol.6 (518–565)
—— (2020e) MHLR Vol.5 (457–518)
—— (2020d) MHLR Vol.4 (425–457)
—— (2020c) MHLR Vol.3 (395–425)
—— (2018) MHLR Vol. 2 (361–395)
—— (2015) MHLR Vol.1 (284–361)
—— (2021b) 'The Three Hephthalite Wars of Peros 574/5–484', *Historia i Swiat 10*, 96–116.
—— (2021a) *Gordian II and Philip the Arab*. Barnsley.
—— (2020b) 'Holy War and a Place in Paradise? Development of the East Roman Holy War from the 4th until the 11th Century', Written with the generous support of the ASMEA Research Grant.
—— (2020a) *Aurelian and Probus. The Soldier Emperors Who Saved Rome.* Barnsley.
—— (2019c) *The Reign of Emperor Gallienus. The Apogee of Roman Cavalry.* Barnsley.
—— (2019b) *Britain in the Age of Arthur. A Military History*. Barnsley.
—— (2019a) 'The Capture of Jerusalem by the Muslims', *Historia i Swiat 8*, 37–58. Written with the generous support of the ASMEA Travel Grant.
—— (2017b) 'Parthian Cataphract vs. the Roman Army 53 BC–AD 224', in *Historia i Swiat 2017*.
—— (2017a) *Caracalla. A Military Biography*. Barnsley. Paperback ed. 2022 with some changes.
—— (2016) 'The Eyes and Ears: The Sasanian and Roman Spies AD 224–450', *Historia i Swiat 2016, Based on research paper of The 8th ASMEA Conference*, Washington DC, 2015. Written with the generous support of the ASMEA Research Grant.
—— (2014) 'Persia, la caida de un imperio', in *Desperta Ferro* 24, Julio 2014.
—— (2013c) 'las campañas de Belisario contra los Sasánidas', in *Desperta Ferro 18*.
—— (2013b) 'Byzantine Military Doctrine', in *Philosophers of War,* 2vols. eds. D. Coetzee and L.W. Eysturlid, Praeger, 331–8.
—— (2013a) 'Arrian, Governor of Cappadocia, Lucius Flavius Arrianus (Arrianos) Xenophon', in *Philosophers of War,* 2 vols. eds. D. Coetzee and L.W. Eysturlid, Praeger, 259–63.
—— (2012) 'La batalla Liegnitz 1241', in *Desperta Ferro* 12.
—— (2011–2) 'East Roman Cavalry Warfare and Tactics 410–641', three parts (based on a lecture) in *Slingshot 279* (Nov. 2011), *280* (Jan. 2012), and *281* (March 2012).
—— (2011c) 'The Reign of Decius 249–251', in *Slingshot* May 2011, 2–8.
—— (2011a-b) An enlarged version of the research paper 'Campaigns of Germanicus, 13–16 AD', *Historicon 2011*. Available online at academia.edu. A research paper 'Campaigns of Germanicus, 14–16 AD', *6th International Fields of Conflict Conference, 15th–18th April 2011 in Osnabrück and Kalkriese.*
—— (2010) 'Macedonian Art of War. The Balkans 335 BC, Granicus River 334 BC, and Gaugamela 331 BC', in *Saga Newsletter 123,* 31–99.
—— (2009c) 'Teutoburg Forest 9 AD', in *Saga Newsletter 119,* June 2009, 9–23.
—— (2009b) 'The Battle of Melitene in AD 576', in *Saga Newsletter 120,* Aug./Sept. 2009, 32–64.
—— (2009a) 'The Campaign and Battle of Sambre in 57 BC', in *Saga Newsletter 117*, March 2009, 5–16.
—— (2008) 'The New Cavalry Formations in the Sylloge Tacticorum, AD 904', in *Saga Newsletter 112*, republished in *Slingshot 291*, 2014. Available online at academia.edu with additional comments.
—— (2004) *The Age of Hippotoxotai*. Tampere.
Warnery, Emanuel von, (1798) *Remarks on Cavalry*. London 1798.

Whitby, Michael (1995) 'Recruitment in the Roman Armies from Justinian to Heraclius (ca. 565–615)', *The Byzantine and Early Islamic Middle East III: States, Resources and Armies*. Princeton, 61–124.

Zuckerman, Constantine (1990) 'The Military Compendium of Syrianus Magister', *JÖB 40*, 209–224.

—— (1994) 'Chapitres peu connus de L'Apparatus bellicus', *TM 12* (1994), 359–389 (contains a partial edition and French tr. of the tenth-century *Apparatus bellicus*).

Index

Dear Reader,

We hope you have enjoyed this book, but why not share your views on social media? You can also follow our pages to see more about our other products: facebook.com/penandswordbooks or follow us on X @penswordbooks

You can also view our products at www.pen-and-sword.co.uk (UK and ROW) or www.penandswordbooks.com (North America).

To keep up to date with our latest releases and online catalogues, please sign up to our newsletter at: www.pen-and-sword.co.uk/newsletter

If you would like a printed catalogue with our latest books, then please email: enquiries@pen-and-sword.co.uk or telephone: 01226 734555 (UK and ROW) or email: uspen-and-sword@casematepublishers.com or telephone: (610) 853-9131 (North America).

We respect your privacy and we will only use personal information to send you information about our products.

Thank you!